Methods in Molecular Biology

Series Editor
John M. Walker
School of Life and Medical Sciences
University of Hertfordshire
Hatfield, Hertfordshire, UK

For further volumes:
http://www.springer.com/series/7651

For over 35 years, biological scientists have come to rely on the research protocols and methodologies in the critically acclaimed *Methods in Molecular Biology* series. The series was the first to introduce the step-by-step protocols approach that has become the standard in all biomedical protocol publishing. Each protocol is provided in readily-reproducible step-by-step fashion, opening with an introductory overview, a list of the materials and reagents needed to complete the experiment, and followed by a detailed procedure that is supported with a helpful notes section offering tips and tricks of the trade as well as troubleshooting advice. These hallmark features were introduced by series editor Dr. John Walker and constitute the key ingredient in each and every volume of the *Methods in Molecular Biology* series. Tested and trusted, comprehensive and reliable, all protocols from the series are indexed in PubMed.

Membrane Trafficking

Methods and Protocols

Edited by

Jingshi Shen

Department of Molecular, Cellular and Developmental Biology,
University of Colorado Boulder, Boulder, CO, USA

Humana Press

Editor
Jingshi Shen
Department of Molecular
Cellular and Developmental Biology
University of Colorado Boulder
Boulder, CO, USA

ISSN 1064-3745 ISSN 1940-6029 (electronic)
Methods in Molecular Biology
ISBN 978-1-0716-2208-7 ISBN 978-1-0716-2209-4 (eBook)
https://doi.org/10.1007/978-1-0716-2209-4

Cover Illustration Caption: See Chapter 10 for more details.

This Humana imprint is published by the registered company Springer Science+Business Media, LLC part of Springer Nature.
The registered company address is: 1 New York Plaza, New York, NY 10004, U.S.A.

Preface

Membrane trafficking research aims to uncover molecular principles governing the movement of proteins and lipids in the endomembrane systems of eukaryotic cells. Over the past four decades, genetic and biochemical studies have identified core engines underlying major membrane trafficking processes, including coat proteins mediating vesicle budding and SNARE proteins driving vesicle fusion. Naturally, a key future research direction is to fully elucidate how the core engines work at the molecular level. Another major research direction is to unravel how the core engines are controlled by regulatory networks to achieve integrated physiological responses. Regulated membrane trafficking pathways play central roles in nearly every aspect of animal physiology, and their imbalances are associated with major forms of human disease. However, most of these pathways remain poorly characterized. With the advent of unbiased genome-scale CRISPR genetic screens, it is now possible to systematically dissect these membrane trafficking pathways. Our and others' CRISPR screens have identified a surprisingly large number of regulators, and many of these molecules have no assigned functions, reinforcing the notion that major gaps still exist in our current knowledge of membrane trafficking.

To facilitate the next phase of membrane trafficking research, this volume presents 25 chapters of methods, ranging from genetics and high-resolution imaging to in vitro biochemical and biophysical assays. Covering virtually all the major trafficking branches, the chapters are organized into three parts: (Part I) the exocytic pathway, which focuses on cargo transport from the ER to the Golgi, through the Golgi cisternae, and to the plasma membrane and the extracellular space; (Part II) the endocytic pathway, which includes cargo endocytosis, endosomal recycling, and lysosomal degradation; and (Part III) emerging topics beyond the conventional exocytic and endocytic pathways. Many of the chapters describe general methods with broad applications, rather than being restricted to specific molecules or pathways.

This volume will be particularly useful for junior researchers new to membrane trafficking as well as established membrane biologists seeking to expand their research programs. The chapters should also be of interest to scientists in other fields who would like to venture into membrane trafficking research.

Boulder, CO, USA *Jingshi Shen*

Contents

Contributors

Syed Mukhtar Ahmed • *Department of Cell and Developmental Biology, Vanderbilt University School of Medicine, Nashville, TN, USA*

Cedric S. Asensio • *Department of Biological Sciences, University of Denver, Denver, CO, USA*

Tamara Basta • *Department of Molecular, Cellular and Developmental Biology, University of Colorado, Boulder, CO, USA*

Xin Bian • *Department of Cell Biology, Yale University School of Medicine, New Haven, CT, USA; State Key Laboratory of Medicinal Chemical Biology, College of Life Sciences, Nankai University, Tianjin, China*

Juan S. Bonifacino • *Neurosciences and Cellular and Structural Biology Division, Eunice Kennedy Shriver National Institute of Child Health and Human Development, National Institutes of Health, Bethesda, MD, USA*

Émeline Bonsergent • *Université de Paris, INSERM, CNRS UMR7057, Paris, France*

Ashley M. Bourke • *Department of Pharmacology, University of Colorado School of Medicine, Aurora, CO, USA*

Shéryl Bui • *Université de Paris, INSERM, CNRS UMR7057, Paris, France*

Marine D. Camus • *Université Paris Saclay, INSERM U1184, Paris, France*

Stephane M. Camus • *Université de Paris, INSERM UMR970, Paris Cardiovascular Research Center, Paris, France*

Julia Dancourt • *Université de Paris, INSERM, CNRS UMR7057, Paris, France*

Sunando Datta • *Department of Biological Sciences, Indian Institute of Science Education and Research Bhopal, Bhopal, India*

Raffaella De Pace • *Neurosciences and Cellular and Structural Biology Division, Eunice Kennedy Shriver National Institute of Child Health and Human Development, National Institutes of Health, Bethesda, MD, USA*

Marijn G. J. Ford • *Department of Cell Biology, University of Pittsburgh School of Medicine, Pittsburgh, PA, USA*

Herbert Y. Gaisano • *Department of Medicine, The University of Toronto and the Toronto General Hospital Research Institute, Toronto, ON, Canada*

Florence Gazeau • *Laboratoire Matières et Systèmes Complexes, CNRS UMR 7057, Université de Paris, Paris, France*

Jinghua Ge • *Department of Cell Biology, Yale University School of Medicine, New Haven, CT, USA*

Jihong Gong • *Key Laboratory of Cognitive Science, Hubei Key Laboratory of Medical Information Analysis and Tumor Diagnosis and Treatment, Laboratory of Membrane Ion Channels and Medicine, College of Biomedical Engineering, South-Central University for Nationalities, Wuhan, China*

Carlos M. Guardia • *Neurosciences and Cellular and Structural Biology Division, Eunice Kennedy Shriver National Institute of Child Health and Human Development, National Institutes of Health, Bethesda, MD, USA*

Daniel R. Gulbranson • *Gladstone Institute of Neurological Disease, Gladstone Institutes, San Francisco, CA, USA*

Wei Guo • *Department of Biology, University of Pennsylvania, Philadelphia, PA, USA*

RUYUE HE • *Jiangsu Key Laboratory for Molecular and Medical Biotechnology, College of Life Sciences, Nanjing Normal University, Nanjing, China*
WON DO HEO • *Department of Life Sciences, KAIST Institute for the BioCentury, Korea Advanced Institute of Science and Technology (KAIST), Daejeon, South Korea*
BOBAE HYEON • *Department of Life Sciences, Korea Advanced Institute of Science and Technology (KAIST), Daejeon, South Korea*
LUHONG JIN • *Department of Biomedical Engineering, Key Laboratory of Biomedical Engineering of Ministry of Education, State Key Laboratory of Modern Optical Instrumentation, Zhejiang Provincial Key Laboratory of Cardio-Cerebral Vascular Detection Technology and Medicinal Effectiveness Appraisal, Zhejiang University, Hangzhou, China*
FEI KANG • *Department of Medicine, The University of Toronto and the Toronto General Hospital Research Institute, Toronto, ON, Canada*
MATTHEW J. KENNEDY • *Department of Pharmacology, University of Colorado School of Medicine, Aurora, CO, USA*
GRÉGORY LAVIEU • *Université de Paris, INSERM, CNRS UMR7057, Paris, France*
JUHYUNG LEE • *Laboratory of Molecular Biology, National Institute of Diabetes and Digestive and Kidney Diseases, National Institutes of Health, Bethesda, MD, USA*
DI-AO LIU • *Department of Biology, University of Pennsylvania, Philadelphia, PA, USA*
YINGHUI LIU • *Jiangsu Key Laboratory for Molecular and Medical Biotechnology, College of Life Sciences, Nanjing Normal University, Nanjing, China*
CONG MA • *Key Laboratory of Molecular Biophysics of the Ministry of Education, College of Life Science and Technology, Institute of Brain Research, Huazhong University of Science and Technology, Wuhan, China*
LU MA • *Department of Cell Biology, Yale University School of Medicine, New Haven, CT, USA; Beijing National Laboratory for Condensed Matter Physics, Institute of Physics, Chinese Academy of Sciences, Beijing, China*
KUNRONG MEI • *School of Pharmaceutical Science and Technology, Tianjin University, Tianjin, China*
FRÉDÉRIC A. MEUNIER • *Clem Jones Centre for Ageing Dementia Research, Queensland Brain Institute, The University of Queensland, Brisbane, QLD, Australia*
GARRY P. MORGAN • *Boulder Electron Microscopy Services, University of Colorado, Boulder, CO, USA*
MAI KHANH NGUYEN • *Abcam Fremont Technology Development Custom Solution, Fremont, CA, USA*
EILEEN T. O'TOOLE • *Boulder Electron Microscopy Services, University of Colorado, Boulder, CO, USA*
SAMEENA PARVEEN • *Department of Biological Sciences, Indian Institute of Science Education and Research Bhopal, Bhopal, India*
MAX PIFFOUX • *Laboratoire Matières et Systèmes Complexes, CNRS UMR 7057, Université de Paris, Paris, France; Interaction Cellules Souches-Niches: Physiologie, Tumeurs et Ré paration Tissulaire, Inserm UMR 1197, Université Paris-Saclay, Villejuif, France; Dé partement d'oncologie Médicale, Centre Léon Bérard, Lyon, France*
BIJETA PRASAI • *Biochemistry and Biophysics Center, National Heart, Lung, and Blood Institute, National Institutes of Health, Bethesda, MD, USA*
AMULYA PRIYA • *Institut Curie, PSL Research University, CNRS, Paris, France*
RAJESH RAMACHANDRAN • *Department of Physiology and Biophysics, Case Western Reserve University School of Medicine, Cleveland, OH, USA*

NALINI R. RAO • *Department of Neurology, Northwestern University Feinberg School of Medicine, Chicago, IL, USA*

AMRA SARIC • *Neurosciences and Cellular and Structural Biology Division, Eunice Kennedy Shriver National Institute of Child Health and Human Development, National Institutes of Health, Bethesda, MD, USA*

JEFFREY N. SAVAS • *Department of Neurology, Northwestern University Feinberg School of Medicine, Chicago, IL, USA*

PRIYANKA SHARMA • *Department of Biological Sciences, Indian Institute of Science Education and Research Bhopal, Bhopal, India*

JINGSHI SHEN • *Department of Molecular, Cellular and Developmental Biology, University of Colorado, Boulder, CO, USA*

AMANDA KARINE ANDRIOLA SILVA • *Laboratoire Matières et Systèmes Complexes, CNRS UMR 7057, Université de Paris, Paris, France*

GALEN T. SQUIERS • *Department of Molecular, Cellular and Developmental Biology, University of Colorado, Boulder, CO, USA*

MICHAEL H. B. STOWELL • *Department of Molecular, Cellular and Developmental Biology, University of Colorado, Boulder, CO, USA; Boulder Electron Microscopy Services, University of Colorado, Boulder, CO, USA*

JUSTIN W. TARASKA • *Biochemistry and Biophysics Center, National Heart, Lung, and Blood Institute, National Institutes of Health, Bethesda, MD, USA*

DAVID TARESTE • *Institut de Psychiatrie et Neurosciences de Paris, Inserm UMR 1266, Université de Paris, Paris, France*

YUAN TIAN • *Department of Biological Science, Florida State University, Tallahassee, FL, USA*

DEREK TOOMRE • *Department of Cell Biology, Yale University School of Medicine, New Haven, CT, USA*

CHUN WAN • *Department of Molecular, Cellular and Developmental Biology, University of Colorado, Boulder, CO, USA*

BING WANG • *Department of Biological Science, Florida State University, Tallahassee, FL, USA*

SHIFENG WANG • *Department of Chinese Medicine Information Science, Beijing University of Chinese Medicine, Beijing, China; Department of Molecular, Cellular and Developmental Biology, University of Colorado, Boulder, CO, USA*

TONG WANG • *Center for Brain Science, School of Life Science and Technology, Shanghaitech University, Shanghai, China*

CHAD D. WILLIAMSON • *Neurosciences and Cellular and Structural Biology Division, Eunice Kennedy Shriver National Institute of Child Health and Human Development, National Institutes of Health, Bethesda, MD, USA*

YINGKE XU • *Department of Biomedical Engineering, Key Laboratory of Biomedical Engineering of Ministry of Education, State Key Laboratory of Modern Optical Instrumentation, Zhejiang Provincial Key Laboratory of Cardio-Cerebral Vascular Detection Technology and Medicinal Effectiveness Appraisal, Zhejiang University, Hangzhou, China*

RUI YANG • *Department of Biological Science, Florida State University, Tallahassee, FL, USA*

XIAOFEI YANG • *Key Laboratory of Cognitive Science, Hubei Key Laboratory of Medical Information Analysis and Tumor Diagnosis and Treatment, Laboratory of Membrane Ion Channels and Medicine, College of Biomedical Engineering, South-Central University for Nationalities, Wuhan, China*

YIHONG YE • *Laboratory of Molecular Biology, National Institute of Diabetes and Digestive and Kidney Diseases, National Institutes of Health, Bethesda, MD, USA*
QIAN YIN • *Department of Biological Science, Institute of Molecular Biophysics, Florida State University, Tallahassee, FL, USA*
HAIJIA YU • *Jiangsu Key Laboratory for Molecular and Medical Biotechnology, College of Life Sciences, Nanjing Normal University, Nanjing, China*
YONGLI ZHANG • *Department of Cell Biology, Yale University School of Medicine, New Haven, CT, USA*
MIN ZHU • *Jiangsu Key Laboratory for Molecular and Medical Biotechnology, College of Life Sciences, Nanjing Normal University, Nanjing, China*

Part I

The Exocytic Pathway

Chapter 1

Characterizing Membrane Traffic in the Early Secretory Pathway Using the RUSH Retention System

Marine D. Camus and Stephane M. Camus

Abstract

The early secretory pathway encompasses the endoplasmic reticulum (ER) and the ER–Golgi intermediate compartment (ERGIC) organelles. The ERGIC is now understood to be a complex cargo sorting hub involved in a variety of cellular and tissue processes, however the traffic pathways to and from the ERGIC are still unclear.

Classical methods employed for the analysis of a cargo's journey along the secretory pathway rely on reversible traffic blocks leading to cargo accumulation in the ER. Although these methods were key to characterize Golgi and post-Golgi traffic routes, their poor specificity to the cargo of interest and limited spatiotemporal resolution make them inadequate for the fine characterization of cargo traffic in the early secretory pathway.

In this chapter, we describe a protocol to study the traffic of cargo proteins in the early secretory pathway using the Retention Using Selective Hook (RUSH) system, a highly specific and sensitive tracking system with a high spatiotemporal resolution. Taking GLUT4 and GLUT1 as examples of unconventionally and conventionally secreted cargo respectively, we describe the steps to clone the cargoes in the RUSH vector and follow and quantify their traffic along the early secretory pathway. This RUSH method can also be used to study the traffic of other cargo proteins in the early secretory pathway.

Key words Glucose transporter, GLUT4, Protein traffic, Early secretory pathway, Endoplasmic reticulum (ER), ER–Golgi intermediate compartment (ERGIC), Retention using selective hook (RUSH)

1 Introduction

The secretory pathway is constituted by functionally interconnected organelles namely the endoplasmic reticulum (ER), the ER–Golgi intermediate compartment (ERGIC) and the Golgi apparatus. The pre-Golgi traffic steps define the so-called early secretory pathway, along which newly synthesized proteins are sorted into specialized transport vesicles that will ultimately allow proteins to successfully reach their final compartment. Since its discovery, the understanding of ERGIC function has

Jingshi Shen (ed.), *Membrane Trafficking: Methods and Protocols*,
Methods in Molecular Biology, vol. 2473, https://doi.org/10.1007/978-1-0716-2209-4_1,

dramatically evolved from a simple intermediate organelle involved in delivering ER-derived cargo to the Golgi apparatus, to a complex cargo sorting hub involved in a variety of cellular and tissue processes encompassing autophagy, blood coagulation, lung or glucose homeostasis [1–4]. However, the traffic pathways from and to the ERGIC are still poorly understood.

Classical biochemical methods laid the foundation for identifying major traffic pathways. These methods are based on synchronization of cargo traffic in the early secretory pathway by blocking membrane traffic from the ER (15 °C temperature block, brefeldin A treatment), leading to cargo accumulation in the ER. The cargo is then released and chased for increasing periods of time to identify traffic routes and evaluate traffic kinetics by biochemical or imaging techniques. However, the poor physiological relevance of these blocks and their low specificity to the cargo being studied are major shortcomings that limit interpretation. Moreover, the inherently low spatiotemporal resolution of these methods makes them ill-suited to the fine characterization of cargo traffic in the early secretory pathway, such as at the ERGIC.

The RUSH system is a versatile, highly specific and sensitive tracking system developed to analyze the kinetics of intracellular traffic pathways [5]. The major refinements of this tool lie in the high spatiotemporal resolution of traffic of a specific cargo, under physiological conditions in live or fixed cells. Indeed, the cargo of interest is fused to a streptavidin-binding protein (SBP) and a fluorescent protein while an ER-resident protein is fused to streptavidin. This allows specific retention of the cargo in the ER without altering the flux of other secreted cargoes, as the classical methods would. The addition of the nontoxic vitamin biotin to the culture medium triggers the instant, synchronous release of the cargo that will resume its traffic to eventually reach its final destination. Taking advantage of the RUSH system, we demonstrated that newly synthesized glucose transporter GLUT4 follows a Golgi independent, unconventional secretion pathway. This unconventional secretion pathway involves a clathrin-mediated sorting step from the ERGIC to form the Glucose Storage Compartment, enabling insulin-mediated GLUT4 translocation [3].

In this chapter, we describe a protocol to characterize the early traffic steps of cargoes by immunofluorescence analysis in fixed cells, taking GLUT4 and GLUT1 as examples of unconventionally and conventionally secreted cargoes respectively. The protocol includes cloning of the cargo (i.e., GLUT4) in the RUSH vector, transfection of the cells and synchronized release of the cargo, and imaging and quantification. The protocol is also amenable to biochemical analysis (*see* **Note 1**) and can be adapted to the studies of other newly synthesized cargoes.

2 Materials

2.1 Molecular Cloning

1. pQBI25-HA-GLUT4-GFP [6] and pCMV6-AC-GLUT1-GFP (OriGene, #RG222696) DNA plasmids.
2. RUSH plasmid encoding Streptavidin-li-CD44-SBP-EFGP (Addgene #65255). The RUSH plasmid is a bicistronic plasmid encoding two genes separated by an internal ribosome entry site (IRES) to ensure high expression level of both genes. The first gene encodes the ER hook, which is the ER resident protein li fused to streptavidin. The second gene encodes the cargo protein (CD44 in the RUSH vector used here for subsequent cloning) fused to the fluorescent protein EGFP [5, 7].
3. TBE 10× buffer: 108 g Tris base, 55 g boric acid, 40 mL of 0.5 M EDTA, pH 8, complete to 1 L with distilled water. Dilute to 1× using distilled water before use. Store TBE 10× and 1× at room temperature.
4. 1% Agarose gel: Mix 1 g of ultrapure agarose with 100 mL of 1× TBE. Using a microwave, boil the solution until the agarose is fully dissolved. Let the solution cool down. Under a fume hood, add nucleic acid detection reagents to the solution before casting the gel.
5. Site-directed DNA mutagenesis kit.
6. SmaI and EcoRI restriction enzymes.
7. Gel extraction kit.
8. T4 DNA ligase.
9. Super Optimized broth with Catabolite repression (S.O.C.) medium: 2% (w/v) tryptone, 0.5% (w/v) yeast extract, 10 mM NaCl, 2.5 mM KCl, 10 mM $MgCl_2$, 20 mM glucose.
10. *E. coli* DH5α competent cells.
11. Leibowitz Broth (LB) agar plates containing 100 μg/mL ampicillin.
12. Plasmid DNA miniprep and maxiprep kits.

2.2 Cell Culture

1. 24-well plates.
2. 2 cm^2 sterile round glass coverslips #1.5.
3. Growth medium for HeLa cells: Dulbecco's Modified Eagle Medium (DMEM) high glucose, 20% (v/v) Fetal Bovine Serum, and 1% (v/v) Penicillin/Streptomycin.
4. DNA transfection reagent. We use jetPRIME® (Polyplus, 3114-01).
5. 4 mM D-biotin stock solution: dissolve 48 mg biotin in 50 mL sterile HeLa cell growth medium. When fully dissolved, pass

the solution through a 0.22 μm filter and collect into a sterile 50 mL conical tube. Protect from light and store at 4 °C for up to 3 months.

6. Phosphate-buffered saline (PBS): 138 mM NaCl, 8 mM Na_2HPO_4, 3 mM KCl, and 1.5 mM KH_2PO_4, pH 7.4.
7. 32% paraformaldehyde (PFA) stock solution, store at −20 °C.
8. 2% PFA: Dilute 32% PFA stock solution to 2% using ice-cold PBS right before use and keep on ice until use.
9. Permeabilization/blocking buffer: 2% (w/v) bovine serum albumin (BSA) and 0.5% saponin (w/v) in PBS.
10. PBS with 2% BSA.
11. Parafilm.
12. Antibodies recognizing the ER protein calreticulin (Stressgen Bioreagents, clone FMC75).
13. Antibodies recognizing ER–Golgi intermediate compartment proteins: LMAN1/ERGIC53 (clone 2B10, OriGene; or clone EPR6979, Abcam), p115 (clone 7D1) [8].
14. Antibodies recognizing the Golgi apparatus proteins: cis-Golgi protein GM130 (P-20, Santa Cruz Biotechnologies).
15. Antibodies recognizing EGFP.
16. Species-specific secondary antibodies coupled to fluorescent dyes: Alexa Fluor 405 (max Ex 405 nm, max Em 421 nm), Alexa Fluor 555 (max Ex 555 nm, max Em 567 nm), Alexa Fluor 647 (max Ex 651 nm, max Em 671 nm).
17. Mounting medium.
18. SuperFrost Plus microscope slides (25 × 75 × 1 mm).
19. Translucent nail polish.
20. Clean cotton swabs.

3 Methods

3.1 Cloning HA-GLUT4 and HA-GLUT1 into the RUSH Vector

To avoid interference with the cargo's natural traffic pathway and kinetics, determine carefully the orientation of the cargo in the RUSH vector in relation to the SBP and the fluorescent proteins. Since fusion of tags to the C-terminus of GLUT4 does not interfere with GLUT4 traffic [6], the GLUT4 gene is inserted upstream of SBP and the fluorescent protein.

1. This step generates the pCMV6-HA-GLUT1-GFP plasmid. Insert a haemagglutinin tag (HA-tag, sequence 5′-ATCGAT TATCCTTATGATGTTCCTGATTATGCTGAG-3′) between amino acids 67 and 68 (at base pair 201) of the GLUT1-GFP sequence in the pCMV6-AC-GLUT1-GFP plasmid using a site-directed mutagenesis kit.

2. **Steps 2–3** modify the pCMV6-HA-GLUT1-GFP plasmid to allow subsequent insertion of HA-GLUT1 in the RUSH backbone. Remove the internal EcoRI restriction site present within the GLUT1-GFP sequence (at base pair 1305) in the pCMV6-HA-GLUT1-GFP plasmid generated in **step 1** using a site-directed mutagenesis kit.
3. Insert an EcoRI restriction site at the 3′ end of the GLUT1-GFP sequence in the plasmid generated in **step 2** using a site-directed mutagenesis kit. When designing the mutagenesis primers, make sure that the gene encoding GLUT1 will be in frame once inserted in the RUSH backbone.
4. This step modifies the pQBI25-HA-GLUT4-GFP plasmid to allow subsequent insertion of HA-GLUT4 into the RUSH backbone. Insert an SmaI restriction site at the 5′ end and an EcoRI restriction site at the 3′ end of HA-GLUT4-GFP sequence in the pQBI25-HA-GLUT4-GFP plasmid using a site-directed mutagenesis kit. When designing the mutagenesis primers, make sure that the gene encoding GLUT4 will be in frame once inserted in the RUSH backbone.
5. Add 10 units of SmaI and EcoRI restriction enzymes to 1 μg of the following plasmids: pRUSH-Str-li-CD44-SBP-GFP plasmid, the pQBI25-HA-GLUT4-EGFP plasmid generated in **step 4**, and the pCMV6-HA-GLUT1-EGFP plasmid generated in **step 3** and incubate between 30 and 60 min at 37 °C.
6. Separate the digestion products by electrophoresis on a 1% agarose gel.
7. Purify the DNA fragments of interest (HA-GLUT1 insert, HA-GLUT4 insert and the linearized pRUSH-Str-li-SBP-EGFP backbone) from the agarose gel using a gel extraction kit.
8. To generate the pRUSH-GLUT4-EGFP and pRUSH-GLUT1-EGFP plasmids, combine 100 ng of the receiving pRUSH-EGFP backbone (Str-li-SBP-EGFP RUSH without CD44) with the HA-GLUT4 insert or the HA-GLUT1 insert, respectively, at a ratio of 1:3. Add T4 DNA ligase and mix thoroughly by pipetting up and down 7–10 times, and incubate for 10 min at room temperature.
9. Add 5 μL of ligation product to a 50 μL vial of competent DH5α *E. coli* competent cells, gently flick the tube and incubate on ice for 30 min.
10. Transform the bacteria by incubating at 42 °C for 30 s.
11. Place the transformed bacteria on ice for 2 min before adding 950 μL of S.O.C. medium and incubate at 37 °C for 1 h with agitation.

12. Spread 50–100 μL of the transformed bacterial suspension on a selective LB agar plate containing 100 μg/mL ampicillin and incubate for 24 h at 37 °C.
13. The next day, pick five isolated colonies and isolate plasmid DNA using a plasmid DNA miniprep kit.
14. Sequence the plasmids to confirm DNA sequences, insert orientation, and that the genes are in frame.
15. Expand the bacterial cultures and isolate plasmids using a plasmid DNA maxiprep kit. The pRUSH-GLUT4-EGFP and pRUSH-GLUT1-EGFP plasmids are available upon request to the authors [3].

3.2 Cell Transfection

Proper traffic of the pRUSH-GLUT4-EGFP and pRUSH-GLUT1-EGFP vectors should be confirmed first before running any experiments (*see* **Note 2**). Determine the most appropriate time points to study your cargo (*see* **Note 3**), and include a $t = 0$ condition as a control.

1. Seed HeLa cells on 2cm^2 sterile coverslips in a 24-well plate at a density of 30,000 cells/cm^2 in 500 μL of warm growth medium and incubate in a humidified atmosphere at 37 °C with 5% CO_2. We recommend seeding cells on the same plate such that biotin incubation can be stopped simultaneously for all time points. We routinely prepare samples in duplicates to ensure we have enough cells for imaging. For ease of imaging, aim for 60–70% cell confluence on the day of transfection.
2. The next day, transfect the cells with pRUSH-GLUT4-EGFP or pRUSH-GLUT1-EGFP plasmid DNA (*see* **Note 4**). For each well, dilute 0.5 μg plasmid DNA in 50 μL of JetPRIME® diluent and add 1 μL of JetPRIME® reagent (1:2 DNA–Jet-Prime ratio). Mix by vortex and incubate for 15 min at room temperature.
3. Add the solution to the cells dropwise and incubate for 6 h at 37 °C.
4. After 6 h, remove the growth media, gently wash the cells with warm PBS, and add fresh growth media.
5. Incubate overnight at 37 °C.

3.3 RUSH Assay

Before starting the experiment, check the transfection yield with a wide-field microscope. With cells amenable to transfection such as HeLa cells, aim for a transfection yield of over 50%.

1. Add fresh growth medium without biotin to the $t = 0$ wells.
2. Add fresh growth medium supplemented with 40 μM of biotin to the appropriate wells. Biotin should be added to the wells from the longest to the shortest time points.

3. At the end of the biotin incubation periods, quickly place the cells on ice and perform three gentle washes with ice-cold PBS to stop intracellular membrane traffic.
4. Fix the cells by adding ice-cold 2% PFA to the wells. Wrap the plate in aluminum foil and incubate for 30 min at room temperature.
5. Wash the fixed cells twice with 25 mM glycine (or NH_4Cl) in PBS to quench unreacted PFA. The cells are now ready to be processed for immunofluorescence.

3.4 Immunofluorescence

From this point onward, steps can be performed at room temperature. Protect the cells from light during incubation times.

1. Incubate cells with the permeabilization/blocking buffer for 1 h.
2. In the meantime, prepare antibody mix solutions by diluting primary antibodies in the permeabilization buffer (*see* **Note 5**). Prepare 40 μL of antibody mix solution per coverslip.
3. Prepare a paraffin film on a plastic tray and place a 30 μL drop of primary antibody mix solution for each coverslip.
4. Place the coverslips upside down onto the antibody solution drops (cells in contact with the antibody solution). Wrap the tray with aluminum foil and incubate overnight at 4 °C.
5. The next day, place the coverslips back in a clean 24-well plate with cells facing upward and wash the cells three times (twice for 5 min then once for 15 min) with PBS at room temperature.
6. Prepare secondary antibody mix solutions by diluting antibodies into the permeabilization/blocking buffer (*see* **Note 6**). Prepare 40 μL of secondary antibody mix solution per coverslip.
7. Prepare a paraffin film on a plastic tray and place a 30 μL drop of secondary antibody mix solution for each coverslip.
8. Place the coverslips upside down onto the drops. Wrap the tray with aluminum foil and incubate for 45 min at room temperature.
9. Place the coverslips back in a 24-well plate and wash the cells three times (twice for 5 min then once for 15 min) with PBS at room temperature.
10. Place an 8 μL drop of mounting medium on a microscope slide.
11. Remove as much PBS as possible from the coverslips by blotting them onto a clean paper towel and lay the coverslip slowly onto the drop of mounting medium. Take extra-care during this step to avoid trapping air bubbles under the coverslips. Up to 3 coverslips can be arranged on one microscope slide.

12. Let the mounting medium cure for 15 h–24 h.
13. Seal the coverslips with translucent nail polish.
14. Gently clean the coverslips with a clean cotton swabs and distilled water in order to remove any salt crystals that may have formed.
15. The samples are now ready for imaging. Sealed coverslips can be stored for a few weeks at 4 °C although the fluorescence intensity of fluorescent proteins (EGFP particularly) decays over time. Thus, imaging should preferably be performed quickly after mounting the coverslips.

3.5 Imaging and Quantification of GLUT4 and GLUT1 Traffic in the Early Secretory Pathway

Perform imaging using confocal microscopy. Superresolution microscopy such as Structured Illumination Microscopy can also be considered (*see* **Note 7**).

1. Among the transfected cells, select healthy-looking cells with intermediate fluorescence expression to determine the expression level associated with proper GLUT4 and GLUT1 subcellular localization (Fig. 1). Too high of an expression level usually results in aberrant subcellular localization and cellular stress.
2. Using a 63× oil-immersion objective, acquire images of a minimum of 10 cells from at least 4 different fields for each coverslip (*see* **Note 8**).
3. Determine subcellular cargo localization by quantifying the degree of colocalization between GLUT4 or GLUT1 and organelle markers by calculating the Pearson's Correlation Coefficient with the coloc2 plugin of ImageJ. Plotting the pixel values over a given distance (line-tracing) within the area of interest can also be considered. The quantification should be performed based on at least 3 independent experiments to ensure reliable quantification.

4 Notes

1. In addition to the spatiotemporal information provided by the RUSH system allowing the identification of a cargo traffic pathway and kinetics, identifying the cargo's interacting proteins can provide valuable and complementary information regarding the molecular mechanisms of the pathway. The particular versatility of the RUSH system allows the specific immunoprecipitation of the cargo using a GFP-trap for subsequent SDS-PAGE analysis. The following adjustments should be implemented to the cell transfection protocol: seed cells in a 6-well plate, transfect 2 μg of DNA per well, scale up the JetPRIME solutions according to the manufacturer's protocol.

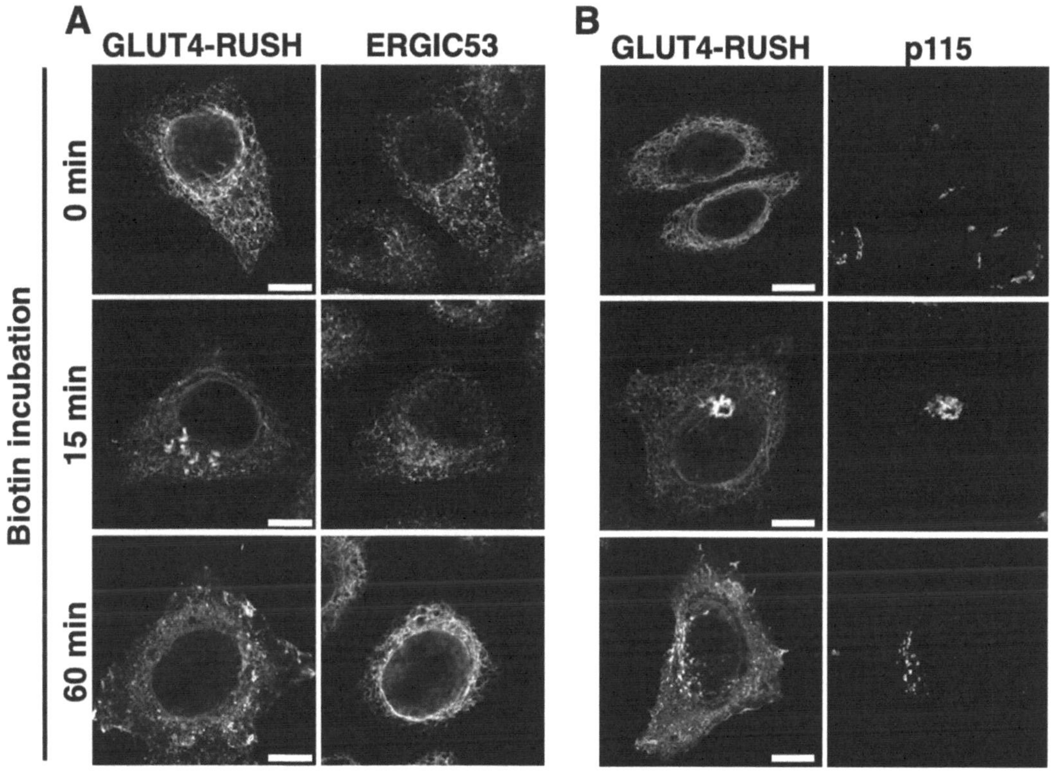

Fig. 1 Representative immunofluorescence staining for pRUSH-GLUT4-EGFP (detected with anti-GFP antibody), ERGIC53 (**a**) or p115 (**b**) in HeLa cells transfected with 0.5 μg of pRUSH-GLUT4-EGFP. Traffic of GLUT4 was tracked at 0, 15, and 60 min after release from the endoplasmic reticulum by biotin. At 15 min postrelease, most of the released GLUT4 is overlapping with ERGIC markers ERGIC53 and p115 in the perinuclear area of the cell. Scale bar: 10 μm

Proceed with the RUSH assay (Subheading 3.3) until **step 4** (place the cells on ice and wash three times) from where the immunoprecipitation procedure should be started.

2. It is advisable to confirm first that the pRUSH-GLUT4-EGFP and pRUSH-GLUT1-EGFP vectors behave similarly to the endogenous GLUT4 and GLUT1 proteins, respectively (intracellular localization). Unfortunately, endogenous GLUT4 immunolabeling is notoriously difficult and most studies describing GLUT4 traffic pathway rely on the use of the GLUT4 fusion protein [6]. This can be achieved by double-transfecting cells with the pRUSH-GLUT4-EGFP along with a plasmid encoding HA-GLUT4 fused to a fluorescent protein such as mCherry. The HA-GLUT4-mCherry fusion protein can be generated by swapping GFP with mCherry in the pQBI25-HA-GLUT4-GFP plasmid. After double-transfecting the cells with pRUSH-GLUT4-EGFP and pQBI25-HA-GLUT4-mCherry, or single transfecting the cells with

pRUSH-GLUT1-EGFP, add 40 μM of biotin to the fresh growth medium in **step 4** of the cell transfection protocol (Subheading 3.2) (after the 6 h incubation with the RUSH plasmids) and incubate overnight to allow for normal traffic of the cargoes to occur. The next day, skip **steps 1–3** of the RUSH assay (Subheading 3.3) and start directly from **step 4** (place the cells on ice and wash three times), and proceed with next steps as described.

3. As a reference, conventionally exocytic cargoes such as GLUT1 rapidly reach the Golgi apparatus in 15 min postrelease (e.g., biotin addition) and exit the Golgi within 30 min. We showed that GLUT4 is delayed in the early secretory pathway and bypasses the Golgi apparatus [3]. To monitor pre-Golgi traffic of GLUT4 and GLUT1, a time point every 5–10 min post-release is appropriate. Time points of 15 min or longer will allow evaluation of the fusion steps with the cis-Golgi and intra-Golgi traffic. These time points are indicative and should be adjusted based on experimental objectives.
4. To simultaneously analyze GLUT4 and GLUT1traffic, the cells can be transfected with both vectors encoding different fluorescent proteins. For example, to generate the pRUSH-GLUT1-mCherry plasmid, swap the GFP tag in the pRUSH-GLUT1-EGFP for mCherry, using the SbfI and FseI restriction enzymes. Make sure to adjust antibody mixes to account for the additional fluorescent protein. Alternatively, the cloning can be done using a RUSH-mCherry vector from Addgene (cat #65256).
5. If RUSH proteins are expressed at low levels with faint GFP fluorescence intensity, use an anti-GFP monoclonal antibody to enhance the detection of the fluorescent protein. The secondary antibody should be coupled to a fluorescent dye with similar excitation and emission wavelengths to the detected fluorescent protein (e.g., Alexa Fluor® 488 for GFP).
6. The emission wavelengths of the fluorescent dyes coupled to secondary antibodies should be as separated as possible to limit signal bleed through between fluorescence channels. If possible, choose weaker fluorescent dyes for primary antibodies with the highest signal-to-noise ratio.
7. Although the method described in this chapter allows accurate localization of the cargo along the early secretory pathway, it can be further improved by combining it with super resolution microscopy such as Structured Illumination Microscopy (SIM). SIM utilizes the same sample preparation protocol as conventional confocal microscopy but will improve lateral resolution of the images by a factor of 2 [9].

8. Aside from basic best practices in fluorescence microscopy imaging [10] it is worth highlighting a few points to consider for refining image acquisition quality. This is especially important when imaging structures that are in close proximity such as in the perinuclear region. Set image capture quality to a minimum of 1024 × 1024 pixels. Set frame averaging: 4 to 6 averaged frames will considerably improve the signal over noise ratio by eliminating background noises. Adjust pinhole diameter: while the default pinhole diameter of 1 airy unit is considered an all-purpose confocal imaging, fine adjustments for each channel can make a substantial improvement of image quality by limiting the detection of contaminating fluorescence from out-of-focus planes. This is especially relevant for signals associated with far-red dyes, which are typically associated with a lower lateral resolution. However, reducing pinhole diameter will result in substantial signal losses so this should be limited to signals associated with the highest signal-to-noise ratio. Once the acquisition settings are optimized, it is preferable to use the same settings across experiments (although minimal adjustments could be needed) to ensure reproducibility.

Acknowledgments

Cell culture studies reported here use the HeLa cell line. Henrietta Lacks, and the HeLa cell line that was established from her tumor cells, have made significant contributions to scientific progress and advances in human health.

S.M. Camus is supported by Marie Skłodowska Curie Actions REMAKIN #846519. This work was partially funded by a grant to F.M. Brodsky from Medical Research Council (MR/S008144/1).

References

1. Ge L, Melville D, Zhang M, Schekman R (2013) The ER-Golgi intermediate compartment is a key membrane source for the LC3 lipidation step of autophagosome biogenesis. Elife 2:e00947. https://doi.org/10.7554/eLife.00947
2. Gilbert A, Jadot M, Leontieva E, Wattiaux-De Coninck S, Wattiaux R (1998) Delta F508 CFTR localizes in the endoplasmic reticulum-Golgi intermediate compartment in cystic fibrosis cells. Exp Cell Res 242(1):144–152. https://doi.org/10.1006/excr.1998.4101
3. Camus SM, Camus MD, Figueras-Novoa C, Boncompain G, Sadacca LA, Esk C, Bigot A, Gould GW, Kioumourtzoglou D, Perez F, Bryant NJ, Mukherjee S, Brodsky FM (2020) CHC22 clathrin mediates traffic from early secretory compartments for human GLUT4 pathway biogenesis. J Cell Biol 219(1). https://doi.org/10.1083/jcb.201812135
4. Nichols WC, Seligsohn U, Zivelin A, Terry VH, Hertel CE, Wheatley MA, Moussalli MJ, Hauri HP, Ciavarella N, Kaufman RJ, Ginsburg D (1998) Mutations in the ER-Golgi intermediate compartment protein ERGIC-53 cause combined deficiency of coagulation factors V and VIII. Cell 93(1):61–70. https://doi.org/10.1016/s0092-8674(00)81146-0
5. Boncompain G, Perez F (2014) Synchronization of secretory cargos trafficking in

populations of cells. Methods Mol Biol 1174: 211–223. https://doi.org/10.1007/978-1-4939-0944-5_14

6. Lampson MA, Racz A, Cushman SW, McGraw TE (2000) Demonstration of insulin-responsive trafficking of GLUT4 and vpTR in fibroblasts. J Cell Sci 113(Pt 22):4065–4076
7. Boncompain G, Divoux S, Gareil N, de Forges H, Lescure A, Latreche L, Mercanti V, Jollivet F, Raposo G, Perez F (2012) Synchronization of secretory protein traffic in populations of cells. Nat Methods 9(5):493–498. https://doi.org/10.1038/nmeth.1928
8. Waters MG, Clary DO, Rothman JE (1992) A novel 115-kD peripheral membrane protein is required for intercisternal transport in the Golgi stack. J Cell Biol 118(5):1015–1026. https://doi.org/10.1083/jcb.118.5.1015
9. Demmerle J, Innocent C, North AJ, Ball G, Muller M, Miron E, Matsuda A, Dobbie IM, Markaki Y, Schermelleh L (2017) Strategic and practical guidelines for successful structured illumination microscopy. Nat Protoc 12(5): 988–1010. https://doi.org/10.1038/nprot.2017.019
10. Lee JY, Kitaoka M (2018) A beginner's guide to rigor and reproducibility in fluorescence imaging experiments. Mol Biol Cell 29(13): 1519–1525. https://doi.org/10.1091/mbc.E17-05-0276

Chapter 2

Monitoring Intra-Golgi Transport with Acute Spatiotemporal Control of a Synthetic Cargo

Julia Dancourt and Grégory Lavieu

Abstract

We hereby describe a method to image cargo trafficking from the cis- to the trans-face of the Golgi apparatus. Briefly, we combine nocodazole treatment that breaks down the Golgi ribbon, temperature blocks that slow down cargo transport, and a drug-controlled aggregation system that controls the size of the cargo and its retention at different stages of the secretory pathway. Using this method, we first position the cargo within the cis-face of the Golgi. When traffic resumes upon temperature block release, kinetics of transport can be assessed by confocal microscopy through colocalization of the cargo with cis- and trans-Golgi markers. This method allows for testing various modes of intra-Golgi transports and can be adapted to investigate other steps of the secretory pathway.

Key words Cargo, Golgi, Endoplasmic Reticulum, Vesicle, Secretory Pathway

1 Introduction

Approximately a third of all proteins produced by a cell travel through the secretory pathway, which starts within the endoplasmic reticulum (ER) and ends up at the plasma membrane or the extracellular space [1]. At the center of the pathway, the Golgi apparatus serves as a triage station, with proteins arriving its cis-face and exiting at its trans face. In mammals, the Golgi apparatus is located at the perinuclear area, where it forms a ribbon structure that results from the lateral fusion of many individual Golgi stacks. Each stack is constituted of 4-to-7 flattened cisternae oriented from a cis-face (the entry side) to the trans-face (exit-face) [2].

The modalities of intra-Golgi transport are still being debated and several models have been proposed [3, 4]. It has been difficult to distinguish between these models because it has not been possible to monitor the movements of both cargoes and Golgi-resident

Julia Dancourt contributed equally with all other contributors.

Jingshi Shen (ed.), *Membrane Trafficking: Methods and Protocols*,
Methods in Molecular Biology, vol. 2473, https://doi.org/10.1007/978-1-0716-2209-4_2,

enzymes concomitantly, in real time, in the Golgi ribbon of living mammalian cells. This has, however, been done in yeasts, in which the Golgi architecture is less complex [5, 6]. In a series of papers [7–13], we and others have used a synthetic biology approach to control: (1) the positioning of a cargo of interest at different stages of the secretory pathways; and (2) the shape and size of the cargo of interest which enable acute spatiotemporal manipulation of the cargo. Combinations of these chemical, physical, cellular and molecular tools are virtually limitless and enable the manipulation of cargos and intracellular organelles behaviors in order to test model predictions within cells [12, 14, 15]. Here, we focus on one particular combination that enables the monitoring of intra-Golgi transport of a model cargo [11].

2 Materials

2.1 Constructs for a Synthetic Secretory Cargo

1. pC4-ssGFPFM4hGH is the prototypal secretory cargo model that we use in this protocol. Briefly, GFP was inserted in the plasmid pC4-ss-FM4hGH using XbaI/SpeI compatibility. pC4-ss-FM4hGH and other plasmids encoding the self-aggregation domain (FM) can be purchased from Takara Bio (Mountain View, CA, catalog number 635066) to generate other cargoes of interest. Signal peptide should be added on the N-terminal side of the protein, fused to a fluorescent protein and one of several repeats of the FM self-aggregation domain. It is important to note that not all FM-containing cargoes behave similarly (*see* **Note 1**). We recommend generating adherent cell lines stably expressing the synthetic cargo, but experiments can also be performed using transient transfection (*see* **Note 2**).

2.2 Chemicals and Cell Culture Medium

1. Cell culture medium appropriate for the cell line of interest. For the cell line used here (HT1080 stably expressing pC4-ssGFPFM4hGH), we use Dulbecco's Modified Eagle's Medium (DMEM) supplemented with 10% Fetal Bovine Serum (FBS, heat inactivated), 0.5 mg/mL Geneticin, and 50 U/mL penicillin-streptomycin.
2. Phosphate Buffered Saline (PBS): 138 mM NaCl, 8 mM Na_2HPO_4, 3 mM KCl, 1.5 mM KH_2PO_4, pH 7.4.
3. Hanks's Balanced Salt Solution (HBSS).
4. Cycloheximide 50 mg/mL in ethanol. Aliquot and store at −20 °C.
5. Nocodazole. Dissolve it in dimethyl sulfoxide to 1 mg/mL, aliquot, and store at −20 °C.
6. D/D solubilizer (Takara Bio, Mountain View, CA, catalog number 635053).

2.3 Immunofluorescence Microscopy and Analysis

1. Fixation solution: 4% paraformaldehyde in PBS stored at 4 °C.
2. Permeabilization solution: 0.1% (w/v) Triton X-100 in PBS.
3. Blocking solution: 1% (w/v) bovine serum albumin in PBS.
4. Primary antibodies: two antibodies from two different species that are raised against cis- or trans-Golgi markers. We recommend anti-gpp130 antibodies for labeling the cis-Golgi and anti-p230 antibodies for labeling the trans-Golgi.
5. Appropriate secondary antibodies coupled to red or far-red dyes (e.g., Alexa Fluor 546 and ATTO647N).
6. Anti-fade mounting medium.
7. Confocal microscope.
8. ImageJ software (NIH).

3 Methods

A FM-fused fluorescent secretory cargo is expressed in the cells of interest (*see* **Note 2**). In the absence of the D/D solubilizer, the cargo is aggregated and retained in the ER. To visually discriminate the cis- and trans-Golgi compartments, treat cells with Nocodazole to breakdown the Golgi ribbon and generate Golgi ministacks. Upon addition of the D/D solubilizer, the disaggregated cargo is loaded within the cis-Golgi at 16 °C. Finally, shift the temperature to 20 or 37 °C so cis→trans Golgi transport is resumed. Perform the whole procedure in the presence of cycloheximide, a protein translation inhibitor, to prevent newly synthetized cargo from interfering with the experiment. Subsequently, fix the cells and label them with antibodies targeting cis- and trans-Golgi markers. Finally, quantify the colocalization between the cargo and the cis- or trans-Golgi marker (Fig. 1a, b). Variations on this protocol can be used to further characterize the transport of the cargo of interest (*see* **Note 3**).

3.1 Cargo Loading in the cis-Golgi and Release to the Trans-Golgi

These steps are performed in temperature-controlled incubators.

1. Plate pC4-ssGFPFM4hGH-expressing HT1080 cells on glass coverslips in the complete culture medium. Let the cells adhere and spread (usually within 24 h).
2. Incubate the cells for at least 16 h in the complete culture medium containing 100 μg/mL cycloheximide under normal culture conditions (37 °C with 5% CO_2). Cycloheximide should be maintained throughout Subheading 3.1.
3. Treat cells with 1 mg/mL Nocodazole for 2 h in HBSS + 0.1% Fetal Bovine Serum.

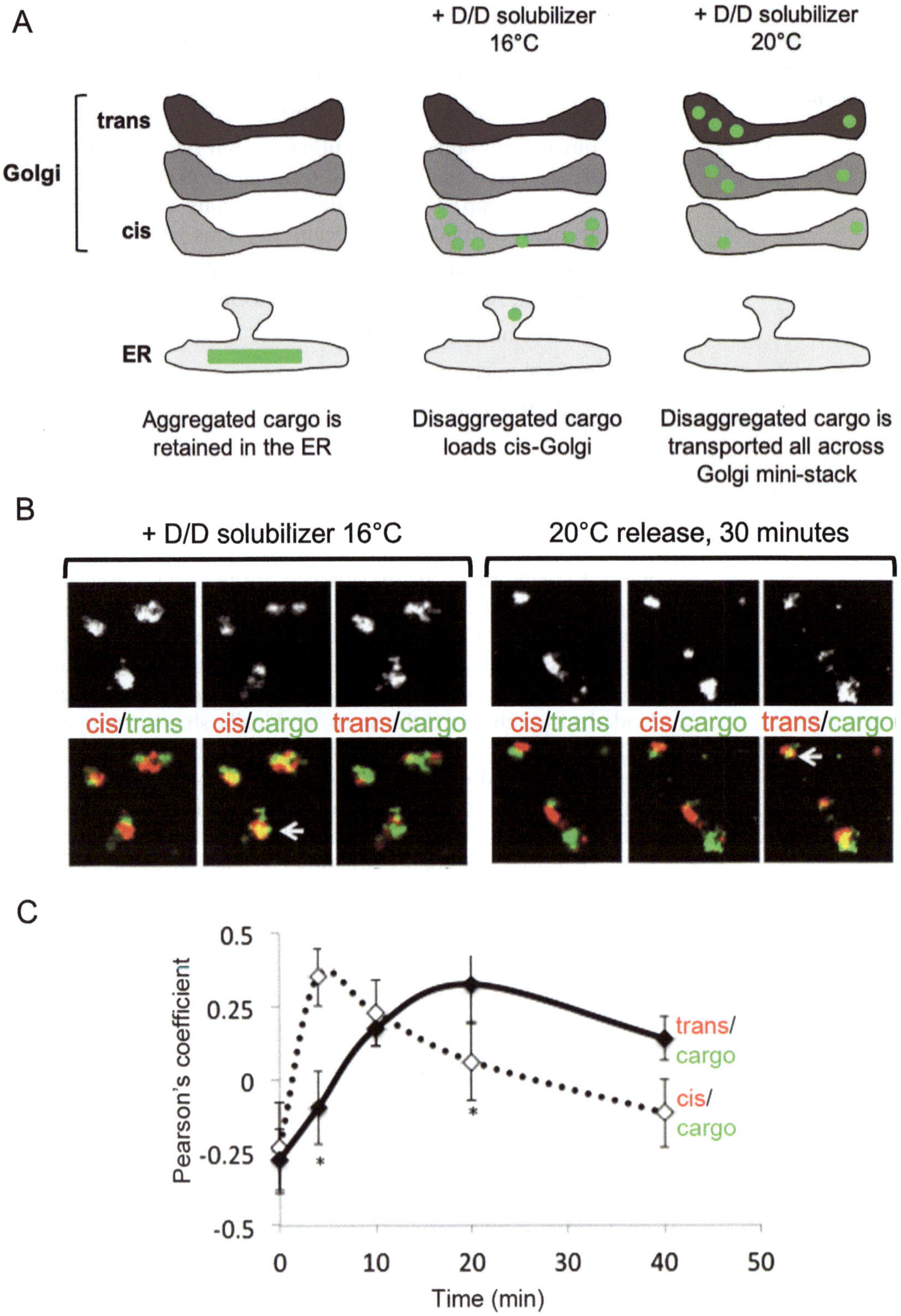

Fig. 1 Method to monitor Intra-Golgi transport with acute spatiotemporal control of a synthetic cargo. (**a**) Schematic description of the protocol. The FM-cargo forms aggregates and is retained in the ER. Nocodazole

4. Place the cells at 16 °C and add D/D solubilizer to a final concentration of 1 mM. Incubate the cells at 16 °C for 1.5 h.
5. Shift the temperature to 20 °C and incubate for various time points to resume cis→trans Golgi transport. We recommend 0, 5, 10, 20, 40 min time points.

3.2 Sample Preparation for Immunofluorescence (See Note 4)

Incubations in the following steps are performed under mechanical agitation.

1. After each time point of 20 °C incubation, terminate the reaction by replacing the culture medium with the fixation solution. Fix the cells for 15 min at room temperature.
2. Rinse twice with PBS and incubate the cells in the permeabilization solution for 10 min at room temperature.
3. Rinse once with PBS and incubate the cells in the blocking solution for 1 h at room temperature.
4. Incubate the cells with primary antibodies diluted in the blocking solution for 16 h at 4 °C.
5. Wash the cells three times for 10 min each using PBS.
6. Incubate the cells with appropriate secondary antibodies diluted in PBS for 1 h at room temperature. If using a GFP-tagged cargo, we recommend labeling gpp130 with a secondary antibody coupled to AlexaFluor 546, and p230 with a secondary antibody coupled to ATTO647N.
7. Wash three times using PBS washes and mount the coverslips on glass slides using an anti-fade mounting medium.

3.3 Immunofluorescence Microscopy and Image Analysis

1. Acquire confocal images using standard procedures with channels for GFP (cargo), red (the cis-Golgi marker), and far red (the trans-Golgi marker).
2. Calculate colocalization metrics using the Colocalization Finder plugin of the ImageJ software. For each ministack, calculate the Pearson coefficient between the cargo and

Fig. 1 (continued) treatment allows for disruption of the Golgi ribbon into mini-stacks. Upon addition of the D/D solubilizer at 16 °C, the cargo disaggregates and loads the cis Golgi. Incubation at 20 °C resumes cis→trans Golgi transport of the cargo. Cycloheximide is maintained throughout the procedure to inhibit cargo neosynthesis. (**b**) Confocal micrographs illustrate representative three-colors images of mini-stacks containing aggregated or disaggregated cargoes. HT1080 cells stably expressing ssGFP-FM4-hGH were subjected to the full protocol. Gpp1305 (cis-Golgi) was labeled with a secondary antibody coupled to Alexa Fluor 546, and p230 (trans-Golgi) with a secondary antibody coupled to ATTO647N. Colocalization of cargo with the cis- or trans-Golgi is indicated by a white arrow. (**c**) Example of intra-Golgi transport kinetic obtained with the protocol, performed here on S2 cells. The graph represents the Pearson's coefficient for each marker combination for each condition. For each condition, we analyzed between 70 and 150 ministacks. *, p values <0.01. ((**b**) and (**c**) are reproduced from [11]) cis-Golgi marker, between the cargo and the trans-Golgi marker, and between the cis- and trans-Golgi markers (as a control).

3. Plot the data in a graph representing the Pearson's coefficient for each marker combination for each condition (Fig. 1c).

4 Notes

1. Type of the cargo (soluble vs membrane), topology of the cargo (Type I vs type II), number of FM repeats influence the physical properties and behaviors of the aggregated cargo. Interestingly, FM-mediated aggregates are electron dense and we strongly recommend characterizing the morphology of the cargo using electron microscopy. Figure 2 shows different types of ER-localized aggregates. Importantly, all these aggregates are retained in the ER when they are newly formed in this organelle but behave differently within the Golgi apparatus.
2. Although the prototypal experiment is based on HT1080 cells stably expressing pC4-ssGFPFM4hGH, any transfection-competent cell line can be used to transiently express this construct.

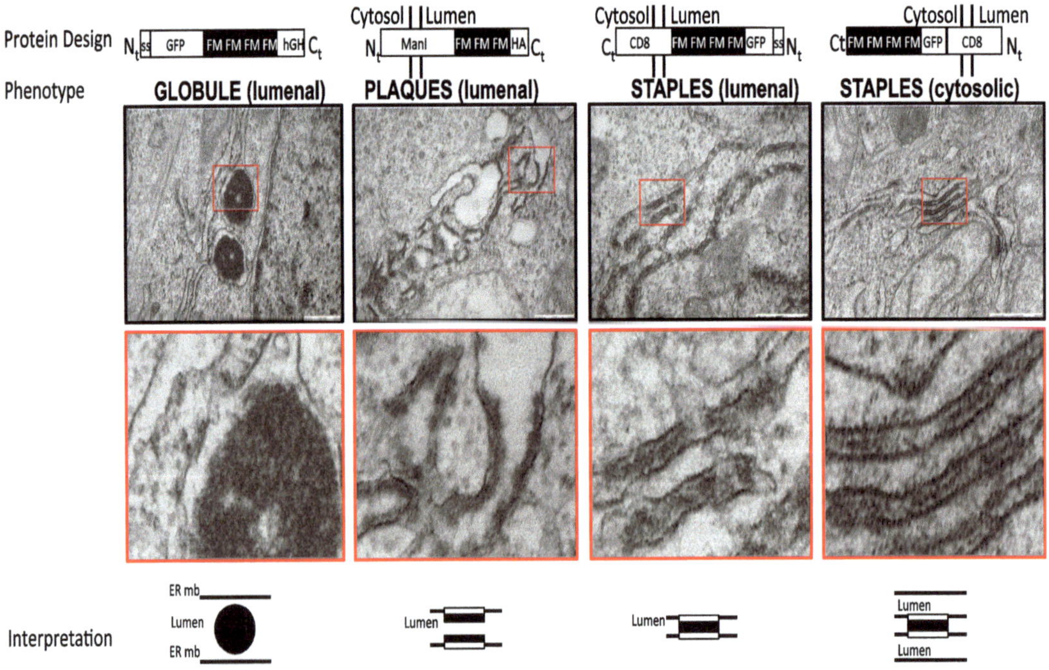

Fig. 2 Different FM-mediated aggregates behave differently. Electron micrographs illustrating different types of FM-mediated aggregates. In the absence of D/D solubilizer, FM-proteins aggregate within the ER

3. Although we hereby provide a protocol to study the rate of intra-Golgi transport, the system is versatile and allows for other studies, below are a few examples.

   By shifting the temperature to 37 °C instead of 20 °C, one can assess the kinetics of cargo exocytosis. In this case, western blotting could be used as a readout rather than fluorescence microscopy in order to detect secreted protein in the culture medium [11].

   Nocodazole is used here to break the Golgi ribbon and allows for visualization of distinct cis- and trans-Golgi compartments by fluorescence microscopy. Omitting Nocodazole allows for the study of the role of an intact Golgi ribbon on cargo secretion. Within the ribbon, cis- and trans- compartments are barely distinguishable by conventional confocal microscopy, but fully resolvable through superresolution microscopy [12].

   Omitting the D/D solubilizer for 30 min prior to the 20 °C leads to cargo reaggregation in the cis-Golgi. Aggregated cargoes require an intact Golgi ribbon to be efficiently transported through the Golgi [11]. The transport rate of an aggregated cargo through an intact Golgi ribbon (no Nocodazole treatment) can also be assessed.

   The abovementioned protocol can be used to measure general alterations to Golgi function in a large range of conditions (drugs, disease, synthetic alteration of Golgi structure, etc.).

   The aggregation system can also be combined with other synthetic approaches [16] to compare two different cargoes side-by-side.

4. We provide an immunofluorescence-based protocol here, but antibody-based detection can be replaced by overexpressing fluorescently tagged Golgi markers, which are compatible with live cell imaging. We recommend Grasp65 and Golgin-97 to label the cis- and trans-Golgi, respectively [9].

Acknowledgments

This work is supported by ANR-18-IDEX-0001, IdEx Université de Paris, ANR-19-CE18-002001, and ANR-20-CE15-002102.

References

1. Dancourt J, Barlowe C (2010) Protein sorting receptors in the early secretory pathway. Annu Rev Biochem 79:777–802
2. Ladinsky MS, Mastronarde DN, McIntosh JR et al (1999) Golgi structure in three dimensions: functional insights from the normal rat kidney cell. J Cell Biol 144:1135–1149
3. Emr S, Glick BS, Linstedt AD et al (2009) Journeys through the Golgi--taking stock in a new era. J Cell Biol 187:449–453

4. Glick BS, Luini A (2011) Models for Golgi traffic: a critical assessment. Cold Spring Harb Perspect Biol 3:a005215
5. Casler JC, Papanikou E, Barrero JJ et al (2019) Maturation-driven transport and AP-1–dependent recycling of a secretory cargo in the Golgi. J Cell Biol 218:1582–1601
6. Kurokawa K, Osakada H, Kojidani T et al (2019) Visualization of secretory cargo transport within the Golgi apparatus. J Cell Biol 218:1602–1618
7. Rivera VM, Wang X, Wardwell S et al (2000) Regulation of protein secretion through controlled aggregation in the endoplasmic reticulum. Science 287:826–830
8. Volchuk A, Amherdt M, Ravazzola M et al (2000) Megavesicles implicated in the rapid transport of intracisternal aggregates across the Golgi stack. Cell 102:335–348
9. Lavieu G, Zheng H, Rothman JE (2013) Stapled Golgi cisternae remain in place as cargo passes through the stack. Elife 2:e00558
10. Pellett PA, Dietrich F, Bewersdorf J et al (2013) Inter-Golgi transport mediated by COPI-containing vesicles carrying small cargoes. Elife 2:e01296
11. Lavieu G, Dunlop MH, Lerich A et al (2014) The Golgi ribbon structure facilitates anterograde transport of large cargoes. Mol Biol Cell 25:3028–3036
12. Dancourt J, Zheng H, Bottanelli F et al (2016) Small cargoes pass through synthetically glued Golgi stacks. FEBS Lett 590:1675–1686
13. Rizzo R, Parashuraman S, Mirabelli P et al (2013) The dynamics of engineered resident proteins in the mammalian Golgi complex relies on cisternal maturation. J Cell Biol 201:1027–1036
14. Wong M, Munro S (2014) The specificity of vesicle traffic to the Golgi is encoded in the golgin coiled-coil proteins. Science 346:1256898
15. Dunlop MH, Ernst AM, Schroeder LK et al (2017) Land-locked mammalian Golgi reveals cargo transport between stable cisternae. Nat Commun 8:432
16. Boncompain G, Divoux S, Gareil N et al (2012) Synchronization of secretory protein traffic in populations of cells. Nat Methods 9:493–498

Chapter 3

Imaging Secretory Granule Budding from the Trans-Golgi Network Using Retention Using Selective Hook (RUSH)

Cedric S. Asensio

Abstract

The retention using selective hook (RUSH) system enables us to synchronize and visualize the movement of cargoes along the secretory pathway. A fluorescently tagged cargo of interest is retained in the endoplasmic reticulum and released in a biotin-dependent manner. Here, we report a detailed protocol describing the steps necessary to perform RUSH experiments to study secretory granule biogenesis at the trans-Golgi network (TGN).

Key words Secretory granules, Insulin, Membrane trafficking, Golgi apparatus, Trans-Golgi network, Vesicle budding, RUSH, Imaging, Cell biology

1 Introduction

Eukaryotic cells rely on a constitutive secretory pathway for the release of newly synthesized proteins. Specialized secretory cells, such as endocrine cells, also express a regulated secretory pathway, which enables the storage of a subset of proteins (e.g., insulin) in intracellular vesicles [1, 2]. These vesicles, named secretory granules (SGs), accumulate intracellularly and undergo exocytosis in response to an appropriate extracellular physiological stimulus (e.g., hyperglycemia for insulin [2]). The retention using selective hook (RUSH) system enables us to synchronize and visualize the movement of cargoes along the secretory pathway [3]. This approach relies on streptavidin acting as a "hook" in the endoplasmic reticulum (ER) for a cargo fused to streptavidin-binding peptide (SBP). Exposure of cells to biotin leads to a rapid, synchronized release of the hooked cargoes from the ER. By using fluorescently tagged reporters, it becomes possible to follow a wave of cargo as it traffics through the secretory pathway.

Jingshi Shen (ed.), *Membrane Trafficking: Methods and Protocols*,
Methods in Molecular Biology, vol. 2473, https://doi.org/10.1007/978-1-0716-2209-4_3,

Here we describe a detailed, optimized protocol to perform RUSH experiments in endocrine INS-1 cells with the goal of synchronizing the trafficking of fluorescently tagged secretory granule cargoes. These experiments provide valuable information regarding the sorting and budding kinetics of these cargoes at the TGN [4]. This method can also be adapted to studies of other cargoes in a variety of cell types.

2 Materials

Prepare all solutions using ultrapure (MilliQ) water and analytical grade reagents. Store them at room temperature, unless indicated otherwise.

2.1 Cell Culture

1. Phosphate-buffered saline (PBS): 138 mM NaCl, 8 mM Na_2HPO_4, 3 mM KCl, and 1.5 mM KH_2PO_4, pH 7.4.
2. Trypsin: 0.25% (w/v) trypsin in PBS.
3. INS-1 cells culture media: biotin-free RPMI, 10% fetal bovine serum, 2.4 mM sodium bicarbonate, 25 mM HEPES (pH 7.4), 1 mM sodium pyruvate, 0.2 μm filtered (*see* **Note 1**).
4. PC12 cells culture media: DMEM, 10% horse serum, 5% cosmic calf serum, 0.2 μm filtered (*see* **Note 1**).

2.2 Transfection

1. Fugene HD (Promega, # E2311).
2. Opti-MEM I (ThermoFisher, # 31985062).
3. Plasmid DNA (endotoxin free) (*see* **Note 2**).

2.3 Imaging

1. Glass bottom imaging chambers or coverslips (#1.5 glass).
2. Poly-L-lysine solution (PLL, 10×): 100 mg in water, 0.2 μm filtered. Store aliquots at −20 °C (*see* **Note 3**).
3. Borate buffer: 0.1 M in water, pH 8.5 (*see* **Note 4**).
4. PLL solution (1×): dilute the 10× PLL stock solution using the borate buffer.
5. Biotin (D-biotin, CAS#: 58-85-5), 1000×): 12 mg in 1 mL of DMSO.
6. PFA: 4% paraformaldehyde in PBS (*see* **Note 5**).
7. Blocking buffer: 2% bovine serum albumin (BSA), 1% fish skin gelatin, 0.02% saponin in PBS, 0.45 μm filtered (*see* **Note 6**).
8. Mounting media.
9. Fluorescent microscope (widefield, confocal, or spinning-disk confocal).

3 Methods

Carry out all procedures at room temperature (RT), unless otherwise specified.

3.1 Transfection

Perform all the steps in a biosafety cabinet. All reagents should be sterile except plasmid DNA. The following protocol is optimized for INS-1 and PC12 cells (*see* **Notes 7** and **8**).

1. On the day before transfection, count cells and plate at 40% confluency in one well of a 24-well plate (*see* **Note 9**). On the following day, prepare the transfection mix: In one 1.5 mL Eppendorf tube (**tube 1**), add plasmid DNA (Table 1) and Opti-MEM I to 100 μL. Flick the tube 10 times and let sit for 5 min at RT (*see* **Note 10**).
2. In a separate 1.5 mL Eppendorf tube (**tube 2**), add Fugene HD and complete with Opti-MEM I to 100 μL. Flick tube 10 times and let sit for 5 min at RT.

   Combine the contents of **tube 1** to **tube 2**, flick the tube 10 times and incubate at RT for 20 min.
3. Add transfection mix dropwise to one well of a 24-well containing 500 μL of cell culture media.
4. Incubate for 4 h in a 37 °C incubator.
5. Change the media and return cells to the incubator.
6. On the next day, add 1xPLL solution to coat imaging chambers or coverslips. Incubate for >1 h at 37 °C.
7. Wash the imaging chambers or coverslips three times with PBS.

Table 1
List of cargoes used to study secretory granule budding from the TGN of endocrine cells, including a negative control (SBP-emdGFP-VSVG) acting as a marker of the constitutive secretory pathway and localizing at the plasma membrane at the steady state. All the cargoes described interact with a lumenal "hook." All the plasmids in this table have been described previously [4] and are available upon request

Plasmid name	Function	Tag	Steady-state localization after release
HA-KDEL	Hook	HA	ER
NPY-emdGFP-SBP	Cargo	emeraldGFP	Secretory granules
Phogrin-emdGFP-SBP	Cargo	emeraldGFP	Secretory granules
VMAT2-emdGFP-SBP	Cargo	emeraldGFP	Secretory granules
NPY-mCherry-SBP	Cargo	mCherry	Secretory granules
SBP-emdGFP-VSVG	Cargo	emeraldGFP	Plasma membrane
Sialyltransferase-TagRFP657	Marker	TagRFP657	TGN

8. Gently dissociate the transfected cells using trypsin and split them onto 4 imaging chambers or coverslips. At this point, the cells will start accumulating fluorescent cargo in the endoplasmic reticulum. This cargo can be released as a synchronized wave upon incubation with biotin.

3.2 Fixed-Cell Imaging and Analysis

The following protocol determines the colocalization of a secretory granule cargo (NPY-GFP-SBP) with insulin at different time points following the addition of biotin.

1. Add biotin to the cell culture media to a final concentration of 50 μM and incubate for an appropriate amount of time. We have routinely used the following time-course for our own experiments (no biotin, 30 min, 1 h, 2 h, 4 h and 24 h).
2. Fix the cells with PFA for 20 min.
3. Rinse the cells three times with PBS.
4. Permeabilize the cells with 0.1% Triton X-100 in PBS. Incubate for 10 min at RT.
5. Rinse the cells three times with PBS.
6. Incubate the cells in the blocking buffer for 1 h.
7. Incubate with primary anti-insulin antibodies (e.g., Millipore-Sigma, #SAB4200691, dilution: 1:1000) in the blocking buffer for 1 h.
8. Wash the cells three times with the blocking buffer (5 min for each wash).
9. Incubate with fluorescent anti-mouse secondary antibodies (e.g., Alexa 555-conjugated) in the blocking buffer for 1 h.
10. Wash the cells three times with the blocking buffer (5 min for each wash).
11. Rinse the cells once with PBS. At this stage, the dishes can be stored in PBS + 0.02% sodium azide at 4 °C up to 1 month.
12. Capture images using a confocal, widefield, or another appropriate fluorescent microscope.
13. Determine colocalization by calculating Manders correlation coefficients (*see* **Note 11**).

3.3 Live-Cell Imaging and Analysis

Perform the live-cell imaging experiments using an environmentally controlled imaging chamber (37 °C, 5% CO_2, 80% humidity). Cotransfect a fluorescent TGN marker (sialyltransferase-TagRFP647) in addition to the hook and cargo (Table 1).

1. Equilibrate the incubator and stage at 37 °C for 30 min before the experiment to avoid temperature drift.
2. Add biotin to the cell culture media to a final concentration of 50 μM.

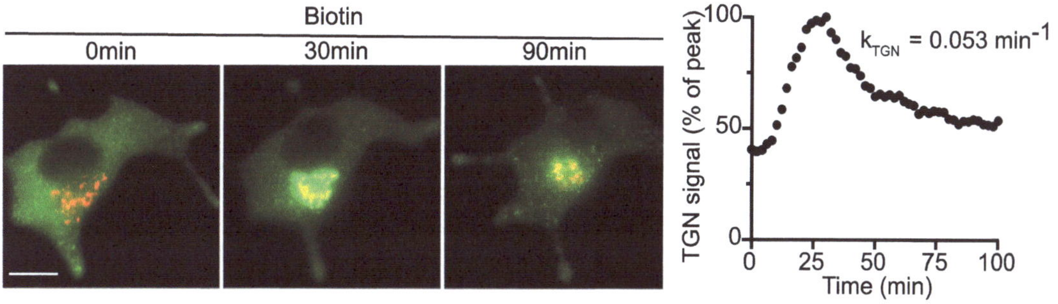

Fig. 1 INS-1 cells transfected with NPY-emdGFP-SBP (green), the TGN marker sialyltransferase-TagRFP657 (red), and HA-KDEL (hook) were incubated with biotin for the indicated times. After fixation, images were captured using a widefield epifluorescence microscope. A typical curve of NPY-emdGFP-SBP fluorescence at the TGN is shown on the right. Data were fitted using a first-order kinetic model to extrapolate the TGN (k_{TGN}) rate constant. Scale bar indicates 10 μm

3. Image the cells every 2 min for 2 h.
4. Process the images using a custom script in the MATLAB software (MathWorks). First, identify regions of interest by manually cropping each frame of the Golgi-labeled channel around the general cell area and by segmenting using an intensity threshold filter. Next, constrain the segmented region of interest by an area filter to reduce noise, generating a binary mask of only the largest objects within the frame. Finally, convolute the mask with the cargo channel to obtain the TGN signal, and convolute the inverse of the mask with the same channel to acquire the background signal.
5. Fit first order decay curves using GraphPad Prism version 8.0.2 (GraphPad Software). To fit k_{TGN}, use the plateau followed by a one phase decay function with a constraint of X0 begin greater than 0. Use the one phase decay function to fit k_{ER}. Use the following criteria to exclude cells from further analysis: (1) cell undergoing division, (2) excessive Golgi fractionation, or (3) extensive out of focus drift. Outside of these criteria, all cells are used for analysis. Typical results are shown in Fig. 1. NPY-emdGFP-SBP reaches the TGN and begins to accumulate into secretory granules within ~30 min of biotin addition.

4 Notes

1. Use a vacuum-filtration unit with a 0.2 μm filter in a biosafety cabinet.
2. Use endotoxin-free Midi or Maxi plasmid DNA of high quality for transfections. We have great success with Macherey-Nagel plasmid purification kits.
3. We typically order a bottle of 100 mg PLL. Add water to the bottle in 20-mL increments and mix to get all powder into

solution. Filter using a 0.2 μm vacuum-filtration unit. Make 1.5 mL aliquots and store at −20 °C.

4. Weigh 3.09 g of boric acid in a beaker and bring the volume to 450 mL with water. While stirring, increase the pH of solution by adding drops of 5 M NaOH until pH reaches 8.5. Bring the final volume to 500 mL with water. Filter using a 0.2 μm vacuum-filtration unit and store at 4 °C.

5. Weigh 4 g paraformaldehyde in a beaker. Working under a chemical hood, bring the volume to 100 mL with PBS. Place the beaker on a hot plate and heat to ~60 °C while stirring. When dissolved, filter the solution using a 0.45 μm vacuum-filtration unit. Use fresh.

6. Weigh 2 g of fatty-acid free BSA, 1 g of fish skin gelatin, and 20 mg of saponin in a beaker. Bring the volume to 100 mL with PBS. Stir and filter using a 0.45 μm vacuum-filtration unit. It is recommended to use fresh but storage at 4 °C for a few days is possible.

7. Use cells up to 10 passages (5–6 weeks) to avoid phenotypic drifts.

8. Standard RPMI has enough biotin to release hooked cargo, it is essential to use biotin-free RPMI for INS-1 cells. We usually grow the cells in biotin-free media for 48 h before starting any RUSH experiments. We did not notice any issues with regular DMEM for PC12 cells. Presence of biotin from the serums was not a problem either in our hands.

9. Cells should be at 60–70% confluency on the day of transfection.

10. We have determined the ideal molar ratio of hook plasmid DNA / cargo plasmid DNA to be 3:1.

11. We have successfully used the Coloc2 Fiji plugin using a region of interest (ROI) excluding the nucleus and Golgi regions. We typically exclude cells from analysis for tM1 or tM2 values of 1 due to thresholding issues as well as cells with no positive correlation.

References

1. Guest PC (2019) Biogenesis of the insulin secretory granule in health and disease. Adv Exp Med Biol 1134:17–32. https://doi.org/10.1007/978-3-030-12668-1_2

2. Hou JC, Min L, Pessin JE (2009) Insulin granule biogenesis, trafficking and exocytosis. Vitam Horm 80:473–506. https://doi.org/10.1016/S0083-6729(08)00616-X

3. Boncompain G, Divoux S, Gareil N et al (2012) Synchronization of secretory protein traffic in populations of cells. Nat Methods 9(5):493–498. https://doi.org/10.1038/nmeth.1928

4. Hummer BH, Maslar D, Soltero-Gutierrez M et al (2020) Differential sorting behavior for soluble and transmembrane cargoes at the trans-Golgi network in endocrine cells. Mol Biol Cell 31(3):157–166. https://doi.org/10.1091/mbc.E19-10-0561

Chapter 4

Spatial and Temporal Control of Protein Secretion with Light

Ashley M. Bourke and Matthew J. Kennedy

Abstract

How newly synthesized integral membrane proteins and secreted factors are sorted and trafficked to the appropriate location in different cell types remains an important problem in cell biology. One powerful approach for elucidating the trafficking route of a specific protein is to sequester it following synthesis in the endoplasmic reticulum and trigger its release with an externally applied cue. Combined with fluorescent probes, this approach can be used to directly visualize each trafficking step as cargo molecules progress through the different organelles of the secretory network. Here, we discuss design strategies and practical implementation of an inducible protein secretion system we recently developed (zapalog mediated ER trap: zapERtrap) that allows one to use light to initiate secretory trafficking from targeted cells or subcellular domains. We provide detailed protocols for experiments using this approach to visualize protein trafficking from the endoplasmic reticulum to the plasma membrane in fibroblast cell lines and primary cultured neurons.

Key words Inducible ER release, Chemo-optogenetic system, Secretory trafficking, Protein secretion

1 Introduction

Nearly all integral membrane and secreted proteins are processed through the cellular secretory network. In Eukaryotic cells, these proteins are synthesized into the endoplasmic reticulum (ER) where they are folded, assembled, and undergo various posttranslational modifications including glycosylation and disulfide bond formation. Following maturation, proteins leave the ER at specialized exit sites and accumulate in the Golgi apparatus (GA) where they are further biochemically modified and ultimately sorted into mobile carrier vesicles that shuttle them to the appropriate subcellular domain (for reviews, *see* [1–3]). Different protein cargoes utilize distinct trafficking routes, with some routes poorly characterized, highlighting the importance of new tools and approaches for investigating secretory protein trafficking.

Jingshi Shen (ed.), *Membrane Trafficking: Methods and Protocols*,
Methods in Molecular Biology, vol. 2473, https://doi.org/10.1007/978-1-0716-2209-4_4,

Classical experiments tracking proteins as they navigate the secretory network were performed by fractionation of newly synthesized proteins pulse labeled with radioactive amino acids and biochemical analysis of posttranslational modifications performed by enzymes residing in different secretory organelles. However, because this method requires cell lysis and solubilization, it yields limited information about the spatial organization of secretory trafficking. The first efforts to directly visualize a protein on its journey through the secretory network in live cells utilized the vesicular stomatitis virus glycoprotein (VSV-G) ts045 temperature-sensitive mutant, which is retained in the ER at elevated temperatures (~40 °C) but traffics normally when cells are cooled (<37 °C) [4–6]. Thus, simply by decreasing the temperature, one could liberate VSV-G (tagged with a fluorescent protein) from the ER and simultaneously visualize its progression through each organelle of the secretory network. While this powerful approach provided the first glimpse into the cellular organization of protein secretion, it was not generalizable to different proteins and lacked fine temporal control. To address these issues, several different approaches emerged that use small molecules (rather than a temperature shift) to trigger forward trafficking of user-defined proteins in cell lines and diverse model organisms [7–12]. While these "trap and release" approaches (i.e., they rely on trapping a protein in the ER and releasing it by addition of a chemical) have proven powerful for studying secretory trafficking in diverse cell types, they lack the spatial regulation required for precision control at the level of individual cells or subcellular domains. The ability to manipulate the surface levels of specific receptors, or secreted proteins within user-defined cell populations would be powerful for studying cell signaling, especially in intact organisms. Furthermore, in large cells of the nervous system such as neurons or glia, whose secretory organelles are broadly dispersed, the ability to initiate secretory trafficking from specific subcellular domains (e.g., dendrites, cell body, or axons) would enable one to address previously intractable questions concerning how (or whether) proteins are spatially compartmentalized in complex cellular environments [13]. Thus, there remains an unmet need for tools that enable control of secretory trafficking in both space and time.

1.1 Why Use Light to Control Cellular Function

New approaches for precise activation, inhibition and observation of cellular processes are continuously deepening and refining our understanding of cell biology. The confluence of chemical genetics (the ability to control the function of an engineered protein with a small molecule that is otherwise biologically inert) and live cell fluorescence microscopy has yielded a powerful toolkit to directly manipulate and observe diverse cellular functions [14–16]. While traditional chemical genetic approaches using small molecule actuators are powerful, they generally lack spatial control for targeting specific groups of cells, individual cells or subcellular domains. This issue is being addressed through development of diverse photosensitive actuator proteins (phytochromes, cryptochromes, and LOV domains, among others) and small molecules that can be activated with a pulse of UV, visible, or infrared light [9, 17–31]. In many ways, light is an ideal actuator for controlling cellular biochemistry since it can be delivered for precise durations at user-defined times over spatial scales ranging from entire organisms to submicrometer subcellular domains. Pioneering approaches used light-sensitive "caged" small molecule second messengers or neurotransmitters that could be photoconverted to a bioactive state [32, 33]. Coupled with real time readouts of cellular function, this "photouncaging" approach has had immense impact for studying compartmentalized signaling pathways. More recent efforts have culminated in photosensitive chemical-induced protein dimerization strategies to leverage the spatial control afforded by light [34–43]. One such approach is based on a compound called zapalog, which is composed of a rapamycin homologue (which binds FK506 binding protein; FKBP), linked to trimethoprim (which binds bacterial dihydrofolate reductase; DHFR) via a photocleavable chemical linker (Fig. 1a) [44]. Zapalog therefore heterodimerizes two target proteins, one tagged with DHFR and the other tagged with FKBP, and this interaction can be subsequently disrupted with a brief pulse of near UV light (405 nm) (Fig. 1b). This approach was originally developed for toggling organelles on and off microtubule-based motor proteins for precise control of organelle localization [44]. Here we describe a detailed method for a new zapalog-based application (zapERtrap) for "uncaging" proteins from the ER, allowing precise spatial and temporal control of secretory trafficking from targeted cells or subcellular domains [45].

2 Materials

2.1 Considcrations for Plasmid Design

We have adapted zapalog to control protein secretion using FKBP fused to a fluorescent protein containing an ER retention motif (the C-terminal amino acid sequence: KDEL) and a target protein fused to DHFR (Fig. 1c) [45]. It is critical to understand the membrane topology of the target protein such that the engineered DHFR tag is presented in the lumen of the ER to make it available for zapalog-mediated binding to the FKBP-KDEL "anchor" which binds to and retains the target protein in the ER (Fig. 2). Brief exposure to 405 nm light, applied either globally (e.g., using full field illumination with a 405 nm laser or CFP filter cube coupled to a full spectrum light source), or locally (using a galvanometer-steered 405 nm laser or digital micromirror device, details below) photolyzes zapalog, liberating the DHFR-tagged target protein from the FKBP-KDEL retention module, allowing forward trafficking to

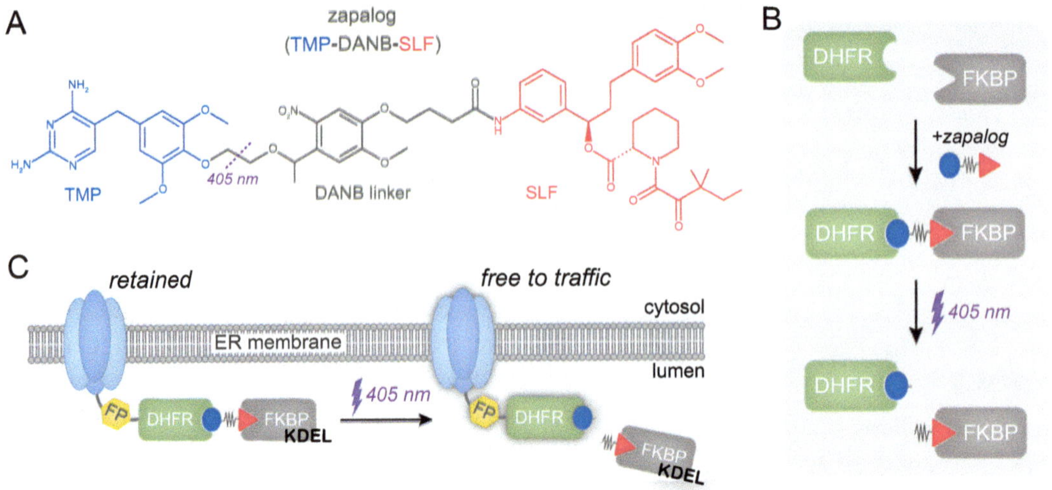

Fig. 1 Photo-inducible protein secretion with zapERtrap. (**a**) Zapalog chemical structure. Zapalog is comprised of the antibiotic trimethoprim (TMP), a synthetic ligand of FKBP (SLF) and a dialkoxynitrobenzyl (DANB) moiety susceptible to photolysis by 405 nm light. The purple dashed line marks the location of photocleavage. (**b**) Photorelease of DHFR/FKBP linked through zapalog. In the presence of zapalog, the proteins DHFR and FKBP become tightly associated through their respective binding to the TMP and SLF moieties of zapalog. Photoexcitation with 405 nm light cleaves the DANB linker and disrupts the association between DHFR and FKBP. (**c**) zapERtrap control of protein secretion. Following synthesis in the ER membrane, DHFR-fused secretory cargo proteins are retained via zapalog-mediated association with FKBP-KDEL present in the ER lumen. Photolysis of zapalog with 405 nm light rapidly disrupts the link between the secretory cargo and the FKBP-KDEL ER anchor, allowing forward trafficking to proceed. The DHFR fusion construct is engineered with a fluorescent protein (FP) tag for visualization of subsequent forward trafficking steps. DHFR, dihydrofolate reductase; FKBP, FK506-binding protein

proceed. This general strategy has worked well for both type I membrane proteins where the DHFR is fused to the N-terminus (e.g., neuroligin 1 and the glutamate receptor subunit, GluA1) and for type II membrane proteins where DHFR is fused to the C-terminus (e.g., transferrin receptor). For N-terminally tagged proteins it is important to preserve the signal peptide that generally directs nascent peptides to the ER (Fig. 2b). In these cases, we have placed the DHFR tag immediately following the signal peptide with good success. While the DHFR tag appears to work well for FKBP-mediated ER retention when it is placed at the C- or N-terminus, we have not yet attempted to integrate DHFR into the middle of a target protein, but this may be an important consideration in cases where N- or C-terminal fusion is known to disrupt protein function and/or localization. Plasmids we have designed that express the FKBP-KDEL ER anchor fused to different fluorescent proteins should be generally effective for retaining diverse targets and are available at the Addgene plasmid repository (along with several DHFR-tagged secretory proteins we have generated).

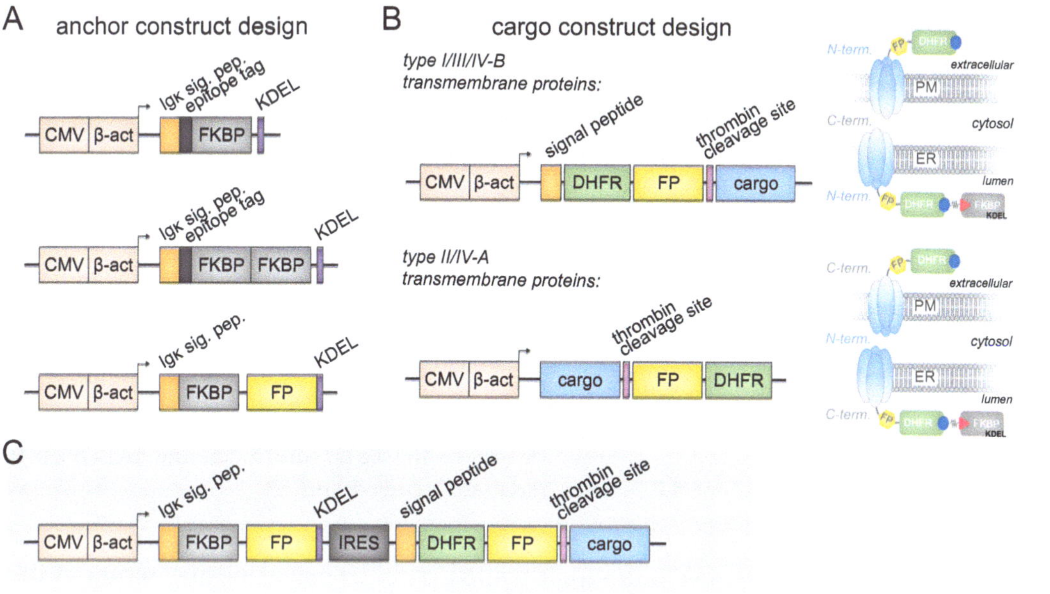

Fig. 2 ZapERtrap construct design. (**a**) Anchor construct design. The essential components of the FKBP anchor construct include an N-terminal signal peptide sequence from mouse immunoglobulin kappa light chain (Igκ signal peptide) for proper ER targeting, an epitope tag to confirm expression/ER localization, and a C-terminal KDEL sequence to retain it within the ER lumen (*top*). Constructs with tandem copies of FKBP (*middle*) enhance ER retention of DHFR-fused cargoes with no apparent effect on the kinetics of cargo release; however, this may not be the case for all cargoes and therefore must be empirically determined. Care should be taken to preserve the N- and C-terminal position of the Igκ signal peptide and KDEL sequence, respectively. Effective anchor constructs include those with an epitope tag (e.g., hemagglutinin) positioned in between the Igκ signal peptide and FKBP (*top and middle*) or with a fluorescent protein (FP) positioned in between FKBP and the KDEL sequence (*bottom*). (**b**) Cargo construct design. N- or C-terminal placement of DHFR depends on the orientation of the cargo of interest in the ER membrane. Transmembrane proteins of types I, III, and IV-B are oriented with the N-terminus on the luminal (extracellular/exoplasmic) face and the C-terminus on the cytosolic face (*top right*). Type II and IV-A transmembrane proteins orient with cytoplasmic N- and luminal (extracellular/exoplasmic) C-terminus (*bottom right*). For cargo retention, DHFR is fused to the terminus within the ER lumen. If the cargo of interest does not contain an internal ER targeting sequence (present in most type II–IV transmembrane proteins), a suitable N-terminal signal peptide sequence must be included for proper targeting to the secretory pathway. Additional considerations are the inclusion of an epitope/FP tag and/or a protease (e.g., thrombin) cleavage site to remove the epitope if necessary. The N- or C-terminal placement of an epitope/FP tag for surface detection depends on the orientation of the cargo in the plasma membrane, and is therefore the same terminus that DHFR is fused to. DHFR should be positioned downstream of the N-terminal signal peptide sequence (as is the case for type I/III/IV-B transmembrane proteins that do not have an internal ER targeting sequence; *top left*) or at the C-terminal end of type II/IV-A transmembrane proteins (*bottom left*). The protease cleavage site is positioned between the epitope/FP tag and the cargo sequence. (**c**) Example of an IRES-containing zapERtrap construct for coexpression of cargo and anchor. IRES-mediated (5′ cap-independent) translation results in a relatively lower expression of the second gene, therefore it is recommended to prioritize expression of the anchor by positioning it upstream of the IRES. Use of a bicistronic zapERtrap construct increases the likelihood for a given transfected cell to express both anchor and cargo and may result in more uniform expression ratios across cells than cotransfection of multiple plasmids. *β-act* β-actin; *CMV* cytomegalovirus; *DHFR* dihydrofolate reductase; *ER* endoplasmic reticulum; *FKBP* FK506-binding protein; *FP* fluorescent protein; *IRES* internal ribosomal entry site; *PM* plasma membrane; *C-term.* C-terminus; *N-term.* N-terminus; *sig. Pep.* signal peptide

2.2 Cell Transfection

1. Zapalog compound (MedChemExpress, cat. no. HY-126316; *see* **Note 1**).
2. Transfection reagent. We use Lipofectamine 2000 (Thermo Fisher Scientific, #11668030).
3. Aluminum foil.
4. Tissue culture biosafety cabinet.
5. Tissue culture incubator set at 37 °C with 5% CO_2.
6. Plasmids encoding FKBP "anchor" (*see* Subheading 2.1 and **Note 2**) and DHFR target cargo (*see* Subheading 2.1 and **Note 3**).

For experiments with heterologous cells

7. COS-7, HeLa or other suitable adherent cell line grown on #1 or #1.5 cover glass (*see* **Notes 4** and **5**).
8. Dulbecco's Modified Eagle Medium (DMEM).
9. DMEM supplemented with 10% fetal bovine serum (FBS), *optional*: 50 units/mL penicillin, 50 μg/mL streptomycin.

For experiments with primary cultured neurons

10. Primary rat hippocampal or cortical cultured neurons plated on poly-D-lysine-coated #1 or #1.5 cover glass (*see* **Note 5**).
11. Neurobasal-A medium.
12. Neurobasal-A medium supplemented with GlutaMAX and B27.

2.3 Live-Cell Imaging

1. HEPES-based imaging solution: 130 mM NaCl, 5 mM KCl, 10 mM HEPES-NaOH, pH 7.4, 30 mM D-glucose, 2 mM $CaCl_2$, 1 mM $MgCl_2$.
2. Cover glass chamber for live cell imaging. We use Ludin chamber type 1 (Life Imaging Services, #10900, *see* **Note 5**).
3. Light source: Any of the following are suitable for activating zapalog: laser light source from confocal microscope, LED/DMD, xenon/mercury/LED lamp with near UV excitation filter (e.g., CFP).
4. Red safelight (LED bicycle tail lights or headlamps with wavelengths ~620 nm work well).
5. Alexa Fluor 647–conjugated polyclonal anti-GFP antibody if simultaneously imaging surface accumulation (Invitrogen, # A31852; RRID:AB_162553).

3 Methods

3.1 Cell Transfection (for One 18 mm Coverslip in a 12-Well Plate)

3.1.1 Transfection of Heterologous Cells

1. Split/plate cells the day before so they are ~50% confluent at the time of transfection (*see* **Note 4**).
2. Add 50 μL of serum-free, antibiotic-free DMEM to a 1.5 mL microtube containing 1.5 μL of Lipofectamine 2000 and incubate at room temperature for 5 min.
3. Add 0.5–1 μg of plasmid DNA (for each CMV-driven expression construct) and/or 100–300 ng of plasmid DNA (for each CAG-driven expression construct) to a 1.5 mL microtube containing 50 μL of serum-free DMEM (*see* **Note 6**). If necessary, adjust the volume of Lipofectamine 2000 to maintain a 1: 0.5 to 1:5 DNA (μg):Lipofectamine 2000 (μL) ratio and scale according to manufacturer's recommendations.
4. Add the diluted Lipofectamine 2000 to the diluted DNA and gently mix by pipetting. Incubate at room temperature for 20 min.
5. During the incubation, remove medium from the well such that only ~800 μL remains, and save this conditioned media. Return cells to incubator.
6. In a tissue culture biosafety cabinet illuminated with red LED safelight, make a 40 μM working stock of zapalog diluted in the saved conditioned media (*see* **Note 1**). Wrap in aluminum foil until ready to use.
7. Add DNA-Lipofectamine transfection mixture to cells dropwise.
8. In a tissue culture biosafety cabinet illuminated with red LED safelight, add 10–20 μL of a 40 μM working stock of zapalog (for a final concentration of 0.5–1 μM) directly to cell media and gently swirl plate to mix (*see* **Note 7**). Wrap plate in aluminum foil.
9. Incubate cells in a 5% CO_2 incubator at 37 °C for 12–24 h to allow expression of the constructs (*see* **Note 8**).

3.1.2 Transfection of Cultured Rat Hippocampal/Cortical Neurons (See **Note 9**)

1. For each well (using standard 12-well dish with cells grown on a 18 mm round cover glass) to be transfected, add 50 μL of serum-free Neurobasal-A media to a 1.5 mL microtube containing 1.5 μL of Lipofectamine 2000 and incubate at room temperature for 5 min.
2. Add 0.5–1 μg of plasmid DNA (for each CMV- or Synapsin-driven expression construct) and/or 100–300 ng of plasmid DNA (for each CAG-driven expression construct) to a 1.5 mL Eppendorf tube containing 50 μL serum-free DMEM (*see* **Note 6**). If necessary, adjust the volume of Lipofectamine

2000 to maintain a 1:0.5 to 1:5 DNA (μg):Lipofectamine 2000 (μL) ratio and scale according to manufacturer's recommendations.

3. Add the diluted Lipofectamine 2000 to the diluted DNA and gently mix by pipetting. Incubate at room temperature for 20 min.
4. During the incubation, remove medium from the well such that 0.8–1 mL remains, and save this conditioned medium. Return cells to incubator.
5. In a tissue culture biosafety cabinet illuminated with red LED safelight, make a 40 μM working stock of zapalog diluted in conditioned media (*see* **Note 1**). Wrap in aluminum foil until ready to use.
6. Add an equal volume of fresh Neurobasal-A supplemented with B27 and GlutaMAX to the saved conditioned media to obtain a 1:1 ratio. Warm in a 37 °C water bath or cell culture incubator.
7. Add DNA-Lipofectamine transfection mixture to cells dropwise. Gently swirl plate to mix and return to incubator.
8. After 1.5 h, remove medium containing DNA-Lipofectamine transfection mixture and replace with 800 μL of warmed 1:1 conditioned–fresh media.
9. In a tissue culture biosafety cabinet illuminated with red LED safelight, add 10–20 μL of a 40 μM working stock of zapalog (for a final concentration of 0.5–1 μM) directly to cell media and gently swirl plate to mix. Wrap plate in aluminum foil.
10. Incubate cells in a 5% CO_2 incubator at 37 °C for 12–24 h (*see* **Note 8**).

3.2 Live Cell Imaging: Global ER Release

1. We use a full enclosure live cell incubator on a spinning disk confocal microscope that requires prewarming to 34–37 °C for at least 1 h prior to use to ensure focal stability during time lapse imaging.
2. *Optional:* If your cargo constructs have incorporated a thrombin protease site to remove extracellular epitopes from cargo that may have escaped zapalog-mediated ER trapping, add 1 unit/mL of thrombin directly to cell medium 10–30 min prior to imaging to eliminate background surface signal (*see* **Note 10**).
3. In a tissue culture biosafety cabinet illuminated with red LED safelight, remove coverslip from the 12-well plate and place in an appropriate live cell imaging chamber (*see* **Note 6**) and add 0.7 mL of prewarmed HEPES imaging solution with or without 0.5–1 μM zapalog (*see* **Note 11**). Gently clean the bottom of the coverslip with a Kimwipe dampened with 70% ethanol.

Transport imaging chamber in a light proof container (an empty tip box wrapped in aluminum foil works well).

4. Place imaging chamber on microscope, taking care to keep cells shielded from light during transfer. Use a red LED safelight for placement. Avoid the use of unfiltered transmitted light. Our full enclosure incubation chamber is made of black plexiglass to avoid ambient light exposure of our samples. We also suggest altering the display mode of computer screens in the imaging area with built-in or third-party blue light filtering programs (we use the f.lux app in “dark mode”).
5. *Optional:* For experiments involving live, real time surface labeling of released cargoes as they appear at the cell surface, add 50 μL of diluted dye-conjugated antibody (typically 1:750 final dilution in the imaging chamber). Gently mix by pipetting. We recommend using fluorescent protein or epitope tag Alexa Fluor 647–conjugated antibodies (*see* **Note 12**).
6. Locate transfected cells with 561 nm excitation light, if possible. Brief exposure to 488 nm excitation light does not appear to appreciably photolyze zapalog; however, it is unknown whether long-term exposure of high intensity 488 nm excitation light causes unwanted photolysis.
7. Establish a 5–10 min baseline by acquiring a series of images obtained with 488, 561 and 640 nm excitation light. We take confocal z-stacks with 0.4 μm spacing and capture 12 optical sections per time point. For purposes of measuring kinetics of cargo accumulation at the Golgi apparatus and/or plasma membrane, it is sufficient to acquire a z-stack image every 1–2 min. Take single-channel images at a frame rate of 2 to 5 Hz to capture fast dynamics of post-ER and/or post-Golgi vesicles as these structures move on the order of 0.5–1 μm/s. Imaging at this frequency should be limited to ~3–5 min as cells can become photodamaged; however this is highly dependent on the wavelength and intensity of excitation light (*see* **Note 13**).
8. After acquiring the baseline images, confirm the absence of background surface signal (640 nm channel) arising from recycling or surface receptors that may have escaped the ER. If appreciable signal is observed, a new cell in a different region of the coverslip should be located and **step 6** repeated (*see* **Note 12**).
9. Initiate global ER release by full-field illumination with 405 nm light of appropriate intensity and exposure time (*see* **Note 14** below). We typically acquire 2 z-stacks with light from a 100 mW fiber-coupled 405 nm laser at 80% laser power (912 μW measured from the objective) delivered from an acousto-optic tunable filter (AOTF) controlled laser launch.

10. Continue imaging to visualize cargo as it progresses from the ER to the Golgi apparatus, to the plasma membrane. It will be obvious if efficient ER release has occurred as cargo should begin to accumulate in the Golgi apparatus within 5 min of the 405 nm light exposure, although kinetics of ER trafficking can vary widely for different cargoes. Subsequent surface signal (detected with dye conjugated anti-extracellular epitope antibody) should be apparent approximately 30 min following 405 nm illumination, although some cargoes may take longer. *See* Fig. 3b for an example.

3.3 Live Cell Imaging: Local ER Release from Single Cells and Subcellular Domains

3.3.1 Local Release in COS-7 Cells

1. Performs **steps 1–8** from Subheading 3.2.
2. Identify a single cell or cells to target.
3. Using a light source capable of targeted illumination (we use the FRAPPA system from ANDOR to steer a diffraction limited spot from a 405 nm laser source), locally stimulate the region of interest. With the FRAPPA system, we typically scan the region of interest with a pixel dwell time of 1 ms with 405 nm power = 23.0 $\mu W/\mu m^2$ measured through a 60× Plan Apochromat NA 1.42 objective). An example of a local release experiment from a single targeted COS-7 cell is shown in Fig. 3d.

3.3.2 Local Release in Neurons

1. Performs **steps 1–8** from Subheading 3.2.
2. Initiate local ER release by targeted illumination with 405 nm light of appropriate intensity. We trigger local zapalog photolysis with 23.0 $\mu W/\mu m^2$ 405 nm illumination (6% total laser power from a 100 mW fiber-coupled laser) using the FRAPPA system to steer a diffraction limited 405 nm spot with a 0.5–1 ms dwell time. Examples of local release experiments where cargo is release either from the soma, or from a defined section of neuronal dendrite is shown in Fig. 4. Once more, efficacy of light-triggered ER release can be assessed by redistribution of cargo to the Golgi apparatus and/or mobile intracellular vesicles following ER exit (Fig. 4).

4 Notes

1. Zapalog is sensitive to ambient light. Zapalog and zapalog-treated cells should be handled in the dark at all times. To prepare 1 M zapalog stocks, dilute in DMSO and make 10 μL aliquots in a tissue culture biosafety cabinet illuminated with red LED safelight. We wrap each aliquot in aluminum foil and store at −20 °C. Use these 1 M stocks to prepare the 40 μM working stock on the day of transfection. The 1 M stocks can undergo multiple freeze–thaw cycles.

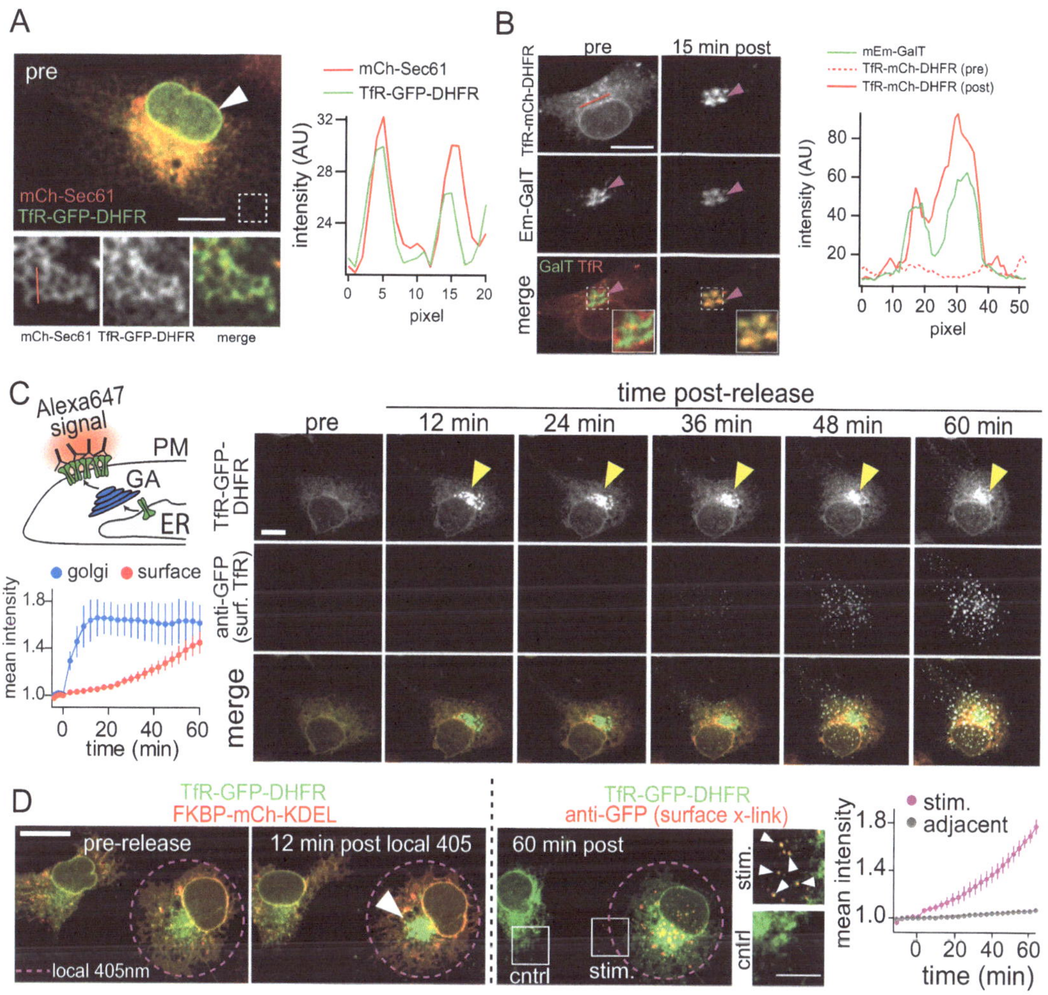

Fig. 3 Inducible and local control of secretory trafficking in COS-7 cells. (**a**) TfR-GFP-DHFR (green) colocalizes with ER marker mCh-Sec61 (red) in COS-7 cells expressing FKBP-KDEL in the presence of zapalog. The white arrowhead denotes the nuclear envelope, which is contiguous with the ER and contains both red and green signal. The bottom panels show enlarged images of mCh-Sec61 and TfR-GFP-DHFR at the cell periphery (denoted by dashed box in top panel) where ER morphology is distinct. A plot of pixel intensities along the red line (in image inset) is plotted on the right. Scale bar, 10 μm; magnified images are 6 × 6 μm. (**b**) Shown is a COS-7 cell expressing TfR-mCh-DHFR, the GA marker mEmerald-GalT and FKBP-KDEL in the presence of zapalog before (left) and 15 min following 405 nm light exposure (right). The magenta arrowheads denote the GA. The insets show colocalization between mEmerald-GalT and TfR-mCh-DHFR. Note the display settings were adjusted for the TfR-mCh-DHFR channel 15 min post-image acquisition to avoid apparent saturation of the signal as it concentrated in the GA. Pixel intensities along the red line shown in the upper left panel are plotted for mEmerald-GalT (green line), TfR-mCh-DHFR before ER release (red dashed line) and 15 min after ER release (solid red line). Scale bar, 10 μm; magnified images are 5 × 5 μm. (**c**) Live-cell antibody surface labeling reveals the spatiotemporal dynamics of cargo surface presentation following ER release. Top Left: Schematic of strategy for real-time visualization of surface accumulation using Alexa 647-anti-GFP, which rapidly binds to TfR-GFP-DHFR as it appears on the cell surface. Right: Image time series of TfR-GFP-DHFR accumulation in the GA (top panels; arrowhead) and on the cell surface (middle panels) following light-triggered ER release at time 0. Scale bar, 15 μm. Bottom Left: Kinetics of TfR-GFP-DHFR accumulation in the

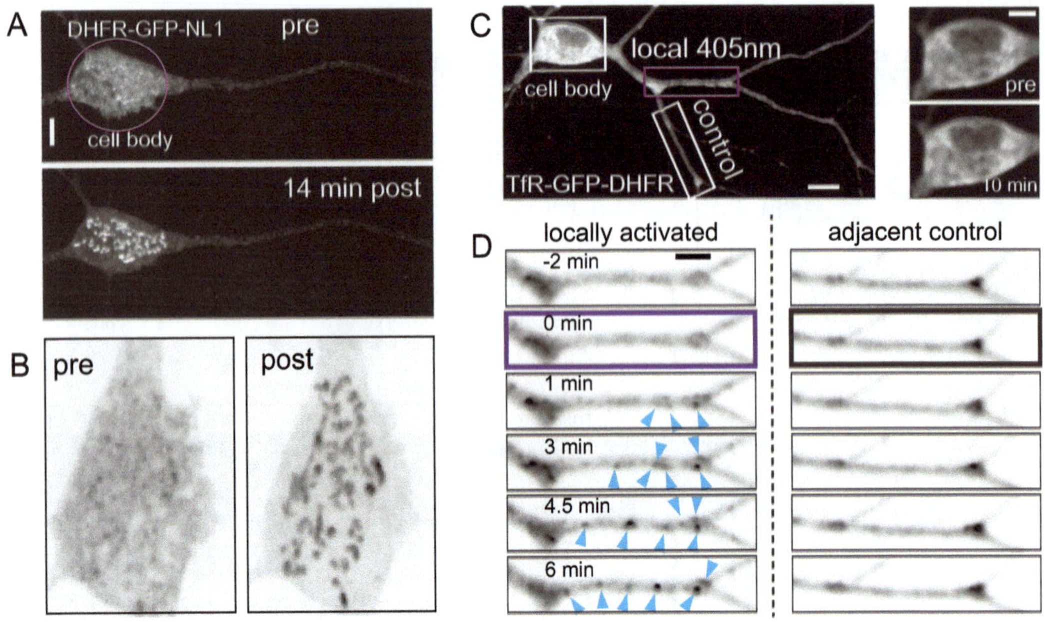

Fig. 4 Local control of secretory trafficking in neurons. (**a**) A hippocampal neuron expressing DHFR-GFP-Neuroligin1 (NL1) along with FKBP-mCh-KDEL (not shown) was imaged before (top) and 14 min following (bottom) local illumination of the cell body with 405 nm excitation light (purple circle). Note the robust redistribution of GFP signal to the Golgi apparatus in the cell body. (**b**) Magnified images of the cell body from (**a**) are shown in gray scale to better visualize the difference in DHFR-GFP-Neuroligin1 localization pre/-postillumination. (**c**) A segment of a dendritic branch on a hippocampal neuron expressing TfR-GFP-DHFR along with FKBP-mCh-KDEL (not shown) was locally illuminated with 405 nm light (purple box). The magnified images to the right demonstrate no detectable accumulation TfR-GFP-DHFR in the cell body Golgi apparatus following local dendritic ER release. (**d**) Magnified images of the targeted (left) and control (right) dendritic branches from (**c**) show intracellular trafficking organelles accumulating TfR-GFP-DHFR in the illuminated branch but not in the neighboring control branch, confirming selective initiation of secretory trafficking within the illuminated region. (Data reproduced with permission from Ref. 45)

Fig. 3 (continued) GA (blue) and on the cell surface (red) following release; mean ± SEM ($n = 6$ cells from 2 independent experiments). (**d**) Local 405 nm excitation triggers ER release of TfR-GFP-DHFR selectively in targeted cells. Left: Focal illumination with 405 nm light (region marked by the pink dashed line) triggers TfR-GFP-DHFR (green) trafficking to the GA (arrowhead) and the surface (right, Alexa 647-anti-GFP puncta shown in red) only in the photoactivated cell. Magnified images (taken from the regions marked by the white boxes) to the right show TfR-GFP-DHFR surface label (arrowheads) in the photoactivated cell ("stim."; top) and neighboring control cell (bottom). Scale bar, 20 μm. Inset scale bar, 10 μm. The plot to the right shows the average time course of TfR-GFP-DHFR surface accumulation for photoactivated cells (pink line) and neighboring control cells that were not photoactivated (gray line); mean ± SEM ($n = 6$ photoactivated cells, 8 control cells from 2 independent experiments). (Data reproduced with permission from Ref. 45)

2. We use the signal peptide sequence from mouse immunoglobulin kappa light chain (METDTLLLWVLLLWVPGSTG) to direct FKBP to the ER. *See* Fig. 2a for more information on anchor construct design.
3. It is critical to know the topology classification of your protein of interest (e.g., types I–IV) for proper construct design. It may be that your protein of interest contains an internal ER targeting signal (as is the case with the majority of type II–IV transmembrane proteins). The classification also refers to the orientation of the N- and C-termini across the ER membrane. For zapERtrap experiments, DHFR must be fused to the cargo at the terminus that faces the ER lumen, which may not be feasible if this is known to disrupt proper localization of the cargo. In all cases, it is critical to confirm proper subcellular targeting of a nonretained DHFR-fused cargo by comparing it to other fusion constructs (preferably to those that have been previously validated/published). *See* Fig. 2b for more information on cargo construct design.
4. For imaging experiments using heterologous cells, we prefer to use COS-7 cells, which have a flat, "spread out" morphology on the coverslip, allowing clear visualization of trafficking organelles.
5. Instead of using 18 mm coverslips and a Ludin chamber, one can use 35 mm glass bottom dishes (#1 or #1.5 cover glass, 14 mm glass diameter, poly-D-lysine-coated).
6. The exact amounts of cargo and anchor plasmid DNA ratios must be empirically determined. Bicistronic expression of cargo and anchor is a possibility (*see* Fig. 2c); however, keep in mind that excess anchor can still lead to background leakiness caused by saturation of endogenous KDEL receptor. Thus, it is critical to assess the efficiency of cargo retention using various cargo/anchor DNA ratios (by surface staining for an extracellular epitope tag in dark vs light treated cells).
7. To conserve zapalog compound, it is possible to "recycle" zapalog on sequential days by applying zapalog-conditioned media from a previous experiment to newly transfected cells.
8. Background "leakiness" of cargo to the plasma membrane increases with expression duration. We find it is best to perform imaging experiments 12–18 h after transfection, especially when expressing proteins driven by CAG and Synapsin promoters.
9. We plate rat hippocampal neurons at a density of 80,000–100,000 cells per well from a 12-well plate. We plate rat cortical neurons at a density of 175,000–200,000 cells per well from a 12-well plate.

10. Incubate cells with 1 unit/mL thrombin 10–30 min prior to imaging in order to cleave GFP (or other affinity tag) and thereby prevent surface signal detection of DHFR-fused cargoes already present on the plasma membrane prior to 405 nm light illumination. Background "leakiness" of cargo to the plasma membrane varies with the target cargo under investigation and from cell to cell in the same experiment, and must be empirically determined. We find thrombin protease cleavage to be faster and more complete than Tobacco Etch Virus (TEV) protease cleavage.
11. We find robust ER retention for at least an hour after washout of zapalog, so inclusion of zapalog in the imaging solution may be unnecessary for short-term imaging experiments. We typically include 0.5–1 μM zapalog in our imaging solution as we typically image for ~2 h after release.
12. Alexa Fluor 647-anti-GFP (Invitrogen, cat. no. A31852) binding to surface cargoes occurs within tens of seconds; however, we recommend waiting 10–15 min after antibody addition before baseline image acquisition to confirm that there is no baseline "leakiness" present in the cell(s) under investigation. Note that presence of dye-conjugated antibody will result in background fluorescence that may obscure weak surface delivery. If this is the case, the antibody can be washed off for imaging at different time points.
13. To minimize cell phototoxicity and maximize detection sensitivity of intracellular cargo trafficking, we perform live-cell labeling of HaloTag-fused cargo constructs using the cell permeant Janelia Fluor 646 HaloTag ligand. Imaging with 640 nm excitation light is less phototoxic to cells than 488 nm excitation light. For combined intracellular trafficking and surface detection, we use DHFR-fused cargoes tagged with both HaloTag and GFP, and Alexa Fluor 555–conjugated GFP antibody for surface detection (GFP antibody binding yields brighter, punctate signaling compared to halo-dye).
14. Zapalog is photosensitive in the near UV spectrum and we have found that 405 nm light delivered either from a laser source or light emitting diode (LED) is effective for photolyzing zapalog and triggering protein trafficking in live cells. A metal halide, mercury, or xenon light source commonly used for fluorescence microscopy should be equally suitable. Importantly, common red and green fluorescent proteins (EGFP, mNeon, mCherry, etc.) can be simultaneously imaged with negligible zapalog photolysis using common laser lines (e.g., 488 and 561 nm) or filter cubes (Cy2/EGFP, Cy3/TRITC, etc.). However, we wish to stress that ambient room light could photolyze zapalog, depending on the duration and intensity of exposure. *We perform all handling of zapalog-treated samples*

(and the zapalog compound itself) under dim red light and filter the transmitted light on our microscope (we use a 45 mm Schott RG610 long pass light filter) when we set up samples for imaging.

For local zapERtrap experiments (e.g., releasing proteins from the ER only in targeted cells or subcellular domains) we generally use galvanometer-steered mirrors (FRAPPA device, Andor Technologies) to scan a diffraction limited 405 nm excitation spot over the region of interest. Typical power settings are 6% laser power from a fiber-coupled 100 mW 405 nm laser (which yields 23.0 $\mu W/\mu m^2$ through a 60× Plan Apochromat NA 1.42 objective) with a scanning dwell time of 1 ms. We have also successfully used an LED source coupled to a digital micromirror device (Mosaic, Andor Technologies) for illumination of a region of interest. Finally, for global ER release (e.g., from all cells within an imaging field), we use full field excitation with a 405 nm laser source directed through a spinning disk scan head with a 50 ms exposure at 80% laser power (100 mW fiber coupled laser), which yields 912 μW at the objective (60×, NA 1.42). One powerful feature of zapERtrap is the ability to titrate the level of protein released simply by adjusting the light intensity. Here we use an acoustooptic tunable filter (AOTF) to control the intensity of 405 nm laser excitation (focused to a diffraction limited spot). Typical powers for triggering ER release are 23.0 $\mu W/\mu m^2$ for 100% release and 6.83 $\mu W/\mu m^2$ triggering ~50% release (measured through a Plan Apochromat NA 1.42 objective; 1 ms dwell time). For titrating the amount of cargo release with full field illumination, we deliver a 50 ms pulse of 405 nm excitation light through a spinning disk scan head. Typical excitation powers measured at the objective are: 912 μW (80% laser power) of full-field excitation for ~100% release; 194 μW (30% laser power) triggers ~50% release and 97 μW (15% laser power) triggers ~25% of release. An intensity-controllable LED or constant power mercury/xenon/metal halide light source should be similarly effective with neutral density filters to achieve the appropriate power. While the light intensities above are a good starting point, a careful light titration is recommended to optimize exposure duration and intensity for effective ER release. At the light exposure times and intensities noted above we have not observed signs of cell toxicity.

Acknowledgments

The work described in this protocol was supported by grants from the National Science Foundation (DGE-1553798; A.M.B.), the Howard Hughes Medical Institute (A.M.B.), and the National Institute of Neurological Disorders and Stroke (R01NS082271, R01NS10755, R35NS116879, UF1NS107710; M.J.K.).

References

1. Lippincott-Schwartz J, Roberts TH, Hirschberg K (2003) Secretory protein trafficking and organelle dynamics in living cells. Annu Rev Cell Dev Biol 16:557–589. https://doi.org/10.1146/annurev.cellbio161557
2. Barlowe CK, Miller EA (2013) Secretory protein biogenesis and traffic in the early secretory pathway. Genetics 193:383–410. https://doi.org/10.1534/GENETICS.112.142810
3. Lee MCS, Miller EA, Goldberg J et al (2004) Bi-directional protein transport between the ER and GOLGI. Annu Rev Cell Dev Biol 20: 87–123. https://doi.org/10.1146/annurev.cellbio20010403105307
4. Presley JF, Cole NB, Schroer TA et al (1997) ER-to-Golgi transport visualized in living cells. Nature 389:81–85. https://doi.org/10.1038/38001
5. Lodish HF, Weiss RA (1979) Selective isolation of mutants of vesicular stomatitis virus defective in production of the viral glycoprotein. J Virol 30:177–189. https://doi.org/10.1128/jvi.30.1.177-189.1979
6. Toomre D, Steyer JA, Keller P et al (2000) Fusion of constitutive membrane traffic with the cell surface observed by evanescent wave microscopy. J Cell Biol 149:33–40
7. Boncompain G, Divoux S, Gareil N et al (2012) Synchronization of secretory protein traffic in populations of cells. Nat Methods 9: 493–498. https://doi.org/10.1038/nmeth.1928
8. Rivera VM, Wang X, Wardwell S et al (2000) Regulation of protein secretion through controlled aggregation in the endoplasmic reticulum. Science 287:826–830
9. Chen D, Gibson ES, Kennedy MJ (2013) A light-triggered protein secretion system. J Cell Biol 201:631–640. https://doi.org/10.1083/jcb.201210119
10. Bowen AB, Bourke AM, Hiester BG et al (2017) Golgi-independent secretory trafficking through recycling endosomes in neuronal dendrites and spines. Elife 6:e27362. https://doi.org/10.7554/eLife.27362
11. Evans AJ, Gurung S, Wilkinson KA et al (2017) Assembly, secretory pathway trafficking, and surface delivery of Kainate receptors is regulated by neuronal activity. Cell Rep 19:2613–2626. https://doi.org/10.1016/j.celrep.2017.06.001
12. Hangen E, Cordeliè Res FP, Petersen JD et al (2018) Neuronal activity and intracellular calcium levels regulate intracellular transport of newly synthesized AMPAR. Cell Rep 24: 1001–1012.e3. https://doi.org/10.1016/j.celrep.2018.06.095
13. Bourke AM, Bowen AB, Kennedy MJ (2018) New approaches for solving old problems in neuronal protein trafficking. Mol Cell Neurosci 91:48–66
14. Spencer DM, Wandless TJ, Schreiber SL, Crabtree GR (1993) Controlling signal transduction with synthetic ligands. Science 262: 1019–1024
15. Ho SN, Biggar SR, Spencer DM et al (1996) Dimeric ligands define a role for transcriptional activation domains in reinitiation. Nature 382: 822–826. https://doi.org/10.1038/382822a0
16. Chen J, Zheng X-F, Brown EJ, Schreiber SL (1995) Identification of an 11-kDa FKBP12-rapamycin-binding domain within the 289-kDa FKBP12-rapamycin-associated protein and characterization of a critical serine residue. Proc Natl Acad Sci U S A 92(11): 4947–4951
17. Kennedy MJ, Hughes RM, Peteya LA et al (2010) Rapid blue-light-mediated induction of protein interactions in living cells. Nat Methods 7:973–975
18. Crefcoeur RP, Yin R, Ulm R, Halazonetis TD (2013) Ultraviolet-B-mediated induction of protein-protein interactions in mammalian cells. Nat Commun 4:1779. https://doi.org/10.1038/ncomms2800
19. Kakumoto T, Nakata T (2013) Optogenetic control of PIP3: PIP3 is sufficient to induce the actin-based active part of growth cones and is regulated via endocytosis. PLoS One 8: e70861. https://doi.org/10.1371/journal.pone.0070861
20. Zhou XX, Chung HK, Lam AJ, Lin MZ (2012) Optical control of protein activity by fluorescent protein domains. Science 338:810–814. https://doi.org/10.1126/science.1226854
21. Toettcher JE, Gong D, Lim WA, Weiner OD (2011) Light control of plasma membrane recruitment using the Phy-PIF system. Methods Enzymol 497:409–423. https://doi.org/10.1016/B978-0-12-385075-1.00017-2
22. Zhang K, Duan L, Ong Q et al (2014) Light-mediated kinetic control reveals the temporal effect of the Raf/MEK/ERK pathway in PC12 cell neurite outgrowth. PLoS One 9:e92917. https://doi.org/10.1371/journal.pone.0092917
23. Toettcher JE, Weiner OD, Lim WA (2013) Using optogenetics to interrogate the dynamic control of signal transmission by the ras/erk

module. Cell 155:1422–1434. https://doi.org/10.1016/j.cell.2013.11.004

24. Levskaya A, Weiner OD, Lim WA, Voigt CA (2009) Spatiotemporal control of cell signalling using a light-switchable protein interaction. Nature 461:997–1001. https://doi.org/10.1038/nature08446
25. Yazawa M, Sadaghiani AM, Hsueh B, Dolmetsch RE (2009) Induction of protein-protein interactions in live cells using light. Nat Biotechnol 27:941–945
26. Lungu OI, Hallett RA, Choi EJ et al (2012) Designing photoswitchable peptides using the AsLOV2 domain. Chem Biol 19:507–517. https://doi.org/10.1016/j.chembiol.2012.02.006
27. Strickland D, Moffat K, Sosnick TR (2008) Light-activated DNA binding in a designed allosteric protein. Proc Natl Acad Sci U S A 105:10709–10714
28. Strickland D, Lin Y, Wagner E et al (2012) TULIPs: tunable, light-controlled interacting protein tags for cell biology. Nat Methods 9: 379–384. https://doi.org/10.1038/nmeth.1904
29. Wang H, Vilela M, Winkler A et al (2016) LOVTRAP: an optogenetic system for photoinduced protein dissociation. Nat Methods 13: 755–758. https://doi.org/10.1038/nmeth.3926
30. Wu YI, Frey D, Lungu OI et al (2009) LETTERS A genetically encoded photoactivatable Rac controls the motility of living cells. Nature 461(7260):104–108. https://doi.org/10.1038/nature08241
31. Zimmerman SP, Hallett RA, Bourke AM et al (2016) Tuning the binding affinities and reversion kinetics of a light inducible dimer allows control of transmembrane protein localization. Biochemistry 55(37):5264–5271. https://doi.org/10.1021/acs.biochem.6b00529
32. Adams SR, Tsien RY (1993) Controlling cell chemistry with caged compounds. Annu Rev Physiol 55:755–784. https://doi.org/10.1146/annurev.ph.55.030193.003543
33. Ellis-Davies GC (2007) Caged compounds: photorelease technology for control of cellular chemistry and physiology. Nat Methods 4: 619–628
34. Karginov AV, Zou Y, Shirvanyants D et al (2011) Downloaded via MPI HIRN-FORSCHUNG on April 8. J Am Chem Soc 133:39. https://doi.org/10.1021/ja109630v
35. Umeda N, Ueno T, Pohlmeyer C et al (2011) A photocleavable rapamycin conjugate for spatiotemporal control of small GTPase activity. J Am Chem Soc 133:12–14. https://doi.org/10.1021/ja108258d
36. Brown KA, Zou Y, Shirvanyants D et al (2015) Light-cleavable rapamycin dimer as an optical trigger for protein dimerization †. Chem Commun 5702:5702. https://doi.org/10.1039/c4cc09442e
37. Schelkle KM, Griesbaum T, Ollech D et al (2015) Light-induced protein dimerization by one- and two-photon activation of gibberellic acid derivatives in living cells. Angew Chem Int Ed Engl 54:2825–2829. https://doi.org/10.1002/anie.201409196
38. Wright CW, Guo ZF, Sen LF (2015) Light control of cellular processes by using photocaged abscisic acid. Chembiochem 16:254–261. https://doi.org/10.1002/cbic.201402576
39. Zhang H, Aonbangkhen C, Tarasovetc EV et al (2017) Optogenetic control of kinetochore function. Nat Chem Biol 13:1096–1101. https://doi.org/10.1038/nchembio.2456
40. Chen X, Venkatachalapathy M, Kamps D et al (2017) "Molecular activity painting": switch-like, light-controlled perturbations inside living cells. Angew Chem Int Ed Engl 56:5916–5920. https://doi.org/10.1002/anie.201611432
41. Ballister ER, Ayloo S, Chenoweth DM et al (2015) Optogenetic control of organelle transport using a photocaged chemical inducer of dimerization. Curr Biol 25:R407–R408. https://doi.org/10.1016/j.cub.2015.03.056
42. Chen X, Wu Y-W (2018) Tunable and Photoswitchable chemically induced dimerization for chemo-optogenetic control of protein and organelle positioning. Angew Chem Int Ed Engl 130:6912–6915. https://doi.org/10.1002/ange.201800140
43. Zimmermann M, Cal R, Janett E et al (2014) Cell-permeant and photocleavable chemical inducer of dimerization. Angew Chem Int Ed Engl 53:4717–4720. https://doi.org/10.1002/anie.201310969
44. Gutnick A, Banghart MR, West ER, Schwarz TL (2019) The light-sensitive dimerizer zapalog reveals distinct modes of immobilization for axonal mitochondria. Nat Cell Biol 21: 768–777. https://doi.org/10.1038/s41556-019-0317-2
45. Bourke AM, Schwartz SL, Bowen AB et al (2021) zapERtrap: a light-regulated ER release system reveals unexpected neuronal trafficking pathways. J Cell Biol 220:e202103186. https://doi.org/10.1083/jcb.202103186

Chapter 5

Generation of Endogenously Tagged Membrane Trafficking Regulators Using CRISPR Genome Editing

Syed Mukhtar Ahmed

Abstract

Vesicle trafficking entails packaging and transport of membrane-associated proteins to their target membranes, and recycling or degradation of endocytosed proteins. Biochemical and cell biological studies of vesicle trafficking often require the introduction of epitope tags or fluorescent protein markers for protein purification and tracking in cells. Previously, such tagging experiments in mammalian cells mainly used overexpression systems, which could lead to artifacts. Abnormally high expression levels also prevent us from studying individual vesicle trafficking events with precision. With the advent of CRISPR technologies, epitope tags and fluorescent proteins can now be introduced into endogenous proteins in almost any cell type that are proliferating in culture. This chapter describes approaches for inserting tags at the endogenous loci of genes, with the vesicle tethering protein complex, exocyst, as an example.

Key words CRISPR-Cas9, Gene editing, Endogenous tagging, Membrane trafficking, Vesicle transport, Tethering complex, Exocyst

1 Introduction

Vesicle transport is a fundamental feature of eukaryotic cells. Newly synthesized proteins are transported from the endoplasmic reticulum and Golgi body to their target membranes, and some of them are destined for exocytosis. Conversely, endocytosed proteins and membranes are routed for recycling or degradation when repurposing is uncalled for [1, 2]. Over the past four decades, we have learned a lot about this crucial cellular process from model organisms, often by tagging proteins for subsequent purification or to identify the localization of proteins of interest in cells. In genetically tractable organisms such as yeasts, the endogenous loci can be easily manipulated to add such tags, but studies in mammalian cell cultures have mostly been limited to the introduction of exogenous cDNA constructs [3]. These strategies, however, frequently lead to uncontrolled expression of proteins - often many-fold greater than

Jingshi Shen (ed.), *Membrane Trafficking: Methods and Protocols*,
Methods in Molecular Biology, vol. 2473, https://doi.org/10.1007/978-1-0716-2209-4_5,

physiological levels even if stable cell lines are generated. It is becoming increasingly evident that abnormally high expression levels may cause serious artifacts such as aggregation, protein degradation, mislocalization, kinetic perturbations, and spurious nonphysiological protein–protein interactions, to name a few [4–7]. Overexpression of proteins of interest also makes it difficult to study molecular events with single-molecule precision, as a high density of fluorescently labeled proteins interferes with their resolution and complicates particle tracking [8].

Now, with the arrival of genome-editing technologies using Clustered Regularly Interspaced Short Palindromic Repeats (CRISPR)-Cas, it has become possible to insert short epitope tags and fluorescent proteins (hereby collectively referred to as tags) directly at the endogenous gene loci of proteins of interest to be expressed in-frame [9, 10]. The CRISPR-Cas system has dramatically simplified the ways we edit genetic information with high precision and flexibility, and is becoming an indispensable tool in bioscience research [11, 12]. Numerous Cas nucleases have been discovered, with distinct properties that can facilitate specific types of editing, but this chapter focuses on the prototypical Type II endonuclease, Cas9 [13, 14]. Cas9 can bind a short RNA species called trans-activating CRISPR RNA (tracrRNA) and a sequence-specific RNA called a crRNA, which interacts with both the Cas9 protein and the tracrRNA. However, in most technical applications, a fusion of tracrRNA and crRNA, termed a single guide RNA (gRNA), is used. A specific genome target site is recognized by the gRNA and the stable Cas9-gRNA ribonucleoprotein (RNP) complex cleaves the DNA with a double-strand break [15, 16]. The RNP can be generated in vitro and introduced into cells by electroporation; or the nuclease and gRNA can be introduced as plasmids using traditional gene transfer methods. The latter method will be described here in detail. Used alone, the dsDNA breaks are most often repaired by nonhomologous end joining (NHEJ), resulting in indel mutations and consequently premature terminations or nonsense-mediated RNA decay [17]. If early exons are targeted, this can cause gene knockouts. One can also simultaneously introduce a donor sequence with flanking homology sequences to seamlessly patch the cleavage site by homology-directed repair (HDR), resulting in specified nucleotide changes or even insertions of longer coding sequences in-frame.

A wide variety of tags exist for biochemical purifications or protein localization studies. For example, the palette of fluorescent proteins and self-labeling protein tags such as Halo-Tag and SNAP-Tag have become the staple for cell biologists [18–20]. The advent of new bright and stable fluorophores coupled to Halo and SNAP ligands with low toxicity in living cells has made them invaluable tools for both fast and long-term live cell imaging. This chapter describes endogenous tagging of membrane trafficking proteins in mammalian cells using CRISPR-Cas9 genome editing. The tagging of the vesicle tethering complex, exocyst, is described as an example.

2 Materials

2.1 Cell Culture

1. NMuMG cells (ATCC) (*see* **Notes 1** and **2**).
2. Dulbecco's Modified Eagle Medium (DMEM) with 4.5 g/L D-glucose, L-glutamine and without sodium pyruvate, supplement with 10% fetal bovine serum (FBS).
3. Phosphate-buffered Saline (PBS; 10×): 1.37 M NaCl, 27 mM KCl, 100 mM Na_2HPO_4, and 18 mM KH_2PO_4, pH 7.4.
4. PBS (1×): Dilute 10× PBS with dH_2O and filter-sterilize before use.
5. 0.25% Trypsin–EDTA (*see* **Note 3**).
6. Cell culture plates: 10 cm dishes, flat-bottom 6-well plates, 12-well plates, 24-well plates, and 96-well plates.
7. 40 μm nylon cell strainers.
8. 15 mL conical tubes.
9. Insulin Solution (*see* **Note 4**).

2.2 Plasmids and Cloning

1. gRNA Cloning vector (Addgene plasmid # 41824) (*see* **Note 5**) [21].
2. pCMVSP6-NLS-hCas9-NLS (*see* **Note 6**) [10, 22].
3. LITMUS29 vector for cloning donor sequences [23].
4. Gene fragments (e.g., gBlock from IDT).
5. Annealing buffer: 10 mM Tris–HCl, pH 8.0, 50 mM NaCl, 1 mM EDTA. NEBuffer 2 (New England Biolabs) can also be used instead.
6. Gibson Assembly Cloning Kit.
7. AflII restriction endonuclease.
8. Choice of restriction endonuclease to linearize LITMUS29 vector.
9. Primers.
10. High-fidelity DNA polymerase (e.g., Phusion High-Fidelity DNA polymerase).
11. T4 Polynucleotide Kinase (PNK).
12. T4 DNA Ligase.
13. T4 Ligase Reaction Buffer (10×): 500 mM Tris–HCl, pH 7.5, 100 mM $MgCl_2$, 100 mM Dithiothreitol, and 10 mM ATP.

14. Deoxynucleotide (dNTP) mix: 10 mM each of dATP, dCTP, dGTP, and dTTP in 0.6 mM Tris–HCl, pH 8.0.
15. Stbl3™ Chemically Competent *E. coli* (ThermoFisher Scientific, Catalog # C7373-03).
16. Plasmid DNA Miniprep kit.
17. PCR purification kit.
18. Gel extraction kit.
19. Plasmid DNA Midi kit.
20. Plasmid DNA Maxi kit.
21. Spectrophotometer.

2.3 Designing gRNAs

One must first identify a region within the gene's coding region to target for tag insertion (*see* **Note 7**). Next, a 23-bp target gRNAs site containing a protospacer adjacent motif (PAM) at the 3′-end of the gRNA should be located (Fig. 2). A Cas9-mediated cleavage site needs not to happen exactly at the site where the tag will be inserted as this will be accomplished by the donor vector. However, one should select the gRNA within ±50 bp from the desired tag insertion site. The exact gRNA sequence can lie on either strand that possesses the PAM sequence. The most commonly used Cas9, derived from *S. pyogenes*, recognizes the PAM sequence *NGG*. The target guide sequence must be searched against the rest of the genomic sequence using NCBI BLAST to ensure none or very little sequence match, especially in the 13-bp region immediately before the PAM site (*see* **Note 8**).

2.4 Transfection Reagent

For NMuMG cells, we found that the Xfect Transfection reagent (Takara Bio, Catalog # 631318) works well with minimal toxicity. Due to the inherently low efficiency of homologous recombination [17], users should select a transfection reagent and optimize it to achieve high transfection efficiencies in their chosen cell types. Electroporation-based gene transfer can be also considered.

2.5 Flow Cytometry Materials to Isolate Positive Clones

1. 5 mL round bottom polystyrene tubes.
2. PBS containing 2 mM EDTA (filter-sterilized).
3. 5% bovine serum albumin (BSA) in PBS buffer (filter-sterilized).
4. Cell sorter with appropriate laser lines (e.g., BD FACSAria III).

2.6 Genotyping Materials for Identifying and Confirming Positive Clones

1. Potential knockin cell lines.
2. Genomic DNA isolation kit.
3. Primers for genotyping.
4. Nuclease-free distilled, deionized water.
5. Taq DNA polymerase.

6. Thermocycler.
7. Electrophoresis apparatus.
8. Agarose, molecular biology grade.
9. TAE Buffer (50×): 2 M Tris base, 1 M acetic acid, and 50 mM EDTA.

3 Methods

3.1 Cloning of Target Guide RNA

A general flow diagram illustrates the steps involved in creating cell lines with genes of interest tagged with fluorescent proteins (Fig. 1). The target gRNA sequence can be cloned into a gRNA cloning vector (*see* **Note 9**).

Once an ideal target gRNA sequence is identified, the 19-bp immediately before the PAM site can be incorporated into two 60-mer oligonucleotides (hereby referred to as oligos) where the flanking arms are homologous to the 5′ U6 promoter and 3′ gRNA scaffold sequences in the cloning vector (Fig. 2). However, a portion of U6 and scaffold are inserted into the oligo design by the user as they do not exist in the vector (*see* **Note 10**). The two synthesized oligos are then annealed and extended using a DNA polymerase with proofreading capability such as Phusion DNA polymerase.

1. Order the two 60-mer oligos described above. All oligos must be designed in the 5′ to 3′ direction and oligos corresponding to the bottom stand must be converted to its reverse complement.
2. Resuspend oligos in the annealing buffer.
3. Mix oligos in equimolar concentrations in a total volume of 50 μL in PCR tubes. It is recommended to use 2 μg of oligos each. Add additional annealing buffer if necessary, to a total of 50 μL.
4. Using a thermocycler programmed to start at 95 °C for 2 min and then ramping down to 25 °C at 5 °C/min to anneal the oligos.
5. Take 2 μL of the annealed oligos and add 2 μL of 10 mM dNTPs, 0.5 μL of high-fidelity DNA polymerase, and the PCR reaction buffer. Bring the total volume to 25 μL using nuclease-free H_2O.
6. For oligo overlap extension, use a thermocycler programmed to start with denaturing at 98 °C for 2 min, followed by annealing at 55 °C for 20 s and extension at 72 °C for 5 min. The two 60-mer oligos with complementary 20 bps are now 100-mer oligos.

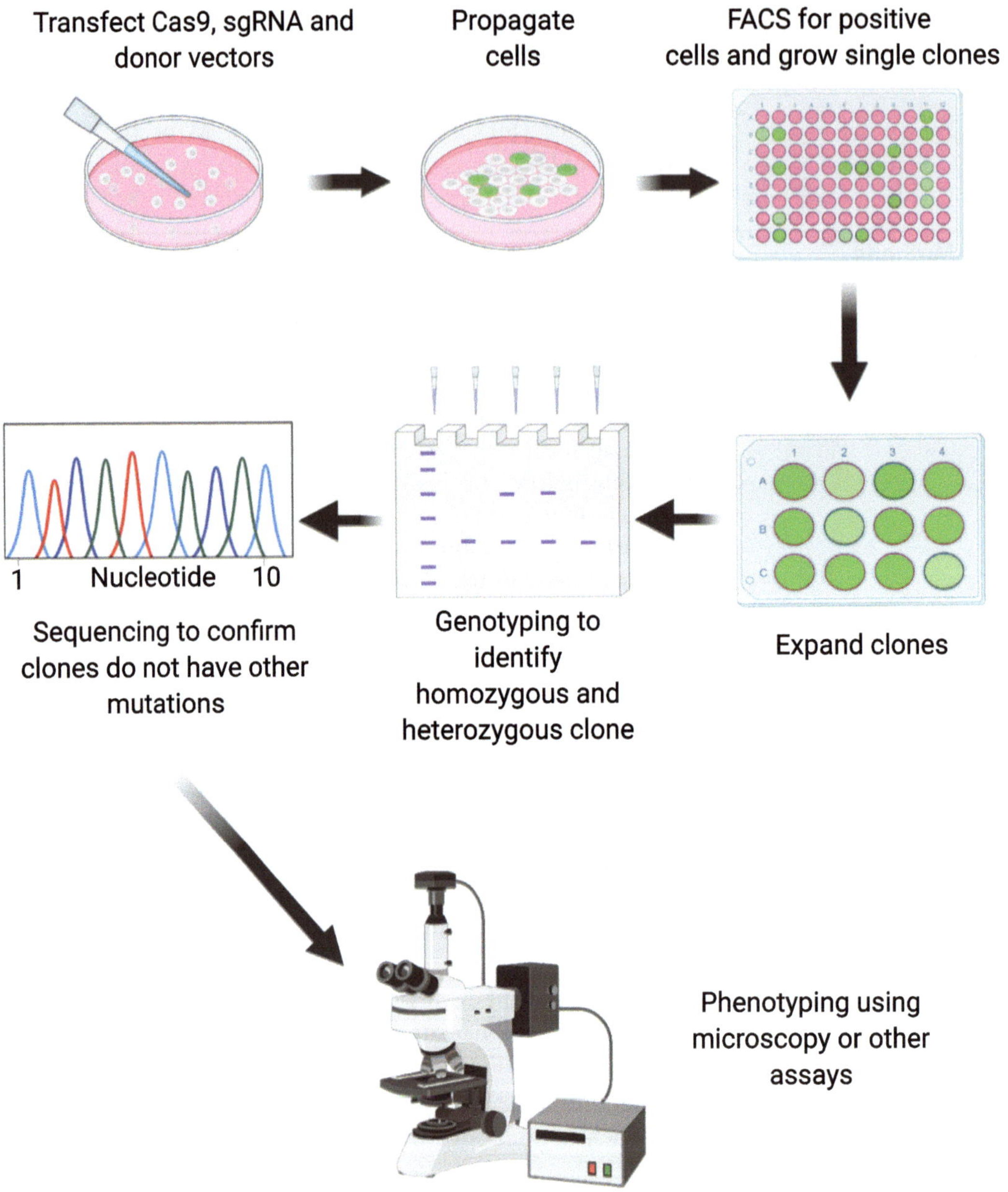

Fig. 1 Procedure flowchart. Flowchart demonstrating the steps involved to generate endogenously tagged proteins in cultured mammalian cells. (Illustration created using BioRender.com)

7. Linearize 1–2 μg of the gRNA Cloning vector using the AflII restriction endonuclease.
8. Purify the linearized vector using a PCR purification kit. Alternatively, resolve the digested vector using agarose gel electrophoresis and purify the desired vector band using a gel extraction kit.

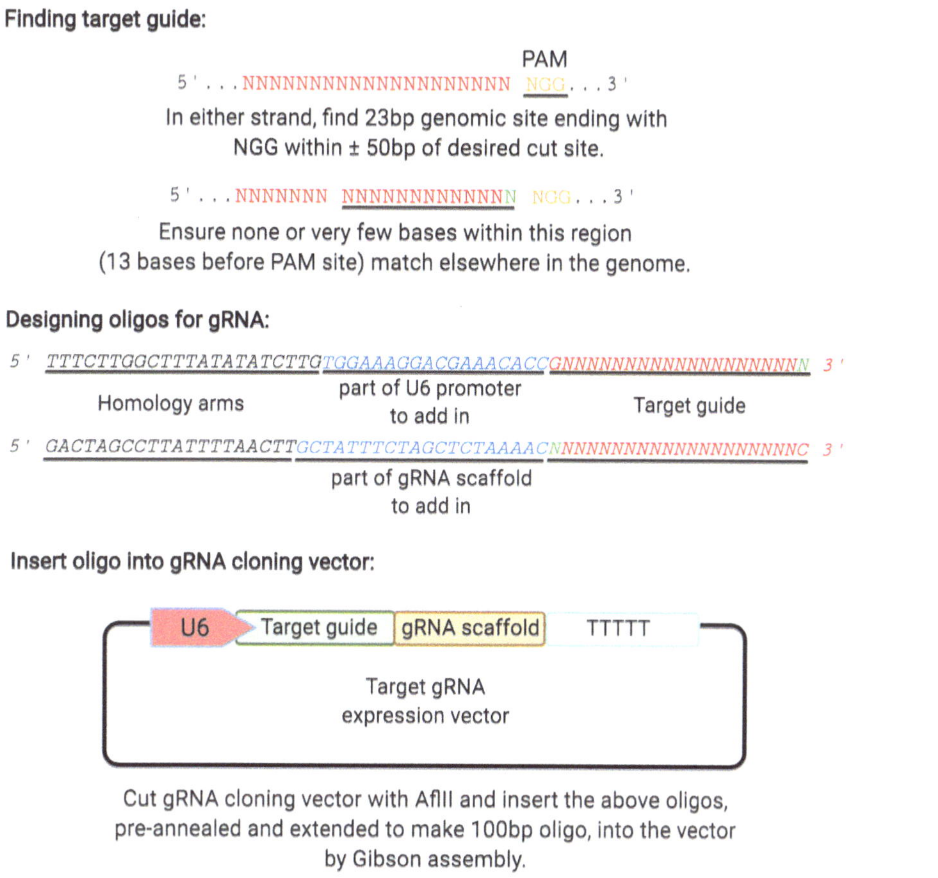

Fig. 2 gRNA oligo design. Schematic showing the procedures to identify and design target guide RNA sequences for cloning into gRNA vector. The green colored "N" is to show the direction of the sequence placement when designing oligos

9. Combine 2.5 μL of 100-mer oligos and 2 μL of purified gRNA cloning vector with the Gibson Assembly Master Mix (*see* **Note 11**). This reaction creates the final gRNA construct.
10. Amplify the cDNA construct by transforming it into Stbl3 *E. coli*. Plasmids should be sequenced using primers near the inserted sequence to confirm accuracy.

3.2 Cloning of the Donor Vector

Similar to gRNA cloning, the tag of choice and the homology arms to the genomic DNA can be synthesized as gBlocks and inserted into the LITMUS29 vector plasmid by Gibson Assembly or using restriction enzyme sites. In the particular example shown in Fig. 3, we used BamHI and XhoI sites to fuse the superfolder-GFP (sfGFP) flanked by 5′ and 3′ homology arms that are synthesized either as gBlocks or amplified directly from genomic sequence by PCR. The homology arms should be 0.5–1.5 kbps long. Longer homology arms have higher HDR efficiency. In this example, we

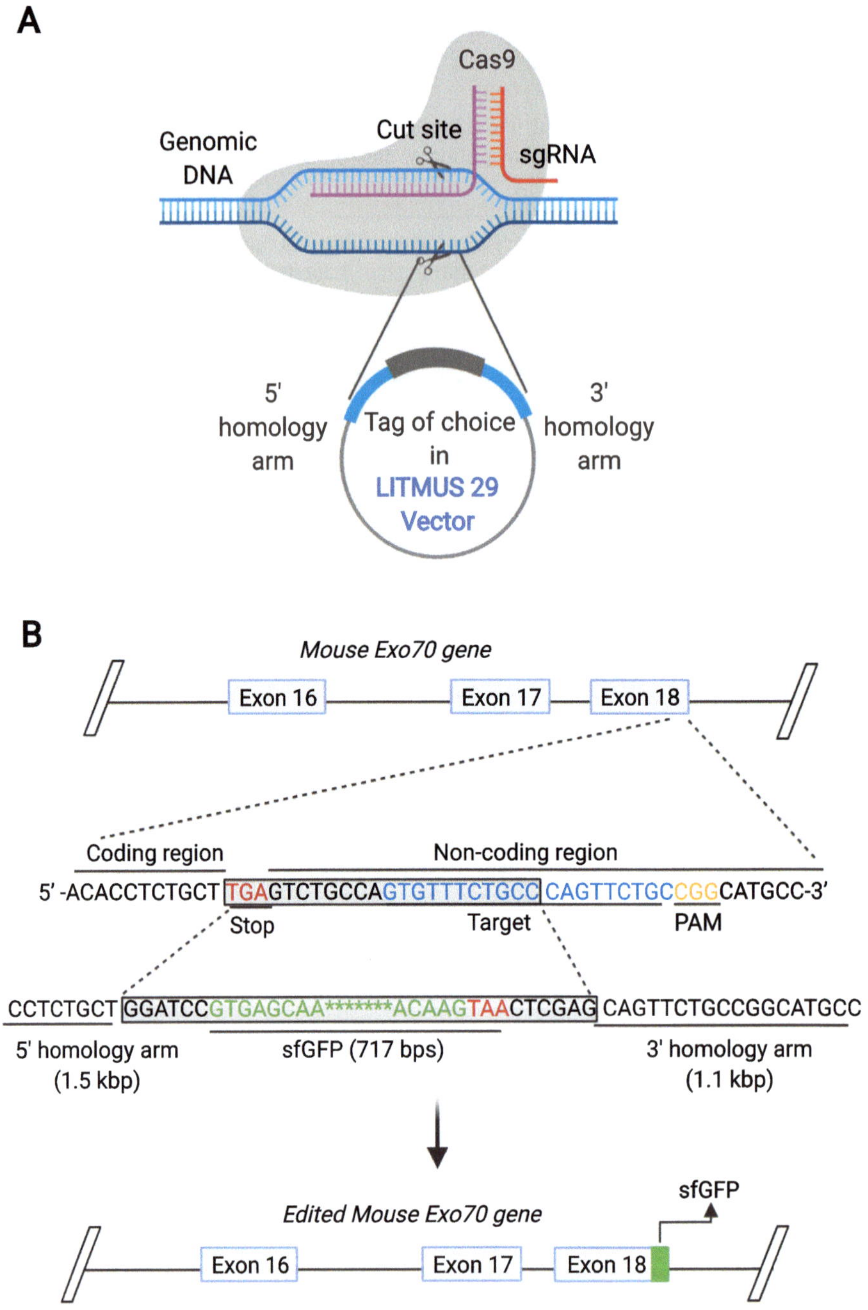

Fig. 3 Donor construct design. (**a**) Schematic shows the design of donor vector containing 5′ homology arm (left blue), the tag of interest (gray), and 3′ homology arm (right blue) as cloned into LITMUS29 vector. (**b**) The diagram shows the design of target guide and donor vector to insert superfolder-GFP (sfGFP) at the C-terminus of the exocyst subunit, Exo70

demonstrate the introduction of sfGFP to the C-terminal end of the *Exo70* (*Exoc7*) gene right before the stop codon (Fig. 3b). The region between the stop codon and part of the target site was replaced with an insertion cassette in frame with the coding region with a stop codon placed after sfGFP. It is important to note that, if the sequence within the target site is unchanged, the PAM site must be removed from the donor sequence to prevent repeated excision by the Cas9-gRNA complex (*see* **Notes 12** and **13**).

1. Order the three gBlocks—5′ homology arm, sfGFP, and 3′ homology arm with 40–60 bps overlap from IDT. The 5′-end of the 5′ homology arm, and the 3′-end of 3′ homology arm should also have 40–60 bp overlaps with the LITMUS29 vector if Gibson Assembly method for cloning is used. Otherwise, the 5′- and 3′-ends should contain the overhangs appropriate for the chosen restriction sites. This entire fragment can be ordered as a single gBlock from IDT.
2. Smaller primer sets should also be ordered to amplify individual gBlocks by PCR as the supplied quantities of the gene fragments from the vendor may not be adequate.
3. If the tag (e.g., sfGFP) is common for multiple cloning, insert restriction enzyme sites at the 5′ and 3′ ends. Ensure that this does not make the tag out of frame with the coding sequence. The 3′ restriction site should be inserted after the stop codon. This helps to stitch the 5′ and 3′ homology arms to sfGFP using traditional restriction endonucleases and T4 Ligase (*see* **Note 14**).
4. Linearize 1–2 μg of the LITMUS29 vector with the chosen restriction enzyme from the multiple cloning sites. The exposed sequence must match the 5′ and 3′ overlapping sequences made above.
5. Purify the linearized LITMUS29 vector using a PCR purification kit or a gel extraction kit.
6. After the 5′ homology arm, sfGFP, and 3′ homology arm are inserted in tandem, and LITMUS29 is linearized, combine 75 ng of the gene fragment and 50 ng of the plasmid (0.02–0.5 pmols) with the Gibson Assembly Master Mix in a total reaction volume of 20 μL, following the manufacturer's protocol.
7. Amplify the construct in Stbl3 *E. coli*.
8. Sequence the plasmids using forward and reverse primers from outside the inserted DNA fragment to ensure undesirable changes are not incorporated.

3.3 Generation of Knockin Cell Lines

As mentioned earlier, end-users should adapt transfection reagents and methods that best suit their cell lines. For NMuMG cells, we use the Xfect transfection reagent with >80% transfection efficiency with low toxicity. Should one decide to generate knockins with multiple genes expressing discrete tags in the same cell we recommend isolation of clones with one label before incorporating the next (*see* **Note 15**).

1. Plate cells in a 10 cm culture dish such that the cells reach 60–70% confluency in 16–24 h.
2. Transfect the cells with equimolar (550 fmol) amounts of Cas9, gRNA, and donor plasmids according to the manufacturer's protocol (*see* **Note 16**).
3. Change cell culture media after transfection as suggested by the manufacturer.
4. Passage the cells for additional 10–14 days to allow gene-edited cells to propagate before sorting by flow cytometry (*see* **Note 17**).
5. A subfraction of cells can be taken for a preliminary pooled genotyping assay to check for the possible positive population. One can also test by fluorescence microscopy for fluorescent cells (*see* Subheading 3.8 and **Note 18**).

3.4 Isolation of Single-Cell Clones

Upon satisfactory expansion of cells, single cell clones must be isolated to generate clonal lines. This can be conveniently done by fluorescence-activated cell sorting (FACS) (*see* **Note 3**). If self-labeling tags are inserted in the genome, cells should be incubated with the appropriate ligands such as a Halo ligand (100 nM) or a SNAP ligand (0.5–1 μM) for 1 h, followed by extensive washing to remove excess dye (*see* **Note 19**). Wild-type cells should also be treated with the fluorescent ligand and used as negative controls for setting gates for FACS.

1. To detach cells for FACS, gently wash cells twice with 1× sterile PBS by pipetting the latter to the side wall of the culture dish.
2. Add 2–5 mL of trypsin to the dish and incubate at 37 °C for 8–10 min or until cells dissociate from the plate and each other to form mostly single cell suspensions.
3. Resuspend the cells in 8 mL of DMEM + 10% FBS and gently pipette up and down 4–5 times to loosen any remaining cell clumps. Collect the media into 15 mL centrifuge tubes.
4. Centrifuge for 2 min at 800 × *g* to pellet the cells.
5. Discard the supernatant and resuspend with 10 mL of sterile PBS.
6. Repeat **steps 4** and **5**.
7. Resuspend cells in PBS containing BSA (5%) and EDTA (2 mM).

8. If necessary, remaining clumps can be removed using a 40 μm cell strainer.
9. Sort desired cells based on fluorescence using FACS.
10. Sorted single cells are placed in each well of 96-well plates after sorting and expanded to 24-well and 6-well plates. One can plate the cells in glass-bottom plates to monitor the fluorescence in the clones during an expansion (*see* **Note 18**). Multiple clones should be tested to avoid clonal phenotypes (*see* **Note 2**). If clonal cells are not required, population of sorted cells can be plated on a 10 cm dish although these cells will have a mix of homozygous and heterozygous knockin cells. Alternatively, one can use a dilution cloning approach to generate single clones without FACS sorting. However, because of the low efficiency of homologous recombination, this would be a cumbersome effort.

3.5 Genotyping and Confirmation of Clones

1. Design a genotyping primer set near the tag insert site. The distance between the primers should be selected in a way so a clear change in molecular weights can be seen on a 2% agarose gel. One primer should be on the genomic DNA sequence with an opposing primer inside the HDR region of the inserted sequence. Two opposing primers flanking the inserted sequence can also be used.
2. Plate cells in 12-well or 6-well flat-bottom cell culture plates.
3. When cells are 90–95% confluent, genomic DNA can be extracted using the DNeasy Blood & Tissue Kit (Qiagen) following the manufacturer's protocol. Users may use an alternative genomic extraction kit or house protocol for this purpose.
4. Perform PCR reactions with the genotyping primers using Taq DNA polymerase according to the manufacturer's guide.
5. Run PCR products on a 2% agarose gel.

   Figure 4a shows representative PCR-based genotyping of Exo70-sfGFP clones where a clear shift in molecular weight is observed for homozygous knockin cells. For heterozygous knockin clones, fragments from both the wild-type and edited alleles can be seen in the PCR products.

3.6 Sequencing

It is recommended to sequence around the region where the tag is inserted before cells are used for downstream biological phenotyping. This ensures mutations are not introduced inadvertently. The PCR genotyping approach described above is also used for sequencing purposes although high-fidelity DNA polymerases should be used to avoid errors incorporated during PCR amplification.

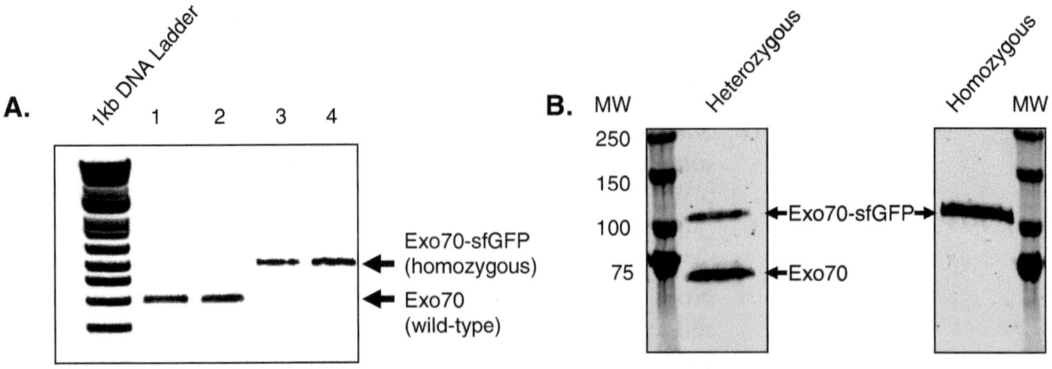

Fig. 4 Validation by genotyping and western blotting. (**a**) PCR based genotyping to verify successful integration of sfGFP into the Exo70 locus. A 5′ primer within Exo70 (exon 18) and a 3′ primer in the intron region flanking the sfGFP insert site were used. Lanes 1 and 2 show no integration of sfGFP, whereas lanes 3 and 4 show insertion of sfGFP at the Exo70 locus in both alleles. (**b**) Western blot analysis using anti-Exo70 antibody show expression of Exo70-sfGFP in heterozygous (left) and homozygous (right) knockin cell lines

1. Amplify the region of interest by PCR with flanking primers using Phusion DNA polymerases.
2. Purify the desired PCR products using agarose electrophoresis and a PCR purification kit.
3. Sequence the purified PCR products using either the 5′ or 3′ primer. Ensure that the entire span of the edited genome sequence is covered.
4. In the sequencing chromatogram, check for seamless integration of the intended coding sequence.

3.7 Validation Using Western Blotting

A mainstay method to verify if the intended tagged protein is expressed properly is by western blot analysis.

1. Prepare cell lysates and run the samples on SDS-PAGE. We recommend 4–20% gradient gels for initial assessment.
2. Transfer the proteins from SDS-PAGE to nitrocellulose or PVDF membranes and probe proteins of interest using anti bodies recognizing the proteins or if available, the tag itself. These experiments determine whether the tagged protein is expressed (Fig. 4b).

3.8 Validation by Microscopy

Upon successfully incorporating a fluorescent tag to the endogenous protein, one should confirm that the subcellular localization of the protein is not affected by tagging. We employ total TIRF microscopy to track the exocyst subunits during vesicle trafficking and fusion with a fast temporal resolution [10]. Figure 5 shows punctate localization of Exo70-sfGFP and Sec5-Halo near the plasma membrane in dual knockin NMuMG cells. This approach can also be used for initial screening of clones upon transfection as described in Subheading 3.3 (*see* **Note** 7).

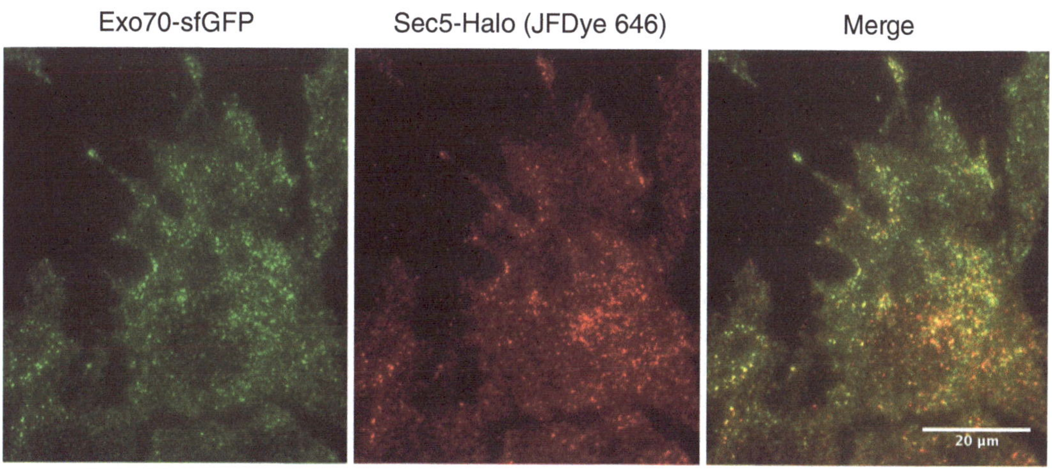

Fig. 5 Visualization of tagged protein localization by microscopy. TIRF images show localization of Exo70-sfGFP and Sec5-Halo in dual knockin NMuMG cells. Sec5-Halo was labeled with Halo-ligand (100 nM) coupled to Janelia Fluor 646 dye. Scale bar: 20 μm

4 Notes

1. Here, we provide examples using the mouse mammary epithelial cell line NMuMG. The parameters described here are optimized for these cells. Other cell-types can be used for these purposes, although cell culture and transfection protocols may vary and need to be optimized by the end-user.
2. To maintain cultures, NMuMG cells should be grown at high density (1:2 or 1:4 sub cultivation dilution is recommended) and passaged before reaching confluency to avoid polarization or selective culture of single cells. When single-cell colony expansion is required, multiple clones should be examined to avert clonal variations in experiments.
3. Cells should be trypsinized for an appropriate duration to generate single-cell suspension, especially before flow cytometry. Trypsin treatment at 37 °C for 7–10 min generally detach NMuMG cells from the plate and each other. The use of cell strainers will reduce clumps. However, if gene-edited cells are trapped in clumps, strainers will reduce overall recovery.
4. Cells can be cultured in the presence or absence of insulin. However, they should be maintained under the same condition unless otherwise required for experimental purposes.
5. There are several plasmids available for the generation of gRNAs and donor vectors. Several mammalian-codon-optimized Cas9 expressing vectors also exist. We have used the three vectors described here successfully [10].

6. In this plasmid, the Cas9 coding sequence is optimized for expression in mammalian cells. Two nuclear localization sequences (NLS), one at the N-terminus and the other at the C-terminus, increase the genome-editing efficiency of Cas9. Other mammalian expression vectors expressing similar Cas9 can also be used.
7. Site for tag insertion should be determined carefully as addition of a tag may perturb the function of the protein. The Allen Cell Biology Institute (https://www.allencell.org) creates heterozygous knockin alleles of genes in induced pluripotent stem cells (iPSCs) that are available for purchase [24], although it is unclear if the tagged allele always functions exactly as the non-tagged allele. To avoid misinterpretations, when possible, we prefer using homozygous knockin cell lines, and rigorously test the functions of tagged proteins. Some of the validation steps we take are presented in this chapter. In the case of the exocyst complex, the analysis is facilitated by the fact that the complex is essential for viability. In other cases, however, it may be necessary to test different tagging strategies, for example, at the N- or C-terminus, or between domains within the protein of interest. For instance, many small GTPases cannot be tagged at the C-terminus because it disrupts post-translational modifications. Sometimes, it may prove impossible to tag a protein while retaining its function. For example, we have been unable to obtain functional Ral GTPases capable of binding exocyst after N-terminal tagging with a GFP, Halo-Tag, or SNAP-Tag.
8. This step is very important to reduce off-target cuts in the genome. Free online platforms such as CHOPCHOP and Benchling provide good resources for designing and optimizing gRNAs.
9. The gRNA can be synthesized as a gene fragment that consists of all components for gRNA expression, namely, U6 promoter, target gRNA, gRNA scaffold, and termination signal. If this method is chosen, subcloning into a target vector is not needed. The gene fragment can be amplified using primers and the purified product is directly used for transfection.
10. If users choose gRNA cloning vectors that already have the entire U6 and scaffold sequence, these regions will be homologous. Users should be mindful of the restriction cleavage generated on the plasmid by the restriction endonuclease used when designing oligos for subsequent Gibson Assembly.
11. The annealing and extension steps described in **steps 4–8** can be done directly in the Gibson Assembly Mix as it contains Phusion DNA Polymerase, Taq DNA Ligase and T5 Exonuclease. However, the protocol described above in our hands produces better and more consistent results.

12. Users should note that splice variants in a gene missing the targeted exon can complicate data interpretations later. If possible, such exons should be avoided. Similarly, the insertion of a tag may affect the stability of proteins or their splice variants.
13. Some exon-intron boundary regions in the genome can be difficult to target making it cumbersome to get successful insertion of desired sequences. In such cases, one should test incorporation by varying the length of the homology arm. Nevertheless, one should note that some proteins may not be possible to tag at all because of difficulties presented by placement of tags and the genomic sequence around the desired tag site.
14. Alternatively, order primers to amplify the 5′ and 3′ homology arms (*see* Fig. 3 for design details). The primers need to contain overlaps as described above if Gibson Assembly will be used to fuse the fragments together. Restriction overhangs may be used instead for traditional cloning methods using restriction endonuclease sites to stitch the individual fragments. However, it is imperative to ensure frameshift is not generated as a result.
15. NHEJ is a major competing DNA repair pathway that limits the efficiency of HDR. HDR efficiency can be therefore promoted by inhibiting NHEJ [25, 26]. Blocking NHEJ can also prevent indels from occurring in one allele while HDR happens in the other allele.
16. Linearizing the donor cDNA using a restriction enzyme outside of the donor region before transfection can yield better results. If this route is chosen, the digested plasmid should be purified using a PCR purification kit before transfection.
17. Due to the low efficiency of HDR, propagation of cells after transfection ensures that sufficient numbers of gene-edited cells are obtained before FACS.
18. Low endogenous protein expression levels may prevent visualization of fluorescent proteins using conventional microscopes. One may have to use high sensitivity sCMOS cameras or high contrast, low background methods, such as Total Internal Reflection Fluorescence microscopy. The use of self-labeling proteins, such as Halo-Tag and SNAP-Tag allow labeling with bright and stable fluorophores to facilitate visualization of low levels of endogenous proteins compared to fluorescent proteins. Properties of dyes and fluorescent proteins can be acquired before experiments using the free and open-source database FPbase (https://www.fpbase.org/).
19. There are cell membrane permeable and impermeable substrates for both Halo-Tag and SNAP-Tag. These substrates covalently bind to Halo-Tag and SNAP-Tag. Users should choose the version depending on their application and

placement of the tag. Cell-permeable substrates are suitable to label both intracellular and cell surface tags, whereas cell impermeable substrates are only suitable for labeling cell surface tags. The user, however, is cautioned that labeled cell surface receptors may be internalized and therefore may be present inside the cells.

Acknowledgments

I thank Dr. Ian Macara, Dr. Loic Fort, and Christian de Caestecker for their comments and insights on the manuscript.

References

1. Rodriguez-Boulan E, Kreitzer G, Müsch A (2005) Organization of vesicular trafficking in epithelia. Nat Rev Mol Cell Biol 6(3):233–247
2. Orlando K, Guo W (2009) Membrane organization and dynamics in cell polarity. Cold Spring Harb Perspect Biol 1(5):a001321
3. Booher KR, Kaiser P (2008) A PCR-based strategy to generate yeast strains expressing endogenous levels of amino-terminal epitope-tagged proteins. Biotechnol J 3(4):524–529. https://doi.org/10.1002/biot.200800012
4. Vavouri T, Semple JI, Garcia-Verdugo R, Lehner B (2009) Intrinsic protein disorder and interaction promiscuity are widely associated with dosage sensitivity. Cell 138(1):198–208. https://doi.org/10.1016/j.cell.2009.04.029
5. Birchler JA, Veitia RA (2012) Gene balance hypothesis: connecting issues of dosage sensitivity across biological disciplines. Proc Natl Acad Sci U S A 109(37):14746–14753
6. Doyon JB, Zeitler B, Cheng J et al (2011) Rapid and efficient clathrin-mediated endocytosis revealed in genome-edited mammalian cells. Nat Cell Biol 13(3):331–337. https://doi.org/10.1038/ncb2175
7. Rivera-Molina F, Toomre D (2013) Live-cell imaging of exocyst links its spatiotemporal dynamics to various stages of vesicle fusion. J Cell Biol 201:673–680. https://doi.org/10.1083/jcb.201212103
8. Yu J (2016) Single-molecule studies in live cells. Annu Rev Phys Chem 67:565–585. https://doi.org/10.1146/annurev-physchem-040215-112451
9. Chen B, Zou W, Xu H et al (2018) Efficient labeling and imaging of protein-coding genes in living cells using CRISPR-tag. Nat Commun 9(1):5065. https://doi.org/10.1038/s41467-018-07498-y
10. Ahmed SM, Nishida-Fukuda H, Li Y et al (2018) Exocyst dynamics during vesicle tethering and fusion. Nat Commun 9:5140. https://doi.org/10.1038/s41467-018-07467-5
11. Doudna JA, Charpentier E (2014) The new frontier of genome engineering with CRISPR-Cas9. Science 346(6213):1258096
12. Sander JD, Joung JK (2014) CRISPR-Cas systems for editing, regulating and targeting genomes. Nat Biotechnol 32(4):347–355
13. Jiang F, Doudna JA (2017) CRISPR-Cas9 structures and mechanisms. Annu Rev Biophys 46:505–529
14. Jinek M, Chylinski K, Fonfara I et al (2012) A programmable dual-RNA-guided DNA endonuclease in adaptive bacterial immunity. Science 337(6096):816–821. https://doi.org/10.1126/science.1225829
15. Deltcheva E, Chylinski K, Sharma CM et al (2011) CRISPR RNA maturation by trans-encoded small RNA and host factor RNase III. Nature 471(7340):602–607. https://doi.org/10.1038/nature09886
16. Chylinski K, Le Rhun A, Charpentier E (2013) The tracrRNA and Cas9 families of type II CRISPR-Cas immunity systems. RNA Biol 10(5):726–737. https://doi.org/10.4161/rna.24321
17. Liu M, Rehman S, Tang X et al (2019) Methodologies for improving HDR efficiency. Front Genet 9:691
18. Rodriguez EA, Campbell RE, Lin JY et al (2017) The growing and glowing toolbox of fluorescent and photoactive proteins. Trends Biochem Sci 42(2):111–129
19. Erdmann RS, Baguley SW, Richens JH et al (2019) Labeling strategies matter for super-resolution microscopy: a comparison between HaloTags and SNAP-tags. Cell Chem Biol

26(4):584–592.e6. https://doi.org/10.1016/j.chembiol.2019.01.003

20. England CG, Luo H, Cai W (2015) HaloTag technology: a versatile platform for biomedical applications. Bioconjug Chem 26(6):975–986. https://doi.org/10.1021/acs.bioconjchem.5b00191
21. Mali P, Yang L, Esvelt KM et al (2013) RNA-guided human genome engineering via Cas9. Science 339(6121):823–826. https://doi.org/10.1126/science.1232033
22. Jao LE, Wente SR, Chen W (2013) Efficient multiplex biallelic zebrafish genome editing using a CRISPR nuclease system. Proc Natl Acad Sci U S A 110(34):13904–13909. https://doi.org/10.1073/pnas.1308335110
23. Evans PD, Cook SN, Riggs PD, Noren CJ (1995) LITMUS(TM): multipurpose cloning vectors with a novel system for bidirectional in vitro transcription. BioTechniques 19(1): 130–135
24. Roberts B, Haupt A, Tucker A et al (2017) Systematic gene tagging using CRISPR/Cas9 in human stem cells to illuminate cell organization. Mol Biol Cell 28(21):2854–2874. https://doi.org/10.1091/mbc.E17-03-0209
25. Maruyama T, Dougan SK, Truttmann MC et al (2015) Increasing the efficiency of precise genome editing with CRISPR-Cas9 by inhibition of nonhomologous end joining. Nat Biotechnol 33(5):538–542. https://doi.org/10.1038/nbt.3190
26. Chu VT, Weber T, Wefers B et al (2015) Increasing the efficiency of homology-directed repair for CRISPR-Cas9-induced precise gene editing in mammalian cells. Nat Biotechnol 33(5):543–548. https://doi.org/10.1038/nbt.3198

Chapter 6

Determine the Function of the Exocyst in Vesicle Tethering by Ectopic Targeting

Kunrong Mei, Di-Ao Liu, and Wei Guo

Abstract

We describe an assay, in which ectopically targeting the exocyst subunit Sec3 to mitochondria is used to determine its role in tethering of post-Golgi vesicles to the plasma membrane. In the assay, we use a plasmid that encodes a fusion protein of the mitochondria protein Tom20 and Sec3 N-terminally tagged with the florescence protein mCherry, and coexpress the plasmid in yeast cells with CIT1-GFP, a marker protein of mitochondria. We then detect the colocalization between Sec3 and CIT1 and other exocyst subunits such as Sec5 on mitochondria using fluorescence microscopy. We further detect the colocalization between Sec3 and Sec4, a Rab protein and a marker of post-Golgi vesicles. Through this assay, we propose that the exocyst subunit Sec3 recruits the other exocyst subunits and secretory vesicles to a target membrane, suggesting that it plays a pivotal role in vesicle tethering. This approach is likely appropriate for studying other tethering complexes at their specific stages of trafficking and may also be used in other eukaryotic cells such as the cultured mammalian cells.

Key words Exocyst, Vesicle, Vesicle tethering, Vesicle fusion, Mitochondria, Ectopic targeting, Multi-subunit tethering complexes

1 Introduction

Membrane trafficking mediates the material exchange between intracellular organelles in eukaryotic cells. Most cellular organelles including endoplasmic reticulum, Golgi, early endosomes, late endosomes, and lysosomes are actively involved in secretory or endocytic trafficking, while mitochondria and peroxisomes are relatively autonomous. There are four continuous steps of vesicle trafficking including vesicle budding, transport, tethering, and fusion [1]. Each step is mediated by specific protein machineries [1]. Tethering proteins mediate the initial interaction and tethering of vesicles to the target membrane, which function after vesicular carrier formation and cargo loading but before SNARE-mediated membrane fusion [2].

Jingshi Shen (ed.), *Membrane Trafficking: Methods and Protocols*,
Methods in Molecular Biology, vol. 2473, https://doi.org/10.1007/978-1-0716-2209-4_6,

Multisubunit tethering complexes (MTCs) are a family of protein complexes consisting of 3–8 subunits implicated in vesicle tethering at different trafficking stages [3]. Ten complexes, TRAPP I, TRAPP II, TRAPP III, HOPS, CORVET, COG, GARP, EARP, Dsl, and the exocyst complex, have been identified as MTCs [3–5]. However, the functions of some MTCs in tethering are quite speculative. Conclusive assays that provide direct evidence for their functions in vesicle tethering have been lacking for MTCs except HOPS and CORVET, mainly due to their complex compositions and difficulties in obtaining enough soluble proteins for in vitro reconstitutions [6].

The exocyst complex is believed to tether the post-Golgi vesicles to the plasma membrane during exocytosis [7]. It consists of eight subunits, Sec3, Sec5, Sec6, Sec8, Sec10, Sec15, Exo70, and Exo84. In yeasts, the exocyst complex interacts with Rab proteins and v-SNARE on the vesicles through Sec15 and Sec6, respectively, and binds to t-SNARE, PI (4,5)P2, and Rho GTPase on the plasma membrane through Sec3 and Exo70. Therefore, exocyst bridges the vesicular membrane and plasma membrane for tethering [8–13]. Here, we describe an ectopic targeting assay to determine the role of exocyst in vesicle tethering. Before the assay, we have prepared a plasmid, in which we fused the N-terminus of Sec3 with TOM20 for mitochondria targeting and mCherry for microscopy detection. The plasmid is cotransformed with CIT1-GFP or Sec5-GFP into yeast cells, and their colocalization is detected via florescence microscopy. We also detect the colocalization between Sec3 and the Rab protein Sec4 on the post-Golgi vesicles through immunofluorescence microscopy. The exocyst subunits are localized at specific sites of plasma membrane such as bud tip and bud neck in log-phase yeast cells in native cells [12, 14, 15]. Using this assay, we have found that Sec3 recruits all other exocyst subunits and the secretory vesicles to mitochondria [16]. We also found that Sec3 truncations containing the CorEx motif could recruit Sec5 and Sec4, while Sec3 truncations without the CorEx motif could not [17]. These findings support a role of exocyst in vesicle tethering and suggest that the exocyst subunit Sec3, particularly its CorEx motif, plays a pivotal role in vesicle tethering.

Since the exocyst is one of the better studied MTCs and bears similar structural and functional features to other MTCs such as COG, GARP, EARP, and Dsl, the study of the exocyst provides helpful insights for the study of all other MTCs [18]. Similarly in mammalian cells, by ectopic targeting to mitochondria, COG4 and COG8 were able to recruit Golgi transport intermediates to the mitochondria [19].

2 Materials

2.1 Plasmids and Yeast Strains

The plasmids and yeast strains in Table 1 are used in the assay. The plasmids were constructed following routine gene cloning strategies (*see* **Notes 1–4**). The map of p415TEF-TOM20-mCherry is shown in Fig. 1.

Table 1
Plasmids and yeast strains

Name	Description	Source
pGV373	p415TEF-TOM20-mCherry	Wei Guo Lab
pG1872	p415TEF-TOM20-mCherry-Sec3	Wei Guo Lab
pG1831	pRS416-CIT1-GFP	Liza Pon Lab
GY1143	*MAT **a** trp1, leu2, his3, ura3, lys2, Sec5-GFP::URA3*	Wei Guo Lab
GY1132	*MAT **a** trp1, leu2, his3, ura3, lys2*	Peter Novick Lab

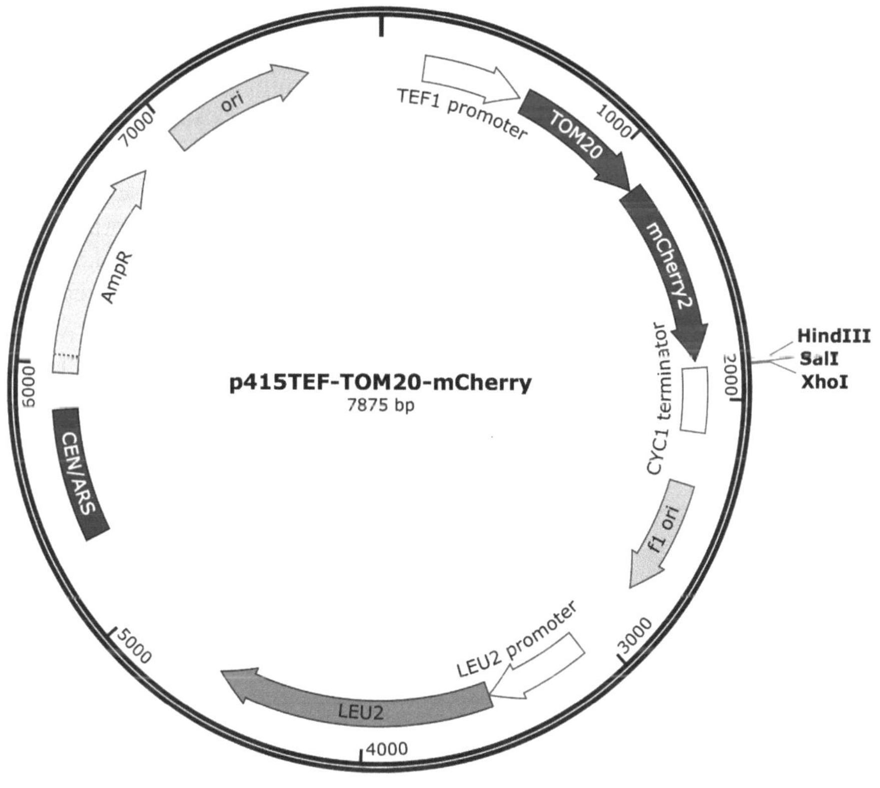

Fig. 1 Map of p415TEF-TOM20-mCherry. The vector backbone is p415TEF (ATCC 87366). Single cleavage sites HindIII, SalI, and XhoI are available for inserting gene of interest

2.2 Yeast Transformation

1. Yeast Extract Peptone Dextrose Medium (YPD): 20 g/L peptone, 20 g/L glucose, 10 g/L yeast extract. Add 800 mL ddH_2O to a glass beaker. Weigh 20 g peptone, 10 g yeast extract and 20 g glucose and transfer to the beaker. Mix with a magnetic stirrer to dissolve the medium well. Make up to 1 L with ddH_2O. Autoclave at 121 °C for 15 min and store at 4 °C before use (*see* **Notes 5** and **6**).
2. Synthetic complete medium without leucine (SC-Leu): 6.7 g yeast nitrogen base, 1.4 g amino acid dropout mix (without uracil, histidine, tryptophan, or leucine), 0.073 g histidine, 0.073 g tryptophan, 0.073 g uracil, 20 g glucose. Weigh and dissolve the reagents in 1 L ddH_2O. Adjust pH to 6.8 by adding 600 μL of 10 M NaOH. Autoclave at 121 °C for 15 min and store at 4 °C.
3. SC-Leu-Ura medium: 6.7 g yeast nitrogen base, 1.4 g amino acid dropout mix (without uracil, histidine, tryptophan, and leucine), 0.073 g histidine, 0.073 g tryptophan, 20 g glucose. Prepare a 1 L solution as in YPD preparation.
4. SC-Leu plates: 6.7 g yeast nitrogen base, 1.4 g amino acid dropout mix (without uracil, histidine, tryptophan, and leucine), 0.073 g histidine, 0.073 g tryptophan, 0.073 g uracil, 20 g glucose, 20 g agar. Weigh and dissolve the reagents in 1 L ddH_2O. Adjust pH to 6.8 by adding 600 μL of 10 M NaOH. Autoclave at 121 °C for 15 min. When the medium drops to 50–70 °C, pour out into petri dishes, with about 20 mL per dish. Store at 4 °C after the medium get solidified (*see* **Note 7**).
5. SC-Leu-Ura plates: 6.7 g yeast nitrogen base, 1.4 g amino acid dropout mix (without uracil, histidine, tryptophan, and leucine), 0.073 g histidine, 0.073 g tryptophan, 20 g glucose, 20 g agar. Prepare plates as in SC-Leu plates preparation.
6. 10 × TE buffer: 100 mM Tris, 10 mM EDTA, pH 8.0. Filter with a 0.22 μm membrane. Store at 4 °C or store 1 mL aliquots at −20 °C.
7. 1 M LiOAc solution: weigh 3.3 g lithium acetate to a 50-mL tube and dissolve in 50 mL ddH_2O, filter-sterilize, and store 1 mL aliquots at −20 °C.
8. TE buffer: dilute 10 mL 10 × TE buffer with 90 mL ddH_2O, filter-sterilize and store at 4 °C.
9. 0.1 M LiOAc–TE buffer: 10 mL 10 × TE buffer, 10 mL 1 M LiOAc, 80 mL ddH_2O. Filter-sterilize and store at 4 °C.
10. 10 mg/mL ssDNA: dissolve 1 g ssDNA in 100 mL TE buffer, stir at 4 °C overnight. Sonicate the ssDNA solution on ice with 300 W power, 10 s on and 10 s off, with a total sonication time of 2 h. Store 1 mL aliquots at −20 °C (*see* **Note 8**).

11. 50% PEG3350 solution: weigh 25 g PEG3350 to a 50-mL tube. Gradually add ddH_2O to dissolve and make up the final volume to 50 mL. Rotate on a rocker at room temperature overnight to help dissolve. Filter-sterilize and store 1 mL aliquots at −20 °C (*see* **Note 9**).
12. 1 M $CaCl_2$ solution: weigh 7.35 g of $CaCl_2{\cdot}2H_2O$ and dissolve with 50 mL ddH_2O. Filter-sterilize and store at room temperature.
13. 5 mM $CaCl_2$ solution: add 250 μL of 1 M $CaCl_2$ to 49.75 mL ddH_2O. Filter-sterilize and store at 4 °C.

2.3 Immunofluorescence Microscopy

1. 16% Paraformaldehyde solution: weigh 16 g paraformaldehyde and transfer to a glass beaker, add 90 mL ddH_2O. Stir at 60 °C for 10–20 min. Add 1 M NaOH dropwise while stirring until the solution becomes clear. Make up to 100 mL with ddH_2O.-Cool the solution to room temperature and filter with a 0.22 μm membrane. Store 10 mL aliquots at −20 °C (*see* **Note 10**).
2. 1 M KH_2PO_4 buffer: weigh 68 g KH_2PO_4 and dissolve with 500 mL ddH_2O. Autoclave and store at room temperature.
3. 1 M K_2HPO_4 buffer: weigh 87 g K_2HPO_4 and dissolve with 500 mL of ddH_2O. Autoclave and store at room temperature.
4. 0.1 M phosphate buffer: take 80.2 mL of 1 M K_2HPO_4, 19.8 mL of 1 M KH_2PO_4, add 900 mL ddH_2O, mix well. The pH is 7.4. Autoclave and store at room temperature.
5. 0.1 M phosphate buffer–4% paraformaldehyde: 2.5 mL of 16% Paraformaldehyde, 802 μL of 1 M K_2HPO_4, 198 μL of 1 M KH_2PO_4, 6.5 mL of ddH_2O.
6. 0.1 M phosphate buffer–1.2 M sorbitol solution: weigh 218.6 g sorbitol, add 80.2 mL of 1 M K_2HPO_4, 19.8 mL of 1 M KH2PO4, add ddH_2O to final volume of 1 L. Autoclave and store at room temperature.
7. 10 mg/mL zymolase: weigh 100 mg zymolase and dissolve in 10 mL of ddH_2O. Store 0.5-mL aliquots at −20 °C (*see* **Note 11**).
8. PBS–BSA solution: weigh 0.5 g BSA into a 50 mL tube, add 0.5 mL of 1 M KH_2PO_4, 2 mL of 1 M K_2HPO_4, 1.5 mL of 5 M NaCl, 0.1 g NaN_3. Add ddH_2O to a final volume of 50 mL. Dissolve and store at 4 °C (*see* **Note 12**).
9. Antibodies: rabbit anti-Sec4 polyclonal antibody (a gift from Dr. Peter Novick's lab). Alexa Fluor 488–conjugated goat anti-rabbit antibody.

3 Methods

3.1 Expression in Yeast Cells

To coexpress TOM20-mCherry-Sec3 and CIT1-GFP, transform pG1872 and pG1831 into the GY1132 cells. To coexpress TOM20-mCherry-Sec3 and Sec5-GFP, transform pG1872 into the GY1143 cells. To test whether TOM20-mCherry-Sec3 could recruit secretory vesicles, transform pG1872 into GY1132 cells. After isolation of the positive strains, grow the yeast cells to log phase and harvest the cells for fluorescence or immunofluorescence microscopy.

3.1.1 Yeast Cell Transformation

1. Inoculate 5 mL of YPD medium with the yeast host strains GY1132 and GY1143, respectively. Grow overnight at 30 °C.
2. The next morning, check OD_{600} of the overnight culture. Inoculate the culture into 5 mL of YPD medium at an OD_{600} of 0.2–0.3.
3. Grow the culture for 3–4 h at 30 °C until OD_{600} reaches 0.6–1.0. For each transformation, use 1.5 mL of cells at OD_{600} of 0.6–1.0 (*see* **Note 13**).
4. Heat ssDNA at 100 °C for 5–10 min and then move the samples on ice quickly (*see* **Note 14**).
5. Transfer 1.5 mL of GY1143 cells and two 1.5 mL of GY1132 cells into sterile 1.5-mL tubes. Centrifuge the cells at 900 × *g* for 1 min and remove the supernatant.
6. Wash the cells with 1 mL ddH_2O. Centrifuge at 900 × *g* for 1 min and remove the supernatant.
7. Wash the cells with 200 μL of 0.1 M LiOAc–TE buffer. Centrifuge the cells at 900 × *g* for 1 min and remove the supernatant.
8. Resuspend the cells in 20 μL of 0.1 M LiOAc–TE buffer. Add 5 μL of pretreated ssDNNA.
9. Add 2 μL of pG1872 into the GY1143 cells. Add 2 μL of pG1872 and 2 μL of pG1831 into the GY1132 cells. Add 2 μL of pG1872 into the GY1132 cells (*see* **Note 15**).
10. For each transformation, add 96 μL of 50% PEG3350 solution, 12 μL of 10× TE buffer, and 12 μL of 1 M LiOAc solution. Mix well with the yeast cells.
11. Incubate at 200 rpm at 30 °C for at least 30 min.
12. For each transformation, add 16 μL DMSO and mix well. Heat shock the cells by incubating in a 42 °C water bath for 15 min.
13. Centrifuge the cells at 900 × *g* for 3 min and remove the supernatant.

14. Wash the cells with 100 μL of 5 mM $CaCl_2$ solution. Centrifuge at 900 × *g* for 1 min and remove the supernatant (*see* **Note 16**).
15. Resuspend the cells with 1 mL YPD medium and transfer into a cell culture tube. Incubate at 30 °C at 200 rpm for 3–6 h (*see* **Note 17**).
16. Transfer the cells into 1.5 mL tubes. Centrifuge the cells at 900 × *g* for 1 min and remove the supernatant. Add 500 μL sterilized ddH_2O to resuspend the cells, and plate 50 μL of cell suspension onto selective plates (*see* **Note 18**). For transforming pG1872 into GY1143 cells and transforming pG1872 and pG1831 into GY1132 cells, plate the cells on SC-Leu-Ura plates. For transforming pG1872 into GY1132 cells, plate the cells on SC-Leu plates.
17. Put the plates in a 30 °C incubator. Colonies will appear after 2–3 days.

3.1.2 Yeast Cell Growth

1. Pick 4–6 colonies from each plate and streak on new selective plates. Grow at 30 °C for 1–2 days.
2. Inoculate the colonies into 2 mL of selective medium. Grow overnight at 30 °C.
3. The next morning, check OD_{600} of the overnight culture. Inoculate the culture to 5 mL of selective medium at an OD_{600} of 0.2–0.3. Grow the culture at 30 °C until OD_{600} reaches 1.0–1.5.
4. Harvest the cells by centrifugation at 900 × *g* for 1 min. Remove the supernatant. Resuspend the cells with 50–100 μL of selective medium and transfer into 1.5 mL tubes.
5. Check the expression of mCherry and GFP via florescence microscopy. For transforming pG1872 into GY1143 cells and transforming pG1872 and pG1831 into GY1132 cells, observation of expression of both mCherry and GFP means the transformation and expression are successful and the colony is suitable for subcellular localization analysis via microscopy. For transforming pG1872 into GY1132 cells, expression of mCherry means transformation and expression are successful and the colony is suitable for subcellular localization analysis of Sec4 via immunofluorescence microscopy (*see* **Note 19**).

3.2 Microscopy Analysis

Perform microscopic analysis using a confocal fluorescence microscope such as the Nikon ECLIPSE Ti confocal microscope equipped with an Apo TIRF 100× 1.49 oil immersion objective lens and a CSU-X1 digital camera. Other fluorescence microscopes are also adequate as long as they can detect mCherry and GFP, and observe yeast cells under the DIC (differential interference contrast) or phase contract mode.

3.2.1 Florescence Microscopy

1. Prepare cells (GY1132 bearing pG1872 and pG1831; GY1143 bearing pG1872) as **steps 2–4** in Subheading 3.1.2.
2. Load 2 μL of cell suspension to the microscope slide.
3. Apply immersion oil onto the 100× objective lens and load the slide on the microscope.
4. Adjust the focus until cells are clearly observed in the bright filed.
5. Shift the visual view until most cells in the view look healthy and are well distributed.
6. Excite the sample with a 488 nm laser to observe the expression and localization of CIT1-GFP or Sec5-GFP. Adjust the exposure time and signal amplification to proper values (*see* **Note 20**).
7. Excite the sample with a 587 nm laser to observe the expression and localization of TOM20-mCherry-Sec3. Adjust the exposure time and signal amplification to proper values (*see* **Note 21**).
8. For the same visual view, take photos under the DIC mode, the 488 nm laser excited mode, and the 587 nm laser excited mode.
9. Shift the visual view to another spot and take photos.
10. Collect 5–10 sets of photos for each sample.
11. Export photos and merge image of each view for data analysis.

3.2.2 Immunofluorescence Microscopy

1. Inoculate cells (GY1132 bearing pG1872) into 2 mL of SC-Leu medium. Grow overnight at 30 °C.
2. The next morning, check OD_{600} of the overnight culture. Inoculate the culture to 5 mL of SC-Leu medium at an OD_{600} of 0.2–0.3. Grow the culture at 30 °C until OD_{600} reaches 1.0–1.5.
3. Add 1.7 mL of 16% paraformaldehyde into the 5 mL of cell culture, mix well, and continue shaking for 10–20 min at 30 °C.
4. Pellet the cells by centrifuging at 900 × *g* for 5 min and remove the supernatant. Resuspend cells with 1 mL of 0.1 M phosphate buffer–4% paraformaldehyde and transfer into a 1.5 mL tube. Keep at 30 °C for 1–2 h or 4 °C overnight.
5. Pellet the cells by centrifuging at 900 × *g* for 2 min. Wash cells 2× with 1 mL of 0.1 M phosphate buffer. Resuspend cells in 500 μL of 0.1 M phosphate buffer–1.2 M sorbitol solution.
6. Add 10 μL of 2-mercaptoethanol, mix well, and keep at room temperature for 30 min.

7. Add 50 μL of 10 mg/mL zymolase, mix gently, and incubate in a 37 °C water bath for 30 min to spheroplast the cells (*see* **Note 22**).
8. Pellet the cells by centrifuging at 2400 × *g* for 5 min. Wash the cells 2× with 1 mL of 0.1 M phosphate buffer–1.2 M sorbitol solution. Resuspend the cells in 50–200 μL of 0.1 M phosphate buffer–1.2 M sorbitol solution (*see* **Note 23**).
9. Wash microscope slide with reaction wells with 95% ethanol and air-dry.
10. Add 15 μL of 0.1% poly-L-lysine solution into each well and spread to cover the whole well. Incubate at room temperature for 5–10 min.
11. Wash 3× with ddH_2O and air-dry.
12. Add 15 μL of spheroplasted cells into each well (*see* **Note 24**). Incubate at room temperature for 5–10 min.
13. Remove the supernatant and immediately plunge the slide into ice-cold methanol and incubate for 5–6 min (*see* **Note 25**).
14. Take the slide out and immediately plunge into acetone for 30 s.
15. Take out the slide and add 25 μL of PBS–BSA solution to each well once the slide is air-dry. Incubate at room temperature for 30 min.
16. Remove the PBS–BSA solution and add 25 μL of rabbit anti-Sec4 polyclonal antibody (1:3000 dilute with the PBS–BSA solution) to each well. Carefully move the slide into a humid chamber and incubate overnight at 4 °C.
17. Remove the antibody solution. Wash 5× with the PBS–BSA solution.
18. Add 25 μL of Alexa Fluor 488 goat anti-rabbit antibody (1:500 dilute with the PBS–BSA solution) to each well. Incubate for 1 h at room temperature in a humid chamber in the dark.
19. Remove the antibody solution. Wash 5× with PBS–BSA solution.
20. Wash 2× with PBS and air-dry the slide.
21. Cover each well with Antifade Mounting Solution and cover-slip. Seal with nail polish oil. Store the slide in the dark at −20 °C (*see* **Note 26**).
22. Run fluorescence microscopy analysis as in Subheading 3.2.1.

4 Notes

1. The pGV373 (p415TEF-TOM20-mCherry) plasmid was constructed by inserting ScTOM20 (NM_001181211) and mCherry into p415TEF (ATCC 87366) via restriction ligation. HindIII, SalI, and XhoI are available for inserting gene of interest into pGV373 to express N-terminal TOM20-mCherry fusion protein. The pG1872 plasmid was constructed by inserting ScSec3 (NM_001178899) into pGV373 via SalI digest and ligation [16].
2. The pG1831 is a gift from Dr. Liza Pon's Lab. It was constructed by insertion of a 2.19 kb fragment containing 5′-UTR of CIT1(−806 to −1 nt), CIT1 leader (1–52 aa), GFP and 3′-UTR of CIT1 regions into the vector pRS416 (ATCC 87521). The product of this plasmid is CIT1 leader directed mitochondria localized GFP in yeast cells [16].
3. GY1143 was constructed via genome integration of Sec5-GFP into the GY1132 strain followed by URA3 selection [17].
4. Alternative vectors [20] and yeast strains could be used to construct and express the TOM20-mCherry fusion proteins and chromosomal GFP fusion proteins as long as they are compatible. Also, mCherry and GFP could be substituted by other fluorescent proteins that are compatible with the fluorescence microscope and could be well distinguished when imaged together.
5. The medium should be well dissolved before autoclaving. Autoclaving should be no more than 15 min at 121 °C. The color of the medium will become dark brown after autoclaving.
6. Alternatively, dissolve 20 g peptone and 10 g yeast extract with 900 mL ddH_2O and autoclave to make YP solution. Dissolve 200 g glucose with 800 mL ddH_2O, make up 1 L with ddH_2O, and autoclave to make 20% glucose solution. Then add 100 mL 20% glucose into 900 mL of YP solution to make 1 L of YPD medium. YPD prepared with this method will be yellow or light brown.
7. pH adjusting is necessary or the plates will not be solidified properly.
8. ssDNA becomes very sticky when dissolved in the TE buffer. After sonication, it will become less sticky. Most of the DNA fragments will be around 200 nt when checked with an agarose gel.
9. The 50% PEG3350 solution is very thick. Filter with a 0.22 μm membrane will be hard but operable.

10. Wear a mask and goggles when weighing paraformaldehyde. Transfer the weighed paraformaldehyde to the beaker inside the fume hood and mix on a stirrer placed inside the hood.
11. Vortex can help dissolve zymolase. When the stock solution is thawed before use, centrifuge at 16200 × *g* for 5 min to pellet the insoluble.
12. The PBS–BSA solution could be store at 4 °C for up to 2 weeks.
13. Cells at OD_{600} of 1.0–2.0 are also usable but not recommended.
14. Single-stranded carrier DNA promotes yeast transformation at higher efficiency than double-stranded carrier DNA [21]. Heating ssDNA at 100 °C breaks down double-stranded DNA. Quick cooling prevents reannealing and keeps the DNA as single-stranded. Pretreating ssDNA before transformation will allow enough cooling time for single-stranded DNA formation.
15. The concentration of the plasmid is 100–1000 ng/μL. If the concentration of the plasmid is lower, 3–20 μL of plasmids could be used.
16. This step is recommended but not required. It likely improves the transformation efficiency for unknown reasons.
17. YPD medium helps the recovery of the yeast cells and is not recommended to be changed into selective medium. Incubation in the cell culture tube help avoid burst of the cells during incubation.
18. The proper numbers of cells to plate vary with the amounts of plasmids to be transformed and the efficiency of transformation. 50 μL could be a starting point. Store the rest of the cells at 4 °C. If the plated cells are too many or too few, recover the stored cells by incubating at 30 °C for 30 min, and then plate again with adjusted numbers of cells.
19. Western blot detection is highly recommended to confirm the expression of the proteins of interest.
20. The exposure time and signal amplification rate are different for different proteins and different microscopes. In our experiment, we set the exposure time for CIT1-GFP 200–500 ms with an amplifier of 300. For Sec5-GFP, which has an endogenous level of expression, the exposure time and amplifier is 0.8–1 s and 800, respectively.
21. TOM20-mCherry-Sec3 is expressed under the TEF promoter, which is strong and constitutively active, resulting in a relatively high expression level. In our experiment, the exposure time for TOM20-mCherry-Sec3 is 100–200 ms with an amplifier of

300. We found that the mitochondria of yeast cells bearing TOM20-mCherry-Sec3 exhibit chunk-like structures, unlike their normal ribbon-like structures. This morphology change may be caused by ectopic-targeting of Sec3 [16].

22. The time of zymolase incubation could be 15–45 min.
23. Resuspending should be very gentle as the spheroplasted cells are fragile. Cells can be stored overnight at 4 °C at this point.
24. In case the recommended volume is not optimal, different amounts of cells could be prepared at this step.
25. Do not let cells dry out before mounting.
26. The slides can be stored for several weeks in the dark at −20 °C.

Acknowledgments

This work was supported by National Institutes of Health grant R35GM141832 to W.G. and National Science Foundation of China (Grant 91954112 and 31900501) to K.M.

References

1. Cai H, Reinisch K, Ferro-Novick S (2007) Coats, tethers, Rabs, and SNAREs work together to mediate the intracellular destination of a transport vesicle. Dev Cell 12: 671–682
2. Guo W, Sacher M, Barrowman J et al (2000) Protein complexes in transport vesicle targeting. Trends Cell Biol 10:251–255
3. Whyte JRC, Munro S (2002) Vesicle tethering complexes in membrane traffic. J Cell Sci 115: 2627–2637
4. Bröcker C, Engelbrecht-Vandré S, Ungermann C (2010) Multisubunit tethering complexes and their role in membrane fusion. Curr Biol 20:R943–R952
5. Schindler C, Chen Y, Pu J et al (2015) EARP is a multisubunit tethering complex involved in endocytic recycling. Nat Cell Biol 17:639–650
6. Brunet S, Sacher M (2014) Are all multisubunit tethering complexes Bona fide tethers? Traffic 15:1282–1287
7. TerBush DR, Maurice T, Roth D et al (1996) The exocyst is a multiprotein complex required for exocytosis in Saccharomyces cerevisiae. EMBO J 15:6483–6494
8. Guo W, Roth D, Walch-Solimena C et al (1999) The exocyst is an effector for Sec4p, targeting secretory vesicles to sites of exocytosis. EMBO J 18:1071–1080
9. Shen D, Yuan H, Hutagalung A et al (2013) The synaptobrevin homologue Snc2p recruits the exocyst to secretory vesicles by binding to Sec6p. J Cell Biol 202:509–526
10. Liu J, Zuo X, Yue P et al (2007) Phosphatidylinositol 4,5-bisphosphate mediates the targeting of the exocyst to the plasma membrane for exocytosis in mammalian cells. Mol Biol Cell 18:4483–4492
11. He B, Xi F, Zhang J et al (2007) Exo70p mediates the secretion of specific exocytic vesicles at early stages of the cell cycle for polarized cell growth. J Cell Biol 176:771–777
12. Zhang X, Orlando K, He B et al (2008) Membrane association and functional regulation of Sec3 by phospholipids and Cdc42. J Cell Biol 180:145–158
13. Yamashita M, Kurokawa K, Sato Y et al (2010) Structural basis for the rho- and phosphoinositide-dependent localization of the exocyst subunit Sec3. Nat Struct Mol Biol 17:180–186
14. TerBush DR, Novick P (1995) Sec6, Sec8, and Sec15 are components of a multisubunit complex which localizes to small bud tips in Saccharomyces cerevisiae. J Cell Biol 130:299–312
15. Finger FP, Hughes TE, Novick P (1998) Sec3p is a spatial landmark for polarized secretion in budding yeast. Cell 92:559–571

16. Luo G, Zhang J, Guo W (2014) The role of Sec3p in secretory vesicle targeting and exocyst complex assembly. Mol Biol Cell 25: 3813–3822
17. Mei K, Li Y, Wang S et al (2018) Cryo-EM structure of the exocyst complex. Nat Struct Mol Biol 25:139–146
18. Yu I-M, Hughson FM (2010) Tethering factors as organizers of intracellular vesicular traffic. Annu Rev Cell Dev Biol 26:137–156
19. Willett R, Kudlyk T, Pokrovskaya I et al (2013) COG complexes form spatial landmarks for distinct SNARE complexes. Nat Commun 4: 1553
20. Mumberg D, Müller R, Funk M (1995) Yeast vectors for the controlled expression of heterologous proteins in different genetic backgrounds. Gene 156:119–122
21. Gietz RD, Schiestl RH, Willems AR et al (1995) Studies on the transformation of intact yeast cells by the LiAc/SS-DNA/PEG procedure. Yeast 11:355–360

Chapter 7

Imaging Insulin Granule Dynamics in Human Pancreatic β-Cells Using Total Internal Reflection Fluorescence (TIRF) Microscopy

Fei Kang and Herbert Y. Gaisano

Abstract

Due to the ultra-thin optical sectioning capability of exclusively illuminating space at the interface where total internal reflection occurs, the TIRF microscope has been indispensable for monitoring biological processes adjacent to the plasma membrane with excellent signal-to-noise ratio. Insulin-containing granules fuse with the plasma membrane to release contents within hundreds of milliseconds, which involves well-orchestrated assembly of SNARE complex and associated proteins. A video-rate multiple-color TIRF microscope offers the unique opportunity to visualize single secretory granule docking and fusion dynamics and can also map its regulators with high spatiotemporal resolution. Here, we describe the basic principles and practical implementation of a fast dual-color TIRF microscope, detailing a how-to guide on imaging and analysis of insulin granule dynamics in human β-cells.

Key words Insulin granules, Total internal reflection fluorescence microscopy, Exocytosis, Live-cell imaging

1 Introduction

Human pancreatic β-cells of the islets of Langerhans are dedicated to secreting insulin to regulate the body's glucose homeostasis. β-cells synthesize and package processed mature insulin into large dense core secretory granules constantly. When sensing elevated blood glucose, the β-cell depolarizes and generates action potentials, which leads to Ca^{2+} influx via voltage-gated Ca^{2+} channels that triggers SNARE complex-mediated fusion of insulin-containing granules with the plasma membrane to release its cargo at millisecond time scales [1]. In order to gain more detailed insights regarding the exocytotic machinery that mediates the sequential steps of granule docking, priming and fusion, and how these steps could be perturbed in diabetes [2, 3], methodologies with high

Jingshi Shen (ed.), *Membrane Trafficking: Methods and Protocols*,
Methods in Molecular Biology, vol. 2473, https://doi.org/10.1007/978-1-0716-2209-4_7,

spatiotemporal resolution are required to visualize the behaviors of individual insulin secretory granules.

TIRF microscopy produces optical sections by generating an evanescent field to selectively excite fluorophores within a few hundred nanometers of the cell–glass contact plane, holding the potential in achieving high spatial and temporal resolution [4, 5]. Principally, the spatial resolution of a fluorescence microscope is determined by the Abbe limit (*see sections below*), which largely depends on the numerical aperture (NA), a measurement of the ability of the objective lens to collect photons emitted by fluorophores [6]. In contrast to point/line-scanning or multiframe reconstruction–based microscopy, TIRF microscopy has the natural advantage in the temporal resolution at a planar-scanning modality. In this chapter, we cover a straightforward approach to adapt existing commercial microscopes to build a fast dual-color TIRF microscope, a step-by-step operating guide on TIR illuminator alignment and achieving TIR illumination. We also present a detailed protocol to prepare human pancreatic β-cells for imaging and give examples to analyze insulin granule fusion events.

2 Materials

2.1 General Materials

1. Coverslips, No. 1½, round, Ø = 25 mm.
2. Sample holder (e.g., SKE, SKU: S-LBCC*-ES or custom built) (*see* Fig. 2a, chamber).
3. Microscope immersion oils, Refractive Index = 1.518. The refraction index of immersion oils should match that of coverslips to maximize the resolution of high NA objectives.
4. Water-resistant markers and green or yellow highlighters with fluorescent water-based ink for illuminator alignment and incident angle adjustment.
5. Phosphate-buffered saline (PBS): 138 mM NaCl, 8 mM Na_2HPO_4, 3 mM KCl, 1.5 mM KH_2PO_4, pH 7.4.
6. 0.1% (w/v) Poly-L-lysine solution. Dilute a 1% (w/v) stock solution with deionized water just prior to using.
7. 0.05% Trypsin-EDTA.
8. Disposable, sterile 15-mL tubes.
9. Deionized water, sterile.
10. Cell culture dishes or Petri dishes, 35 mm, sterile.
11. 37 °C incubator with humidified atmosphere of 5% CO_2.
12. Krebs-Ringer bicarbonate buffer (KRBB): 120 mM NaCl, 5 mM KCl, 2 mM $CaCl_2$, 1 mM $MgSO_4$, 25 mM $NaHCO_3$, 10 mM HEPES, and 0.1% BSA with low or high concentrations of glucose, pH 7.4 at 5% CO_2.

13. Culture medium for human islets and dispersed β-cells: CMRL-1066 medium containing 5.5 mM glucose, 100 units/mL penicillin, 100 μg/mL streptomycin, 2 mM GlutaMAX, 250 μg/mL Gentamycin, and 10% Fetal Bovine Serum (FBS).

2.2 TIRF Microscope

The optical train of a fast dual-color TIRF is schematically illustrated in Fig. 2a. The basic components include the microscope, light source, TIRF illuminator, OptoSplit II, and the detector. Listed below is an overview of two common commercial fluorescence microscopes. We discuss the implementation of video-rate dual-color TIRF imaging (*see* **Note 1**).

2.2.1 Nikon TIRF Microscope

Microscope: Inverted Nikon Eclipse Ti microscope.

Objectives: Apo TIRF 60×/1.49, oil, Apo TIRF 100×/1.49, oil.

Camera: Andor iXon897, 512 × 512, pixel size, 16 μm.

Lasers: 405 nm, 488 nm (50 mW), 561 nm (70 mW), 640 nm.

Emission Filters: 530/50, 641/75.

2.2.2 Olympus TIRF Microscope

Microscope: Olympus IX81.

Objectives: 60×/1.49 (O), 150×/1.45 (O).

Camera: Hamamatsu C9100-13 EM-CCD, 1024 × 1024, pixel size, 13 μm.

Lasers: 491 nm (50 mW), 561 nm (50 mW).

Emission Filters: 520/35, 624/40.

2.2.3 Basic Requirements for the Objective and Camera to Optimize the Spatial Resolution

Objectives capable of TIRF illumination must have an NA larger than 1.38, which is determined by the refractive indexes of the coverslip n_1 and culture medium n_2 at the interface where total internal reflection occurs:

$$\mathrm{NA} = n_1 \times \sin\theta = n_1 \times \frac{n_2 \times \sin\frac{\pi}{2}}{n_1} = n_2 = 1.38$$

In order to match the resolution of a high NA objective, the magnification should be larger than 60×. The spatial resolution is given by Abbe limit, $d = 0.5 \times \lambda/\mathrm{NA}$. According to the Nyquist limit:

$$d \times \text{Magnification} > 2 \times \text{pixel size}$$

where pixel size is the physical dimension of CCD, λ is the maximum wavelength of the emission of a fluorophore. In general, a magnification of at least 60× is required to yield maximum spatial resolution.

3 Methods

3.1 Preparation of Coverslips

1. Immerse coverslips in acetone and sonicate 15 min.
2. Transfer the coverslips to a 1 M KOH solution and sonicate again for 20 min.
3. Wash the coverslips extensively with deionized water and sonicate 10–15 min. Repeat 4–5 times.
4. Store coverslips in 95% ethanol. Never touch the coverslips with your fingers.

3.2 Preparation of Dispersed Human Pancreatic β-Cells

1. Tilt one coverslip at the edge of each 35 mm petri dish wall and leave it in biosafety cabinet for 10 min to allow ethanol to evaporate completely.
2. Place the coverslip on the bottom of the dish and coat it with the poly-L-lysine solution for 15 min. Wash twice with sterile deionized water.
3. Pick islets out one by one with a P-200 μL pipette under a stereoscope to make sure that no acinar or other adjacent tissues are collected with the islets (*see* Fig. 1). Generally, 200–250 islets are sufficient to seed 10–12 coverslips for imaging.
4. Wash twice with PBS. Transfer islets into a sterile 15 mL centrifuge tube containing 10 mL of PBS. When inverting the tube, islet particles are clearly visible by eyes. Centrifuge for 2 min at $200 \times g$ and then aspirate and discard the supernatant. Repeat once more.

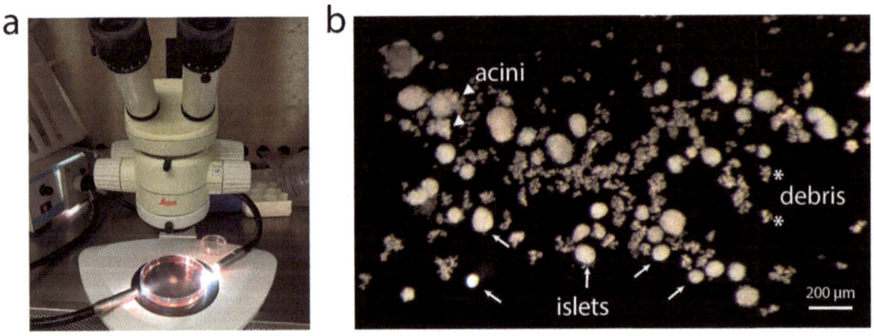

Fig. 1 Picking islets using a stereo microscope. (**a**) Setup of a stereo microscope and illuminating islets with fiber optic illuminators from both sides. (**b**) Picking of islets using a P-200 μL pipette. Healthy islets are easily recognized under a stereo microscope by their smooth surfaces and compact spherical appearances, and they are approximately 50–200 μm in diameter. By contrast, debris and exocrine acini look more transparent and aggregate loosely

5. Add 500 μL of trypsin solution into a tube of ~200 islets and gently shake the tube with hand to resuspend the islets. Incubate the tube at 37 °C for ~5–6 min and shake vigorously by hand every 2 min. Observe frequently during this critical step and stop the enzymatic dissociation by adding 10 mL of culture medium when most of the islet particles have become invisible and the solution has become turbid.
6. Centrifuge for 2 min at 200 × *g* and carefully aspirate and discard the supernatant.
7. Resuspend the cell pellet in 200 μL of culture medium by pipette. Seed and spread 20 μL of the dispersed cells evenly on the centre of the coverslip within a dish. The sample should be enough for 10 coverslips. Place dishes into the incubator for 15 min to allow the cells to attach onto the coverslips.
8. Add 2 mL of prewarmed culture medium gently to the side of the dish to avoid disturbing the cells attached on the coverslip. Culture the cells overnight at 37 °C under a 5% (v/v) CO_2 atmosphere.
9. Change the culture medium, then infect with approximately 1×10^{10} adenovirus particles per millilitre medium for 4 h. Change the medium and culture the cells overnight. Imaging should be performed between 24 and 48 h after infection.

3.3 Optimizing the Temporal Resolution

In order to achieve fast dual-color imaging, the system should avoid the use of any mechanical switches.

1. Switch excitation laser within microseconds by an electronically controlled Acousto-optic tunable filter (AOTF).
2. Focus the excitation laser at the rear focal plane of the objective by an easy-to-adjust way. This can be achieved with either a commercial illuminator or a home-built one. A number of commercial TIR illuminators are available as shown in Fig. 2b.
3. Perform simultaneous spatially equivalent imaging at different optical wavelengths on a single camera. An epifluorescence microscope often selects the required fluorescence by switching to a matched filter set installed in each filter cube, which is the major bottleneck of high-speed imaging. To overcome this constraint, the OptoSplit II is introduced, aiming at dividing the image into two separate components based on wavelength, and projecting both simultaneously onto a single camera chip. This is achieved by the beam splitter (Fig. 2 a ⑤) inside the filter cube of the OptoSplit II. Components with longer wavelength of the emission pass through the beam splitter while the other components are reflected, both of which are focused side-by-side on the CCD chip after being filtered by an appropriate set of filters (e.g., Semrock, 520/35-25, bandpass filter for GFP; 617/73 bandpass filter for RFP; FF580, beam

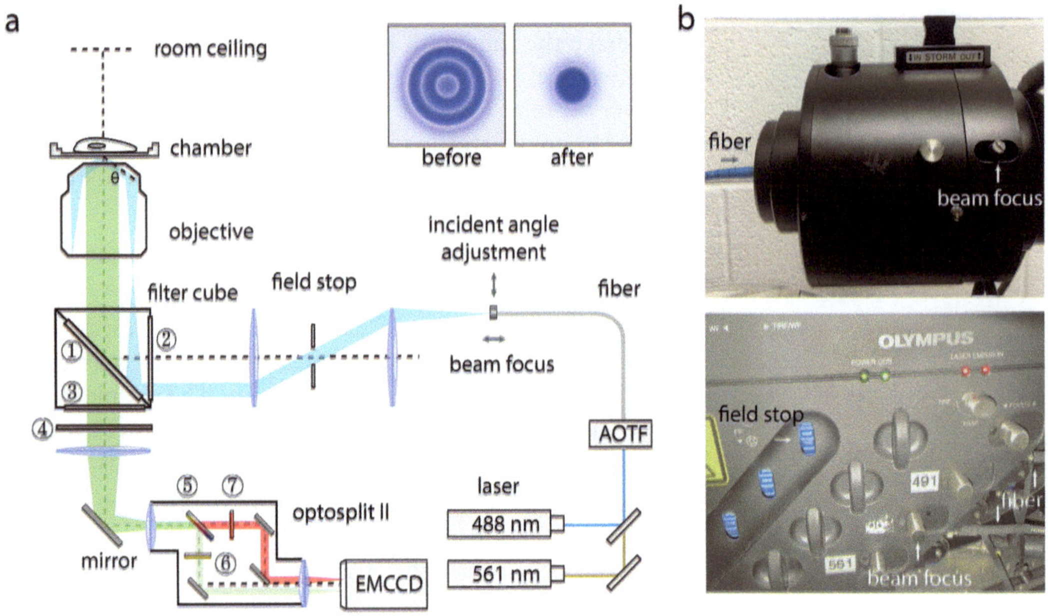

Fig. 2 Schematic of the fast dual-color objective-based TIRF microscope. (**a**) optical pathway and principal components of fiber-coupled objective type TIRF microscope. ① Dichroic mirror, ② excitation filter, ③emission filter (leave blank), ④ filter turret (leave blank), ⑤ beam splitter, ⑥ bandpass emission filter for shorter wavelength, ⑦ bandpass emission filter for longer wavelength. (**b**) Nikon (top) and Olympus cellTIRF (bottom) illuminator

splitter). We recommend commercial solutions that provide a compatible interface to couple with the microscopes and readily interchangeable filter cubes (e.g., Cairn Research).

3.4 Illuminator Alignment

1. Turn the system on. Laser and mercury arc lamp need several minutes to reach optimal operating conditions. Electron-multiplying charge-coupled devices (EMCCD) requires cooling down to −80 °C and approximately 10 min are required to achieve a stable output.
2. Mount a new coverslip onto the chamber and adjust the focus. Draw an “x” on the upper surface of the coverslip with water-resistant marker and fill chamber with 500 μL PBS. Drop a little glob of immersion oil onto the TIRF objective and place the chamber on the microscope stage and fasten it with clips. Move the stage and turn the focus knob to bring “x” into focus.
3. Slowly move the microscope stage to get the “x” out of the field of view. Open the shutters and allow laser beam to pass through the objective and project onto the room ceiling. Adjust the longitudinal position of the fiber-output such that the laser beam is collimated along the optical axis (Fig. 2a, dash line) and hits the ceiling right above the objective.

4. Focus the laser spot on the ceiling from the more diffuse pattern (Fig. 2a *before* image) until an Airy disc pattern is achieved.
5. Tune the lateral position of fiber-output along the optical axis to minimize laser spot on the ceiling (Fig. 2a *after* image).
6. Discard PBS in the chamber and draw another "x" on coverslip beside the previous one with a highlighter pen. Pipette 500 μL of PBS again into the chamber. Look through the eyepiece, you would be able to observe a large number of fluorescent particles moving randomly. Once again adjust fiber-output longitudinal position continually. The laser beam is expected to shoot out at an increasing angle. Keep doing this carefully until you see the laser beam exiting horizontally and eventually totally reflecting back toward the objective lens. Only a small number of settled-down fluorescent particles could be observed at this time, indicating TIRF mode has been achieved.

3.5 Coverslip Correction

High NA objectives used for TIRF microscopy are sensitive to minute thickness variation of coverslips and the operating temperature. To avoid image degradation, perform compensation before imaging by adjusting the objective correction collars.

1. Turn the correction collar coarsely to align the indicator labeled on the objective barrel with 0.17 mm at 23 °C or 0.17 mm at 37 °C if you use a lens heater.
2. Look at the live image through the camera and rotate the collar right and left slightly to find a position to achieve the optimal resolution and contrast of the image.

3.6 Imaging and Analysis at the Single-Granule Level

1. Twenty minutes before imaging, mount the coverslip on the chamber, wash it twice and incubate it in 500 μL of KRBB buffer with a low-glucose concentration (i.e., 2.8 mM).
2. Record images at 10 frames per second or faster, which can be achieved by enabling the frame transfer mode or crop mode of the EMCCD. Image acquisition for analysis of glucose-stimulated biphasic insulin secretion usually requires 15 min or longer, with the first 4–5 min considered the peak phase, and the subsequent 10 min or more as the second phase. A stage-top incubator may be employed to maintain the required temperature (i.e., 37 °C).
3. Save the raw images as a stack of thousands of frames. Before analyses, basic post-processing should be done using the Fiji software [7], an open-source platform for biological image analysis, such as setting image intervals and scales. Drag and drop image stack or folder directly on the Fiji toolbar to open image files as stacks. Select *"Image > Properties..."* to set the

frame interval. Select *"Analyze > Set Scale..."* to set the pixel size (*see* **Note 2**). The scale bar can be stamped by select *"Analyze > Tools > Scale Bar..."*.

4. Analyze granule dynamics by the fluorescence intensity trace. Center circle region-of-interest (ROI) on each granule and add them to the ROI manager by hitting the hotkey "t". Plot the fluorescence intensity trace with the *"Image/Stacks/Plot Z-axis Profile"* command and analyse endocytosis behaviors (*see* **Note 3**).

4 Notes

1. Although this protocol is dedicated to dual-color imaging, it can also be adapted to three-color imaging with OptoSplit III or four-color imaging with two side-by-side OptoSplit II. Since the configuration allows switching laser rapidly and continuously without mechanical shutters, the image acquisition rate is only largely limited by the laser power, sensitivity of EMCCD, and brightness of fluorophores. Therefore, this method has the capability of imaging at the millisecond range or faster.
2. Image pixel size = physical pixel dimension of CCD/total magnification. For example, considering an image stack acquired with a 60× objective, magnified with a 1.5× Optovar, on a microscope equipped with the Andor EMCCD DU897, we can calculate the image pixel size:

$$\text{pixel size} = 16\ \mu\text{m}/(60 \times 1.5) = 0.178\ \mu\text{m/pixel}$$

3. Secretory insulin granules fuse with plasma membrane and release contents into the extracellular space. Upon fusion pore opening, the substance exchange between granule lumen and extracellular environment occurs instantly. Various strategies have been developed to monitor specific fusion events of interest in the β cell, including labeling of granule cargos (e.g., insulin, NPY or syncollin) with fluorophores or tagged with sensors (e.g., pH and Zn^{2+} sensors), which can record the precise behaviors of the granules emptying their contents [8, 9].

   Recognizing the patterns of fusion events is the crucial step for the analysis of insulin secretion. For cargo-labeling probes, such as mTurquoise2- or mCherry-tagged NPY and syncollin, the fluorescence intensity of granule puncta initially increases, followed by an abrupt spreading upon exocytosis. The increase is mainly induced by granule approaching the exponentially decaying field of TIR illumination whereas the sudden spreading is caused by cargo diffusion into surrounding region after granule fusion. These are the criteria to identify fusion events. However, for some atypical fusion events, the trademark fluorescence decline could be confused with a granule becoming undocked or just a short visit without actual docking. In these cases, pH-sensitive fluorescent probes are particularly useful since there is a characteristic transient sharp increase of fluorescence intensity due to rapid neutralization of the acidic granule lumen resulting from the opening of the fusion pore opening to the alkaline cell exterior, followed by the release of the cargo into the cell exterior that leads to rapid dissipation of a fluorescence cloud [10]. This strategy can also be used to monitor fusion pore dynamics [11]. Figure 3 gives an example

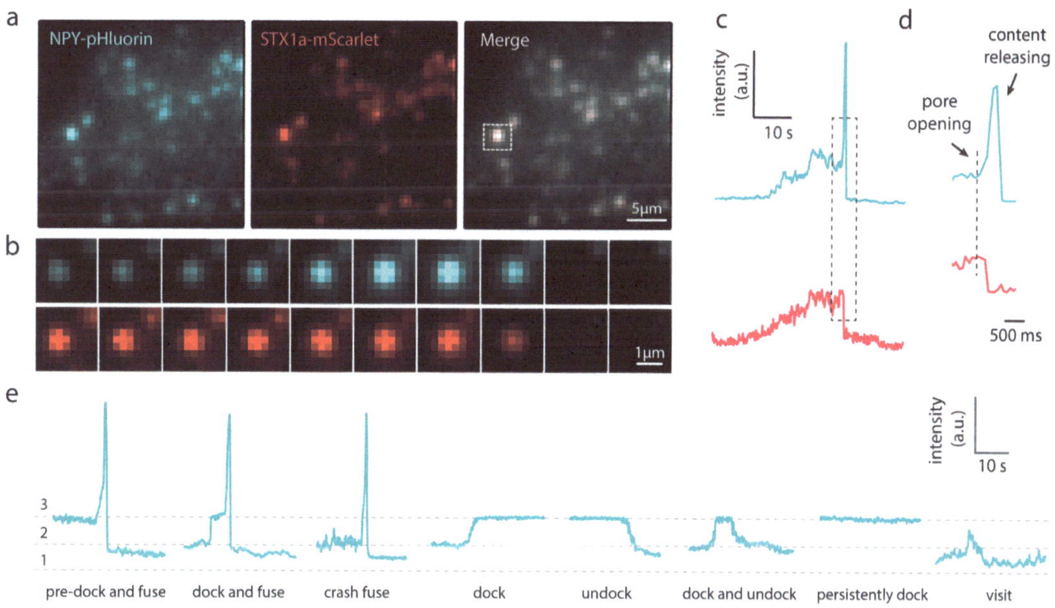

Fig. 3 Imaging insulin granule dynamics in human pancreatic β-cells with fast dual-color TIRF microscopy. (**a**) Dual-color TIRFM images of insulin granule exocytosis in human β-cell expressing pH-sensitive probes, NPY-pHluorin and STX1a-mScarlet. (**b**) Montage of the fusion events indicated in dashed box in (**a**), which shows that the granule docks precisely on the Stx1A molecule where it eventually fuses with the plasma membrane and empties its contents. (**c**) Time course of the fusion events indicated in dashed box in (**a**). The time course of typical fusion events monitored by luminal pH-based granule probe are characterized by an instantaneous rise to a peak amplitude, followed by a rapid decay to the baseline. (**d**) Magnification of the window (the dashed box) indicated in (**c**). At the very beginning of fusion pore opening, a narrow aqueous channel bridging the granule lumen and extracellular space only allows protons to pass through, which leads to a sharp increase in the fluorescence of the pH-sensitive probes. As the fusion pore expands and granule membrane collapses, insulin crystals are leaked and eventually released completely into the cell exterior. (**e**) Schematics of typical insulin granule dynamics. Individual granule behavior can be learned through analysis of the intensity traces of a single granule. Initially, granules staying away from the TIRF evanescent field can not be observed and the intensity remain at the baseline (line 1). When approaching the plasma membrane, the labeled granules are gradually illuminated by the decaying intensity of the TIRF field (line 2). After translocating to the target plasma membrane, some of the granules are trapped and stabilized in a highly restricted region by docking receptors [1, 12, 13], referred to as being "docked" (line 3). Granules that fail to stably dock can either move randomly on the same plane or go back inside the cell interior. Secretion-competent granules that are fully "primed" fuse with the plasma membrane with a characteristic flash as shown in (**c**)

of TIRF microscopy imaging of insulin granules monitored by a pH-sensitive probe and summarizes the different modes of insulin granule dynamics.

Acknowledgments

This work was supported by operating grants from the Canadian Institute for Health Research PJT-159741 and PJT-148652, equipment and human islets from the Canadian Foundation for Innovation Projects no 19442 (Center for Diet, Digestive Tract and Disease) and 33603 (Centre for Islet Research and Therapeutics), respectively, and a Tier 1 Canada Research Chair.

References

1. Gaisano HY (2017) Recent new insights into the role of SNARE and associated proteins in insulin granule exocytosis. Diabetes Obes Metab 19(Suppl 1):115–123. https://doi.org/10.1111/dom.13001
2. Thurmond DC, Gaisano HY (2020) Recent insights into Beta-cell exocytosis in type 2 diabetes. J Mol Biol 432(5):1310–1325. https://doi.org/10.1016/j.jmb.2019.12.012
3. Liang T, Qin T, Kang F, Kang Y, Xie L, Zhu D, Dolai S, Greitzer-Antes D, Baker RK, Feng D, Tuduri E, Ostenson CG, Kieffer TJ, Banks K, Pessin JE, Gaisano HY (2020) SNAP23 depletion enables more SNAP25/calcium channel excitosome formation to increase insulin exocytosis in type 2 diabetes. JCI Insight 5(3): e129694. https://doi.org/10.1172/jci.insight.129694
4. Axelrod D (2008) Chapter 7 Total internal reflection fluorescence microscopy. In: Biophysical tools for biologists, Volume Two: In Vivo Techniques. Methods in Cell Biology. Elsevier Science, Amsterdam, pp 169–221. https://doi.org/10.1016/s0091-679x(08)00607-9
5. Axelrod D (2001) Total internal reflection fluorescence microscopy in cell biology. Traffic (Copenhagen, Denmark) 2(11):764–774. https://doi.org/10.1034/j.1600-0854.2001.21104.x
6. Lipson A, Lipson SG, Lipson H (2010) Optical physics. Cambridge University Press, Cambridge
7. Schindelin J, Arganda-Carreras I, Frise E, Kaynig V, Longair M, Pietzsch T, Preibisch S, Rueden C, Saalfeld S, Schmid B, Tinevez J-Y, White DJ, Hartenstein V, Eliceiri K, Tomancak P, Cardona A (2012) Fiji: an open-source platform for biological-image analysis. Nat Methods 9(7):676–682. https://doi.org/10.1038/nmeth.2019
8. Gandasi NR, Vesto K, Helou M, Yin P, Saras J, Barg S (2015) Survey of red fluorescence proteins as markers for secretory granule exocytosis. PLoS One 10(6):e0127801. https://doi.org/10.1371/journal.pone.0127801
9. Li WH (2017) Probes for monitoring regulated exocytosis. Cell Calcium 64:65–71. https://doi.org/10.1016/j.ceca.2017.01.002
10. Zhu D, Koo E, Kwan E, Kang Y, Park S, Xie H, Sugita S, Gaisano HY (2013) Syntaxin-3 regulates newcomer insulin granule exocytosis and compound fusion in pancreatic beta cells. Diabetologia 56(2):359–369. https://doi.org/10.1007/s00125-012-2757-0
11. Aoki R, Kitaguchi T, Oya M, Yanagihara Y, Sato M, Miyawaki A, Tsuboi T (2010) Duration of fusion pore opening and the amount of hormone released are regulated by myosin II during kiss-and-run exocytosis. Biochem J 429(3):497–504. https://doi.org/10.1042/bj20091839
12. Gandasi NR, Barg S (2014) Contact-induced clustering of syntaxin and munc18 docks secretory granules at the exocytosis site. Nat Commun 5:3914. https://doi.org/10.1038/ncomms4914
13. Xie L, Zhu D, Dolai S, Liang T, Qin T, Kang Y, Xie H, Huang YC, Gaisano HY (2015) Syntaxin-4 mediates exocytosis of pre-docked and newcomer insulin granules underlying biphasic glucose-stimulated insulin secretion in human pancreatic beta cells. Diabetologia 58(6):1250–1259. https://doi.org/10.1007/s00125-015-3545-4

Chapter 8

Monitoring Synaptic Exocytosis Using a Whole-Cell Patch Clamp Technique

Jihong Gong, Xiaofei Yang, and Cong Ma

Abstract

Whole-cell patch clamping is a standard method to monitor the secretion of synaptic vesicles. In this chapter, we describe the basic steps of whole-cell patch clamping for measuring synaptic exocytosis, aiming to provide reference for researchers who are new to this field.

Key words Synaptic exocytosis, Whole-cell patch clamping, Electrophysiology, Synaptic transmission, Vesicle, Vesicle fusion, Neurotransmitter release

1 Introduction

Calcium-dependent synaptic vesicle exocytosis mediates synaptic transmission between neurons [1]. Neurotransmitter-loaded synaptic vesicles experience the process of docking to the presynaptic active zone, priming to a "ready-to-fuse" state, and Ca^{2+}-triggered fusion with the presynaptic membrane to release neurotransmitters into synaptic cleft [2, 3]. Subsequently, the released neurotransmitters bind to their receptors and induce the changes of postsynaptic membrane current and voltage. Patch clamping can measure the membrane current or voltage, and has been widely used to study the function and mechanism of synaptic vesicle exocytosis [4]. A classical patch-clamp system includes the following components: mechanical part (i.e., shockproof platform, electrostatic shield, and equipment rack), optical part (i.e., microscope, video monitor), electronic components (i.e., amplifier, stimulator, data acquisition equipment, computer system), and micromanipulator. In addition, the micropipette puller instrument is needed for preparation of microelectrode.

Jingshi Shen (ed.), *Membrane Trafficking: Methods and Protocols*,
Methods in Molecular Biology, vol. 2473, https://doi.org/10.1007/978-1-0716-2209-4_8,

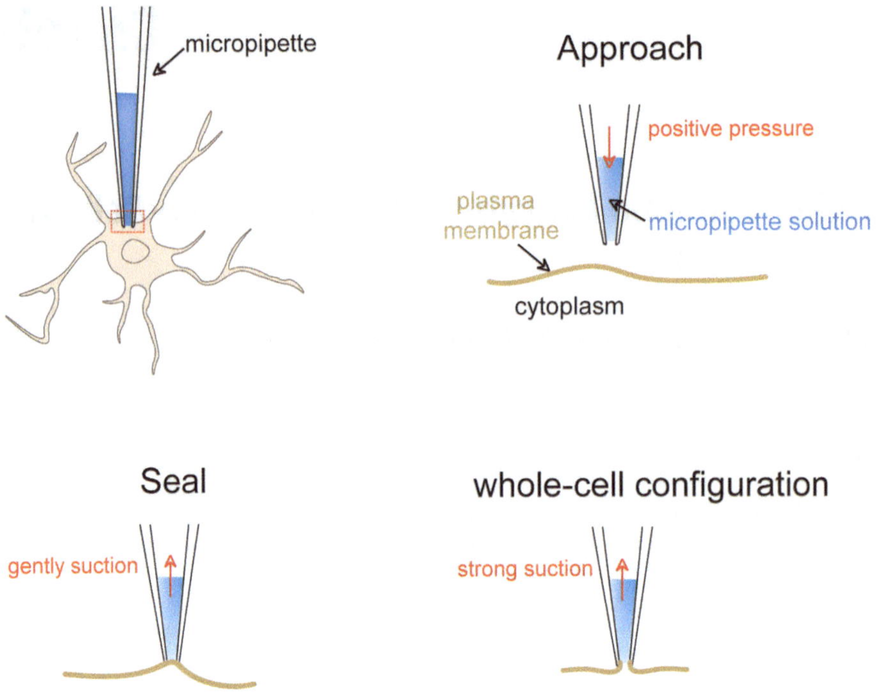

Fig. 1 Schematic showing basic procedures of whole-cell configuration. Neurons are immersed in the bath solution, and the micropipette are filled with the micropipette solution. Approach means putting the micropipette close to the plasma membrane of neurons. Then, gentle suction is applied to create a tight contact between the micropipette and the membrane. If it was operated properly, the resistance is expected to increase up to 1 GΩ or higher, which achieves "Seal" formation. Apply a short and relatively strong suction to rupture the small patch of the membrane covered by the micropipette, leaving the micropipette sealed with the rest of the cell

Patch clamping contains four classical measuring modes: "cell-attached" mode, "whole-cell" mode (Fig. 1), "inside-out" mode, and "outside-out" mode [5]. The whole-cell voltage clamping method is typically used to monitor synaptic vesicle exocytosis [6, 7]. Here, we introduce how to use this method to monitor excitatory and inhibitory spontaneous release (mini-release), action potential-evoked release (evoked-release), and readily release pool (sucrose-release). This protocol includes: (1) preparation of patch-clamp solutions and microelectrode; (2) sealing microelectrode with the neuron membrane; (3) whole-cell patch clamp recording; and (4) data processing. Please note that the method described here is applicable for recording cultured neurons. The methods for neuron isolation and culturing are described previously [8–10].

2 Materials

2.1 Solutions

1. AMPA-receptor mediated mini EPSCs (Excitatory PostSynaptic Currents) recording bath solution: 150 mM NaCl, 4 mM KCl, 1 mM $MgCl_2$, 2 mM $CaCl_2$, 10 mM HEPES–NaOH, and 10 mM glucose, pH 7.2–7.4. Adjust the osmotic pressure to about 315 mOsm with dd H_2O. Add 100 μM picrotoxin, 50 μM AP-5, and 1 μM tetrodotoxin before use.
2. NMDA-receptor mediated mini EPSCs recording bath solution: 150 mM NaCl, 4 mM KCl, 4 mM $MgCl_2$, 2 mM $CaCl_2$, 10 mM HEPES–NaOH, and 10 mM glucose, pH 7.2–7.4. Adjust the osmotic pressure to about 315 mOsm with dd H_2O. Add 1 μM tetrodotoxin, 20 μM CNQX, 100 μM picrotoxin, and 15 μM glycine before use.
3. Mini IPSCs (Inhibitory PostSynaptic Currents) recording bath solution: 150 mM NaCl, 4 mM KCl, 1 mM $MgCl_2$, 2 mM $CaCl_2$, 10 mM HEPES–NaOH, and 10 mM glucose, pH 7.2–7.4. Adjust the osmotic pressure to about 315 mOsm with dd H_2O. Add 20 μM CNQX, 50 μM AP-5, and 1 μM tetrodotoxin before use.
4. AMPA-receptor mediated evoked EPSCs recording bath solution: 150 mM NaCl, 4 mM KCl, 1 mM $MgCl_2$, 2 mM $CaCl_2$, 10 mM HEPES–NaOH, and 10 mM glucose, pH 7.2–7.4. Adjust the osmotic pressure to about 315 mOsm with dd H_2O. Add 100 μM picrotoxin, 50 μM AP-5 before use.
5. NMDAR-receptor mediated evoked EPSCs recording bath solution: 150 mM NaCl, 4 mM KCl, 4 mM $MgCl_2$, 2 mM $CaCl_2$, 10 mM HEPES–NaOH, and 10 mM glucose, pH 7.2–7.4. Adjust the osmotic pressure to about 315 mOsm with dd H_2O. Add 20 μM CNQX, 100 μM picrotoxin, and 15 μM glycine before use.
6. Evoked IPSCs recording bath solution: 150 mM NaCl, 4 mM KCl, 1 mM $MgCl_2$, 2 mM $CaCl_2$, 10 mM HEPES–NaOH, and 10 mM glucose, pH 7.2–7.4. Adjust the osmotic pressure to about 315 mOsm with dd H_2O. Add 20 μM CNQX and 50 μM AP-5 before use.
7. Sucrose-evoked EPSCs bath solution: 150 mM NaCl, 4 mM KCl, 1 mM $MgCl_2$, 2 mM $CaCl_2$, 10 mM HEPES–NaOH, and 10 mM glucose, pH 7.2–7.4. Adjust the osmotic pressure to about 315 mOsm with ddH_2O. Add 100 μM PTX and 1 μM tetrodotoxin before use.
8. Sucrose-evoked IPSCs bath solution: 150 mM NaCl, 4 mM KCl, 1 mM $MgCl_2$, 2 mM $CaCl_2$, 10 mM HEPES–NaOH, and 10 mM glucose, pH 7.2–7.4. Adjust the osmotic pressure to about 315 mOsm with dd H_2O. Add 20 μM CNQX, 50 μM AP-5, and 1 μM tetrodotoxin before use.

9. Micropipette solution for mini/Sucrose-evoked EPSCs/IPSCs recording: 140 mM CsCl, 5 mM NaCl, 10 mM HEPES–CsOH, 5 mM EGTA, 0.3 mM Na_2GTP, 3 mM Mg_2ATP, pH 7.2–7.4. Adjust the osmotic pressure to about 305 mOsm with dd H_2O.
10. Micropipette solution for evoked EPSCs or evoked IPSCs recording: 140 mM CsCl, 5 mM NaCl, 10 mM HEPES–CsOH, 5 mM EGTA, 0.3 mM Na_2GTP, 3 mM Mg_2ATP, pH 7.2–7.4. Adjust the osmotic pressure to about 305 mOsm with dd H_2O. Add 5 mM QX-314 before use.

    See **Notes 1** and **2** for the solution preparation.

2.2 Supplies, Equipment, and Software

1. Borosilicate tubes.
2. Shockproof platform, electrostatic shield, and equipment rack.
3. Microscope with a video monitor.
4. Micromanipulator (e.g., Sutter MP-225).
5. Micropipette puller (P-97, Sutter instrument).
6. Amplifier: EPC10 (HEKA) with the software PatchMaster.
7. Data analysis software such as Mini Analysis (Synaptosft), Clampfit (Molecular Devices), and Igor software (HEKA).

3 Methods

3.1 Patch-Clamp Solutions and Microelectrode Preparation

1. Take out the bath solution from the refrigerator and warm it up to room temperature.
2. Take out the micropipette solution from the −20 °C freezer and allow it to melt slowly on ice (*see* **Note 3**).
3. Pull a glass microelectrode from the borosilicate glass capillary tubes using the micropipette puller. The resistance of the microelectrode is about 3–5 MΩ (*see* **Notes 4** and **5**).

3.2 Forming a Seal of Between the Microelectrode and the Membrane of Neurons

1. Turn on the patch-clamp system, including the amplifier, the microscope, the micromanipulation, and the computer in this sequence. Next, turn on the PatchMaster software. A convenient way to set the desirable adjustments of the amplifier (HEKA EPC10) is achieved by using the SETUP mode, a button on the software PatchMaster. Click "Voltage-clamp" in the item "Whole-cell" and choose 0 mV in item "V-Membrane". Set "Filter 1" to 10 kHz and "Filter 2" to 2.9 kHz. Set "Rs Comp" and "C-slow" to "Off."
2. Create a new folder to store the data.
3. Take out one coverslip with cultured neurons and wash it with the bath solution. Put the coverslip into the chamber.

4. Fill the microelectrode with the microelectrode solution and flick it gently to remove air bubbles (*see* **Note 6**).
5. Install the microelectrode into the pipette holder that connects to the amplifier and ensure that the microelectrode solution passes through the silver wire in the pipette holder.
6. Before moving the tip of the microelectrode into the bath solution, apply a positive pressure to the microelectrode with a syringe or your mouth. Positive pressure can prevent impurities from blocking the microelectrode tip, and avoids other solutions from entering the tip.
7. Move the microelectrode tip into the bath solution (note that this may lead to the current trace going off-scale). Click on the "Auto V_0" button until the trace reappears. Find the microelectrode tip and the cell of interested under the microscope. Record the microelectrode resistance with the amplifier. The values of microelectrode resistance are displayed in "R-memb" (*see* **Notes** 7 and **8**).
8. Adjust the microelectrode close to a cell using micromanipulation. Under the microscope, slowly approach the cell with the microelectrode. If the resistance increases by 0.1–0.3 MΩ, the tip is likely well-touched the cell (*see* **Note 9**).
9. To form a seal, gently and rapidly switch the pressure from positive to negative in the microelectrode using your mouth or a syringe. Hyperpolarizing the voltage pulses from −10 mV to −70 mV often helps seal formation. The resistance will increase rapidly.
10. Release pressure and wait for a while. When the resistance becomes greater than 1 GΩ, the giga-seal is accomplished, and the current trace becomes flat (Fig. 2).

3.3 Whole-Cell Patch-Clamp Recording

1. Adjust C-fast compensation (fast capacitance compensation, mainly compensates the microelectrode capacitance) by clicking the "SEAL" button) on the PatchMaster software. This will decrease the spike size of current traces on PatchMaster. Now the cell-attached configuration has been set up.
2. On the basis of cell-attached configuration, apply a suction or high voltage pulse (600,800 mV) at the tip to break the cell membrane covered by the microelectrode. Now the "whole-cell" configuration is established (Fig. 3). With the breaking of the cell membrane, a sudden increase of the capacitive transients can be obtained.
3. Lower values of resistance are desirable. Sometimes additional suction is required to achieve lower values of resistance, which will enlarge the amplitudes of capacitive transients.

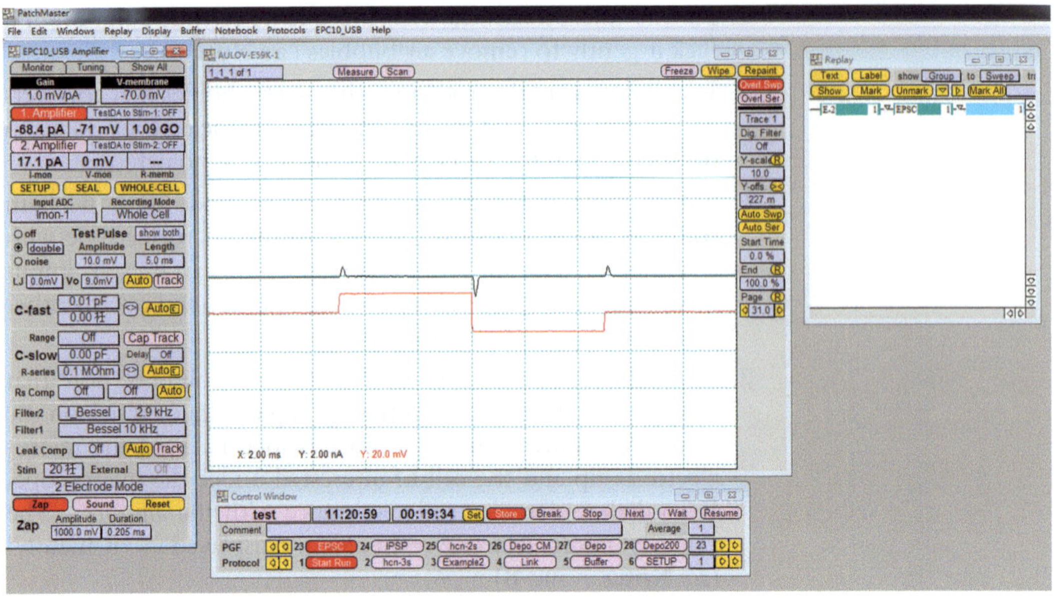

Fig. 2 Forming a seal of microelectrode with the membrane of neurons. The giga-seal is established as the resistance becomes greater than 1 GΩ

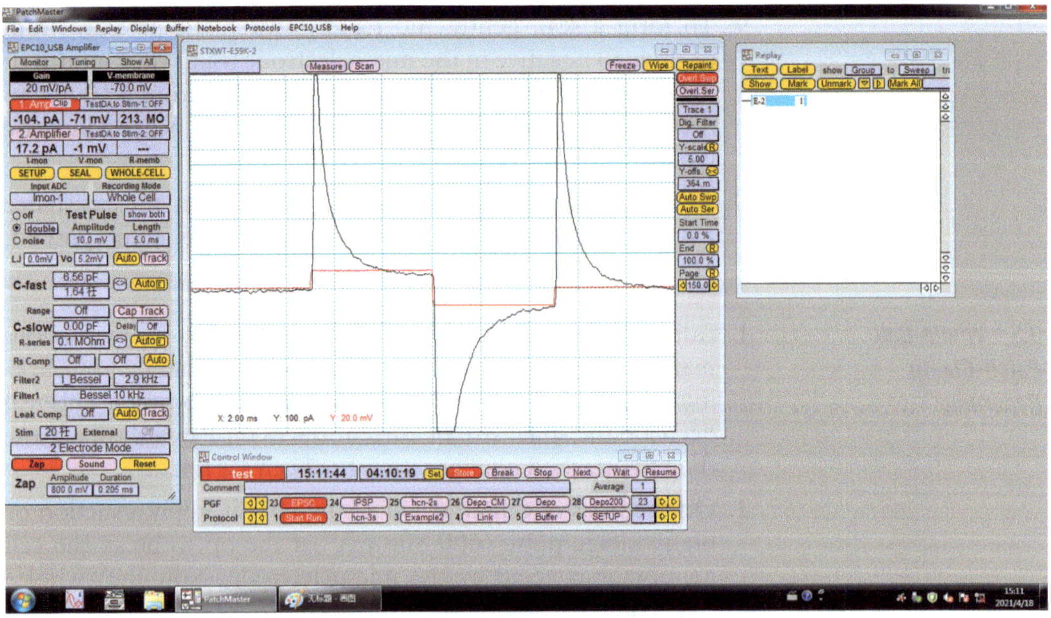

Fig. 3 "Whole-cell" configuration displayed on the PatchMaster. With the "whole-cell" configuration being established, a sudden increase appears in the capacitance transients

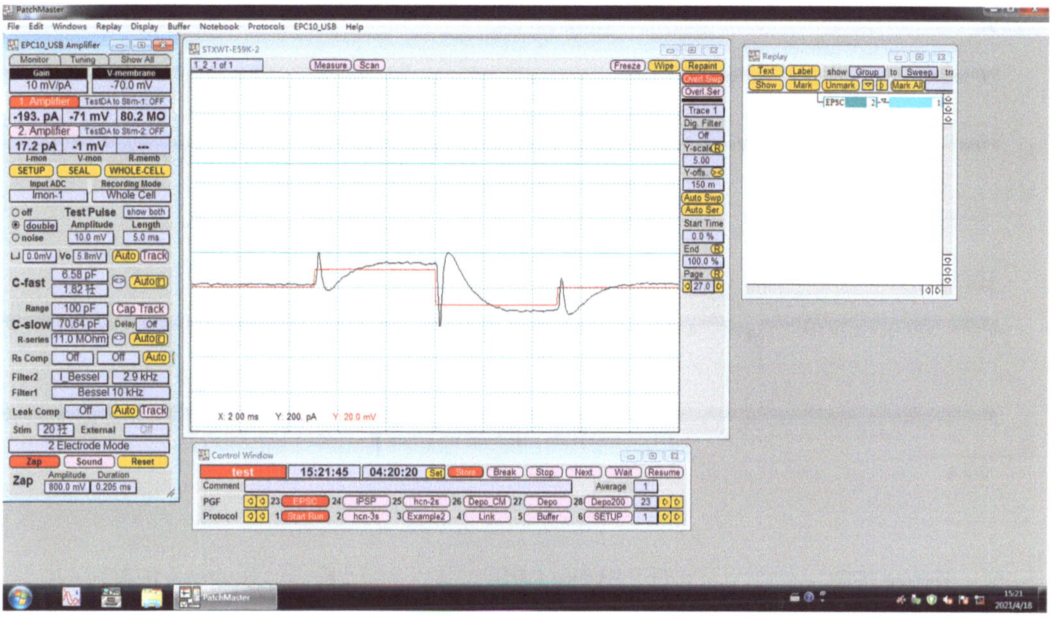

Fig. 4 "Whole-cell" configuration after compensation. With the "whole-cell" configuration established, photos of PatchMaster after Rs compensation, C-slow compensation and whole-cell compensation

4. Click the "Auto" button (near the "Rs" of PatchMaster) for Rs (series resistance) compensation (*see* **Note 10**). To compensate for slow capacitance, click on the "Auto" button (close to "C-slow") and the "Whole Cell" button on PatchMaster (Fig. 4) (*see* **Note 11**).
5. After approximately 1 min, record AMPA receptor-mediated mini-EPSCs and evoked-EPSCs as well as mini-IPSCs and evoked-IPSCs with a clamping potential of −70 mV. Record NMDA-receptor mediated mini-EPSCs and evoked-EPSCs at a holding potential of +40 mV. Record action potential-evoked releases using single extracellular stimulus pulses (90 μA, 1 ms). Record sucrose-evoked vesicle release with 0.5 M sucrose added to bath solution.

3.4 Data Processing

3.4.1 Mini-Release

1. Examples of mini-release recordings (single vesicle spontaneous releases) are shown in Fig. 5a, b (*see* **Note 12**). Usually, the frequency and amplitude of mini-releases are calculated. Open the data using the Clampfit software, and click the "event detection" button (*see* **Note 13**).
2. Open the created corresponding template in the small window. Select the "negative-going" from pull down menu and set the template match threshold to "4". After setting up, click the "OK" button (Fig. 6).

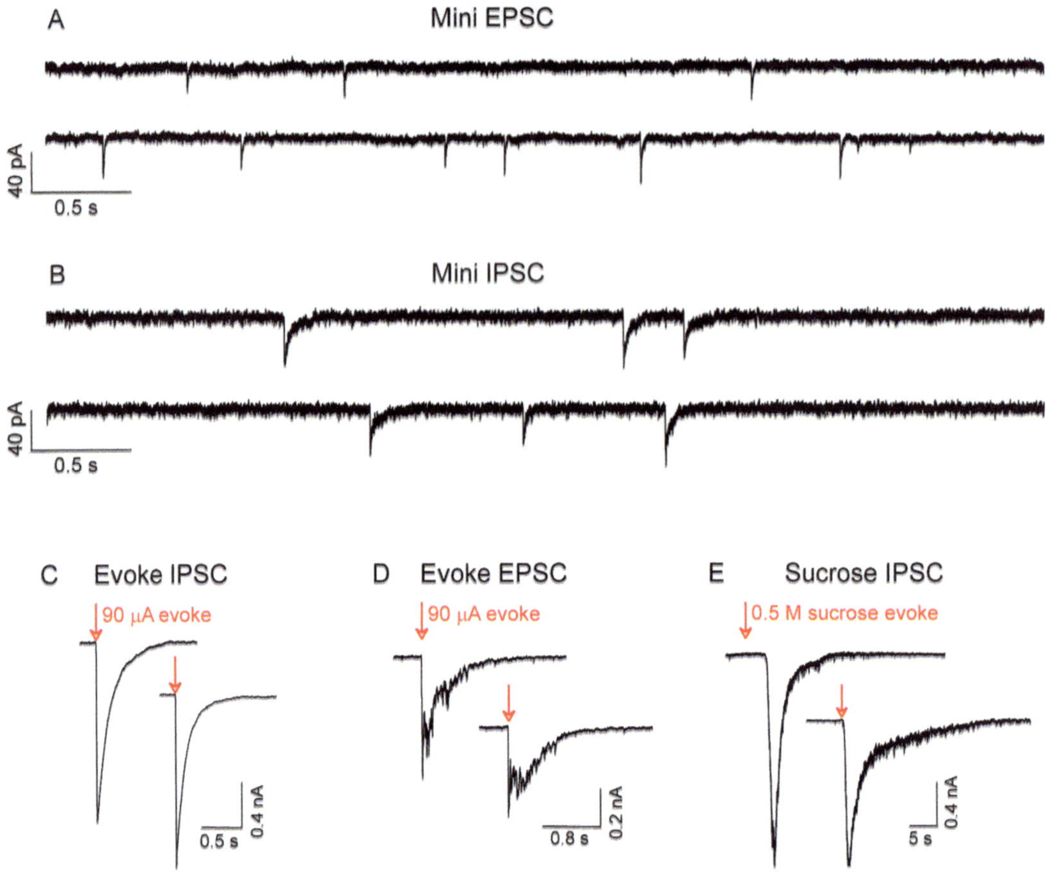

Fig. 5 Representative traces in synaptic exocytosis. Sample trances of mini EPSCs (**a**) and mini IPSCs (**b**) recorded in cortical neurons. Sample traces of action potential-evoked IPSCs (**c**) and EPSCs (**d**) recorded in cortical neurons. (**e**), Sample traces of IPSCs evoked by 0.5 M sucrose for 30 s

3. The mini-release events are automatic identified by the software, as shown in Fig. 7. The parameters of each mini-release event are shown in the "results" table. The frequency and average amplitude can be calculated.

3.4.2 Evoke-Release

Examples of "evoke" release are shown in Fig. 5c, d (*see* **Note 14**). Generally, the amplitude and the charge transfer of evoke IPSCs are displayed.

1. Open the evoked-release data. Right click and select "bring cursors" to show the four cursors.
2. Use cursor 1 and 2 to pick "non-signal" trace (the baseline), and use cursor 3 and 4 to pick signal trace for processing data. Be careful not to pick the stimulus trail.
3. Find the "statistics" button in the toolbar, click it and choose the "Peak amplitude" and "Area" to calculate them (*see* Fig. 8).

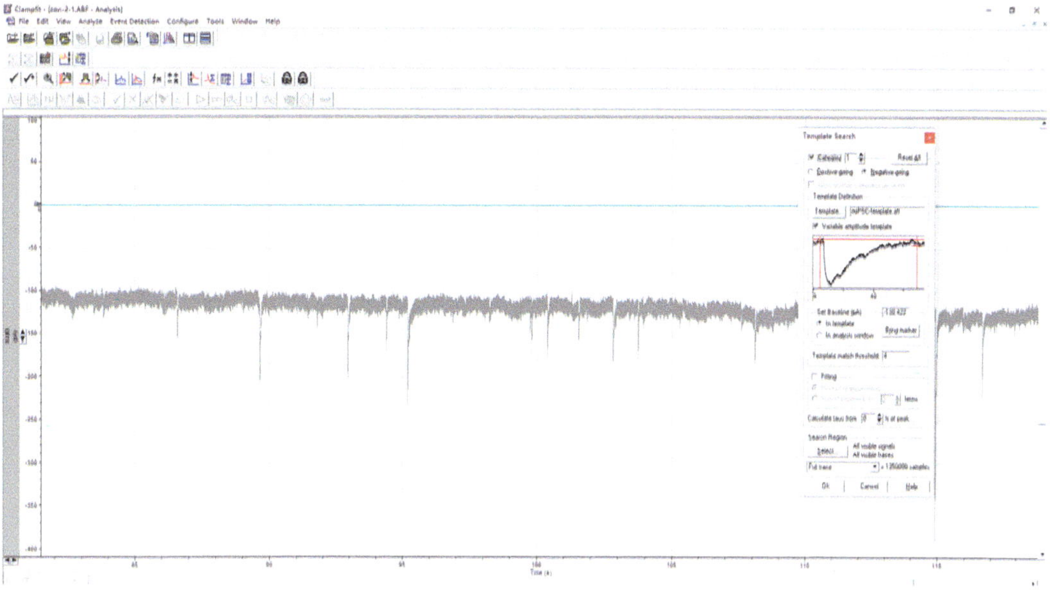

Fig. 6 The photo of Clampfit showing the function of "mini" template search

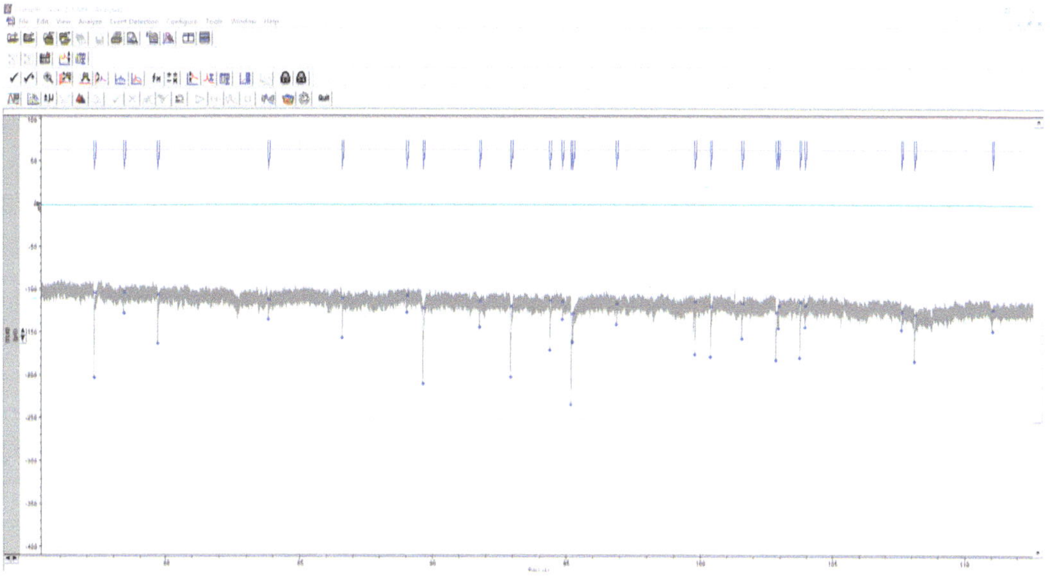

Fig. 7 The "mini" event data are identified by the Clampfit

Select the "negative-going" or "positive-going" from pull down menu of the software to match AMPA receptor-mediated EPSCs and NMDA receptor-mediated EPSCs, respectively.

4. The calculation results will be shown in the "results" table after clicking "OK".

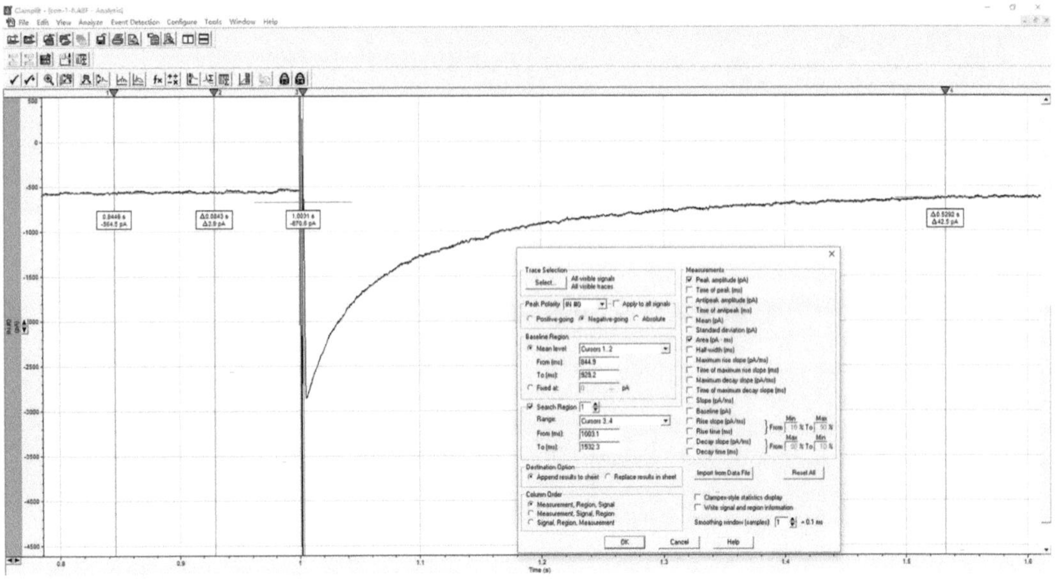

Fig. 8 Analysis of evoked-release with Clampfit

3.4.3 Sucrose-Release

1. When detecting the synaptic readily release pool (Fig. 5e), the recovery of postsynaptic current induced by sucrose is slow. The area of the peak, which represents the charge transfer, is calculated. The shapes of excitatory and inhibitory postsynaptic currents evoked by 0.5 M sucrose are similar, although the excitatory currents are usually smaller than inhibitory currents. Open the sucrose-release data, right click and select the "bring cursors".
2. Use cursor-1 and -2 to pick "non-signal" trace (the baseline), and use cursor-3 and -4 to pick signal trace for processing data. Find the "statistics" button in toolbar, click it and choose the "Area" to calculate the charge transfer (*see* Fig. 9).
3. The calculation results will be shown in the "results" table after clicking "OK".

4 Notes

1. Use a 0.22 μm filter to remove impurities in the bath solutions and micropipette solution, which will help to improve experimental success rate.
2. To achieve a perfect seal, the osmolarity of the micropipette solution should be 15–20 mOsm lower than that of the bath solution.
3. During the experiment, the micropipette solution needs to be kept on ice because it contains GTP and ATP.

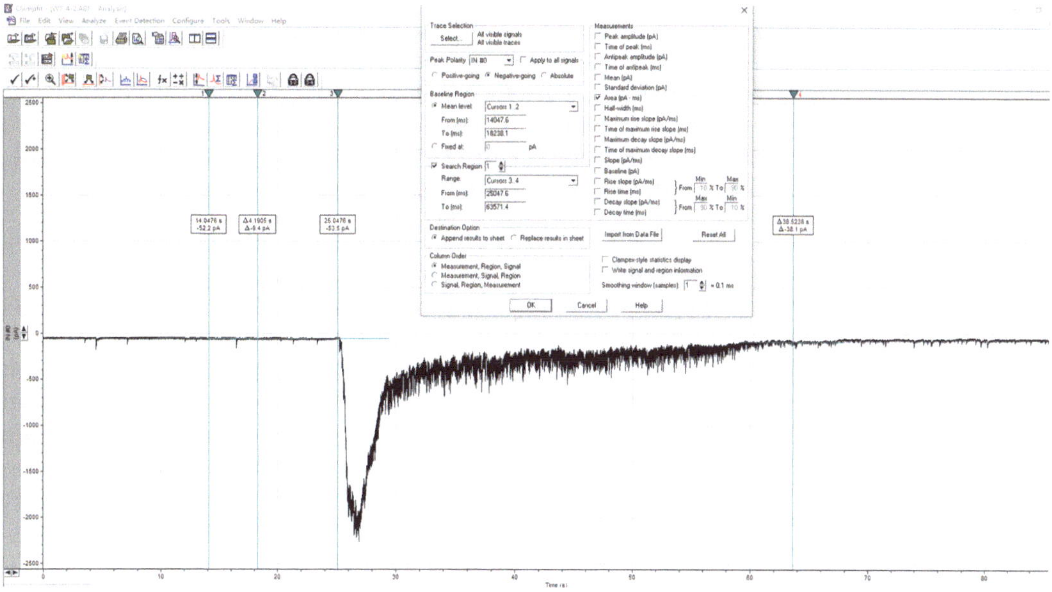

Fig. 9 Analysis of sucrose-release with Clampfit

4. Microelectrodes with different thickness and lengths can be prepared by using different settings of the micropipette puller. The desired resistance of the microelectrode is 3–5 MΩ for clamping neurons.
5. You can also use the pipette microforge to polish the tip of the micropipette with high temperature to obtain desired diameter and resistance. For neurons with relatively large cell volume, it is much easier to be clamped without polishing. The polishing steps can be selected according to your experimental requirements.
6. Keeping the glass microelectrode clean is very important for the success of the experiment. We recommend to use the glass microelectrode immediately after being pulled by the micropipette puller.
7. When moving the microelectrode tip into the bath solution, a decrease or disappearance of the reaction current indicates that the resistance of the microelectrode increases. Possible reasons are: (1) there are bubbles at the tip of the microelectrode; (2) the bath electrode is not put in the bath solution such that the circuit is disconnected; (3) there is insufficient solution in the microelectrode; (4) the silver wire does not contact with the micropipette solution.
8. If the resistance value is less than 1 MΩ after the microelectrode is placed into the bath solution, the microelectrode tip is likely to be broken.
9. Cell health is also a prerequisite for successful experiments.

10. It is important that the resistance is stable for "Rs" compensation.
11. The slow capacitance is mainly the capacitance between the microelectrode tip and the bath solution. After establishing a whole-cell configuration, the slow capacitance is almost equivalent to the cell-membrane capacitance.
12. Generally speaking, one signal in the "mini" signal recording represents the fusion of a vesicle with the cell membrane. The amplitude of different vesicle fusion events in one cell is rather different (Fig. 5a, b). The current of a single vesicle release is small. For example, the average amplitude of mini-IPSCs is about 40–60 pA, while the amplitude of mini-EPSCs is about 20–40 pA. Currents from different strains of mice or even different cells from the same animal may exhibit differences. It is worth noting that the current triggered by an excitatory transmitter differs from that triggered by an inhibitory transmitter. The rate of excitatory signal decline and recovery is faster than that of inhibitory (Fig. 5a, b). Moreover, the amplitude of mini spontaneous inhibitory current is larger.
13. "Mini Analysis" can convert the "itx" data format from PatchMaster to "ABF" format, which can be opened by Clampfit.
14. The evoked postsynaptic current is larger than spontaneous current and varies among different cells. The peak of inhibitory current decreases and rises more cleanly, while the peak of the excitatory current rises with a delay, and the signal appears in a zigzag manner (Fig. 5c, d).

References

1. Südhof TC (2014) The molecular machinery of neurotransmitter release (Nobel lecture). Angew Chem Int Ed Engl 53:12696–12717
2. Südhof TC (2004) The synaptic vesicle cycle. Annu Rev Neurosci 27:509–547
3. Südhof TC (2012) The presynaptic active zone. Neuron 75:11–25
4. Jahn R, Fasshauer D (2012) Molecular machines governing exocytosis of synaptic vesicles. Nature 490:201–207
5. Segev A, Garcia-Oscos F, Kourrich S (2016) Whole-cell patch-clamp recordings in brain slices. J Vis Exp 112:54024
6. Wang S, Li Y, Gong J et al (2019) Munc18 and Munc13 serve as a functional template to orchestrate neuronal SNARE complex assembly. Nat Commun 10:69
7. Wang X, Gong J, Zhu L et al (2020) Munc13 activates the Munc18-1/syntaxin-1 complex and enables Munc18-1 to prime SNARE assembly. EMBO J 39:e1036318
8. Hilgenberg LG, Smith MA (2007) Preparation of dissociated mouse cortical neuron cultures. J Vis Exp:562
9. Kaech S, Banker G (2006) Culturing hippocampal neurons. Nat Protoc 1:2406–2415
10. Yang X, Cao P, Südhof TC (2013) Deconstructing complexin function in activating and clamping Ca^{2+}-triggered exocytosis by comparing knockout and knockdown phenotypes. Proc Natl Acad Sci U S A 110:20777–20782

Chapter 9

Live-Cell Superresolution Imaging of Retrograde Axonal Trafficking Using Pulse–Chase Labeling in Cultured Hippocampal Neurons

Tong Wang and Frédéric A. Meunier

Abstract

The entanglement of long axons found in cultured dissociated hippocampal neurons restricts the analysis of the machinery underlying directed axonal trafficking. Further, hippocampal neurons exhibit "en passant" presynapses that may confound the analysis of long-range retrograde axonal transport. To solve these issues, we and others have developed microfluid-based methods to specifically follow the fates of the retrograde axonal cargoes following pulse–chase labeling by super-resolution live-cell imaging, and automatically tracking their directed transport and analyzing their kinetical properties. These methods have allowed us to visualize the trafficking of fluorescently tagged signaling endosomes and autophagosomes derived from axonal terminals and resolve their localizations and movements with high spatial and temporal accuracy. In this chapter, we describe how to use a commercially available microfluidic device to enable the labeling and tracking of retrograde axonal carriers, including (1) how to culture and transfect rat hippocampal neurons in the microfluidic device; (2) how to perform pulse–chase to label specific populations of retrograde axonal carriers; and (3) how to conduct the automatic tracking and data analysis using open-source software.

Key words Axon trafficking, Retrograde trafficking, Long-range cargo transport, Superresolution microscopy, Live-imaging microscopy, Microfluidic device, Pulse–chase labeling, Activity-dependent

1 Introduction

1.1 Compartmentalization of Cultured Neurons Using Microfluidic Devices

Microfluid based methods were developed to compartmentalize neurons into somatodendritic regions, axon shafts and axon terminals [1]. Polydimethylsiloxane (PDMS) is the base of the commercialized devices, reversibly bound to acid-washed glass coverslips via electrostatic interactions [2, 3], as shown in Fig. 1a. The cell bodies with the diameter of ~20–30 μm [4] are physically restricted by the length (450 μm) and narrow diameter (10 μm) of the microfluidic channels, as shown in Fig. 1b. Only the thin and long axons can cross these channels to the terminal chambers, where the nerve terminals are formed and enriched with presynaptic machinery

Jingshi Shen (ed.), *Membrane Trafficking: Methods and Protocols*,
Methods in Molecular Biology, vol. 2473, https://doi.org/10.1007/978-1-0716-2209-4_9,

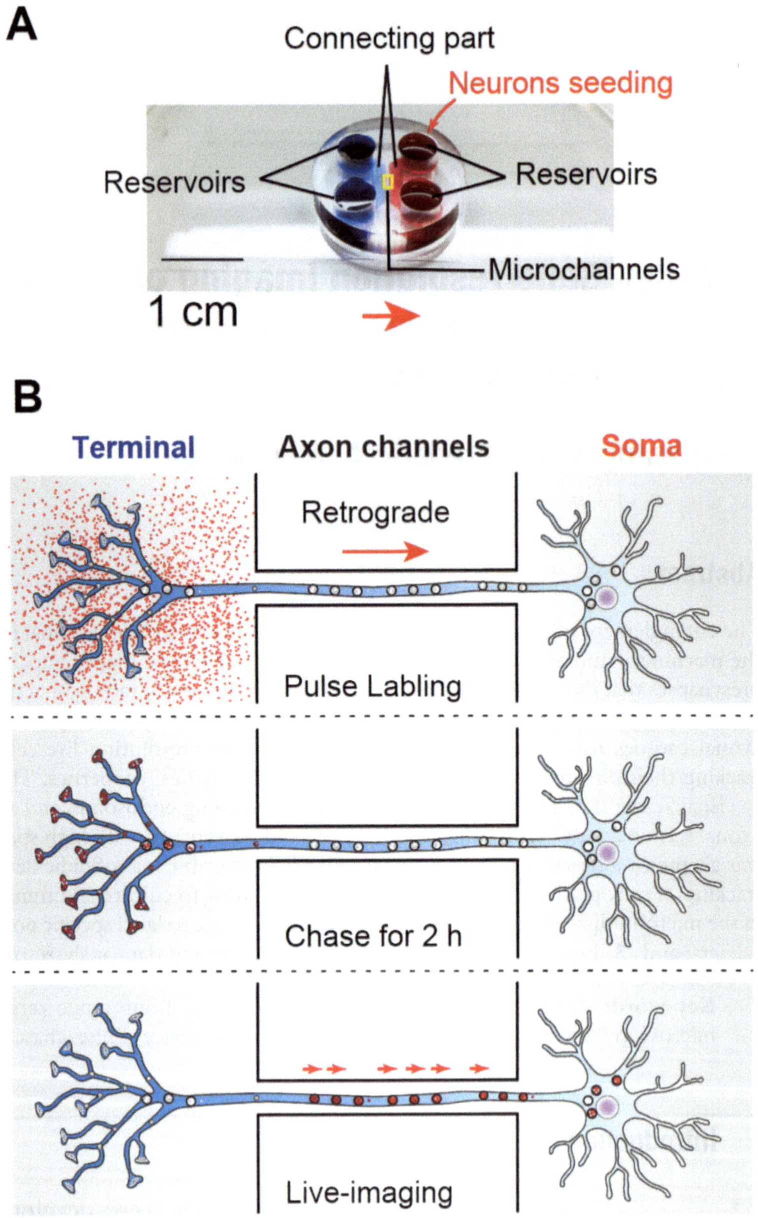

Fig. 1 The diagram showing the pulse–chase labeling steps for retrograde axonal cargoes. (**a**) Structure of the commercially available microfluidic device for compartmentalization of cultured neurons. Neurons are only seeded in the soma reservoirs (red). The connecting part (red) between the soma reservoirs is filled with the suspension of neuronal cells. The reservoirs and connecting part in the terminal side (blue) are filled with plating medium only. Axons extend through the microchannels between the two connecting parts (red and blue). Bar = 1 cm. (**b**) Cartoon showing the amplification of the boxed region showing how neurons are growing in the microchannels. In DIV 12–14 neurons, axons extend to the terminal chamber (blue) with their soma restricted to the soma chamber. The nerve terminals are pulse-labeled with fluorescently tagged reagents (e.g., BoNT/A-Hc-Atto647N and CTB-Alexa Fluor 488) for 5 min, conjugated with the depolarizing stimulation by high K^{+}. Following a 2 h chase time, a flux of retrograde axonal cargoes labeled with tagged reagents is detected (chased) along the axon channels using a time-lapse microscope

even in the absence of innervating postsynaptic components [5], represented by the VAMP2-positive axonal regions in the terminal chamber (Fig. 2a). Significantly, inside the 10 μm axon channels, axons form unidirectional bundles, similar to the actual anatomic structure of fascicles composed of multiple paralleled axons originating from the same nuclei or ganglia [6]. This thin layer of axons in the microfluidic channel can be easily imaged alive, providing an ideal model for observing simultaneous passages of long-range trafficking events within multiple axons [7–9]. These regions of interest (ROIs) along the axon channels ensure the segments are roughly placed at a similar distance to the soma, as shown in the boxed regions of Fig. 2a. This design therefore facilitates the detection of synchronized trafficking events in multiple parallel axons (Fig. 2b–d), offering a highly efficient way to quantity the passages of labeled cargoes within a given time window (Fig. 3a).

1.2 Labeling of the Long-Range Retrograde Axonal Cargoes with Fluorescently Tagged Reagents

The retrograde trafficking machinery plays a critical role in bringing neurotrophic signaling endosomes [10, 11] and degradative cargoes such as autophagosomes [12–16] back to the soma of neurons. Surface receptors are organized in functional nanodomains on the plasma membrane of presynaptic nerve terminals [17–20]. Applications of the tagged ligands to these nerve terminals allows precise labeling of diverse groups of receptors [21–23]. Following endocytosis, those ligand-bound receptors are then separated into different pools of vesicles, some of which are tagged with Dynein via adaptors, forming the axon carriers destined for long-range transport. For instance, BoNT/A-Hc hijacks a specific subset of the endocytic synaptic vesicle by binding to their surface receptors [8, 9, 24–26]. Following intracellular uptake in nerve terminals, some of the BoNT/A are targeted to autophagosomes destined for retrograde trafficking to the soma [8, 9]. Long-range retrograde trafficking of those large cargoes is mainly driven by dynein motors, which are associated with the cargoes via various adaptors [7, 8, 27, 28]. Recent studies using time-lapse super-resolution microscopy revealed that the radial contraction mediated by the axonal periodic actomyosin also significantly impacts the axonal vesicle trafficking [29–32].

Similarly, Cholera toxin subunit-b (CTB) binds to lipid rafts, where the tyrosine kinase receptor B (TrkB) is located [33]. Brain-derived neurotrophic factor (BDNF) is the primary ligand of TrkB receptors. Following BDNF binding, TrkB receptors dimerize and are activated, leading to their internalization into a group of endosomes called signaling endosomes. Signaling endosomes are labeled by CTB and undergo long-range retrograde axonal trafficking to bring the associated survival signals back to cell bodies, eliciting survival responses [7, 21–23]. In addition, specific antibody–antigen recognition can also be used to trace the intracellular trafficking of the surface proteins following their specific antibodies binding to the extracellular antigen domains. For instance, the intracellular

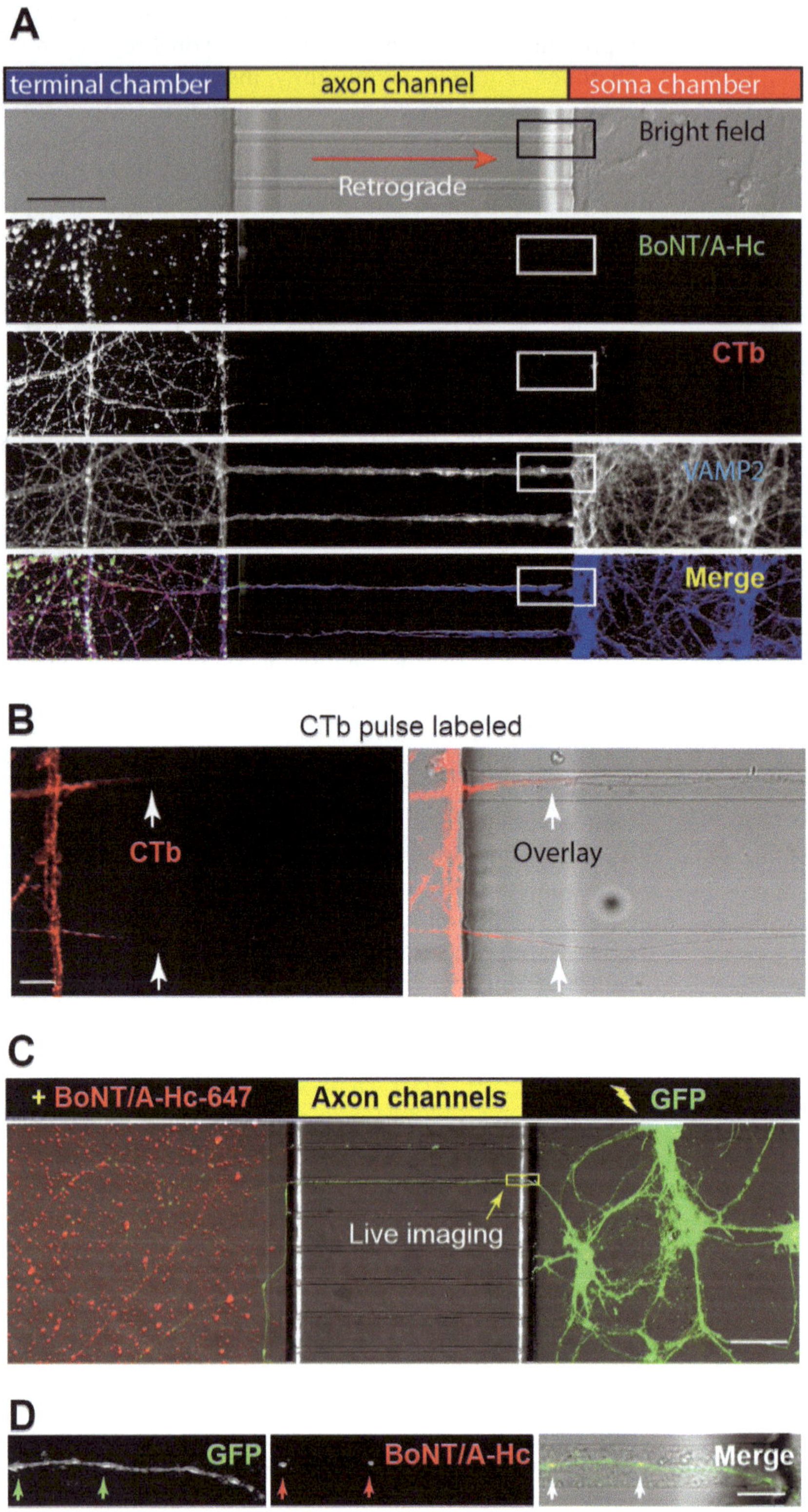

Fig. 2 The pulse–chase labeled cargoes derived from nerve terminals are carried by the retrograde axonal trafficking to the soma of the neuron. (**a**) The cell bodies of neurons cultured in microfluidic devices are restricted to the soma chamber, as shown in the bright field images. In contrast, the axons crossed the

fate of surface-stranded VAMP2-pHluorin can be tracked following activity-dependent endocytosis of anti-GFP nanobodies recognizing its extracellular domain [19, 34]. Importantly, we found that most of the fluorescently tagged ligands/toxins used as probes in axon terminal chamber separate to two pools with different fates: one pool undergoes retrograde trafficking via axon [7], while the other pool remains local recycling [19, 34].

1.3 Pulse–Chase Labeling to Investigate the Flux of Retrograde Cargoes

Most of the tagged axonal carriers imaged at the end of the microchannels have an overall retrograde direction (Fig. 3b, c) [7, 8, 31]. As axonal trafficking is tightly regulated by neuronal activity [27], examining the modulation of these retrograde cargoes labeled under different stimulation paradigms can help us to understand how synaptic activity controls various aspects of axonal trafficking. We found that raising synaptic activity via a short pulse of depolarization with high potassium significantly increases the flux of autophagosomes undergoing retrograde transport [8]. Interestingly, we also found that the retrograde flux of signaling endosomes is also up-regulated by synaptic activity, suggesting that the sorting and retrograde trafficking machinery are tightly coupled to the levels of neuronal activity. Importantly, in both cases, the impacts of the short 5 min pulsed stimulation on retrograde trafficking fluxes are quickly observed but only become significant following a 2 h-delay [7, 8]. This delay might reflect the time required for the initiation, formation and sorting of autophagosomes [35] or endosomes from the nerve terminals [10, 23], and the loading of these organelles to the activated cytoplasmic dyneins. These delayed activity-dependent retrograde fluxes may function to maintains homeostasis in remote synapses by adjusting the frequency of retrograde carriers delivering the signaling [7] or/and degradative [8] cargoes back to the cell bodies.

1.4 Resolving the Long-Range Axonal Trafficking Using Live-Cell Superresolution Microscopy

Combining the pulse–chase labeling method with single-molecule tracking (SPT), we could image single molecules of Atto647N-labeled BoNT/A-Hc (BoNT/A-Hc-Atto647N) undergoing retrograde transport for the first time [36]. The photoactivated localization microscopy (PALM)-based single-particle tracking and analysis are conducted using open-source software [37, 38]. To specifically

Fig. 2 (continued) 450 μm-channel and reached the terminal chamber, showing strong VAPM2 staining. Fluorescently tagged BoNT/A-Hc and CTB are both redistricted to the terminal chamber following the 5 min pulse labeling step. Bar = 100 μm. (**b**) Zoomed-in view of the CTB labeling is restricted in the terminal portion of axons crossing the microfluidic devices. Arrows indicate the front ends of the CTB signal in distal axons. Bar = 10 μm. (**c**) Rat hippocampal neurons are transfected with Lipofectamine 2000 on DIV 12 and live-imaged on DIV14. The BoNT/A-Hc labeled cargoes are delivered retrogradely via the axons of the transfected neuron to the soma. The live-imaging windows are placed in the axon channels near the soma chamber. Bar = 100 μm. (**d**) Amplification of the boxed regions in (**a**), showing two BoNT/A-Hc tagged cargoes (arrows), are transported in the axon of a GFP expressing neuron. Bar = 10 μm

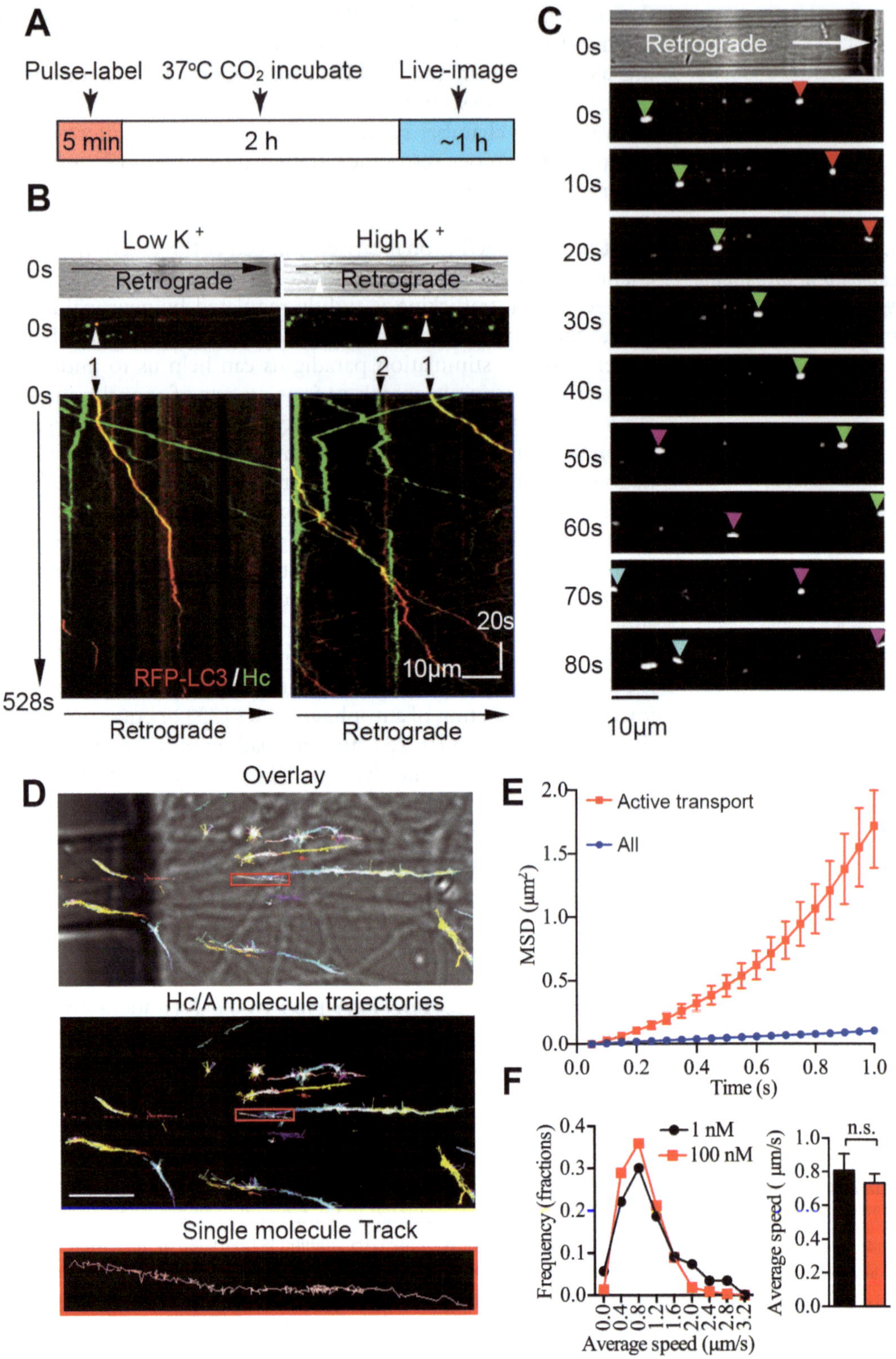

Fig. 3 Time-lapse images of pulse–chase labeled retrograde cargoes in one axon channel. (**a**) Scheme showing the steps of pulse–chase labeling steps for the live-imaging experiments of the retrograde axonal

isolate trajectories of BoNT/A-Hc-Atto647N carriers undergoing directed motion are filtered based on both their duration (>25 consecutive frames) and mean square displacement (MSD) fitting to a polynomial equation [8, 39] (Fig. 3d–f). The nature of different types of long-range carriers are further characterized using immunofluorescent staining with organelle markers such as TrkB (signaling endosome) or VAMP2 (synaptic vesicle) (Fig. 4a, b). Importantly, structured illumination microscopy (SIM), another super-resolution method, allowed us to detect several closely aligned small vesicles that would otherwise appear as a large tubular structure in conventional light microscopy [7]. Such carriers are therefore likely to be linked together during long-range retrograde trafficking via an unknown mechanism (Fig. 4b).

1.5 Conclusion

In this chapter, three types of techniques have been combined to analyze the trafficking of retrograde axonal carriers at the single-molecule level with subdiffractional precision in live cells. First, pulse–chase labeling allows the accurate temporal labeling of certain pools of axonal carriers in response to pulse stimulation. Then, spatial separation using microfluidic devices facilitates specific labeling of retrograde carriers. Finally, single-molecule tracking precisely segregates the targeted retrograde axonal carriers based on the kinetics of their movement, with different fluorescent codes identifying their sources. Using this method, labeling, tracking, and analyzing diverse types of retrograde axonal cargoes could be streamlined, opening up a new avenue to understand the molecular basis of membrane endocytosis, sorting and retrograde trafficking in live neurons.

Fig. 3 (continued) cargoes. Briefly, BoNT/A-Hc (Hc-647) pulse labeling of synapses followed by 2 h chase to observe the retrograde transport. Briefly, DIV 14 hippocampal neurons grown in microfluidic devices were pulsed 5 min with BoNT/A-Hc-Atto647N (100 nM for confocal; 1 nM for PALM). Following a 2 h chase, the live-imaging microscopic images are acquired. (**b**, **c**) Along the axons within the microfluidic channels, the retrograde carriers labeled with BoNT/A-Hc-Atto647N were analyzed. The kymograph shows the overlapping tracks of Hc-Atto647N and LC3-positive autophagosome. xBar = 10 μm; yBar = 20 s. (**c**) The flux of retrograde cargoes is captured using the live-imaging confocal microscope. Different cargoes are marked out with triangles of different colors. Bar = 10 μm. (**d**) Single-molecule trajectories of BoNT/A-Hc-Atto647N trapped in synaptic vesicles were traced over 200 s, with a track duration filter of 2.5 s. Bar = 10 μm. Trajectories of BoNT/A-Hc-Atto647N single molecules were tracked, overlaid with a bright-field image (top panel). High power view of a single molecule trajectory (bottom panel). Bar = 5 μm. (**e**) MSD reflecting the mobility of all BoNT/A-Hc-Atto647N molecules and those undergoing active transport (directed motion). (**f**) Left panel: the frequency distribution of the average speeds of BoNT/A-Hc-Atto647N labeled molecules (1 nM) and vesicular cargoes (100 nM), respectively. Right panel: the statistical comparison of their average speeds. ((**d**, **e**) are reproduced from Ref. [8] with the publisher's permission)

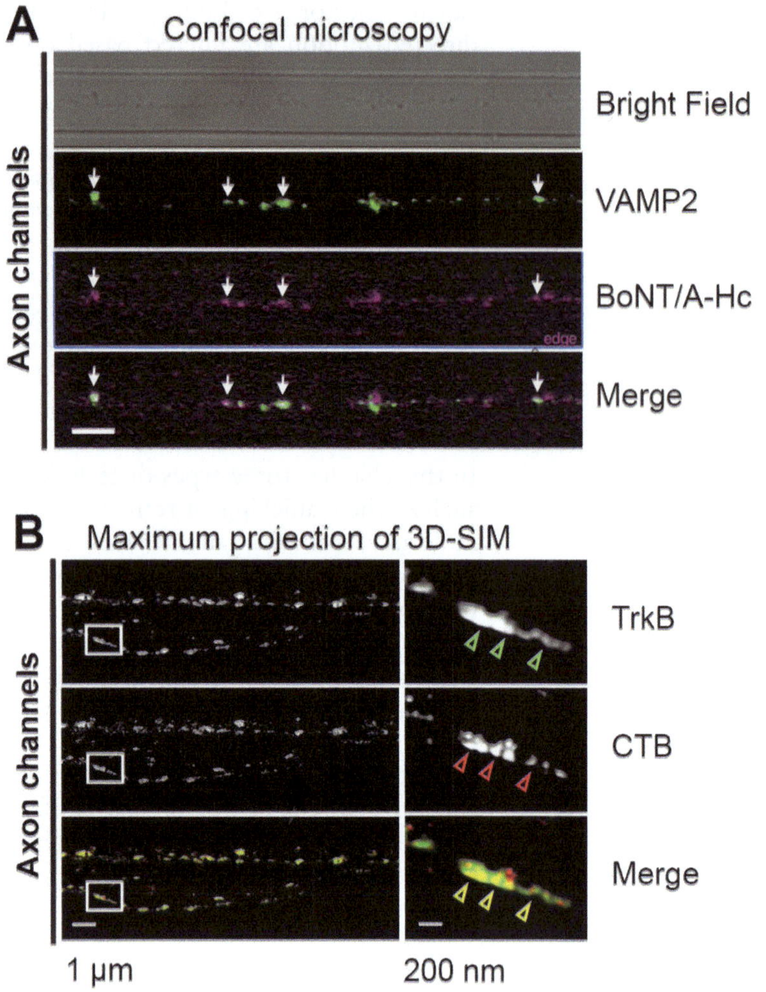

Fig. 4 Immunofluorescent staining of retrograde cargoes in axons. Axons of DIV 14 hippocampal neurons in the microfluidic devices were fixed and immunofluorescently stained after pulse–chase labeling steps. (**a**) Confocal microscopic image showing that the BoNT/A-Hc-Atto647N endocytosed in the synapses colocalize with the synaptic vesicle marker VAMP2 along the axonal shafts. Bar = 10 µm. (**b**) 3D-SIM showing that CTB labeled endosomes colocalize with the signaling endosomal marker TrkB receptor. Bar = 1 µm (left); 200 nm (right)

2 Materials

All buffers and solutions are prepared using ultrapure water and tissue culture grade reagents, sterile-filtered through 0.22 µm filters, and stored at 4 °C unless indicated otherwise. All waste materials and reagents are disposed of according to the institution's waste disposal regulations.

2.1 Preparation of the Microfluidic Devices

1. Microfluidic devices: for live-imaging and immunofluorescent staining, commercially available devices from Xona (RD 450) or equivalent can also be used in conjugation with 24 × 60 × 0.17 mm microscope coverslips (*see* **Note 1**).
2. Glass coverslips: #1.5 thickness glass microscope coverslips, 24 × 60 × 0.17 mm.
3. Petri dishes for tissue culture.
4. Sterilized forceps: Adson forceps, 1.2 mm tip, 125 mm or similar type.
5. 70% (v/v) ethanol.
6. 0.1 M borate buffer (pH 8.5): weigh 1.24 g boric acid and 1.90 g sodium tetraborate dissolved in 450 mL of water in the beaker. Mix and adjust the pH to 8.5. Make up the final volume to 500 mL.
7. Coating solution (100×): make the 100 × poly-L-lysine hydrobromide (PLL) coating solution by dissolving 10 mg PLL into 10 mL of 0.1 M borate buffer (pH 8.5) (*see* **Note 2**). The resultant 10 mg/mL coating stock solution is passed through 0.2 μm syringe filter into a sterile 15 mL conical tube and divided to 1 mL aliquots. Store at −20 °C.

2.2 Set up the Primary Rat Neuronal Culture

1. Dissecting medium: Hanks's buffered salt solution (HBSS) without Ca^{2+}, without Mg^{2+} and no phenol red, sterile-filtered. For each 500 mL bottle of HBSS, add in 5 mL of Penicillin-Streptomycin antibiotics (10,000 U/mL). Store at 4 °C.
2. Trypsin digesting medium: 2.5% Trypsin (w/vol)-EDTA with phenol red.
3. Serum supplement: supplement reagent such as MITO+ Serum extender (Corning, #355006) or equivalent. Dissolve the reagent powder in 10 mL of ultrapure water, filtered through the 0.2 μm filter, and aliquoted to 1 mL and stored at 20 °C.
4. Plating medium: Dulbecco's modified Eagle medium (DMEM), with high glucose, without L-Glutamine, with sodium pyruvate, with phenol red. In a 500 mL bottle of DMEM, add 1 mL of MITO+ Serum extender, 25 mL of Fetal Bovine Serum (FBS) and 0.5 mL of gentamicin (50 mg/mL) (*see* **Note 3**).
5. Anatomical microscope with the light source in the bottom plate.
6. Tools for dissection: iris scissors (straight, 11.5 cm), dissecting forceps (1.2 mm tip, 125 mm), 45° bent thin-tipped tweezers (Dumont tweezers, styles 5 and 5/45 or equivalent). All instruments should be autoclaved before dissection. Dissection should be conducted in a laminar flow hood equipped with UV light.

2.3 Maintenance and Transfection of Cultured Neurons

1. Cell culture incubator with 5% CO_2, 37 °C and humidity levels >80% RH.
2. Maintaining medium: In 500 mL of Neurobasal™ medium (with phenol red), add 10 mL of B27 supplements (50×) and 5 mL of GlutaMAX supplement (100×) (*see* **Note 4**). The medium is aliquoted into 50 mL conical tubes and stored at 4 °C.
3. 5-fluoro-2′-deoxyuridine (FDU) stock: 100 mg of 5-Fluoro-2′-deoxyuridine and 100 mg of Uridine are dissolved with 40 mL of Neurobasal™ medium to make the 1000 × FDU stock. Filter-sterile and divide the stock solution into 1 mL aliquots and store them at −20 °C.

2.4 Pulse–Chase Labeling and Stimulation

1. 1 M HEPES (pH 8.0) stock buffer: Dissolve 119.0 g of HEPES in 400 mL of ultrapure water, adjust the pH to 8.0 with NaOH. Make the total volume to 500 mL with ultrapure water. The stock solution could be stored at room temperature for up to 6 months.
2. 5 M NaCl solution: 29.22 g NaCl is dissolved in ultrapure water to a total volume of 100 mL. The stock solution can be stored at room temperature for up to 6 months.
3. 1 M KCl solution: 23.8 g KCl is dissolved in ultrapure water to a total volume of 100 mL.
4. 1 M $CaCl_2$ solution: 14.70 g of $CaCl_2$ are dissolved in ultrapure water to a total volume of 100 mL. The stock solution could be stored at room temperature for up to 6 months.
5. 1 M $MgCl_2$ solution: 20.34 g $MgCl_2$ are dissolved in ultrapure water to a total of 100 mL.

   The stock solution could be stored at room temperature for up to 6 months.
6. 560 mM D-glucose stock: 5.04 g of D-glucose are dissolved in ultrapure water to make a total volume of 50 mL. Divide the stock into 10 mL aliquots and store them at −20 °C for up to 6 months.
7. 0.5 M ascorbic acid stock: 0.88 g of fresh ascorbic acid is dissolved in ultrapure water to make the total volume of 10 mL. Divide the stock into 1 mL aliquots and store them at −20 °C for up to 6 months.
8. 10% (w/vol) BSA: 1.0 g of BSA is dissolved in ultrapure water to a total volume of 10 mL. Divide the stock into 1 mL aliquots and store them at −20 °C for up to 6 months.
9. Low K^+ and High K^+ buffer: To prepare 100 mL low K^+ buffer (0.5 mM $MgCl_2$, 2.2 mM $CaCl_2$, 5.6 mM KCl, 145 mM NaCl, 5.6 mM D-glucose, 0.5 mM ascorbic acid, 0.1% BSA and 15 mM HEPES, pH 7.4), add 1.5 mL of 1 M HEPES

(pH 8.0) stock, 2.9 mL of 5 M NaCl stock, 560 μL of 1 M KCl stock, 220 μL of 1 M $CaCl_2$ stock, 50 μL of 1 M $MgCl_2$ stock, 1 mL of 560 mM D-glucose stock, 100 μL of 0.5 M ascorbic acid stock and 1 mL of 10% BSA stock, adjust the pH to 7.4 and make the final volume to 100 mL. The high K^+ buffer is prepared similarly, with 56 mM KCl (add 5.6 mL of 1 M KCl stock) and 95 mM NaCl (add 1.9 mL of 5 M NaCl stock) instead of the indicated amount in Low K^+ buffer. The osmolarity of both buffers should be adjusted to 290–310 mOsm. Resultant High K^+ and Low K^+ buffers are filter-sterilized and stored at 4 °C for up to 2 weeks (*see* **Note 5**).

2.5 BoNT/A-Hc Expression and Purification

1. Bacteria homogenizer, French Press G-M™ high-pressure cell press homogenizer (Glen Mills) or equivalent equipment.
2. Luria–Bertani (LB) broth bacterial growth medium: Add 10 g peptone, 5 g yeast extract, and 5 g sodium chloride in 950 mL of ddH_2O, stir until all the reagents are fully dissolved. Adjust the pH to 7.0 with 5 N NaOH. Make the final volume to 1 L with distilled water. Autoclave for 20 min at 15 psi.
3. 1000 × Kanamycin stock: 50 mg/mL kanamycin in ddH_2O. Add 1 g of kanamycin to 20 mL of distilled water to make a 50 mg/mL stock. Filter-sterilize and store in 1 mL aliquots at −20 °C.
4. 1000 × Isopropylthio-β-D-galactoside (IPTG) Stock: 200 mM IPTG in ddH_2O. Add 0.5 g of IPTG in 8 mL of ddH_2O. Adjust the volume of the solution to 10 mL with ddH_2O. Filter-sterilize the solution and store in 1 mL aliquots at −20 °C.
5. Iminodiacetic acid (IDA) agarose resins: IDA cross-linked affinity chromatography resin for affinity purification of 6 × HIS-tagged BoNT/A-Hc. Store at 4 °C.
6. 1 M Tris–HCl (pH 8.0): weigh out 121.1 g Tris into 1 L clean beaker, add 800 mL of ddH_2O, dissolve it and add in about 42 mL of HCl to adjust the pH to 8.0. Make the final volume to 1 L with ddH_2O. Autoclave for 20 min at 15 psi.
7. Protease inhibitors: the protease inhibitor cocktail set III (#539138, Merck) or similar products.
8. Lysis buffer: 10 mM Tris–HCl, pH 7.5, 300 mM NaCl, 1% Triton X-100, add in 1 U DNase I and protease inhibitors. To make 100 mL of lysis buffer, add 1 mL 1 M Tris–HCl (pH 8.0) stock, 6 mL 5 M NaCl stock, 1 mL Triton X-100, 100 μL DNase I (1 U/μL), and 100 μL protease inhibitors (*see* **Note 6**), adjust the pH to 7.5. Make up the final volume to 100 mL, store at 4 °C for up to 2 weeks.
9. Wash buffer: 10 mM Tris–HCl, pH 7.5, 300 mM NaCl, 5 mM imidazole. To make 100 mL of wash buffer, add 1 mL 1 M Tris–HCl (pH 8.0) stock, 6 mL of 5 M NaCl stock, 1 mL of

500 mM imidazole stock. Make up the final volume to 100 mL and store at 4 °C for up to 2 weeks.

10. Elution buffer: 10 mM Tris–HCl (pH 7.5), 300 mM NaCl, 500 mM Imidazole. To make 100 mL of wash buffer, add 1 mL of 1 M Tris–HCl (pH 8.0) stock, 6 mL of 5 M NaCl stock, 3.4 g of imidazole, adjust the pH to 7.5 using HCl. Make up the final volume to 100 mL store at 4 °C for up to 2 weeks.

2.6 Conjugation of Purified BoNT/A-Hc to Atto647N Maleimide

1. Bradford reagent to determine protein concentration: reagents should be stored at 4 °C.
2. 1 M DL-dithiothreitol, anhydrous (DTT) stock: dissolve 1.5 g of DTT in 8 mL of ddH_2O (*see* **Note** 7). Adjust the final volume to 10 mL, divide the solution into 1 mL aliquots, wrap the tube with foil and store in −20 °C.
3. β-mercaptoethanol (*see* **Note** 7).
4. Sephadex column with desalt capacity. We use the PD Mini-Trap G-25 Column (GE Healthcare).
5. Atto647N maleimide (#05315, Merck) or equivalent products are used. Dissolve each dye in DMSO to the final concentration of 10 mM, aliquot to 5 μL stocks, wrap with foil and store in −80 °C.
6. 10% (w/v) BSA in 1 × PBS. Divide the stock into 1 mL aliquots and store them at −20 °C for up to 6 months (*see* **Note 8**).
7. Cholera Toxin Subunit B (CTB) Alexa Fluor 488 and 555 conjugates (#C34775 and #C34776, Thermo Fisher) or equivalent products to label retrograde signaling endosomes.

2.7 Live Imaging of Retrograde Axon Cargoes

1. Inverted confocal microscope: live imaging was performed using 63 × 1.4 NA/190 μm WD/0.132 μm/pixel (1024 × 1024) objective on a Zeiss LSM 710 inverted point-scanning confocal microscope equipped with high-sensitivity BiG (GaAsP) detectors, argon 488 nm, 561 nm and 637 nm lasers, gallium arsenide phosphide (GaAsP) detectors and CO_2 and temperature incubating chamber.
2. Inverted SIM microscope: SIM images are acquired using an Alpha Plan-Apochromat 100×/1.46 NA oil-immersion objective on the Zeiss ELYRA PS.1 SIM inverted microscope (Carl Zeiss) equipped with an sCMOS camera (PCO AG) for SIM acquisition.
3. PALM microscope: a commercial inverted microscope (Nikon, TiE) equipped with a perfect focus system and the 100 × 1.49 numerical aperture CFI Apo objective (Nikon) are needed for single-molecular imaging experiments. Under slightly oblique illumination ($ILas^2$; Roper Scientific), movements of

ATTO647N-labeled molecules on the axon carriers are captured at least 16 frames/s through an EM-CCD (Evolve512 delta; Photometrics).

4. Live-imaging buffer: 25 mM HEPES, 145 mM NaCl, 5.6 mM KCl, 0.5 mM $MgCl_2$, 2.2 mM $CaCl_2$, and 5.6 mM D-glucose, pH 7.4. Add 2.5 mL of 1 M HEPES (pH 8.0) stock, 2.9 mL of 5 M NaCl stock, 560 μL of 1 M KCl stock, 220 μL of 1 M $CaCl_2$ stock, 50 μL of 1 M $MgCl_2$ stock, 1 mL of 560 mM D-glucose stock, adjust the pH to 7.4 and make the final volume to 100 mL. The osmolarity of this buffer should be adjusted to 290–310 mOsm. The resultant imaging buffer is filter-sterilized and stored at 4 °C for up to 2 weeks.

2.8 Fixation and Immunofluorescent Staining

1. Fixation solution: 4% Paraformaldehyde (PFA), 4% sucrose in 1 × phosphate buffer (PBS). To make 40 mL of fixation solution, add a 10 mL vial of methanol free 16% PFA, 1.6 g of sucrose, 4 mL of 10 × PBS into one beaker, add 20 mL of ultrapure water, adjust the pH to 7.4 and make up the final volume to 40 mL. The fixation solution is filtered through a 0.22 μm filter and stored at 4 °C for up to 2 weeks (*see* **Note 9**).
2. Perforation buffer: 0.1% Triton X-100 in 1 × PBS, freshly made. Mix 1 mL, Triton X-100 with 9 mL of ultrapure water, rotate on rotors for over 10 min to ensure it is fully resolved, store the 10% Triton X-100 solution at 4 °C. Before each experiment, dilute 50 μL of 10% Triton X-100 stock to 5 mL 1 × PBS to make the fresh perforation buffer (*see* **Note 10**).
3. Blocking and antibody dilution buffer: 1 × PBS, 1% (w/v) BSA, 0.1% saponin, 0.2% Gelatin. To prepare 50 mL, add 5 mL 10 × PBS, 0.5 g BSA and 30 mL Milli-Q water and mix well. While stirring, add 500 μL 10% saponin and 5 mL 2% Gelatin, add to 50 mL with Milli-Q, then filtered through a 0.22 μm filter and stored the 5 mL aliquots at −20 °C (*see* **Note 11**).
4. Primary antibodies are stored and diluted as recommended by the manufacturer (*see* **Note 12**).
5. Fluorescently tagged secondary antibodies are stored, avoiding light, and diluted as recommended by the manufacturer (*see* **Note 13**).
6. Antifade mounting mediums: antifade mounting medium not interfering with GFP and can solidify and should be used for confocal microscopy imaging.
7. Liquid antifade mounting medium: antifade mounting medium not interfering with GFP and cannot solidify. This should be used for 3D-SIM imaging.

3 Methods

All the pressures are carried out at room temperature, unless otherwise specified.

3.1 Preparation of the 60 × 24 mm Glass Coverslip

1. In the fume hood, soak the whole package of 100-piece glass coverslips in a clean 2 L glass triangle bottle. Pour 200 mL of 40% nitric acid to soak all the coverslips (*see* **Note 14**). Cover the container with thick aluminum foil, clearly label it with warning signs. Soak the coverslips in nitric acid for over 24 h in the fume hood at room temperature with a very slow shake (<20 rpm).
2. Remove all the nitric acid (*see* **Note 15**). Add in 500 mL of ddH_2O, shake with a slow speed of 60 rpm for 10 min. Repeat the wash seven times.
3. Repeat **step 2** using ultrapure water three times.
4. Remove ultrapure water, rinse the coverslips with 200 mL of 99.7% ethanol, shake the container at 60 rpm for 5 min. Repeat the ethanol wash twice.
5. Remove the ethanol as clean as possible from the beakers containing the glass coverslips.
6. Transfer the coverslips to a clean 2 L glass beaker. Cover the top of the beaker with thick aluminum foil, use forceps to pinch 10–15 small holes on the foil to allow evaporation of ethanol.
7. Adjust the temperature setting of an empty air-dry oven to 120 °C. Place the foil-covered beaker with coverslips into the oven, dry for 1 h to completely evaporate all traces of liquid. After cooling down, store the cleaned coverslips in a sterile 10 cm petri dish and seal the dish with parafilm.

3.2 Preparation of the Microfluidic Devices for Neuronal Culture

1. Soak the new devices into 70% ethanol for over 5 min in an autoclaved beaker. Shake the beaker gently 10–15 times.
2. Soak the acid-cleaned coverslips from Subheading 3.1 into 70% ethanol for over 5 min in an autoclaved beaker. Shake the beaker gently 10–15 times.
3. In a new 10 cm tissue culture dish, place the soaked microfluidic device with groove side facing up, use 10 mL pipet to add in 10–15 mL of sterile ultrapure water to rinse the device. Repeat the rinse twice. Vacuum the ultrapure water as clean as possible. Left the petri dish lid open and dry the device in the biosafety hood with airflow on.
4. In a new 10 cm culture dish, place the cleaned coverslips and use a 10 mL pipet to add 10–15 mL of sterile ultrapure water to rinse them. Repeat the rinse twice. Vacuum the ultrapure water as clean as possible. Left the petri dish lid open and dry the coverslips in the biosafety hood with airflow on.

5. Press the device to join it with the coverslip with bent forceps. Before pressing down, make sure the etched side of the device is facing the coverslip. Combined devices and coverslips are shown in Fig. 1a (*see* **Note 16**).
6. For device coating, dilute the 10 mg/mL PLL stock solution to a final concentration of 0.1 mg/mL by pipetting 1 mL aliquot into 99 mL of 0.1 M borate buffer (pH 8.5). The diluted working solution could be wrapped with aluminum foils and store at 4 °C for half a year. It can also be divided into aliquots and stored at −20 °C indefinitely.
7. Add 200 μL of 0.1 mg/mL PLL coating solution diagonally to two reservoirs in the opposite direction (*see* **Note 17**). Place two 24 × 60 mm coverslips with devices attached to one 10 cm tissue culture dish. Cover the lid of the dish, put back the device with coating solutions into the cell culture incubator and incubate overnight. The minimal coating time is 2 h for microfluidic devices.
8. After coating, wash the device with 200 μL of sterile ultrapure water by adding it diagonally to the reservoirs on both sides (*see* **Note 18**).
9. Preincubate the coated device with 200 μL of plating medium by adding the medium diagonally to the reservoirs on both sides. Return the dish with devices to the cell culture incubator for preincubation.

3.3 Set up the Rat Hippocampal Neuron Culture in the Microfluidic Device

1. Before dissection, put the following tools (anatomical microscope, ophthalmic surgical scissors × 1, ophthalmic bent forceps × 1, fine-tip 45° forceps × 2, and two frozen 500 mL ice bags) into the dissection hood, spraying them thoroughly with 70% ethanol, and placing them under UV for 30 min to sterilize (*see* **Note 19**).
2. Add 10 mL of dissection buffer into two sterile 6 cm Petri dishes and placed them on top of the ice bags.
3. Place two aliquots of 1 mL of digestion buffer in the cell culture incubator.
4. Take 4–5 freshly recovered embryonic day 18 (E18) Sprague-Dawley rat embryos from the animal house (or other suppliers) (*see* **Note 20**). Transfer the embryos to a 50 mL sterile conical tube filled with dissection buffer (*see* **Note 21**).
5. Take the embryos out from the tube using the ophthalmic bent forceps, cut the heads of embryos with ophthalmic surgical scissors, and place them in the 6 cm sterile petri dish. Submerge the heads into cold HBSS dissection buffer. Use fine-tip 45° forceps to fix the heads for surgical scissors to cut along the midline of the skull with the other fine-tip 45° forceps. Fully expose the brain.

6. Take out the brains with the fine-tip 45° forceps, place them in another new 6 cm dish filled with cold dissection buffer. Place the brain with the ventral side up, use the fine-tip forceps to pinch off the two cortical hemispheres. Remove the meninges by tearing them off from the rostral end.
7. Further dissected out the hippocampal pieces with the fine 45° forceps. Submerge them in cold dissection buffer in a sterile 6 cm Petri dish.
8. Add 1 mL of digestion buffer to a sterile 15 mL centrifuge tube, prewarm it in the cell culture incubator for 10 min. Collect all the hippocampal pieces into the 15 mL tube filled with digestion buffer. Tighten the cap, gently shake it, then put it in the cell culture incubator and incubate for 18 min.
9. After digestion, remove the excess digestion buffer from the tube, add 5 mL of prewarmed plating medium into the 15 mL tube containing the tissue to terminate the digestion. Gently shake the tube and remove the plating medium as much as possible. Repeat this **step 3** times.
10. Add 3 mL of plating medium, dissociate the tissue by gently pipetting up and down 10–15 times with a 1 mL pipet.
11. Spin down the dissociated cells at 500 × *g* for 5 min.
12. After centrifugation, remove the supernatant and resuspend the pellet cells with 2 mL of plating medium. Pass the dissociated cells through a 40 μm cell strainer to filter out undigested tissue debris.
13. Count the cells and adjust to a final concentration of 5×10^6 cells/mL (*see* **Note 22**).
14. Seed the suspended neurons into the microfluidic device. Remove the preincubation medium from all the reservoirs. Using 200 μL pipet tips, add 100 μL of the cell suspension to one single side of the device. This side is termed the soma side and clearly labeled, as shown in red in Fig. 1a (*see* **Note 23**). Gently squirt it through the connecting part between two reservoirs on the soma side (Fig. 1a). Repeat the operation of adding 100 μL of cell suspension from the other reservoirs on the same side (the soma side) by gentle squirting through the connecting part again (*see* **Note 24**).
15. Remove all the cell suspensions from the reservoirs, but make sure the connecting section is still filled with cell suspension. Incubate the devices to cell culture incubator for 10 min (*see* **Note 25**).
16. After neurons have attached to the coverslip in the connecting section, gently fill all the reservoirs with a prewarmed plating medium. Return the dish containing the devices to the cell culture incubator and incubate overnight.

3.4 Maintenance and Transfection of Neurons Cultured in the Device

1. On the next day after seeding (DIV 2), remove 100 μL of plating medium from each reservoir and add 200 μL of pre-warmed maintaining medium to each reservoir. Return the devices to the cell culture incubator (*see* **Note 26**).
2. After the neurons are cultured in vitro for 3–5 days (DIV 3), kill glial cells by changing the culture medium to the maintaining medium with 1 × FDU.
3. Feed the neurons in the device twice a week by removing 100 μL of the conditioned medium in the reservoirs and replace it with 200 μL of fresh maintaining medium.
4. On DIV 10–12 (*see* **Note 27**), prepare the transfection using Lipofectamine 2000 (Invitrogen) or equivalent reagent. In a sterile 1.5 mL microtube, add 100 μL of Neurobasal medium and 1 μL of Lipofectamine 2000 reagent, mix them well and incubate for 5 min at room temperature. In another sterile 1.5 mL microtube, add 100 μL of Neurobasal medium and 1 μg of plasmid DNA, mix well. Combine these two complexes and mix well by gently pipetting up and down 3–5 times. Incubate this 200 μL mixture at room temperature for 10 min (*see* **Note 28**).
5. Take out the device from the cell culture incubator. Remove all the medium from two reservoirs in the soma side of the device (about 200 μL). Store the conditioned medium into a new sterile 1.5 mL microtube (*see* **Note 29**).
6. In one reservoir on the soma side, add the 200 μL of transfection mixture. Wait until the liquid in the two reservoirs reached the same level. Then, return the dish with the devices to the cell culture incubator for 2–4 h.
7. Add 200 μL fresh maintaining medium to the stored conditioned medium and put this mixed medium with a loose lid into the cell culture incubator.
8. Remove all the transfection medium from the reservoirs on the soma side. Add back the mixed conditioned medium into these reservoirs. Transfection efficiency can be detected after 24 h, as shown in Fig. 2c, d.

3.5 Expression and Purification of His-Tagged BoNT/A-Hc

1. pET28–6 × His-BoNT/A-Hc, which encodes the C-terminus of the botulinum neurotoxin heavy chain (BoNT/A-Hc) [24, 40], is used here as an example for labeling probes used in the following steps.
2. Add 5 μL of antibiotics into 5 mL of LB in a sterile bacterial test tube. Seed the glycerol stock of pET28–6 × His-BoNT/A-Hc bacteria into this test tube. Shake the test tube at 37 °C for 12 h in a shaker set at 225 rpm.
3. Add all 5 mL overnight bacterial culture into the autoclaved triangle bottle filled with 800 mL of LB medium supplemented

with 50 μg/mL kanamycin. Shake at 37 °C at 250 rpm for 3 h, until the OD 600 reaches 0.6–0.9.

4. Induce the expression of tagged protein with 200 μL of IPTG stock. Continue to shake the bacterial culture overnight at 20 °C at 250 rpm.
5. Spin down all the bacteria at 7000 rpm 10 min at 4 °C and remove the supernatant (*see* **Note 30**).
6. Resuspend the bacteria pellet in 40 mL of lysis buffer. The bacterial suspension is homogenized by passing through the French Press homogenizer at 1500 psi (*see* **Note 31**).
7. Centrifuge the bacterial lysis for 30 min at 150,000 × *g*, 4 °C. Store the supernatant at 4 °C.
8. In a 5 mL tube, wash 1 mL cobalt beads (stored in ethanol) with 1 mL of wash buffer by centrifugating at 6000 × *g* for 60 s. Repeat the wash five times. Remove the wash buffer.
9. Add the 40 mL bacterial supernatant to the washed beads, slowly rotate the mixture for 4–12 h at 4 °C.
10. Wash the chromatography column with ddH_2O. Load the supernatant and beads into the column.
11. Wash the column with beads with 5 mL of wash buffer (*see* **Note 32**).
12. Elute the tagged protein with 1 mL of elution buffer and collect the 1 mL of eluted solution. Repeat the elution step twice and store the eluted fractions at 4 °C (*see* **Note 33**).
13. Run SDS-PAGE gel to check the expression of the tagged proteins.
14. Use IgG as the standard to determine the protein concentration of the elution by using the Bradford assay.
15. The solution is deoxygenated by bubbling nitrogen gas for 30 min on ice. Add about tenfold molar excess of DTT (about 300–400 μM of DTT) to the protein solutions, incubate for 30 min on ice.
16. Dialyze the protein solution for 24 h in 2 L of precooled 1× PBS at 4 °C, change the 1 × PBS once (*see* **Note 34**).
17. Measure the concentration of the dialyzed protein using the Bradford assay.
18. Add twofold excess of Atto647N-maleimide dye into the dialyzed protein solution. Leave the conjugation reaction to happen for 4–12 h at 4 °C.
19. Add about tenfold molar excess of β-mercaptoethanol (compared to the amount of dye) to the solution to stop the reaction.
20. Remove unbound dyes using a Spandex desalt column.

21. Determine the concentration of the Atto647N conjugated protein using the Bradford assay.
22. Add 1% total volume of 10% (w/v) BSA into the BoNT/A-Hc-Atto647N solution to stabilize the protein (*see* **Note 35**). Aliquot the conjugated protein solution to 20–50 μL and store at −80 °C.

3.6 Pulse–Chase Labeling and Stimulation of Neurons

1. Add 200 μL of prewarmed Low K^+ buffer to each of two microtubes. Dilute the BoNT/A-Hc-Atto647N to the final concentration of 100 nM in one tube of Low K^+ buffer. This concentration is for the experiment using confocal microscopy. For the single-particle tracking (SPT) experiments, 1 nM BoNT/A-Hc-Atto647N is used to label at the single molecular level. Prewarm these two tubes with Low K^+ buffer in a 37 °C water bath for 10 min.
2. Take the device with neurons from the cell culture incubator and examine it under the microscope. Open the dish's lid to mark the soma side with a marker pen on the coverslip bearing the device.
3. In a microtube rack, place the 1.5 mL microtube with BoNT/A-Hc-Atto647N in Low K^+ buffer near the soma side and the other with Low K^+ buffer only near the terminal side. Remove the conditioned medium from all 4 reservoirs and store it in a sterile 1.5 mL microtube.
4. Add 140 μL of Low K^+ with BoNT/A-Hc-Atto647N to the two reservoirs (70 μL each) on the soma side and 200 μL of Low K^+ buffer to the two reservoirs (100 μL each) on the terminal side of the device.
5. Return the device and the stored conditioned medium to the cell culture incubator, and wait for 5 min for the pulse labeling to happen in the terminal chamber. The same pulse labeling step for High K^+ replaces the Low K^+ buffer with the High K^+ buffer (*see* **Note 36**).
6. The same pulse labeling step can label the nerve terminals with the CTB under either Low K^+ or High K^+ conditions. The restricted labeling of both BoNT/A-Hc and CTB to the terminals is shown in Fig. 2a, b.
7. Take the device out from the cell culture incubator and remove the labeling medium from all four reservoirs. Wash the reservoirs three times by adding prewarmed Low K^+ medium diagonally to the reservoirs at opposite sides.
8. Refill the device with the stored and prewarmed conditioned medium and return it to the cell culture incubator for an additional 2 h. The flux of labeled retrograde cargoes can be observed as shown in Fig. 2d (*see* **Note 37**).

3.7 Live Imaging of Retrograde Axonal Trafficking Events

1. During the 2 h incubation time, prepare the inverted microscope for live-imaging of the labeled cargos. Prewarm and turn on the CO_2 in the imaging chamber by setting the temperature to 37 °C and CO_2 level to 5%, at least 30 min before imaging. The time scale of the pulse–chase experiment is shown in Fig. 3a.
2. Replace the culture medium with 400–500 μL of the prewarmed live-imaging buffer. This method is applicable to multiple live-imaging microscopic approaches. Here, we list the steps for time-lapse confocal microscopy (**steps 3** and **4**) and SPT detection using PALM microscope (**steps 5** and **6**).
3. Add a drop of immersion oil to the 63× oil inverted objective of the confocal microscope. Pinch the coverslips with devices tightly to the versatile adaptor. Place the adaptor to the microscope and adjust the focal plane to ensure the coverslip touches the oil drop. Look for axon channels under the bright-field light path. Then turn on the fluorescent light to look for the axons expressing GFP. Save the positions of GFP-expressing axons with the multiposition function of the microscope.
4. Set up the image acquisition parameters and image the saved ROIs sequentially. Next, select the ROIs in axon channels proximal to the soma chamber (*see* **Note 38**). Each channel is imaged for about 5 min, and a time resolution higher than 1 Hz (1 frame per second) should be used to ensure accurate cargo tracing and trajectory analysis (*see* **Note 39**).
5. For SPT detection in axon terminal and channels, an inverted PALM microscope equipped with a 100 × oil-immersion objective with a numerical aperture (NA) 1.49 should be used. The high-performance EMCCD camera with high-speed acquisition software is needed to capture the BoNT/A-Hc-Atto647N single molecule at a speed of at least 16 ms/frame.
6. Place a drop of immersion oil on the 100 × objective of the inverted single-molecule-imaging microscope and place the coverslip with the device on top of the slide holder. Adjust the focus of the objective lens to achieve oblique illumination of axons in the channels and terminal chamber. In the acquisition software, set up the parameter to acquire time-lapse images in several ROIs with a high acquisition speed. 4000 to 20,000 frames should be acquired within a time frame of a few minutes using the high-speed recording mode of the software (*see* **Note 40**).
7. Fixation and staining of the device should be conducted immediately after the live-imaging steps (Go to Subheading 3.9).

3.8 Analysis of Retrograde Axon Trafficking Events Using Live-Imaging Movies

1. Time-lapse images of the retrograde cargoes are analysed for the kinetics using the open-source software FIJI (NIH, version 1.53a) (*see* **Note 41**).
2. Use the Bio-Formats plugin to import the movie to FIJI. Adjust the brightness and contrast to allow the clear visualization of the labeled axon carriers.
3. Use "Image → colour → channels tool → colour" to define colors for different types of carriers. Then, use the composite mode to view the movement of the various types of cargoes.
4. Kymographs were generated using the plugin Multi-Kymograph, set the line thickness according to the diameter of the axon, as shown in Fig. 3b (*see* **Note 42**).
5. Use the FIJI TrackMate (v 6.1) plugin to automatically trace the movement of retrograde axonal cargos. Open the time-lapse acquisitions of retrograde axon cargoes using Bio-Format. Then, run the TrackMate Plugin ("Plugin → tracking → TrackMate"). Make sure the pixel size and time interval of the imported movie is correct. Draw the ROI and saved it to the ROI manager.
6. Select "LoG detector" go next to detect spots. On the LoG detector page, adjust the pixel and threshold value to detect most of the spots. Next, click the "Preview" to examine the accuracy of detection whenever the values are changed. Click next until the "select a tracker" page is shown.
7. Set the linking algorithm as "Simple LAP tracker" and the "Gap" to 0 (*see* **Note 43**).

   Finish the setup steps, and run the trajectory tracker. Filter the traced trajectories using the "track duration" filter to exclude the trajectories last shorter than 25 frames, then go next.
8. Click the "next" until the last page. In this, select an action page, choose "Export tracks to Icy," save one "_Icy.xml" file.
9. Open open-sourced software Icy (v.2.0.3.0, Institut Pasteur), click "Detection & Tracking - Track Manager". In the jump out window, click "file-load", choose the "_Icy.xml" file generated in the last step and load it to the track manager. In the "add Track Processor..." button, choose "motion profile processor." It will generate an Excel file containing the average track speed, duration, and displacement values of all selected tracks, which can be used to analyze the directional movement of these axon trafficking cargoes. Moving cargoes in the retrograde direction are shown in Fig. 3c.
10. For the analysis of SPT tracks, in the "motion profile processor," select "Track processor MSD." This calculates the Mean square displacement (MSD) of the loaded tracks. It will

generate an Excel file contains the MSD of each track. MSD values of the first 20 steps per trajectory are used to analyze the overall mobility of the SPT in retrograde axonal trafficking. Mobility analytic results are shown in Fig. 3d–f.

3.9 Fixation and Immunofluorescent Staining of the Compartmentalized Neuron

1. Fixation and staining of the device should be conducted immediately after the live-imaging (Subheading 3.8). Remove all the imaging buffer from the 4 reservoirs and add 250 μL of cold fixation buffer to each reservoir (*see* **Note 44**). Fix the neurons in the device overnight at 4 °C (*see* **Note 45**).
2. Remove the fixation buffer, wash the device by adding 200 μL of 1× PBS diagonally to two opposite reservoirs, incubate at room temperature for 5 min. Repeat the wash five times (*see* **Note 46**).
3. Block the neurons by adding 200 μL of blocking buffer diagonally to two opposite reservoirs, incubate at room temperature for 60 min. Dilute the primary antibodies in the blocking buffer. Next, add 200 μL of antibody solution diagonally to two opposite reservoirs, incubate overnight at 4 °C.
4. Wash the device by adding 200 μL of 1× PBS diagonally to two opposite reservoirs, incubate at room temperature for 5 min. Repeat the wash 3–5 times.
5. Dilute the secondary antibodies in the blocking buffer. Add 200 μL of secondary antibody solution diagonally to two opposite reservoirs, incubate at room temperature for 1 h.
6. Wash the device by adding 200 μL of 1× PBS diagonally to two opposite reservoirs, incubate at room temperature for 5 min. Repeat the wash 3–5 times.
7. If the sample is prepared for confocal microscopy imaging, mount the coverslip with a drop of antifade mounting medium, which does not interfere with GFP and can solidify. If the sample is prepared for 3D-SIM, use the liquid antifade mounting medium [41, 42].

4 Notes

1. For live-imaging in a glass-bottom 3.5 cm dish, custom-made microfluidic devices are designed and fabricated as detailed in the previously published protocol paper [34].
2. As PLL is very sticky and hard to weigh out, we usually use 10 mL of 0.1 M borate buffer (pH 8.5) to wash out all the content from a new bottle containing 10 mg PLL. The solution needs to be sterile-filtered through the 0.22 μm filter and immediately wrapped with foil, as PLL is light-sensitive.

3. Variation of the plating medium is common, and here is another recipe we tested that also works well: 10% FBS, 10% F12 supplement in DMEM.
4. We usually do not include any antibiotics in the maintaining medium. However, gentamicin or penicillin–streptomycin can be included as the components of the maintaining medium when necessary.
5. Low K^+ and High K^+ buffers are the keys to the success of the following pulse–chase labeling. We found the freshly made ones produce more repeatable results than frozen-thawed aliquots.
6. Protease inhibitors from different companies have different dilution rate. Follow the one available in your lab and make sure there is 1 mM phenylmethanesulfonyl fluoride (PMSF) in the lysis buffer.
7. Both DTT and β-mercaptoethanol are toxic if inhaled or absorbed through the skin. Therefore, avoid contact with the skin, eyes or clothing. Handle them with gloves in a chemical fume hood.
8. We usually use up the whole aliquoted dye to avoid freeze and thaw. To prevent degradation of purified BoNT/A-Hc, a final concentration of 0.1% (w/v) of BSA is added to stabilize this homemade reagent. BoNT/A-Hc should be stored, avoiding light in aliquots of 50 μL at −80 °C. Thaw the frozen aliquots on ice and make BoNT/A dilutions fresh for each experiment.
9. The 4% PFA fixation solution and all other solutions used in the immunofluorescent staining procedure following the fixation step do not need to be sterile. Instead, the purpose of filtering is to remove any small floating object that could potentially affect imaging.
10. We do not store the perforation buffer and always make it fresh during the experiment.
11. The leftover of each aliquot of blocking buffer can be frozen and stored for subsequent use.
12. We suggest storing the primary antibodies at −20 °C. For this purpose, 50% (v/v) pure glycerol should be added to some of the primary antibodies initially labeled as storing at 4 °C. Unless listed in the product information sheet of the antibody, the starting antibody diluting is 1–500 for our experiments.
13. We usually use the dilution factor between 1/1000 and 1/5000 for most of the fluorescently tagged secondary antibodies.
14. Nitric acid is highly corrosive and causes burns by all exposure routes. Therefore, use it only in the chemical fume hood with proper personal protective equipment (PPE).

15. The recycled nitric acid can be stored in the dark at room temperature. It can be reused at least three times.
16. Always check under a light microscope to see whether the device is combined tightly to the coverslip. Remember to put back its lid when removing the dish outside of the biosafety hood. Before coating, make sure the device is (1) tightly combined with the coverslip and (2) both parts are completely dried. If there is leakage, the coating medium would appear outside of the device very soon. Therefore, we suggest not to continue with the leaking device and restart from Subheading 3.2.
17. The reason to only add in one reservoir on each side is to avoid bubbles.
18. Coated and washed dishes can be wrapped by aluminum foils and stored at 4 °C for up to 2 weeks.
19. These two ice bags are prepared to keep the temperature low while dissecting the brain tissue, making sure they are flat and can hold two 6 cm dishes on the top side.
20. The dam should be anesthetized and sacrificed following the institute's animal ethics regulations. Likewise, all animal experiments should be performed following the institute's animal ethics codes, with approval from the animal ethics committee.
21. The embryos can be stored on ice for up to 4 h before dissection.
22. The counted cell suspension can be stored on ice for up to 4 h before plating.
23. At this stage, there is no need to add any medium or cell suspensions to the reservoirs on the other side of the microfluidic channels.
24. Always use the light microscope to check the condition of the neurons seeded in the device. Make sure that there is no bubble in the connecting section on the soma side. It is also crucial to check whether there are enough evenly distributed neurons in the connecting section.
25. This incubation step allows the neurons in the connecting section of the device attached to the coverslips. To avoid the dry out of the medium from this device, make sure the device is contained in a 10 mL culture dish with the lid closed and do not exceed the incubation time of 10 min.
26. Due to the liquid surface tension, each reservoir can hold more liquid than its designed volume. Therefore, we usually add an excessive amount of maintaining medium to avoid the rapid dry-out of the device.

27. We choose to transfect neurons on DIV 12 because presynaptic machinery matures around DIV14. Transfection on DIV12 allows us to perform the activity-dependent pulse–chase labeling on DIV14. Hippocampal neurons in the device can be successfully transfected from DIV 3 to DIV 12 with an overall efficiency between 1% and 5%. This transfection ratio is ideal for live-imaging approaches. Transfection outside this window reduced the transfection rate in our hands.
28. Plasmid DNA should be prepared using an endotoxin-free plasmid extraction kit, and the DNA concentration should be higher than 1 μg/μL for best transfection efficiency.
29. We do not add any transfection reagent to the reservoirs at the nerve terminal side.
30. This bacterial pellet can be stored at −80 °C for years.
31. All parts of the French press machine should be precooled at 4 °C.
32. Beads can be split here and stored at −80 °C to keep unlabeled protein more stable.
33. We usually only use the first eluted fraction of the purified protein for the following steps.
34. The pore size of the dialytic bag is determined by the molecular weight of the target tagged protein. For example, for BoNT/A-Hc (50 kD), the size of the dialytic membrane is 50 kD.
35. For 200 μL of conjugated protein solutions, add 2 μL of 10% (w/v) BSA stock.
36. The aim of adding more buffer to the soma side is to create pressure toward the terminal side. This pressure is used against the diffusion, restricting the Atto647N-BoNT/A-Hc labeling to the terminal side.
37. The 2 h incubation time aims to chase the retrograde axonal trafficking of the internalized toxins (e.g., BoNT/A-Hc and CTB), which happens in 2 h after their uptake at the axonal terminals.
38. The width of the ROI should be bigger than 10 μm, the length of the ROI should be longer than 40 μm. However, the size of the ROI should be optimized with the scanning speed of the confocal microscope.
39. We usually image 8–10 channels for each device, which takes roughly 1 h to finish the whole live-imaging session for each device.
40. The number and length of the time-lapse movies should be determined by the number of single molecular fluorophores in the live cell. If the fluorophores are low in abundance, elongate the acquisition time is helpful to accumulate enough SPT for

further analysis. Moreover, it is also beneficial to run a pre-bleach section of several minutes to quench the nonspecific fluorophores attached to the coated surface of the dish or PDMS device.

41. These steps (Subheading 3.8) can be adapted to the analysis for single-particle tracking of Atto647N-BoNT/A-Hc internalized at the presynaptic nerve terminals.
42. Only single-channel images can be analysed with Multi-Kymograph. Therefore, multiple channel movies should be split first using the "Image →colour → split channels."
43. The size of the linking distance should be experimentally determined by the speed/mobility of fluorescently tagged cargoes/ single particles. We usually start by putting in a small value and slowly increasing it until the axonal cargoes' trajectories are mostly linked.
44. Never use harsh vacuum suction to remove liquid in any step in the fixation and immunofluorescent staining process.
45. PFA is a toxic and hazardous reagent. Use appropriate personal protective equipment, handle the fixation buffer in the fume hood, and dispose of wastes properly.
46. We found the axons in the middle part of the axon channels are sometimes difficult to stain, likely due to slowed or blocked liquid flow inside. Should this occur, the device can be removed before the fixation step.

Acknowledgments

This work was supported by the National Natural Foundation of China (31871036), the Science and Technology Commission of Shanghai Municipality grant 20ZR1436600 and Pujiang Fellowship 20PJ1410300 to TW and the Australian Research Council grant LE130100078, and the National Health and Medical Research Council grants GNT1155794 and GNT1120381 to FAM.

Data availability statement: The unpublished data used in this study are available from the corresponding author upon reasonable request.

References

1. Taylor AM, Blurton-Jones M, Rhee SW, Cribbs DH, Cotman CW, Jeon NL (2005) A microfluidic culture platform for CNS axonal injury, regeneration and transport. Nat Methods 2(8):599–605. https://doi.org/10.1038/nmeth777
2. Taylor AM, Jeon NL (2010) Micro-scale and microfluidic devices for neurobiology. Curr Opin Neurobiol 20(5):640–647. https://doi.org/10.1016/j.conb.2010.07.011
3. Wang J, Ren L, Li L, Liu W, Zhou J, Yu W et al (2009) Microfluidics: a new cosset for

neurobiology. Lab Chip 9(5):644–652. https://doi.org/10.1039/b813495b
4. Spencer RL, Bland ST (2019) Chapter 5—Hippocampus and hippocampal neurons. In: Fink G (ed) Stress: physiology, biochemistry, and pathology. Academic Press, Cambridge, Massachusetts, pp 57–68
5. Nguyen QT, Son YJ, Sanes JR, Lichtman JW (2000) Nerve terminals form but fail to mature when postsynaptic differentiation is blocked: in vivo analysis using mammalian nerve-muscle chimeras. J Neurosci 20(16):6077–6086. https://doi.org/10.1523/jneurosci.20-16-06077.2000
6. Grimm A, Winter N, Rattay TW, Härtig F, Dammeier NM, Auffenberg E et al (2017) A look inside the nerve—morphology of nerve fascicles in healthy controls and patients with polyneuropathy. Clin Neurophysiol 128(12): 2521–2526. https://doi.org/10.1016/j.clinph.2017.08.022
7. Wang T, Martin S, Nguyen TH, Harper CB, Gormal RS, Martinez-Marmol R et al (2016) Flux of signalling endosomes undergoing axonal retrograde transport is encoded by presynaptic activity and TrkB. Nat Commun 7: 12976. https://doi.org/10.1038/ncomms12976
8. Wang T, Martin S, Papadopulos A, Harper CB, Mavlyutov TA, Niranjan D, et al. (2015) Control of autophagosome axonal retrograde flux by presynaptic activity unveiled using botulinum neurotoxin type a. J Neurosci. 35 (15):6179–6194.
9. Restani L, Giribaldi F, Manich M, Bercsenyi K, Menendez G, Rossetto O et al (2012) Botulinum neurotoxins A and E undergo retrograde axonal transport in primary motor neurons. PLoS Pathog 8(12):e1003087. https://doi.org/10.1371/journal.ppat.1003087
10. Zweifel LS, Kuruvilla R, Ginty DD (2005) Functions and mechanisms of retrograde neurotrophin signalling. Nat Rev Neurosci 6(8): 615–625. https://doi.org/10.1038/nrn1727
11. Harrington AW, Ginty DD (2013) Long-distance retrograde neurotrophic factor signalling in neurons. Nat Rev Neurosci 14(3): 177–187. https://doi.org/10.1038/nrn3253
12. Maday S, Twelvetrees AE, Moughamian AJ, Holzbaur EL (2014) Axonal transport: cargo-specific mechanisms of motility and regulation. Neuron 84(2):292–309. https://doi.org/10.1016/j.neuron.2014.10.019
13. Gowrishankar S, Yuan P, Wu Y, Schrag M, Paradise S, Grutzendler J et al (2015) Massive accumulation of luminal protease-deficient axonal lysosomes at Alzheimer's disease amyloid plaques. Proc Natl Acad Sci U S A 112(28):E3699–E3708. https://doi.org/10.1073/pnas.1510329112
14. Yue Z (2007) Regulation of neuronal autophagy in axon: implication of autophagy in axonal function and dysfunction/degeneration. Autophagy 3(2):139–141. https://doi.org/10.4161/auto.3602
15. Farfel-Becker T, Roney JC, Cheng X-T, Li S, Cuddy SR, Sheng Z-H (2019) Neuronal Soma-derived degradative lysosomes are continuously delivered to distal axons to maintain local degradation capacity. Cell Rep 28(1): 51–64.e4. https://doi.org/10.1016/j.celrep.2019.06.013
16. Lee S, Sato Y, Nixon RA (2011) Lysosomal proteolysis inhibition selectively disrupts axonal transport of degradative organelles and causes an Alzheimer's-like axonal dystrophy. J Neurosci 31(21):7817–7830. https://doi.org/10.1523/jneurosci.6412-10.2011
17. Zhou R, Han B, Xia C, Zhuang X (2019) Membrane-associated periodic skeleton is a signaling platform for RTK transactivation in neurons. Science 365(6456):929–934. https://doi.org/10.1126/science.aaw5937
18. Li H, Yang J, Tian C, Diao M, Wang Q, Zhao S et al (2020) Organized cannabinoid receptor distribution in neurons revealed by super-resolution fluorescence imaging. Nat Commun 11(1):5699. https://doi.org/10.1038/s41467-020-19510-5
19. Joensuu M, Padmanabhan P, Durisic N, Bademosi AT, Cooper-Williams E, Morrow IC et al (2016) Subdiffractional tracking of internalized molecules reveals heterogeneous motion states of synaptic vesicles. J Cell Biol 215(2): 277–292. https://doi.org/10.1083/jcb.201604001
20. Bademosi AT, Lauwers E, Padmanabhan P, Odierna L, Chai YJ, Papadopulos A et al (2017) In vivo single-molecule imaging of syntaxin1A reveals polyphosphoinositide- and activity-dependent trapping in presynaptic nanoclusters. Nat Commun 8:13660. https://doi.org/10.1038/ncomms13660
21. Steketee MB, Moysidis SN, Jin XL, Weinstein JE, Pita-Thomas W, Raju HB et al (2011) Nanoparticle-mediated signaling endosome localization regulates growth cone motility and neurite growth. Proc Natl Acad Sci U S A 108(47):19042–19047. https://doi.org/10.1073/pnas.1019624108
22. Zhou B, Cai Q, Xie Y, Sheng ZH (2012) Snapin recruits dynein to BDNF-TrkB signaling endosomes for retrograde axonal transport and is essential for dendrite growth of cortical neurons. Cell Rep 2(1):42–51. https://doi.org/10.1016/j.celrep.2012.06.010

23. Marlin MC, Li G (2015) Biogenesis and function of the NGF/TrkA signaling endosome. Int Rev Cell Mol Biol 314:239–257. https://doi.org/10.1016/bs.ircmb.2014.10.002
24. Harper CB, Martin S, Nguyen TH, Daniels SJ, Lavidis NA, Popoff MR et al (2011) Dynamin inhibition blocks botulinum neurotoxin type A endocytosis in neurons and delays botulism. J Biol Chem 286(41):35966–35976. https://doi.org/10.1074/jbc.M111.283879
25. Harper CB, Papadopulos A, Martin S, Matthews DR, Morgan GP, Nguyen TH et al (2016) Botulinum neurotoxin type-A enters a non-recycling pool of synaptic vesicles. Sci Rep 6:19654. https://doi.org/10.1038/srep19654
26. Harper CB, Popoff MR, McCluskey A, Robinson PJ, Meunier FA (2013) Targeting membrane trafficking in infection prophylaxis: dynamin inhibitors. Trends Cell Biol 23(2): 90–101. https://doi.org/10.1016/j.tcb.2012.10.007
27. Guedes-Dias P, Holzbaur ELF (2019) Axonal transport: driving synaptic function. Science 366:6462. https://doi.org/10.1126/science.aaw9997
28. Reck-Peterson SL, Redwine WB, Vale RD, Carter AP (2018) The cytoplasmic dynein transport machinery and its many cargoes. Nat Rev Mol Cell Biol 19(6):382–398. https://doi.org/10.1038/s41580-018-0004-3
29. Li P, Bademosi AT, Luo J, Meunier FA (2018) Actin remodeling in regulated exocytosis: toward a mesoscopic view. Trends Cell Biol 28(9):685–697. https://doi.org/10.1016/j.tcb.2018.04.004
30. Pan X, Zhou Y, Hotulainen P, Meunier FA, Wang T (2021) The axonal radial contractility: structural basis underlying a new form of neural plasticity. BioEssays 43:e2100033. https://doi.org/10.1002/bies.202100033
31. Wang T, Li W, Martin S, Papadopulos A, Joensuu M, Liu C et al (2020) Radial contractility of actomyosin rings facilitates axonal trafficking and structural stability. J Cell Biol 219(5):e201902001. https://doi.org/10.1083/jcb.201902001
32. Costa AR, Sousa SC, Pinto-Costa R, Mateus JC, Lopes CD, Costa AC et al (2020) The membrane periodic skeleton is an actomyosin network that regulates axonal diameter and conduction. Elife 9:e55471. https://doi.org/10.7554/eLife.55471
33. Assaife-Lopes N, Sousa VC, Pereira DB, Ribeiro JA, Sebastião AM (2014) Regulation of TrkB receptor translocation to lipid rafts by adenosine A2A receptors and its functional implications for BDNF-induced regulation of synaptic plasticity. Purinergic Signalling 10(2): 251–267. https://doi.org/10.1007/s11302-013-9383-2
34. Joensuu M, Martínez-Mármol R, Padmanabhan P, Glass NR, Durisic N, Pelekanos M et al (2017) Visualizing endocytic recycling and trafficking in live neurons by subdiffractional tracking of internalized molecules. Nat Protoc 12(12):2590–2622. https://doi.org/10.1038/nprot.2017.116
35. Maday S, Wallace KE, Holzbaur ELF (2012) Autophagosomes initiate distally and mature during transport toward the cell soma in primary neurons. J Cell Biol 196(4):407–417. https://doi.org/10.1083/jcb.201106120
36. Betzig E, Patterson GH, Sougrat R, Lindwasser OW, Olenych S, Bonifacino JS et al (2006) Imaging intracellular fluorescent proteins at nanometer resolution. Science 313(5793): 1642–1645. https://doi.org/10.1126/science.1127344
37. Schindelin J, Arganda-Carreras I, Frise E, Kaynig V, Longair M, Pietzsch T et al (2012) Fiji: an open-source platform for biological-image analysis. Nat Methods 9(7): 676–682. https://doi.org/10.1038/nmeth.2019
38. Chaumont F, Dallongeville S, Olivo-Marin J-C (2011) ICY: A new open-source community image processing software
39. Maucort G, Kasula R, Papadopulos A, Nieminen TA, Rubinsztein-Dunlop H, Meunier FA (2014) Mapping organelle motion reveals a vesicular conveyor belt spatially replenishing secretory vesicles in stimulated chromaffin cells. PLoS One 9(1):e87242. https://doi.org/10.1371/journal.pone.0087242
40. Couesnon A, Shimizu T, Popoff MR (2009) Differential entry of botulinum neurotoxin a into neuronal and intestinal cells. Cell Microbiol 11(2):289–308. https://doi.org/10.1111/j.1462-5822.2008.01253.x
41. Olivier N, Keller D, Rajan VS, Gönczy P, Manley S (2013) Simple buffers for 3D STORM microscopy. Biomed Opt Express 4(6): 885–899. https://doi.org/10.1364/BOE.4.000885
42. Szczurek A, Contu F, Hoang A, Dobrucki J, Mai S (2018) Aqueous mounting media increasing tissue translucence improve image quality in structured illumination microscopy of thick biological specimen. Sci Rep 8(1): 13971. https://doi.org/10.1038/s41598-018-32191-x

Chapter 10

Electron Tomographic Methods for Studying Organelles of the Murine Chemical Synapse

Tamara Basta, Garry P. Morgan, Eileen T. O'Toole, Nalini R. Rao, Jeffrey N. Savas, and Michael H. B. Stowell

Abstract

Electron tomography of the chemical synapse provides important architectural information regarding the organization of synaptic organelles including synaptic vesicles, Nissl bodies, and early endosomes. Here, we describe methods for the preparation of select murine brain regions for high-pressure freezing, freeze substitution, and EM tomographic analysis of synaptic structures. The method uses fresh brain slices prepared using a vibratome and biopsy punches to collect specific brain regions of interest suitable for subsequent preservation and EM tomographic imaging.

Key words Synapse, Murine, Brain, Electron microscopy, Electron tomography, High-pressure freezing, Freeze substitution, Subcellular organelles

1 Introduction

The chemical synapse represents the primary cell-cell junction responsible for rapid information transfer in the brain and is essential for fundamental processes such as learning, memory, and consciousness. Since the early light microscopy investigations of Cajal [1], researchers have strived for enhanced techniques for increasingly high resolution methods of microscopy necessary for the investigation of synaptic organization. The use of EM tomography to study the synapse has primarily investigated the neuromuscular junction [2, 3] and the ribbon synapse from saccular hair cells [4]. EM tomographic studies of the synapse and the various methods employed to date have been reviewed [5] and similar methods employing recently developed optogenetic activation have been described [6].

Here, we describe the methods for preparation of murine brain tissue that result in high quality preservation for subsequent tomographic analysis. This embedding method uses freshly prepared

Jingshi Shen (ed.), *Membrane Trafficking: Methods and Protocols*,
Methods in Molecular Biology, vol. 2473, https://doi.org/10.1007/978-1-0716-2209-4_10,

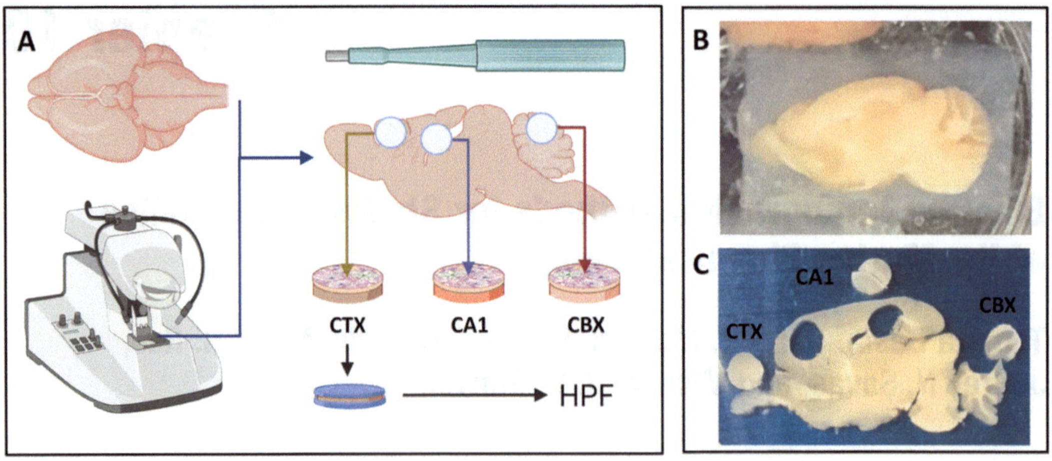

Fig. 1 Schematic flow diagram of methods for sectioning, punching and HPF of murine brain samples. (**a**) Flow diagram of sample preparation steps. Freshly dissected whole murine brains are embedded in agarose on ice and 100 μm sagittal sections are cut using a chilled vibratome. Subsequently a biopsy punch is utilized to excise specific brain regions in diameters (2 mm) suitable for placing in the HPF planchet. (**b**) Agarose embedded murine brain hemisphere on ice. (**c**) Brain section following biopsy punch excision of targeted regions. Brain regions are labelled cortex (CTX), hippocampus CA1 region (CA1), and cerebellum (CBX). (Panel **a** was produced using Biorender (biorender.com))

brain slices collected after vibratome sectioning followed by punch excision of select brain regions for high pressure freezing, freeze substitution, and finally embedding in resin, Fig. 1. These procedures have recently been used to investigate synaptic changes in Alzheimer's disease model mice [7]. This method is advantageous as the preparation of tissue does not require perfusion fixation of the animal and yields well-preserved material for investigation of chemical synapses in situ from various brain regions of interest and is less technically demanding than higher-resolution cryo-ET methods used to study the synapse [8]. The procedures described in this chapter can also be used to study organelle structures in other tissues.

2 Materials

2.1 Vibratome Sectioning of Brain

1. Vibrating-blade Microtome. We use Leica VT1000 S.
2. Double edge razor blades.
3. 2 mm biopsy punches.
4. Phosphate buffer (20 mM Na_2PO_4, 100 mM NaCl, pH 7.4).
5. Agarose 2–4% (w/v) Low melting (*see* **Note 1**).
6. Plastic molds 12 × 12 mm.
7. 45–50 °C Warming plate.

8. Cover Glass No. 1 1/2, 24 × 60 mm for fishing brain slices from the vibratome buffer tray.
9. Clean, soft, superfine art brush for moving brain slices.
10. Super Glue. We use Loctite super glue (catalog #1647358).

2.2 High-Pressure Freezing (HPF), Freeze-Substitution (FS), and Resin Embedding

1. High-Pressure Freezer. We use Wohlwend Compact 02 (Engineering Office M. Wohlwend GmbH CH-9466 Sennwald/Switzerland).
2. Cryoprotectant solution: 2% sucrose, 150 mM D-mannitol, in Dulbecco's Modified Eagle Medium (DMEM).
3. HPF specimen carriers: type A and type B aluminum HPF planchettes.
4. FS cocktail: 1% osmium tetroxide and 0.1% uranyl acetate in EM grade anhydrous acetone. Use extreme care when working with freeze substitution solution (1% osmium tetroxide and 0.1% uranyl acetate in acetone) and always use a fume hood and protective gloves. Osmium tetroxide and uranyl acetate are hazardous materials and must be disposed of according to hazardous disposal regulations.
5. 2.0 mL cryo-vials.
6. FS vessel: medium Styrofoam box that fits in a standard freezer, metal block to hold cryo-vials, and dry ice to fill the box (~3 kg).
7. Resin: 27.5 g of Embed-812 (Epon substitute), 17.0 g of Araldite-502, 55.0 g of dodecenyl succinic anhydride (DDSA), DMP-30 (accelerator, *see* **Note 2**).
8. Embedding capsules, flat bottom (*see* **Note 3**).
9. Embedding capsule holder/rack.
10. 1.0 mL disposable, graduated transfer pipettes.
11. Single edge razor blades.
12. Fine-tipped tweezers/forceps (No. 3 or 5).
13. Dissecting needle or very fine paint brush.
14. Liquid nitrogen (LN).
15. 1-Hexadecene.
16. Colored cryo-vial lid inserts.
17. Lab oven, heated to 60 °C.
18. 15 mL and 50 mL conical tubes.

2.3 Sectioning and Electron Tomography

1. Ultramicrotome.
2. Glass slides.
3. 0.7% formvar in ethylene dichloride.

4. Copper slot grids.
5. 2% uranyl acetate in dH_2O.
6. Reynolds's lead citrate stock solution: 1.33 g lead nitrate, 1.76 g sodium citrate, 8 mL 1 N NaOH, and boiled dH_2O to make 50 mL.
7. 15 nm colloidal gold.
8. Intermediate voltage electron microscope (200–300 kV) equipped with a compustage.
9. High tilt specimen holder.
10. 2k × 2k digital camera.
11. Image acquisition software (SerialEM; http://bio3d.colorado.edu).
12. Tomography reconstruction software (IMOD; http://bio3d.colorado.edu).

3 Methods

3.1 Preparation of Mouse Brain Slices and High-Pressure Freezing (HPF)

1. Sacrifice mice by cervical dislocation (in accordance with IACUC approved protocol), remove brains from the skull and wash in ice-cold PBS.
2. Cut brains into two halves along longitudinal–sagittal axis and place into a soft plastic mold with the sagittal plain facing the bottom.
3. Prepare agarose blocks by pouring a 2% low melting agarose in PBS (~40–42 °C) over the brain hemispheres in the plastic molds, covering the whole hemisphere and 2 mm over the top of the brain.
4. Incubate on ice for 2–3 min, until the is agarose is solidified.
5. Gently remove the agarose block from the plastic mold.
6. Attach agarose blocks onto a specimen disc stage (sagittal plain on the top) using super glue.
7. Prepare a buffer tray containing ice-cold PBS submerged in a cooling bath filled with ice.
8. Collect brain slices with a vibrating-blade Microtome. Set the vibratome parameters: speed of blade (advancement) at 3.5, frequency at 8 and blade angle at 180°. The exact settings may need to be determined empirically. Cut the slices with a thickness of 0.2 μm. When the hippocampus becomes visible, use the mouse brain atlas for help if necessary (http://atlas.brain-map.org/atlas/), switch to 0.1 μm thickness and slice again. Typically, between 4-5 slices of 0.1 μm thickness per hemisphere (6 months old mouse or older) can be collected. Collect

tissue slices as they are cut and use a fine brush to carefully transfer single cut sections from the knife surface to a microscopic slide. Transfer the microscopic slide to a plate containing PBS.

9. Micro-dissect specific brain regions using 2 mm biopsy punches to obtain 2 mm diameter punches of CA1-hippocampus, CBX-cerebellum, and CTX-frontal cortex regions.
10. Quickly submerge biopsy punches in a cryoprotectant solution.
11. Select either the 100 μm or 200 μm well of the type A planchette (whichever more closely matches the tissue thickness) and load the planchette into the HPF specimen rod with the selected depth facing upward (*see* **Note 4**).
12. Fill the planchette well with the cryoprotectant solution (0.2–0.4 μL) and use fine tweezers to hold the planchette in place.
13. Use either a dissecting needle or a very fine tipped paint brush to transfer the biopsy punch into the planchette well. Use the forceps to hold the planchette in place (*see* **Note 5**).
14. Precoat the flat surface of a type B planchette with 1-hexadecene, and then place this flat side down on top of the tissue and type A planchette, making a closed chamber.
15. Close the tip of the HPF specimen rod, transfer/insert into the HPF, freeze the sample (as per the manufacturer's instructions), and rapidly transfer to a LN vessel.
16. Remove the planchettes/tissue from the specimen rod, split apart, and transfer the type A planchette with the tissue into a cryo-vial with a frozen 1.0 mL aliquot of the FS cocktail. Duplicate samples can be added to the same cryo-vial. Use colored lid inserts to distinguish between different samples.
17. Store samples in LN until ready to start FS protocol. Samples can be stored indefinitely in LN.

3.2 FS and Embedding of Frozen Brain Slices

1. Precool metal block in LN, then transfer cryo-vials with samples into the block.
2. Partially fill a Styrofoam box with dry ice (approximately one-third), place metal block with samples into the box, then fill with the remaining dry ice, completely covering the block and cryo-vials.
3. Place lid on box and put into −20 °C freezer.
4. FS process begins as the sample warms to dry ice temperature, −78.5 °C. Hold at this temperature for 72–96 h.

5. Remove box lid and break apart any remaining dry ice, leaving sample in box for 8 h (very slow warm up begins).
6. Remove metal block/samples from box and leave in a −20 °C freezer for 16 h.
7. Transfer metal block/samples to a 4 °C refrigerator and incubate for 4 h.
8. Transfer metal block/samples to room temperature (fume hood) and incubate for 1 h.
9. Working in fume hood, remove FS cocktail from vial using a 1-mL transfer pipette) and replace with 1.5 mL acetone. Incubate for 15 min at room temperature.
10. Repeat **step 9** twice.
11. Fill a shallow watch glass dish or plate with acetone, and transfer planchettes (with samples) into watch glass using forceps.
12. Use a dissecting needle to separate/remove sample from the planchette well (*see* **Note 6**).
13. Fill a cryo-vial with 25% resin–75% acetone mix (*see* **Note 2**) and transfer the sample into the cryo-vial.
14. Transfer cryo-vials to tube rotator wheel, rotate overnight (~16 h) at a low speed at room temperature.
15. Remove the 25% resin–75% acetone mix, replace with 50% resin–50% acetone, and return to rotator for ~8 h.
16. Remove the 50% resin–50% acetone mix, replace with 75% resin–25% acetone, return to rotator, and leave overnight (~16 h).
17. Remove the 75% resin–25% acetone mix, replace with 100% resin, and return to rotator for 4 h.
18. Remove the 100% resin, replace with fresh 100% resin, and return to rotator for 4 h.
19. Remove the 100% resin, replace with fresh 100% resin, return to rotator, and leave overnight (~16 h).
20. Remove the 100% resin, replace with fresh 100% resin+DMP-30, return to rotator, and leave for 4 h.
21. Remove the 100% resin+DMP-30 and replace with fresh 100% resin+DMP-30.
22. Transfer each piece of tissue into embedding capsule (*see* **Note 3**), fill capsule with 100% resin+DMP-30, and close lids.
23. Place capsules into a holder/rack, and transfer to a 60 °C oven to polymerize for 24–48 h.
24. Remove samples/rack from the oven.

3.3 Sectioning and Preparation of Grids for Electron Tomography

1. Remove blocks from rubber molds or capsules. Use a razor blade to trim a trapezoid to include the cross or en face tissue.
2. Place the sample block in a ultramicrotome and collect serial, semithick (250–300 nm) sections onto formvar-coated copper slot grids (*see* **Note 7**).
3. Poststain grids using 2% aqueous uranyl acetate followed by Reynolds's lead citrate (*see* **Note 8**).
4. Affix 15 nm colloidal gold particles to either side of the sections to serve as fiducials for alignment (*see* **Note 9**).

3.4 Electron Tomography

The Boulder EM lab has developed image acquisition (SerialEM) [9] and reconstruction software (IMOD) [10] that is used worldwide for tomography studies. The software is freely available and can be found at http://bio3d.colorado.edu. An emphasis on the details for using this software is described below although other commercial and academic software packages are available.

1. Place grid in a high tilt holder and insert in an intermediate voltage electron microscope (200–300 kV) equipped with a digital camera (*see* **Note 10**).
2. Open SerialEM image acquisition software.
3. Find regions with clear synapses that include the presence of synaptic vesicles and sharp pre- and postsynaptic membranes. An example of this is shown in Fig. 2a.
4. Go to imaging magnification (typically 1–1.2 nm pixel) and adjust eucentric height. Preirradiate the section for a total electron dose of 2000 e/A^2.
5. Use the image acquisition software to automatically collect a series of tilted views from ±60° (or ±70°) at 1.5° increments.
6. After the first tilt series is collected, rotate the grid 90° and collect a second tilt series about the orthogonal axis.
7. Open a terminal window on your computer and type etomo to start the eTomo graphical user interface of the IMOD software package. The eTomo graphical user interface manages the IMOD programs in a user-friendly workflow. In addition, a batch interface in etomo can be used to run the reconstruction steps automatically on multiple data sets. The software will go through a series of steps to first create a prealigned stack that is used for the generation of a fiducial model using the positions of the colloidal gold particles. The fiducial model is then used for aligning the tilt series images. Once the aligned stack is generated, a reconstruction is computed. The program will align and compute reconstructions from both axes and then combine the information to produce a final, dual-axis reconstruction. Figure 2b is an example of a tomographic slice computed from the tilt series shown in Fig. 2a.

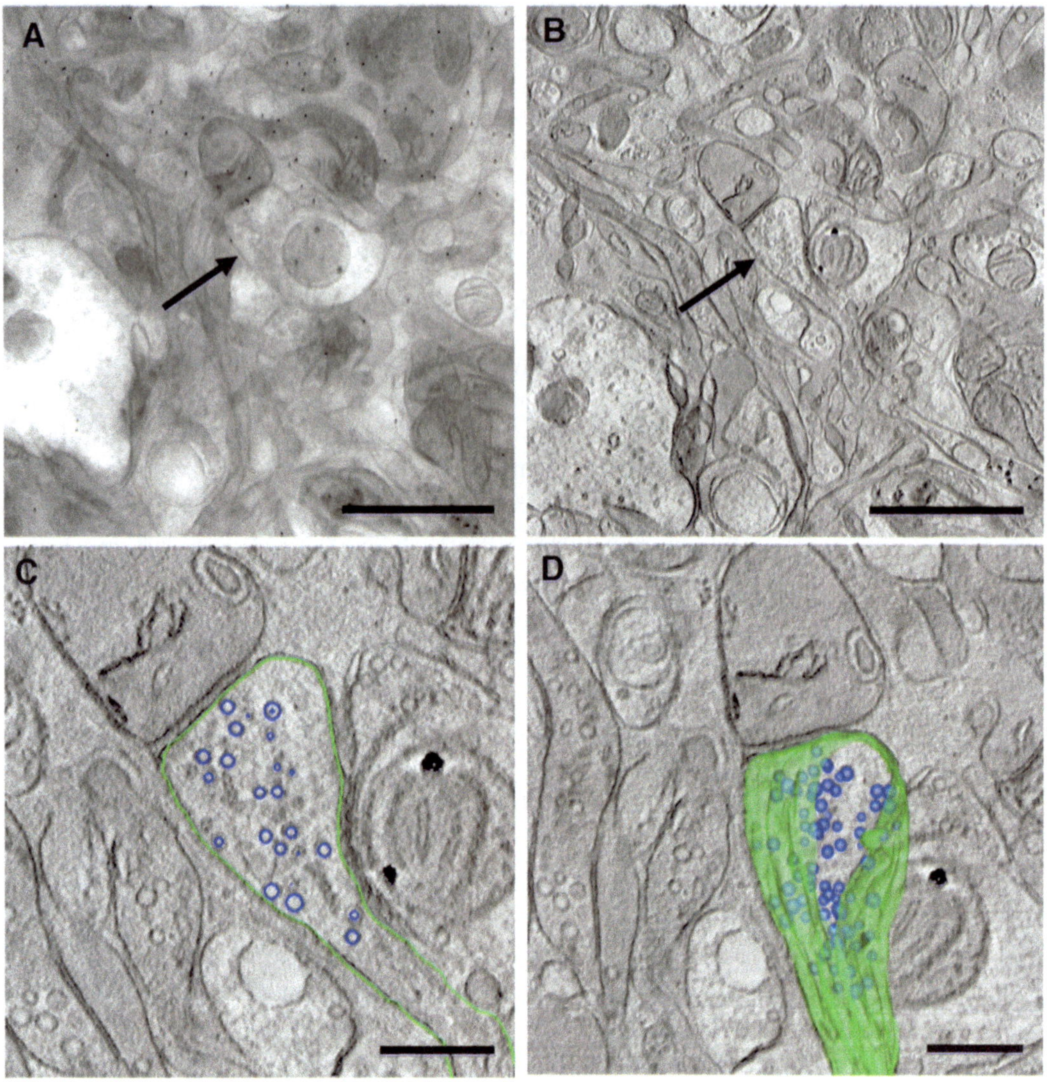

Fig. 2 Reconstruction of a synapse from a mouse cortex using electron tomography. (**a**). Example of a 300 nm thick section used for tilt series collection. Gold particles (dark spheres) are used as alignment markers. Arrow points to a synapse. (**b**) A selected tomographic slice illustrates the detail present in the computed 3D volumes. Arrow points to a synapse modeled in (**c**, **d**). (**c**) The membrane of the synapse is modeled in green and synaptic vesicles (blue) are modeled as a series of scattered points. (**d**) The complete 3D model projected on a selected tomographic slice. Numerous synaptic vesicles (blue spheres) can be seen suspended in the region of the synapse. Bar = 1 μm (**a**, **b**); Bar = 250 nm (**c**, **d**)

8. Open reconstructions using the 3dmod program in the IMOD package and model cell structures by hand, tracing the contours of synapse membranes and modeling synaptic vesicles as scattered points (Fig. 2c). The model contours are then projected in 3D and are subsequently meshed to give a precise rendering of the three-dimensional information (Fig. 2d) (*see* **Note 11**).

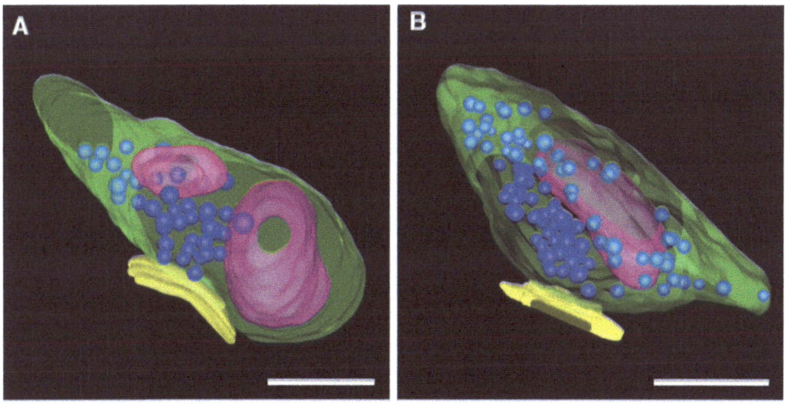

Fig. 3 Colored 3D models created in IMOD from selected synapses. Comparison of synaptic vesicle abundance and organization between amyloid precursor protein (APP) knockin (KI) mouse models of amyloid pathology. (**a**) Representative 3D reconstruction of axon terminal from App KI control Swedish KM670/671NL(App$^{NL/NL}$) mouse model that has no amyloid pathology. (**b**) Representative 3D reconstruction of axon terminal from App KI Arctic (E693G), Iberian, and Swedish mutation App$^{NL-G-F/NL-G-F}$ mouse model that presents with severe amyloid pathology. Comparison of axon terminals shows an increase in abundance of synaptic vesicles (blue) at the presynaptic membrane (green) in App$^{NL-G-F/NL-G-F}$ compared to App$^{NL/NL}$. The postsynaptic density is modeled in yellow and mitochondria in magenta. This figure illustrates the value of electron tomography for understanding the 3D architecture of complex cellular structures. Bar = 250 nm. (This figure was modified from Hark et al. [7])

9. 3D volumes and models from synapses of wild-type and experimentally treated or knock-out animals can then be used for quantification and visualization of structural differences in synaptic organization. *See* Fig. 3, modified from Hark et al. [7] (*see* **Note 12**).

4 Notes

1. It is very important to use agarose with gel strength >1200 g/cm^2 and a low melting point (less than 38 °C).
2. Because resin components are highly viscous, it is best to add components based on weight rather than volume. DMP-30 accelerator is not added to resin initially (for **steps 12–18**) in order to avoid increased resin viscosity, thereby improving the infiltration process. Resin with DMP-30 will slowly polymerize at room temperature. Thus, once mixed, store at 4 °C to slow this reaction. Add 0.15 g DMP-30 per 10 g resin mix in a 15 mL conical tube, or as needed.
3. Flat bottomed embedding capsules allow the tissue biopsy to sit flat to facilitate cutting a face-on section of the entire biopsy with the ultramicrotome. If an edge-on section of the biopsy is preferred, a flat embedding mold can be used instead.

4. Vibratome section thickness can vary to some extent, and it is important that tissue punch closely fits the depth of the well. If the punch appears to be too thick for the 100 μm well, use the 200 μm on the opposite side instead.
5. Be sure to flatten tissue biopsy and carefully center it in the planchette well. If tissue sticks to a needle or brush, try dragging it along the rim of the planchette, or use forceps in a similar fashion. Any tissue sitting outside of the well will be lost. Placing the flat side of the type B planchette onto the type A containing the sample will create a closed chamber for freezing. If tissue is not lying flat, it will be deformed and most likely not usable after freezing.
6. Samples will be somewhat brittle by this step and can easily break apart. Sometimes samples will simply fall out of the planchette well when inverted, but more often the sample will need to be "coaxed" out. To avoid breaking tissue, hold the planchette in place with forceps while using a dissecting needle to scrape around the outer edge of the well (some tissue will be damaged), and then slowly work the needle under the tissue to gently pry it out.
7. It is important to use slot grids to ensure that the tilted views are not impeded by grid bars during data collection. We formvar-coat our grids by dipping a glass slide in 0.7% formvar in ethylene dichloride, allowing the slide to dry, then casting the formvar film onto the surface of a water dish. Copper grids are placed on the floating formvar film and the entire film with grids is picked up using a glass cover slip. The grids are then be removed by tapping around the grid edge with the tip of a fine forceps. Commercially prepared formvar-coated grids can also be purchased.
8. Poststaining is done using 2% aqueous uranyl acetate followed by Reynolds's lead citrate. These are standard electron microscopy poststains, but the staining times are increased due to the thickness of the sections. Stain for 5 min in 2% uranyl acetate, rinse 2× in dH_2O, and stain 4 min in Reynolds's citrate followed by rinsing 2× in dH_2O. Wear protective gloves and dispose of these heavy metal stains according to hazardous waste protocols.
9. Affix colloidal gold to the sections by placing ~20 μL of 15 nm colloidal gold on a square of parafilm and float the grid on top of a drop for several minutes. Blot the excess gold by touching the tip of a triangle of filter paper to the metal edge of the grid, taking care not to touch the formvar. Turn the grid over and repeat on the other side. Rinse grid in a drop of dH_2O and blot.
10. A high tilt specimen holder is required to collect tilt series over a wide angular range (±60 or 70°). For dual-axis tomography, it is useful to load the slot grid with the slot in a vertical

orientation (parallel to the rod) for the first tilt series, then rotating the grid so that the slot is horizontal (perpendicular to the rod) for the second tilt series.

11. Tomographic volumes can be viewed, and cellular structures can be modeled using the 3dmod program in the IMOD software package. Other commercial software package for visualization and modeling are also available (e.g., Amira).
12. Quantitative measurements such as membrane surface area and volume as well as synaptic vesicle size and number can be extracted from model contour data using the IMODINFO program in the IMOD software package. This information is useful for testing ideas about structural relationships among organelles within the synapse in 3D.

Acknowledgments

This work was supported in part by R01 AG061787 to J.N.S. and the MCDB Neurodegenerative Disease Fund to M.H.B.S. N.R. is supported by a Mechanisms of Aging and Dementia T32AG20506. Electron microscopy was done at the University of Colorado, Boulder EM Services Core Facility in the MCDB Department.

References

1. Ramón y Cajal S (1899) Textura del Sistema-Nervioso del Hombre y de los Vertebrados. Moya, Madrid, Spain
2. Harlow ML, Ress D, Stoschek A, Marshall RM, McMahan UJ (2001) The architecture of active zone material at the frog's neuromuscular junction. Nature 409:479–484. https://doi.org/10.1038/35054000
3. Rizzoli SO, Betz WJ (2004) The structural organization of the readily releasable pool of synaptic vesicles. Science 303:2037 2039. https://doi.org/10.1126/science.1094682
4. Lenzi D, von Gersdorff H (2001) Structure suggests function: the case for synaptic ribbons as exocytotic nanomachines. BioEssays 23:831–840. https://doi.org/10.1002/bies.1118
5. Perkins GA, Jackson DR, Spirou GA (2015) Resolving presynaptic structure by electron tomography. Synapse 69:268–282. https://doi.org/10.1002/syn.21813
6. Imig C, López-Murcia FJ, Maus L, García-Plaza IH, Mortensen LS, Schwark M, Schwarze V, Angibaud J, Nägerl UV, Taschenberger H, Brose N, Cooper BH (2020) Ultrastructural imaging of activity-dependent synaptic membrane-trafficking events in cultured brain slices. Neuron 108:843–860.e8. https://doi.org/10.1016/j.neuron.2020.09.004
7. Hark TJ, Rao NR, Castillon C, Basta T, Smukowski S, Bao H, Upadhyay A, Bomba-Warczak E, Nomura T, O'Toole ET, Morgan GP, Ali L, Saito T, Guillermier C, Saido TC, Steinhauser ML, Stowell MHB, Chapman ER, Contractor A, Savas JN (2021) Pulse-chase proteomics of the app Knockin mouse models of Alzheimer's disease reveals that synaptic dysfunction originates in presynaptic terminals. Cell Syst 12(2):141–158.e9. https://doi.org/10.1016/j.cels.2020.11.007
8. Fernández-Busnadiego R, Zuber B, Maurer UE, Cyrklaff M, Baumeister W, Lučić V (2010) Quantitative analysis of the native presynaptic cytomatrix by cryoelectron tomography. J Cell Biol 188:145 156. https://doi.org/10.1083/jcb.200908082
9. Mastronarde DN (2005) Automated electron microscope tomography using robust prediction of specimen movements. J Struct Biol 152:36–51. https://doi.org/10.1016/j.jsb.2005.07.007
10. Kremer JR, Mastronarde DN, McIntosh JR (1996) Computer visualization of three-dimensional image data using IMOD. J Struct Biol 116:71–76. https://doi.org/10.1006/jsbi.1996.0013

Chapter 11

In Vitro Reconstitution Studies of SNAREs and Their Regulators Mediating GLUT4 Vesicle Fusion

Yinghui Liu, Ruyue He, Min Zhu, and Haijia Yu

Abstract

The GLUT4 vesicle fusion is mediated by soluble N-ethylmaleimide-sensitive factor attachment protein receptors (SNAREs) and a variety of regulatory proteins. For example, synip and tomosyn negatively regulate GLUT4 SNARE-mediated membrane fusion. Here we describe in vitro reconstituted assays to determine the molecular mechanisms of SNAREs, synip, and tomosyn. These methods can also be extended to the studies of other types of membrane fusion events.

Key words GLUT4, Synip, Tomosyn, Vesicle fusion, Liposome reconstitution, Lipid mixing, Content mixing, SNARE assembly, Docking

1 Introduction

The insulin-stimulated exocytosis of glucose transporter 4 (GLUT4) plays a central role in maintaining blood glucose homeostasis. The fusion of GLUT4-containing vesicles with the plasma membrane is the last key step of GLUT4 exocytosis [1]. It is well known that the core machinery mediating intracellular vesicle fusion is soluble *N*-ethylmaleimide-sensitive factor attachment protein receptors (SNAREs) [2–4]. The vesicular (v-) SNARE pairs with the target membrane (t-) SNAREs to form a four-helix *trans*-SNARE complex, which provides energy to drive membrane fusion [3–13]. In GLUT4 vesicle fusion, syntaxin-4 and SNAP-23 constitute the t-SNAREs, whereas VAMP2/synaptobrevin serves as the primary v-SNARE protein [14–17].

In response to insulin signals, the SNARE assembly is highly regulated in GLUT4 exocytosis. A group of regulators are required to interact with SNAREs to achieve the spatial and temporal regulation of vesicle fusion [18–20]. Synip and tomosyn are two SNARE regulators, negatively regulating GLUT4 vesicle fusion [21, 22]. Both regulators bind to the syntaxin-4 monomer and

Jingshi Shen (ed.), *Membrane Trafficking: Methods and Protocols*,
Methods in Molecular Biology, vol. 2473, https://doi.org/10.1007/978-1-0716-2209-4_11,

inhibit GLUT4 vesicle fusion [21, 22]. The fusion brake could be due to the inhibition of either the t-SNARE assembly or the ternary complex formation. It is challenging to delineate the molecular mechanism of a negative regulator in the complex membrane trafficking process in vivo.

By the in vitro reconstituted liposome fusion assays, we identified the inhibitory roles of synip and tomosyn in GLUT4 SNARE-dependent membrane fusion reactions [23–25]. However, the results generated from lipid mixing or content mixing assays still could not tell in which step these inhibitory factors block the fusion. We developed the sequential SNARE assembly assay and identified both regulators have little effect on the formation of the binary t-SNARE complex. Instead, they interact with the t-SNARE complexes stoichiometrically as clamps to exclude the v-SNARE [23, 24]. The assembly of the ternary SNARE complex bridges the opposed membranes and membrane docking. Using the SNARE-dependent liposome docking assay, we demonstrated that these inhibitory regulators efficiently block membrane docking caused by the t-SNARE and v-SNARE assembly [23, 24].

In the in vitro defined system, we reconstituted SNARE proteins into the membrane by detergent dilution. The reactions designed in a membrane environment mimic the physiological situations inside the cell. The functional reconstitution system enabled us to establish fundamental features of these regulatory proteins in GLUT4 SNARE-dependent vesicle fusion. Here we review these protocols of protein generation, sequential SNARE assembly, lipid mixing, content mixing, and the docking assay for analysis of GLUT4 vesicle fusion. We anticipate these assays will pave the way to study other regulated vesicle fusion systems beyond GLUT4 exocytosis, such as cytolytic granule secretion or insulin secretion [26, 27].

2 Materials

2.1 DNA Plasmids

1. Untagged syntaxin-4: gene encoding untagged rat syntaxin-4 is cloned into pET28a vector [23, 24].
2. His_6-SNAP-23: the His_6-tagged mouse SNAP-23 sequence is cloned into pET15b vector [23, 24].
3. His_6-SUMO-syntaxin-4: the rat *syntaxin-4* gene is cloned into a pET28a-based SUMO vector [23, 24].
4. His_6-SUMO-VAMP2: the mouse *Vamp2* gene is cloned into a pET28a-based SUMO vector [23, 24].
5. GST-VAMP2 CD (cytosolic domain): the mouse *Vamp2* gene (encoding residues 1–95) is cloned into a pGEX4T-1 vector [23, 24].

6. His_6-SUMO-synip: the mouse *synip* gene is cloned into a pET28a-based SUMO vector [23].
7. His_6-MBP-tomosyn-1: the human tomosyn-1 gene is cloned into a baculovirus transfer vector pFastBacMBP [24]. The construct encodes a His_6-MBP-tomosyn-1 fusion protein with a Tobacco Etch Virus (TEV) cleavage site between the His_6-MBP tag and tomosyn-1.

2.2 Protein Expression and Purification

1. BL21-Gold (DE3) competent cells.
2. Rosetta (DE3) competent cells.
3. DH10Bac competent cells.
4. Luria-Bertani (LB) medium: Dissolve 10 g Tryptone, 5 g yeast extract and 10 g NaCl in ~950 mL water. Adjust the pH to 7.0 using NaOH. Adjust the final volume to 1 L using water, autoclave for 121 °C, 30 min.
5. 100 mg/mL ampicillin: Dissolve 5 g ampicillin in 50 mL water. Filter-sterilize the solution through a 0.22 μm syringe filter.
6. 100 mg/mL kanamycin: Dissolve 5 g kanamycin in 50 mL water. Filter-sterilize the solution through a 0.22 μm syringe filter.
7. LB agar plate: Add 15 g agar in 1 L LB medium. Autoclave for 121 °C, 30 min. Cool down the autoclaved medium, add antibiotics to prepare agar plates. For BL21-Gold (DE3) competent cells, add 1 mL 100 mg/mL ampicillin or/and 1 mL 100 mg/mL kanamycin. For Rosetta (DE3) competent cells, add 1 mL 100 mg/mL kanamycin and 500 μL 50 mg/mL chloramphenicol. For DH10Bac competent cells, add 500 μL 100 mg/mL kanamycin, 1 mL 7 mg/mL gentamicin, 1 mL 10 mg/mL tetracycline, 5 mL 20 mg/mL Bluo-Gal, and 168 μL 1 M isopropyl β-D-1-thiogalactopyranosid (IPTG).
8. 0.5 M Tris(2-carboxyethyl)phosphine (TCEP): Dissolve 7.15 g TCEP in 50 mL water. Filter sterilize the solution through a 0.22 μm syringe filter.
9. 1 M isopropyl β-D-1-thiogalactopyranoside (IPTG): Dissolve 11.92 g IPTG in 50 mL water. Filter-sterilize the solution through a 0.22 μm syringe filter.
10. 20% Triton X-100: Dilute 10 mL Triton X-100 with water to a final volume of 50 mL.
11. Bacterial cell lysis buffer: 25 mM HEPES (pH 7.4), 400 mM KCl, 10% glycerol (v/v), 20 mM imidazole, 2 mM β-mercaptoethanol (freshly added), and protease inhibitor cocktail (Roche, one tablet per 100 mL buffer). Dissolve 5.96 g HEPES, 29.82 g KCl, and 1.36 g imidazole in ~700 mL water. Add 100 mL glycerol and adjust the pH to 7.4 using HCl. Add water to adjust the final volume to 1 L (*see* **Note 1**).

12. Insect cell lysis buffer: 25 mM HEPES (pH 7.4), 400 mM KCl, 10% glycerol, 20 mM imidazole, 1% Triton, 2 mM β-mercaptoethanol (freshly added), and protease inhibitor cocktail (one tablet per 100 mL buffer). Dissolve 5.96 g HEPES, 29.82 g KCl, and 1.36 g imidazole in ~700 mL water. Add 100 mL glycerol and adjust the pH to 7.4 using HCl. Add 50 mL 20% Triton X-100 and use water to adjust the volume to 1 L (*see* **Note 1**).
13. Membrane protein storage buffer: 25 mM HEPES (pH 7.4), 400 mM KCl, 1% *n*-octyl-β-D-glucoside (OG), 10% glycerol, and 0.5 mM TCEP (freshly added). Dissolve 5.96 g HEPES, 29.82 g KCl and 10 g OG in ~700 mL water. Add 100 mL glycerol and adjust the pH to 7.4 using HCl. Adjust the final volume to 1 L with water (*see* **Note 1**).
14. Soluble protein storage buffer: 25 mM HEPES (pH 7.4), 150 mM KCl, 10% glycerol, and 0.5 mM TCEP. Dissolve 5.96 g HEPES and 11.18 g KCl in ~700 mL water. Add 100 mL glycerol and adjust the pH to 7.4 using HCl. Add water to achieve a final volume of 1 L (*see* **Note 1**).
15. Wash buffer A (for membrane proteins): 25 mM HEPES (pH 7.4), 400 mM KCl, 10% glycerol, 20 mM imidazole, 1% OG, 0.5 mM TCEP (freshly added). Dissolve 5.96 g HEPES, 29.82 g KCl, and 1.36 g imidazole in ~700 mL water. Add 100 mL glycerol and adjust the pH to 7.4 using HCl. Add 100 mL 10% OG and use water to adjust the volume to 1 L (*see* **Note 1**).
16. Wash buffer B (for membrane proteins): 25 mM HEPES (pH 7.4), 400 mM KCl, 10% glycerol, 60 mM imidazole, 1% OG, 0.5 mM TCEP (freshly added). Dissolve 5.96 g HEPES, 29.82 g KCl, and 4.08 g imidazole in ~700 mL water. Add 100 mL glycerol and adjust the pH to 7.4 using HCl. Add 100 mL 10% OG and use water to adjust the volume to 1 L (*see* **Note 1**).
17. Elution buffer A (for membrane proteins): 25 mM HEPES (pH 7.4), 400 mM KCl, 10% glycerol, 300 mM imidazole, 1% OG, 0.5 mM TCEP (freshly added). Dissolve 5.96 g HEPES, 29.82 g KCl, and 20.4 g imidazole in ~700 mL water. Add 100 mL glycerol and adjust the pH to 7.4 using HCl. Add 100 mL 10% OG and use water to adjust the volume to 1 L (*see* **Note 1**).
18. Wash buffer C (for soluble proteins): 25 mM HEPES (pH 7.4), 150 mM KCl, 10% glycerol, 20 mM imidazole, 0.5 mM TCEP (freshly added). Dissolve 5.96 g HEPES, 11.18 g KCl, and 1.36 g imidazole in ~700 mL water. Add 100 mL glycerol and adjust the pH to 7.4 using HCl. Use water to adjust the volume to 1 L (*see* **Note 1**).

19. Wash buffer D (for soluble proteins): 25 mM HEPES (pH 7.4), 150 mM KCl, 10% glycerol, 60 mM imidazole, 0.5 mM TCEP (freshly added). Dissolve 5.96 g HEPES, 11.18 g KCl, and 4.08 g imidazole in ~700 mL water. Add 100 mL glycerol and adjust the pH to 7.4 using HCl. Use water to adjust the volume to 1 L (*see* **Note 1**).
20. Elution buffer B (for soluble proteins): 25 mM HEPES (pH 7.4), 150 mM KCl, 10% glycerol, 300 mM imidazole, 0.5 mM TCEP (freshly added). Dissolve 5.96 g HEPES, 11.18 g KCl, and 20.4 g imidazole in ~700 mL water. Add 100 mL glycerol and adjust the pH to 7.4 using HCl. Use water to adjust the volume to 1 L (*see* **Note 1**).
21. Elution buffer C (for soluble GST fusion proteins): 25 mM HEPES (pH 7.4), 150 mM KCl, 10% glycerol, 10 mM reduced glutathione, 0.5 mM TCEP (freshly added). Dissolve 0.6 g HEPES, 1.12 g KCl, and 0.31 g Glutathione in ~70 mL water. Add 10 mL glycerol and adjust the pH to 7.4 using HCl. Use water to adjust the volume to 1 L (*see* **Note 1**).

2.3 Liposomes

1. 15 mM unlabeled lipid mixture: Dissolve all lipids in chloroform. Prepare the unlabeled lipid mixture by mixing 1-palmitoyl-2-oleoyl-*sn*-glycero-3-phosphocholine (POPC), 1-palmitoyl-2-oleoyl-*sn*-glycero-3-phosphoethanolamine (POPE), 1-palmitoyl-2-oleoyl-*sn*-glycero-3-phosphoserine (POPS), and cholesterol at a molar ratio of 60:20:10:10.
2. 3 mM fluorescence-labeled lipid mixture: Dissolve all lipids in chloroform. Prepare the fluorescence-labeled lipid mixture by mixing POPC, POPE, POPS, cholesterol, (*N*-(7-nitro-2,1,3-benzoxadiazole-4-yl)-1,2-dipalmitoyl phosphatidylethanolamine (NBD-DPPE), and *N*-(Lissamine rhodamine B sulfonyl)-1,2-dipalmitoyl phosphatidylethanolamine (rhodamine-DPPE) at a molar ratio of 60:17:10:10:1.5:1.5.
3. 15 mM Biotin-labeled lipid mixture: Dissolve all lipids in chloroform. Prepare the Biotin-labeled lipid mixture by mixing POPC, POPE, POPS, cholesterol, and 1,2-distearoyl-*sn*-glycero-3-phosphoethanolamine-*N*-[biotinyl(polyethyleneglycol)-2000] at a molar ratio of 60:18:10:10:2.
4. Liposome reconstitution buffer: 25 mM HEPES (pH 7.4), 100 mM KCl, 10% Glycerol, and 0.5 mM TCEP. Dissolve 5.96 g HEPES and 7.45 g KCl in ~700 mL water. Add 100 mL glycerol and adjust the pH to 7.4 using HCl. Use water to adjust the volume to 1 L (*see* **Note 1**).
5. 30% and 80% Nycodenz solution (w/v). Dissolve 0.298 g HEPES, 0.372 g KCl, and 15 g (30%) or 40 g (80%) Nycodenz in ~40 mL water. Add 5 mL glycerol and adjust the pH to 7.4 using HCl. Use water to adjust the volume to 50 mL.

3 Methods

3.1 Recombinant Protein Expression and Purification

3.1.1 Binary t-SNAREs

1. Transform the untagged syntaxin-4 and His-SNAP-23 expression plasmids into BL21-Gold (DE3) competent cells and plate the cells on an LB agar plate with 100 μg/mL kanamycin and 100 μg/mL ampicillin.
2. Pick a single colony and inoculate it into 100 mL LB preculture with 100 μg/mL kanamycin and 100 μg/mL ampicillin. Grow the preculture overnight at 37 °C in a shaker set at 220 rpm.
3. Add 50 mL of the preculture into 2 L LB culture with 100 μg/mL kanamycin and 100 μg/mL ampicillin. Grow the culture at 37 °C in a shaker set at 220 rpm.
4. When the OD_{600} of the culture reaches 0.6, add IPTG to the final concentration of 0.2 mM.
5. After overnight induction at 16 °C, centrifuge the culture at 3600 × *g* for 20 min.
6. Resuspend the bacteria pellet in 40 mL bacterial cell lysis buffer.
7. Add 10 mL of 20% Triton (v/v in water) to the lysate and mix well.
8. Lyse the cells using an AVESTIN-EmulsiFlex-C3 homogenizer.
9. Centrifuge the lysate for 30 min at 4 °C at 180,000 × *g* in a Beckman ultracentrifuge (Type 45 Ti rotor).
10. Prepare Ni-NTA beads in bacterial cell lysis buffer. Add 0.5 mL Ni-NTA beads to the supernatant and rotate the mixture at 4 °C for 2 h.
11. Spin down the Ni-NTA beads at 500 × *g* for 2 min. Wash the beads with 5 mL wash buffer A and wash buffer B, respectively.
12. Collect the protein by adding the elution buffer A. Use SDS-PAGE and Bradford assay to check the purified protein. Aliquot and store at −70 °C.

3.1.2 His_6-SUMO-Syntaxin-4

1. Transform the His_6-SUMO-Syntaxin-4 plasmid into BL21-Gold (DE3) competent cells and plate the cells on an LB agar plate with 100 μg/mL kanamycin.
2. Pick a single colony and inoculate it into 100 mL LB preculture with 100 μg/mL kanamycin. Grow the preculture overnight at 37 °C in a shaker set at 220 rpm.
3. Add 50 mL of the preculture into 2 L LB culture with 100 μg/mL kanamycin. Grow the culture at 37 °C in a shaker set at 220 rpm.

4. When the OD_{600} of the culture reaches 0.6, add IPTG to the final concentration of 1 mM.
5. After another 3.5 h at 37 °C, centrifuge the culture at 3600 × *g* for 20 min.
6. Purify the proteins as described in Subheading 3.1.1. Use SDS-PAGE and Bradford assay to check the protein. Aliquot and store at −70 °C.

3.1.3 VAMP2

1. Transform the His_6-SUMO-VAMP2 plasmid into BL21-Gold (DE3) competent cells and plate the cells on an LB agar plate with 100 μg/mL kanamycin. Express the protein in a similar way as His_6-SUMO-Syntaxin-4.
2. Purify the proteins following the same procedure as His_6-SUMO-Syntaxin-4 except for the protein collection step. After Washing the beads with 5 mL wash buffer A and wash buffer B, add 60 μg His_6-tagged SUMO protease to the protein-bound Ni-NTA beads in 1 mL of the membrane protein storage buffer.
3. After incubation at 4 °C overnight, collect the cleaved VAMP2 proteins in solution. Use SDS-PAGE and Bradford assay to check the purified protein. Aliquot and store at −70 °C.

3.1.4 His_6-SNAP-23

1. Transform the His_6-SNAP-23 plasmid into BL21-Gold (DE3) competent cells and plate the cells on an LB agar plate with 100 μg/mL ampicillin.
2. Pick a single colony and inoculate it into 100 mL LB preculture with 100 μg/mL ampicillin. Grow the preculture overnight at 37 °C in a shaker set at 220 rpm.
3. The next morning, add 50 mL of the preculture into 2 L LB culture with 100 μg/mL ampicillin. Grow the culture at 37 °C in a shaker set at 220 rpm.
4. When the OD_{600} of the culture reaches 0.6, add IPTG to the final concentration of 1 mM.
5. After another 2 h at 37 °C, centrifuge the culture at 3600 × *g* for 20 min.
6. Resuspend the bacteria pellet in 50 mL bacterial cell lysis buffer. Lyse the cells using a homogenizer.
7. Centrifuge the lysate for 30 min at 4 °C at 180,000 × *g* in a Beckman ultracentrifuge (Type 45 Ti rotor).
8. Prepare Ni-NTA beads in bacterial cell lysis buffer. Add 0.5 mL Ni-NTA beads to the supernatant and rotate the mixture at 4 °C for 2 h.
9. Spin down the Ni-NTA beads at 500 × *g* for 2 min. Wash the beads with 5 mL wash buffer C and wash buffer D, respectively.

10. Collect the His_6-SNAP-23 proteins by adding the elution buffer B. Dialyze the protein (molecular weight cutoff: 6–8 kDa) overnight at 4 °C against 2 L soluble protein storage buffer.
11. Check the protein using SDS-PAGE and Bradford assay. Aliquot the purified proteins and store them at −70 °C.

3.1.5 GST-VAMP2 CD

1. Transform GST-VAMP2 CD plasmid into BL21-Gold (DE3) competent cells and plate the cells on an LB agar plate with 100 μg/mL ampicillin. Express the protein in a similar way as His_6-SNAP-23.
2. Purify the proteins following the same procedure as His-SNAP-23 except for the protein collection step. After Washing the beads with 5 mL wash buffer C, add elution buffer C to collect the GST tagged VAMP2 CD proteins.
3. Dialyze the protein (molecular weight cutoff: 6–8 kDa) overnight at 4 °C against 2 L soluble protein storage buffer. Check the protein using SDS-PAGE and Bradford assay. Aliquot the purified proteins and store them at −70 °C.

3.1.6 Synip

1. Transform the His_6-SUMO-synip plasmid into Rosetta (DE3) competent cells and plate the cells on an LB agar plate with 100 μg/mL kanamycin and 25 μg/mL chloramphenicol. Express the protein in a similar way as binary t-SNAREs in Subheading 3.1.1 (*see* **Note 2**).
2. Purify synip following the same procedure as His_6-SNAP-23.
3. Add 40 μg SUMO protease to the eluted proteins and dialyze (molecular weight cutoff: 6–8 kDa) at 4 °C overnight against 2 L soluble protein storage buffer.
4. Use gel filtration chromatography to remove His_6–SUMO tags from synip proteins. Check the protein using SDS-PAGE and Bradford assay. Aliquot the purified proteins and store them at −70 °C.

3.1.7 Tomosyn-1

1. Transform the His_6-MBP-tomosyn-1 plasmid into DH10Bac competent cells and plate the cells on an LB plate with 50 μg/mL kanamycin, 7 μg/mL gentamicin, 10 μg/mL tetracycline, 100 μg/mL Bluo-Gal, and 40 μg/mL IPTG.
2. After 36–48 h, pick one white colony and inoculate into 100 mL LB culture supplemented with 50 μg/mL kanamycin, 7 μg/mL gentamicin, and 10 μg/mL tetracycline.
3. Prepare recombinant Bacmid DNA using a plasmid DNA maxiprep kit.
4. In a 6-well culture plate, seed 0.9 million *Sf9* cells in 2 mL of Sf-900 III SFM media without antibiotics and allow the cells to attach for at least 1 h.

5. Mix 0.5 μg Bacmid DNA and 8 μL Cellfectin II transfection reagent in 200 μL Sf-900 III media without antibiotics and incubate at room temperature for 20 min.
6. Replace with 0.8 mL fresh Sf-900 III media without antibiotics. Add the transfection mixture and keep the cells at 27 °C for 5 h.
7. Remove the media. Add 2 mL fresh media and continue to grow the cells for 96 h.
8. Harvest the P1 virus by centrifugation at 200 × *g* for 5 min. Add 1 mL P1 virus to 5 × 10^7 *Sf9* cells in 30 mL Sf-900 III media supplemented with 100 μg/mL penicillin and 100 μg/mL streptomycin to amplify the P2 virus stock.
9. Infect *Sf9* cells by P2 virus to get the P3 virus stock, use 10 mL P3 virus to infect 2 × 10^9 *Sf9* cells in 1 L Sf-900 III media supplemented with 100 μg/mL penicillin and 100 μg/mL streptomycin.
10. Grow the infected *Sf9* cells for 48 h in a shaker set at 130 rpm. Centrifuge the cell culture at 200 × *g* for 5 min.
11. Use 50 mL insect cell lysis buffer to suspend the pellet and rotate at 4 °C for 1 h.
12. Centrifuge the cell lysate at 40,000 × *g* for 30 min in a Beckman ultracentrifuge (Type 45 Ti rotor).
13. Purify tomosyn-1 protein following the same procedure as His_6-SNAP-23 and synip.
14. Add 60 μg TEV protease to the eluted protein and dialyze (molecular weight cutoff: 6–8 kDa) at 4 °C overnight against 2 L soluble protein storage buffer.
15. Use gel filtration chromatography to remove His_6-MBP tags from tomosyn-1 proteins. Check the protein using SDS-PAGE and Bradford assay. Aliquot the purified proteins and store them at −70 °C.

3.2 Reconstitution of t-SNARE Proteins into Unlabeled or Biotin-Labeled Liposomes

1. Add 100 μL unlabeled or Biotin-labeled lipid mixtures to 12 × 75 mm glass (*see* **Note 3**).
2. Dry the lipids by Argon air flow for 15 min, then continue to dry them using vacuum pump for one more hour.
3. Add 500 μL t-SNARE proteins (125 μg t-SNAREs or 100 μg His_6-SUMO-Syntaxin-4) to the dried unlabeled lipid film. Include 10 μg SUMO protease in the samples of His_6-SUMO-Syntaxin-4. Add 500 μL t-SNAREs (125 μg t-SNAREs) to the dried Biotin-labeled lipid film.
4. Vortex the samples continuously at room temperature for 15 min to dissolve the lipid films.

5. Add 1 mL liposome reconstitution buffer to the sample during shaking on a vortex mixer.
6. Dialyze the liposome samples overnight at 4 °C in dialysis tubes (molecular weight cutoff: 12–14 kDa) against 2 L reconstitution buffer.
7. Add equal volumes of 80% Nycodenz solution to the liposome samples and mix well.
8. Transfer the t-SNARE liposomes into an 11 × 60 mm ultracentrifuge tube. Sequentially overlay the sample with 750 μL 30% Nycodenz solution and 250 μL liposome reconstitution buffer without glycerol.
9. Centrifuge the t-SNARE liposome samples for 3 h and 40 min at 400,000 × *g* in a Beckman ultracentrifuge (SW60 Ti rotor). Collect 400 μL t-SNARE liposomes from the 0/30% Nycodenz interface.

3.3 Reconstitution of v-SNARE Proteins into Liposomes for Lipid Mixing and Docking Assay

1. Add 100 μL labeled lipid mixtures to 10 × 75 mm glass tubes.
2. Dry the lipids by Argon air flow for 15 min, then continue to dry them using vacuum pump for one more hour.
3. Add 100 μL v-SNARE protein (37.5 μg VAMP2) to the dried labeled lipid film.
4. Vortex the samples continuously at room temperature for 15 min to dissolve the lipid films.
5. Add 200 μL liposome reconstitution buffer to the v-SNARE sample during shaking on a vortex mixer.
6. Dialyze the liposome samples overnight at 4 °C in dialysis tubes (molecular weight cutoff: 12–14 kDa) against 2 L reconstitution buffer.
7. Add equal volumes of 80% Nycodenz solution to the liposome samples and mix well.
8. Transfer the v-SNARE liposomes into a 5 × 41 mm ultracentrifuge tube. Sequentially overlay the sample with 300 μL 30% Nycodenz solution and 50 μL liposome reconstitution buffer without glycerol.
9. Centrifuge the v-SNARE liposome samples for 4 h at 320,000 × *g* in a Beckman ultracentrifuge (SW55 Ti rotor). Collect 75 μL v-SNARE liposome samples from the 0/30% Nycodenz interface.

3.4 Reconstitution of SNARE Proteins into Liposomes for Content Mixing Assays

1. Prepare unlabeled binary t-SNARE liposomes as described in Subheading 3.2. To prepare t- or v-liposomes containing sulforhodamine B, add 100 μL unlabeled lipid mixtures to 12 × 75 mm glass tubes and 20 μL unlabeled lipid mixtures to 10 × 75 mm glass tubes.

2. Dry the lipids by Argon air flow for 15 min, then continue to dry them using vacuum pump for one more hour.
3. Add 500 μL t-SNARE proteins (125 μg t-SNAREs) or 100 μL v-SNARE protein (37.5 μg VAMP2) to the dried unlabeled lipid film, respectively.
4. Vortex the samples continuously at room temperature for 15 min to dissolve the lipid films.
5. Add 1 mL liposome reconstitution buffer containing 75 mM sulforhodamine B to the t-SNARE sample during shaking on a vortex mixer. Add 200 μL liposome reconstitution buffer containing 75 mM sulforhodamine B to the v-SNARE sample during shaking on a vortex mixer.
6. Dialyze the samples and isolate liposomes by centrifuge as described in Subheadings 3.2 and 3.3.
7. Use the sulforhodamine B-loaded liposomes immediately after preparation.

3.5 t-SNARE Assembly Assay

1. Mix 50 μL Syntaxin-4 liposomes with soluble SNAP-23 proteins in the presence or absence of SNARE regulators. Rotate the samples at 4 °C for 1 h.
2. Add equal volumes of 80% Nycodenz solution to the liposome samples and mix well.
3. Transfer the sample into a 5 × 41 mm ultracentrifuge tube. Sequentially overlay the sample with 200 μL each of 35% and 30% Nycodenz solution, add 20 μL liposome reconstitution buffer without glycerol on the top.
4. Centrifuge the samples for 4 h at 320,000 × *g* in a Beckman ultracentrifuge (SW55 Ti rotor). Collect 40 μL floated liposome samples from the 0/30% Nycodenz interface. Analyze the samples using SDS-PAGE (Fig. 1).

3.6 Ternary SNARE Assembly Assay

1. Mix 50 μL binary t-SNARE liposomes with soluble GST-VAMP2 CD proteins in the presence or absence of SNARE regulators. Rotate the samples at 4 °C for 1 h (*see* **Note 4**).
2. Add equal volumes of 80% Nycodenz solution to the liposome samples and mix well.
3. Transfer the sample into a 5 × 41 mm ultracentrifuge tube. Sequentially overlay the sample with 200 μL each of 35% and 30% Nycodenz solution, add 20 μL liposome reconstitution buffer without glycerol on the top.
4. Centrifuge the samples for 4 h at 320,000 × *g* in a Beckman ultracentrifuge (SW55 Ti rotor). Collect 40 μL floated liposome samples from the 0/30% Nycodenz interface. Analyze the samples using SDS-PAGE (Fig. 1).

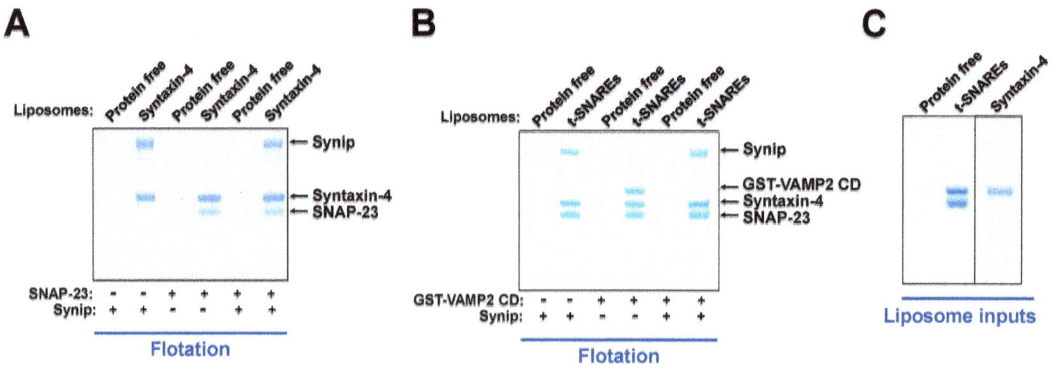

Fig. 1 Synip inhibits the formation of the ternary SNARE complex but not the formation of the t-SNARE complex. (**a**) Coomassie Blue–stained SDS-PAGE gel showing the binding of synip to syntaxin-4 liposomes does not block the t-SNARE assembly. Syntaxin-4 liposomes were incubated with or without synip at 4 °C for 1 h before SNAP-23 was added. After another 1 h, the samples were floated up on a Nycodenz gradient. Protein-free liposomes were used in the control experiments. (**b**) Coomassie Blue–stained SDS-PAGE gel showing the binding of synip to t-SNARE liposomes efficiently inhibits the formation of the ternary SNARE complex. The binary t-SNARE liposomes containing the heterodimer of syntaxin-4 and SNAP-23 were incubated with or without synip at 4 °C for 1 h, before GST-VAMP2 CD was added. After another 1 h, the samples were floated up on a Nycodenz gradient. Protein-free liposomes were used in the control experiments. (**c**) Coomassie Blue–stained SDS-PAGE gels showing the input liposomes. (Adapted with permission from Ref. 23. Copyright The American Society for Biochemistry and Molecular Biology, Inc. (2013))

3.7 Liposome Lipid Mixing Assay

1. Mix 45 μL unlabeled binary t-SNARE liposomes and 5 μL fluorescence-labeled v-SNARE liposomes in the presence or absence of 5 μM SNARE regulators. In the control experiments, add 20 μM GST-VAMP2 CD to block SNARE zippering.
2. Load the mixed samples into a 96-well microplate preheated to 37 °C to initiate reactions.
3. Measure the NBD fluorescence (excitation: 460 nm; emission: 538 nm) of the samples every 2 min in a BioTek Synergy HT microplate reader.
4. At the end of the reaction, add 10 μL 10% CHAPS to each sample. Continue to measure NBD fluorescence every 2 min for another 20 min.
5. Normalize the lipid mixing data as the percentage of maximum fluorescence change after CHAPS treatment (Fig. 2).

3.8 Liposome Content Mixing Assay

1. Mix 45 μL unlabeled t-SNARE liposomes and 5 μL sulforhodamine B-loaded v-SNARE liposomes in the presence or absence of 5 μM SNARE regulators. In the leakage control experiments, replace unlabeled t-SNARE liposomes with sulforhodamine B-loaded t-SNARE liposomes (*see* **Note 5**).
2. Load the mixed samples into a 96-well microplate preheated to 37 °C.

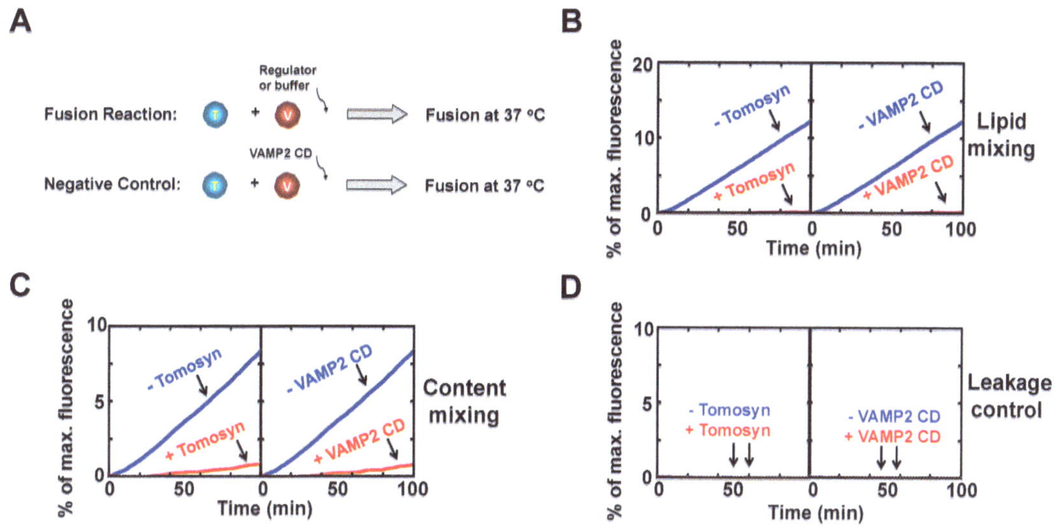

Fig. 2 Tomosyn inhibits the GLUT4 SNARE-mediated lipid mixing and content mixing reactions. (**a**) Illustrations of the reconstituted liposome fusion procedures. The t-SNARE liposomes were reconstituted with syntaxin-4/SNAP-23, while the v-SNARE liposomes were prepared using VAMP2. (**b**) Lipid mixing of the reconstituted proteoliposomes in the absence or presence of 5 μM tomosyn. In negative controls, VAMP2 CD was added to the reactions to the final concentration of 20 μM. The fusion reactions were performed as in (**a**). (**c**) Content mixing of the reconstituted fusion reactions. The sulforhodamine B-loaded v-SNARE liposomes were directed to fuse with unlabeled t-SNARE liposomes in the absence or presence of 5 μM tomosyn. The fusion reactions were performed as in (**a**). (**d**) The leakage controls of the content mixing reactions. The fusion reactions were performed as in (**c**), except the sulforhodamine B was included in both v- and t-liposomes. Little fluorescence increase was observed, indicating that no detectable content leakage occurred during the fusion reactions. (Adapted with permission from Ref. 24. Copyright The American Society for Biochemistry and Molecular Biology, Inc. (2014))

3. Measure sulforhodamine B fluorescence (excitation: 565; emission: 585 nm) of the samples every 2 min in a BioTek Synergy HT microplate reader.
4. At the end of the reaction, add 10 μL 10% CHAPS to each sample. Continue to measure sulforhodamine B fluorescence every 2 min for another 20 min.
5. Normalize the content mixing data as the percentage of maximum fluorescence change after CHAPS treatment (Fig. 2).

3.9 Docking Assay

1. Prepare avidin-conjugated agarose beads in liposome reconstitution buffer.
2. Incubate 45 μL biotin-labeled t-SNARE liposomes with 10 μL avidin-conjugated agarose beads at room temperature for 1 h.
3. Wash the beads three times with the reconstitution buffer. Mix t-SNARE liposomes-bound beads with v-SNARE liposomes in the presence or absence of 5 μM SNARE regulators. Rotate the samples at 4 °C for 30 min. In the negative control reaction, add VAMP2 CD to obtain the background fluorescence signal.

4. After washing three times with the reconstitution buffer, add CHAPS to the final concentration of 1% to solubilize the bead-bound liposomes.
5. Centrifuge the samples for 2 min at 1500 × *g*, collect the supernatant.
6. Measure the rhodamine fluorescence in the supernatant in a BioTek Synergy HT microplate reader, which reflect the amounts of docked v-SNARE liposomes (Fig. 3).

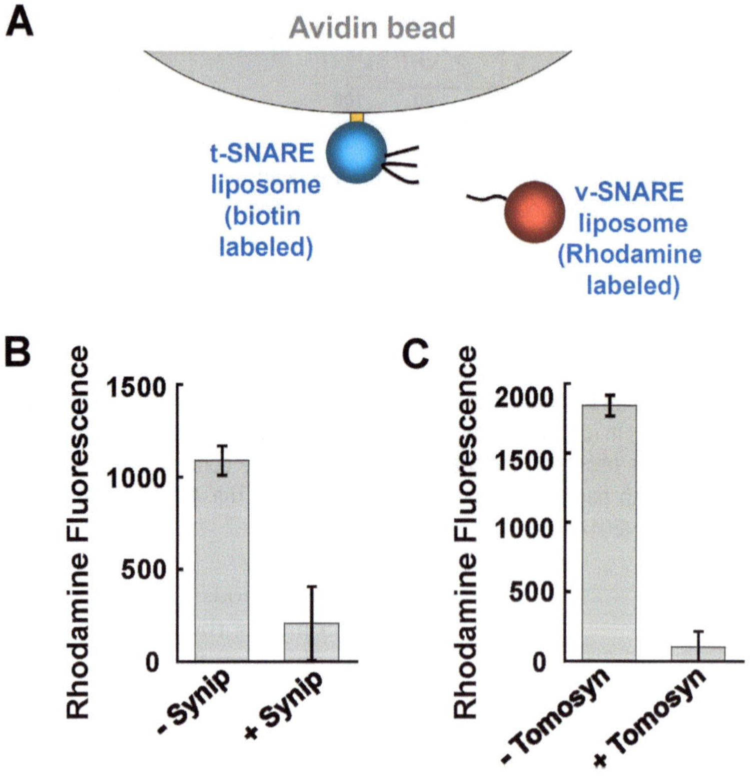

Fig. 3 GLUT4 SNARE-mediated membrane docking is blocked by synip or tomosyn. (**a**) Diagram of the avidin bead pull-down assay. (**b**, **c**) Measurements of docking of t- and v-SNARE liposomes. Biotin-labeled t-SNARE liposomes were anchored to avidin beads and were used to pull down fluorescence-labeled v-SNARE liposomes. The binding reactions were performed in the absence or presence of 5 μM synip (**b**) or tomosyn (**c**). The binding reaction containing 20 μM VAMP2 CD was used as a negative control to obtain the background fluorescent signal. The background fluorescence was subtracted from other binding reactions to reflect specific SNARE-dependent liposome docking. The data are presented as rhodamine fluorescence intensity. Error bars indicate standard deviation. (Adapted with permission from Ref. 23, 24. Copyright The American Society for Biochemistry and Molecular Biology, Inc. (2013, 2014))

4 Notes

1. In all steps, β-mercaptoethanol, TCEP, and protease inhibitor cocktail need to be freshly added into the buffer.
2. The expression of synip proteins requires Rosetta (DE3) competent cells and overnight low-temperature inductions.
3. To maximumly excludeGlucose transporter-4 (GLUT4) the nonspecific soluble protein-membrane association, use the neutral lipid phosphatidylcholine to prepare the liposomesLiposomes in t-SNARE assembly or ternary SNARE assemblySNARE assembly assays.
4. In the ternary SNARE assemblySNARE assembly assay, include a GST tag at the N-terminus to visualize VAMP2 CD better. The GST-tagged VAMP2 CD is fully competent for SNARE complex assembly.
5. To get better activities in the content mixing assay, conduct the experiments immediately after preparing the sulforhodamine B-loaded liposomes.

Acknowledgments

We thank Dr. Chun Wan and Dr. Jingshi Shen for the comments and critically reading the manuscript. This work was supported by the National Natural Science Foundation of China (NSFC) grants (Nos. 31871425 and 91854117) and Natural Science Foundation of Jiangsu Province (BK20200036).

References

1. Klip A, McGraw TE, James DE (2019) Thirty sweet years of GLUT4. J Biol Chem 294: 11369–11381
2. Sollner T, Whiteheart SW, Brunner M, Erdjument-Bromage H, Geromanos S, Tempst P, Rothman JE (1993) SNAP receptors implicated in vesicle targeting and fusion. Nature 362:318–324
3. Weber T, Zemelman BV, McNew JA, Westermann B, Gmachl M, Parlati F, Sollner TH, Rothman JE (1998) SNAREpins: minimal machinery for membrane fusion. Cell 92: 759–772
4. Sudhof TC, Rothman JE (2009) Membrane fusion: grappling with SNARE and SM proteins. Science 323:474–477
5. Jahn R, Scheller RH (2006) SNAREs--engines for membrane fusion. Nat Rev Mol Cell Biol 7: 631–643
6. Wickner W, Schekman R (2008) Membrane fusion. Nat Struct Mol Biol 15:658–664
7. Gao Y, Zorman S, Gundersen G, Xi Z, Ma L, Sirinakis G, Rothman JE, Zhang Y (2012) Single reconstituted neuronal SNARE complexes zipper in three distinct stages. Science 337: 1340–1343
8. Melia TJ, Weber T, McNew JA, Fisher LE, Johnston RJ, Parlati F, Mahal LK, Sollner TH, Rothman JE (2002) Regulation of membrane fusion by the membrane-proximal coil of the t-SNARE during zippering of SNAREpins. J Cell Biol 158:929–940
9. Xu T, Rammner B, Margittai M, Artalejo AR, Neher E, Jahn R (1999) Inhibition of SNARE complex assembly differentially affects kinetic components of exocytosis. Cell 99:713–722
10. Pobbati AV, Stein A, Fasshauer D (2006) N- to C-terminal SNARE complex assembly

promotes rapid membrane fusion. Science 313: 673–676
11. Li F, Pincet F, Perez E, Eng WS, Melia TJ, Rothman JE, Tareste D (2007) Energetics and dynamics of SNAREpin folding across lipid bilayers. Nat Struct Mol Biol 14:890–896
12. Zhou P, Bacaj T, Yang X, Pang ZP, Sudhof TC (2013) Lipid-anchored SNAREs lacking transmembrane regions fully support membrane fusion during neurotransmitter release. Neuron 80:470–483
13. Xu H, Zick M, Wickner WT, Jun Y (2011) A lipid-anchored SNARE supports membrane fusion. Proc Natl Acad Sci U S A 108: 17325–17330
14. Zhao P, Yang L, Lopez JA, Fan J, Burchfield JG, Bai L, Hong W, Xu T, James DE (2009) Variations in the requirement for v-SNAREs in GLUT4 trafficking in adipocytes. J Cell Sci 122:3472–3480
15. Olson AL, Knight JB, Pessin JE (1997) Syntaxin 4, VAMP2, and/or VAMP3/cellubrevin are functional target membrane and vesicle SNAP receptors for insulin-stimulated GLUT4 translocation in adipocytes. Mol Cell Biol 17:2425–2435
16. Bostrom P, Andersson L, Vind B, Haversen L, Rutberg M, Wickstrom Y, Larsson E, Jansson PA, Svensson MK, Branemark R, Ling C, Beck-Nielsen H, Boren J, Hojlund K, Olofsson SO (2010) The SNARE protein SNAP23 and the SNARE-interacting protein Munc18c in human skeletal muscle are implicated in insulin resistance/type 2 diabetes. Diabetes 59: 1870–1878
17. Stockli J, Fazakerley DJ, James DE (2011) GLUT4 exocytosis. J Cell Sci 124:4147–4159
18. Bryant NJ, Govers R, James DE (2002) Regulated transport of the glucose transporter GLUT4. Nat Rev Mol Cell Biol 3:267–277
19. Leto D, Saltiel AR (2012) Regulation of glucose transport by insulin: traffic control of GLUT4. Nat Rev Mol Cell Biol 13:383–396
20. Yu H, Rathore SS, Davis EM, Ouyang Y, Shen J (2013) Doc2b promotes GLUT4 exocytosis by activating the SNARE-mediated fusion reaction in a calcium- and membrane bending-dependent manner. Mol Biol Cell 24: 1176–1184
21. Widberg CH, Bryant NJ, Girotti M, Rea S, James DE (2003) Tomosyn interacts with the t-SNAREs syntaxin4 and SNAP23 and plays a role in insulin-stimulated GLUT4 translocation. J Biol Chem 278:35093–35101
22. Min J, Okada S, Kanzaki M, Elmendorf JS, Coker KJ, Ceresa BP, Syu LJ, Noda Y, Saltiel AR, Pessin JE (1999) Synip: a novel insulin-regulated syntaxin 4-binding protein mediating GLUT4 translocation in adipocytes. Mol Cell 3:751–760
23. Yu H, Rathore SS, Shen J (2013) Synip arrests soluble N-ethylmaleimide-sensitive factor attachment protein receptor (SNARE)-dependent membrane fusion as a selective target membrane SNARE-binding inhibitor. J Biol Chem 288:18885–18893
24. Yu H, Rathore SS, Gulbranson DR, Shen J (2014) The N- and C-terminal domains of tomosyn play distinct roles in soluble N-ethylmaleimide-sensitive factor attachment protein receptor binding and fusion regulation. J Biol Chem 289:25571–25580
25. Wang S, Liu Y, Crisman L, Wan C, Miller J, Yu H, Shen J (2020) Genetic evidence for an inhibitory role of tomosyn in insulin-stimulated GLUT4 exocytosis. Traffic 21: 636–646
26. Hong W (2005) Cytotoxic T lymphocyte exocytosis: bring on the SNAREs! Trends Cell Biol 15:644–650
27. Thurmond DC, Gaisano HY (2020) Recent insights into Beta-cell exocytosis in type 2 diabetes. J Mol Biol 432:1310–1325

Chapter 12

Imaging Single-Vesicle Exocytosis with Total Internal Reflection Fluorescence Microscopy (TIRFM)

Yingke Xu, Luhong Jin, and Derek Toomre

Abstract

Total internal reflection fluorescence microscopy (TIRFM) provides extremely thin optical sectioning with excellent signal-to-noise ratios, which allows for visualization of membrane dynamics at the cell surface with superb spatiotemporal resolution. In this chapter, TIRFM is used to record and analyze exocytosis of single glucose transporter-4 (GLUT4) containing vesicles in 3T3-L1 adipocytes.

Key words TIRFM, Vesicle, GLUT4, Exocytosis, Vesicle fusion, Live-cell imaging

1 Introduction

Vesicle trafficking plays a pivotal role in the formation and maintenance of intracellular compartments as well as in the communications between cells and the environment. Defects in vesicle trafficking have been linked to the pathogenesis of many human diseases, such as cancer, diabetes, and neurodegenerative diseases [1–4].

Exocytosis, the fusion of intracellular membrane vesicles with the plasma membrane, occurs constitutively in all eukaryotic cells but is regulated in some cell types in response to extrinsic stimuli (e.g., hormones and growth factors). Previous studies by using biochemical, molecular biological and genetic approaches have identified many of the machineries involved in the regulation of distinct steps of vesicle exocytosis, such as vesicle transportation along the cytoskeleton and vesicle tethering/docking and fusion with the plasma membrane [5, 6]. Take the insulin-regulated glucose transporter-4 (GLUT4) trafficking as an example; although it has been known that insulin regulates multiple steps of GLUT4 trafficking and the plasma membrane is the major site of insulin action, the spatiotemporal coupling of insulin signaling and GLUT4 translocation remains to be elucidated [7].

Jingshi Shen (ed.), *Membrane Trafficking: Methods and Protocols*,
Methods in Molecular Biology, vol. 2473, https://doi.org/10.1007/978-1-0716-2209-4_12,

Total internal reflection fluorescence microscopy (TIRFM) is a powerful technique for studying of biological processes near the cell surface [8, 9]. TIRFM as its name implies relies on an optical phenomenon called "total internal reflection" (TIR) and the generation of an evanescent field. TIR occurs when light travels from a higher refractive media to a lower one (e.g., from glass coverslip to the cell), at an angle greater than the critical angle the incident light undergoes total refection; that is it is reflected out of the high refractive index media. However, the powerful part that is exploited by TIRFM is that some of the light still penetrates a small distance (e.g., <200 nm) into the lower refractive media, which generates an electromagnetic field propagating at the *Z* direction called an evanescent wave. The evanescent wave decreases exponentially with distance, illuminating only a few hundred nanometers beyond the coverslip/cell interface. Thus, TIRFM has exceptional signal-to-noise which can visualize single molecules and is ideally suited to visualize the dynamics of single-vesicle exocytosis [10, 11]. The latest development of TIRFM incudes the multiangle or variable-angle TIRFM that takes the advantage of supercritical angle illumination for nanometer axial sectioning imaging [12, 13]. In this chapter, we describe the application of TIRFM in imaging and analyzing single GLUT4 vesicle exocytosis in 3T3-L1 adipocytes.

2 Materials

2.1 Setup of TIRFM System

The multiangle ring-illumination TIRFM system is shown in Fig. 1. This system is based on a classical objective-type TIRFM. The manual micrometer laser adjustment module is replaced by a two-dimensional galvanometer (2D-GM) system to automatically control the tilting and radial angle of the illumination laser beam (*see* **Note 1**). In addition, spinning the azimuthal angle by ring illumination eliminates uneven illumination and interference fringes introduced in traditional TIRFM (Fig. 2).

1. Fluorescence is excited by 488 nm laser (Melles Griot), and the laser is controlled with electronic shutters (Unibliz) for reducing light exposure. Optical filters and dichroic mirrors for excitation light are BrightLine full-multiband laser filters (LF405/488/561/635; Semrock).
2. TIRF is achieved through an Olympus 100× 1.49 numerical aperture (NA) oil immersion objective lens (UAPON 100 × OTIRF) (*see* **Note 2**). The focal point of the illuminating laser beam was controlled by the galvo mirrors, which move the spinning ring-pattern light at the peripheral position of the back focal plan of the TIRF objective. Images were collected by a back-illuminated EMCCD camera (Andor iXon Life 888), which is controlled by Andor iQ software.

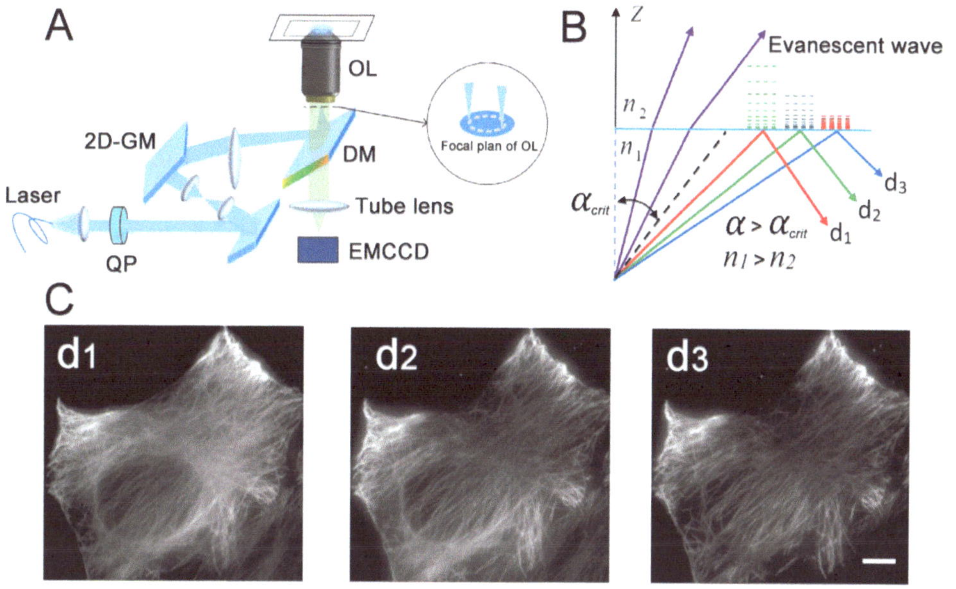

Fig. 1 The setup of the multiangle ring-illumination TIRFM system (**a**), the diagram depicts multiangle TIRF (**b**) and fluorescent labeled microtubules in a COS-7 cell imaged at different TIRF angles (**c**). The COS-7 cells were labeled with mouse anti-β-tubulin monoclonal antibody, followed by the Alexa Fluor 488–conjugated goat anti-mouse IgG secondary antibody. *QP* quarter wave plate; *2D-GM* two-dimensional galvanometers; *DM* dichroic mirror; *OL* objective lens; α_{crit} critical angle. Scale bar, 10 μm

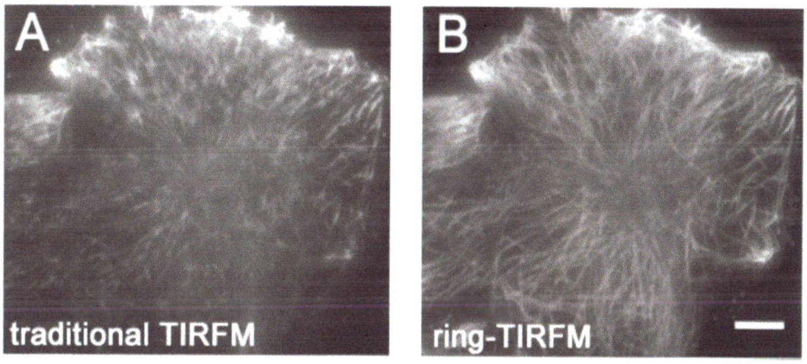

Fig. 2 Fluorescent labeled microtubules within the same COS-7 cell were Imaged by traditional single point illumination TIRFM (**a**) and ring-illumination TIRFM (**b**). The cells were labeled with mouse anti-β-tubulin monoclonal antibody, followed by the Alexa Fluor 488-conjugated goat anti-mouse IgG secondary antibody. The image quality is greatly improved by spinning ring illumination. Scale bar, 10 μm

2.2 Cell Culture and Transfection

1. 3T3-L1 preadipocytes.
2. Cell culture media: Dulbecco's modified Eagle's medium (DMEM, high glucose) supplemented with 10% fetal bovine serum (FBS), 50 IU/mL penicillin, and 50 μg/mL streptomycin.

3. Cell differentiation media: DMEM containing 10% FBS, 500 μM isomethylbutylxanthine (IBMX; Sigma #I-5879), 0.25 μM dexamethasone (DEX; Sigma #D-1756), and 160 nM insulin (Sigma #I-5523).
4. Trypsin–EDTA (solution: 0.25% trypsin, 1 mM ethylenediaminetetraacetic acid (EDTA).
5. Phosphate-buffered saline (PBS; pH 7.4, without $CaCl_2$ and $MgCl_2$).
6. Electroporation solution: Cell Line Nucleofector Kit L (Lonza) (*see* **Note 3**).
7. Electroporator. We use the Nucleofector 2b device (Lonza).
8. Glass bottom microwell dishes: 35 mm petri dish, 14 mm microwell, No. 1.5 coverglass.
9. The VAMP2-pHluorin plasmid was provided by Dr. J. Rothman [14].

2.3 TIRFM Imaging

1. KRH imaging buffer: 25 mM HEPES-KOH, pH 7.4, 125 mM NaCl, 5 mM KCl, 1.3 mM $CaCl_2$, 1.2 mM $MgSO_4$, 20 mM D-glucose, and 0.2% bovine serum albumin (BSA).
2. Immersion oil ($n = 1.518$).
3. Live cell imaging chamber (*see* **Note 4**).

3 Methods

3.1 Culture and Differentiation of 3T3-L1 Preadipocytes

1. Culture 3T3-L1 preadipocytes in cell culture media. Passage the cells every 2 or 3 days.
2. Two days after the 3T3-L1 preadipocytes reach 100% confluency, switch the cells the differentiation media.
3. After 3 days in differentiation media, switch the cells back to cell culture media. The media are changed every 2 days.
4. Use the differentiated adipocytes for experiments between 8 and 10 days after the initiation of differentiation.

3.2 Electroporation of 3T3-L1 Adipocytes

1. Dissociate 3T3-L1 adipocytes using 0.25% trypsin–EDTA.
2. Add cell culture media to stop trypsinization.
3. Centrifuge the cells at 1000 rpm for 10 min at room temperature and carefully discard the supernatant.
4. Count the cells and determine cell density.
5. Centrifuge the required number of cells (2×10^6 cells per sample) at 1000 rpm for 10 min at room temperature and then remove the supernatant completely.
6. Resuspend the cell pellet in 100 μL Nucleofector solution L.

7. Add 2 μg of the VAMP2-pHluorin plasmid. Mix the plasmids with the cell solution gently (*see* **Note 5**).
8. Transfer the cell–DNA suspension into 0.2-cm cuvette. Close the cuvette with a cap.
9. Insert the cuvette into the Nucleofector cuvette holder and choose the program "A-033".
10. Take the cuvette out of the holder and immediately transfer the cells into a 15 mL falcon tube filled with 5 mL of fresh cell culture media.
11. Mix the cells gently and aspire 1 mL of the cell solution into a 35 mm glass bottom dish filled with 1 mL of cell culture media.
12. Incubate the cells in a humidified incubator at 37 °C, 5% CO_2 for 18–24 h.

3.3 Imaging of Single GLUT4 Vesicle Exocytosis

1. Starve 3T3-L1 adipocytes with the KRH imaging buffer in the incubator for at least 2 h.
2. Turn on the TIRFM system and the live cell imaging chamber (Olympus) at least 30 min before the imaging experiment.
3. Apply a drop of immersion oil on the TIRF objective. Mount the glass bottom dish onto the microscope stage.
4. Find VAMP2-pHluorin expressing cells in the epifluorescence mode (*see* **Note 6**).
5. Gradually increase the angle of the illuminating laser beams to reach the TIRF illumination (*see* **Note 7**).
6. Simultaneously adjust the focus to the footprint of the cell.
7. Use the Andor iQ software to choose the appropriate parameters for the EMCCD camera (window size, gain, exposure time, binning, etc.).
8. Acquire time-lapse images before and after insulin (100 nM) stimulation (Fig. 3). We typically record the 3T3-L1 adipocytes in the basal state for 2 min, and insulin stimulation for another 5 min with acquisition rate at 5–10 Hz.
9. Save the images to the hard disk as multitiff files or single-tiff files.

3.4 Analyzing Single-Vesicle Fusion Events

1. Import the image sequences into ImageJ.
2. Using the Grid tool to divide images into small grids and manually review multiple times to visually identify putative vesicle fusion events.
3. Mark the coordinates of the putative fusion sites and choose a small region of interest around each putative fusion site for further analysis.
4. Use circular regions with diameters of 10 pixels (~1.8 μm) to calculate the intensity of single vesicles.

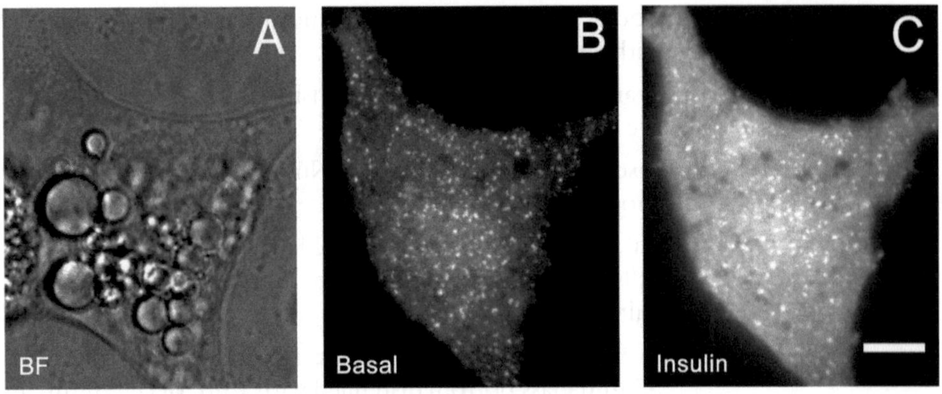

Fig. 3 TIRFM imaging of the insulin-stimulated GLUT4 vesicle translocation. 3T3-L1 adipocytes were electroporated with VAMP2-pHluorin. The brightfield (BF) image of a 3T3-L1 adipocyte (**a**), and the footprint of the same cell were shown before (**b**) and after 5 min of insulin stimulation (**c**). The increase of VAMP2-pHluorin signals under the TIRFM demonstrates the relocation of GLUT4 to the cell surface after insulin treatment. Scale bar, 10 μm

5. Use the average intensity of a ring just outside the circle for local background subtraction.
6. Use the integrated intensity, peak intensity, and full-width at half-maximum (FWHM) of Gaussian distribution to validate and distinguish between full fusion and kiss-and-run fusion events (Fig. 4).

4 Notes

1. There are a few advantages of using galvanometer mirrors for controlling the angle and position of the illumination laser light, as these mirrors (1) enable rapid (millisecond) switching between different modes of light illumination (e.g., epifluorescence, highly inclined and luminated optical sheet [HILO], and TIRFM by varying the incident beam angle); (2) enable ring-shaped illumination (by steering the beam in a ring in the back focal plane of the objective) to eliminate the inhomogeneous evanescent excitation artifacts caused by fixed-azimuth illumination (Fig. 2); and (3) allow for multiangle TIRFM imaging and axial nanometer reconstruction.
2. The NA of an objective describes its light-gathering power and the maximum angle at which the excitation light can emerge from the objective. To achieve TIRF, the NA of the objective must be greater than the refractive index of the cells (n = 1.36–1.38), preferably by a substantial margin. The most common commercially available objectives for TIRFM are 1.45 NA and 1.49 NA, and are both available in 60×, 100×, and 150× magnification. Currently, the 100× 1.7 NA

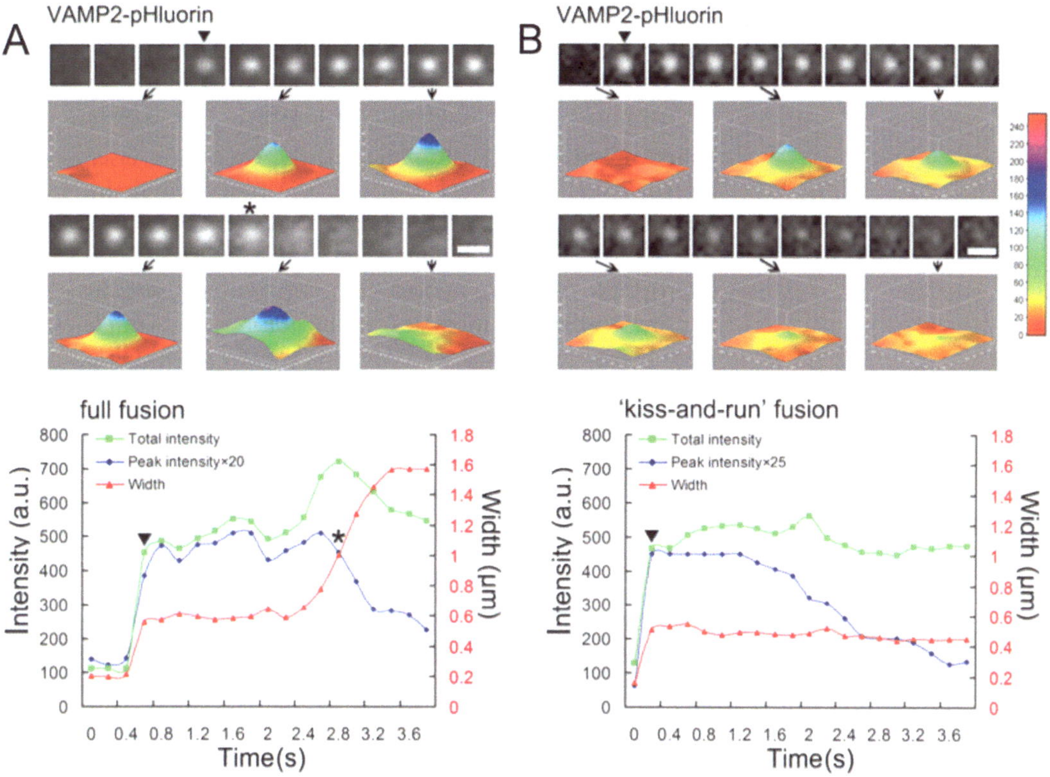

Fig. 4 Analysis of single-vesicle fusion events. Examples of a VAMP2-pHluorin full fusion (**a**) and kiss-and-run (**b**) events in 3T3-L1 adipocytes. The 3D fusion profiles of selected frames (arrows) are shown with the peak intensity, total intensity, and FWHM plotted underneath. Note that the full vesicle fusion has an increase in the FWHM, whereas the kiss-and-run event does not (and in the latter, the slope of the peak intensity decay is shallower). Arrowheads indicate opening of the fusion pore; asterisks indicate full vesicle collapse. All intensities were normalized to their initial values. Interval: 200 ms. Scale bars, 1 μm

objective is also available in the market, but it requires a special toxic immersion oil and expensive nonstandard coverslips with a higher matching refractive index.

3. Lonza provides optimized electroporation buffer for different cell lines, which can be searched at the "Cell & Transfection Database" on its website.

4. The live cell imaging chamber is a miniaturized version of an incubator which can be mounted on top of the stage of a microscope while performing the imaging experiments.

5. Avoid leaving the cells in Nucleofector solution for extended periods of time (e.g., longer than 15 min), as this would reduce cell viability. Mix plasmids with the cell solution gently. The samples must cover the entire bottom of the cuvette and avoid generating any air bubbles.

6. pHluorin, a pH-sensitive variant of green fluorescent protein, is quenched in an acidic environment (e.g., lumen of a vesicle) and becomes much brighter in a neutral environment (e.g., cell surface). This can help identify true vesicle fusion events.
7. When nearing TIRF illumination, the excitation laser is total internally reflected back through the objective again, so the laser light emitted from the objective is greatly reduced. If you observe through the CCD camera (or microscope eyepieces if appropriate filters are in place for safety) when gradually adjusting the angle, the intensity of the fluorescent sample would have a sudden increase and decrease greatly should you further increase the angle of the illuminating light. This indicates that the TIRF illumination is close.

Acknowledgments

This work was supported by the National Key Research and Development Program of China (2018YFE0119000), Zhejiang Provincial Natural Science Foundation (LR18H180001), the Fundamental Research Funds for the Central Universities (2021XZZX022), and the Zhejiang University Education Foundation Global Partnership Fund.

References

1. Xu R, Rai A, Chen M, Suwakulsiri W, Greening DW, Simpson RJ (2018) Extracellular vesicles in cancer—implications for future improvements in cancer care. Nat Rev Clin Oncol 15: 617–638
2. Bogan JS (2012) Regulation of glucose transporter translocation in health and diabetes. Annu Rev Biochem 81:507–532
3. Jahn R, Fasshauer D (2012) Molecular machines governing exocytosis of synaptic vesicles. Nature 490:201–207
4. Olkkonen VM, Ikonen E (2006) When intracellular logistics fails--genetic defects in membrane trafficking. J Cell Sci 119:5031–5045
5. Wu B, Guo W (2015) The exocyst at a glance. J Cell Sci 128:2957–2964
6. Jahn R, Sudhof TC (1999) Membrane fusion and exocytosis. Annu Rev Biochem 68: 863–911
7. Zhou X, Shentu P, Xu Y (2017) Spatiotemporal regulators for insulin-stimulated GLUT4 vesicle exocytosis. J Diabetes Res 2017: 1683678
8. Steyer JA, Almers W (2001) A real-time view of life within 100 nm of the plasma membrane. Nature 2:268–276
9. Toomre D, Manstein DJ (2001) Lighting up the cell surface with evanescent wave microscopy. Trends Cell Biol 11:298–303
10. Rivera-Molina F, Toomre D (2013) Live-cell imaging of exocyst links its spatiotemporal dynamics to various stages of vesicle fusion. J Cell Biol 201:673–680
11. Xu Y, Rubin BR, Orme CM, Karpikov A, Yu C, Bogan JS, Toomre DK (2011) Dual-mode of insulin action controls GLUT4 vesicle exocytosis. J Cell Biol 193:643–653
12. Jin L, Wu J, Xiu P, Fan J, Hu M, Kuang C, Xu Y, Zheng X, Liu X (2017) High-resolution 3D reconstruction of microtubule structures by quantitative multi-angle total internal reflection fluorescence microscopy. Opt Commun 395:16–23
13. Yang Q, Karpikov A, Toomre D, Duncan JS (2011) 3-D reconstruction of microtubules from multi-angle total internal reflection fluorescence microscopy using Bayesian framework. IEEE Trans Image Process 20: 2248–2259
14. Miesenbock G, De Angelis DA, Rothman JE (1998) Visualizing secretion and synaptic transmission with pH-sensitive green fluorescent proteins. Nature 394:192–195

Part II

The Endocytic Pathway

Part II

Chapter 13

Localizing Proteins on Single Trafficking Organelles in 3D with Semisynthetic Gold Labeling and Platinum Replica Electron Microscopy

Bijeta Prasai and Justin W. Taraska

Abstract

The three-dimensional structures of organelles can be visualized at high resolutions using electron microscopy and tomography. Combining genetically encoded tags with tomography enables the specific targeting and detection of identified proteins inside cells. Here, we describe a method for attaching metal-binding gold nanoparticles to proteins genetically tagged with hexa-histidine sequences. We apply this strategy to visualize the position of intracellular proteins on single organelles in unroofed cells with platinum replica electron microscopy at the nanoscale in three dimensions. We have found that this combined method can label and localize proteins with isotropic high precision to generate quantitative maps of protein positions in and around trafficking organelles at the inner plasma membrane of mammalian cells.

Key words Electron microscopy, Electron tomogram, Gold nanoprobe, Unroofing, PREM, Platinum replica, Correlative fluorescence, Histidine tags

1 Introduction

The architectures of multicomponent molecular machines and organelles that drive cellular events are largely unknown. Therefore, the link between the molecular organization of organelles, their dynamics, their regulation, and their physiology remains unclear. Advanced technologies such as superresolution fluorescence and electron microscopy are at the forefront of addressing these gaps in our understanding. Electron microscopy surpasses fluorescence techniques in terms of resolution, however, the shortage of contrast agents that can label specific cellular objects for EM limits the application of this imaging modality. Immunogold labeling is the classic method to visualize targets of interest with EM [1, 2]. However, it is challenging to use and implement globally for several reasons. First, there is commonly a loss of antigenicity of target proteins due to the sometimes-harsh EM sample preparation

Jingshi Shen (ed.), *Membrane Trafficking: Methods and Protocols*,
Methods in Molecular Biology, vol. 2473, https://doi.org/10.1007/978-1-0716-2209-4_13,

process [3, 4]. Moreover, the limited and unreliable set of commercial antibodies hinders widescale applicability of immunolabeling to all targets of interests within the cell. Finally, the large size of the antigen complex can position gold targets upward of 30 nm away from a protein of interest. Thus, the development and use of genetically encoded labeling approaches that are targetable, specific, and small are needed to enable detection with high precision inside cells [5].

Here, we describe a genetically encoded platinum replica electron microscopy (PREM) labeling method that uses histidine-based affinity-tags and NTA-metal (nickel)-binding gold-nanoparticles and tomography to label and detect proteins associated with endocytic and exocytic vesicles. This method can be used to mark a variety of proteins genetically fused with common short six-histidine tags. High affinity nickel-histidine interactions can position gold nanoprobes less than 5 nm from the target. Thus, unlike antibodies, this strategy allows for high precision localizations [6, 7]. Yet, access of gold probes into living or intact cells and tissues is challenging. To overcome this issue, we perform labeling in unroofed cells that allows for unhindered access and labeling of the nanoprobes to cytosolic and membrane proteins [8]. Specifically, we generate six-histidine and GFP tagged proteins associated with either endocytic clathrin-coated, caveolae, or exocytic dense-core vesicles (Fig. 1). In unroofed cells, we use commercially available Nickel-NTA conjugated 10 nm gold particles to label histidine-tagged proteins (Fig. 1) [6, 9]. To ensure that the cells imaged are expressing histidine-tagged proteins of interest, we use correlative fluorescence to locate and target the cells for EM. The pipeline for this method is shown in Fig. 2. First, a fluorescent map of GFP-expressing cells is generated. After metal coating of the sample for PREM, phase-contrast maps are made before and after platinum replica is produced and compared with fluorescence signals. EM imaging and tomography is then done only on cells that exhibit GFP fluorescence.

To confirm the applicability of this method, we imaged well studied proteins known to organize on specific regions of vesicle subtypes. For example, clathrin light chain A and cavin1 are known coat proteins of endocytic clathrin-coated and caveolae vesicles, respectively [10, 11]. With EM and tomography, structures encased with gold nanoparticles are visible as shown in Fig. 4. Likewise, EPS15, a protein known to be located at the base of the clathrin vesicle is marked by the gold nanoprobe exclusively at the vesicle edge [12, 13]. We have used this method to successfully explore the previously unknown three dimensional axial location of exocytic vesicle proteins such as Rab3a, Rab27a, Rabphilin3a, Granuphilin-a, and Rim2 on docked dense core vesicles [13]. This labeling method allows for the global nanoscale marking of proteins in and around organelles in 3D inside cells.

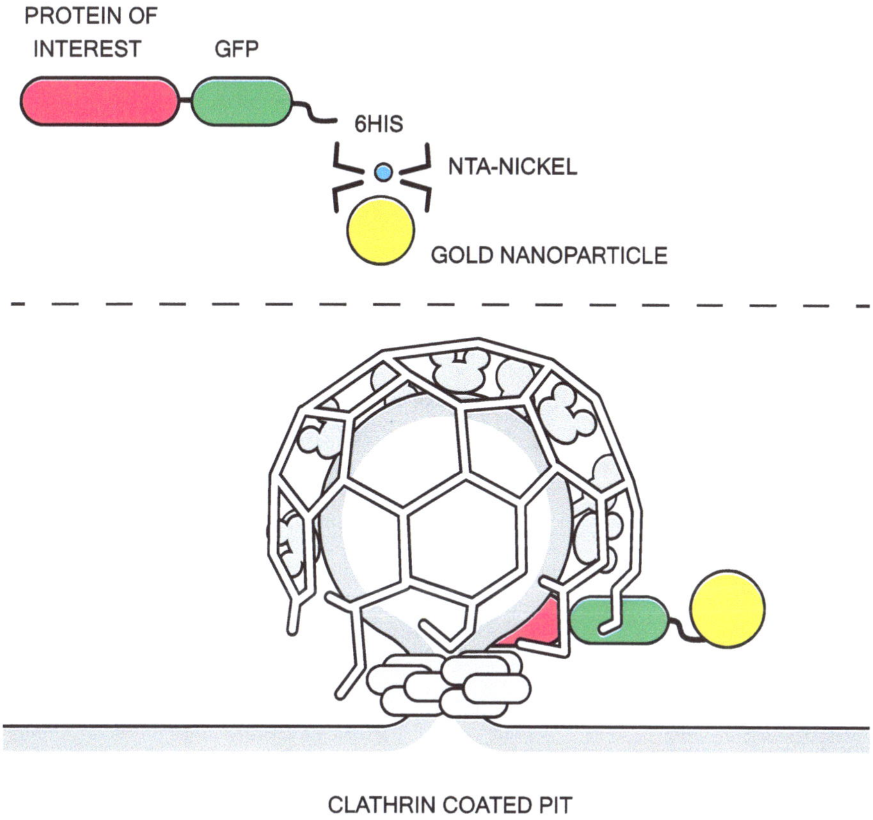

Fig. 1 A schematic (upper panel) of a protein of interest (POI) tagged with GFP and six-histidines (His). The His tag is the target for the Ni-NTA-Au (gold nanoprobe). The cartoon in the lower panel shows a gold nanoprobe-labeled clathrin coated pit that is expressing His-GFP-POI

2 Materials

2.1 Cell Culture and Transfection

1. Cells of interest: HeLa (American Type Culture Collection), and PC12-GR5 cells line—originated from R. Nishi, P. Stork, and W. Almers (OHSU, Portland, Oregon).
2. Plasmids expressing Six-Histidine- and GFP-tagged proteins of interest.
3. 200 proof ethanol.
4. For PC12 cells: DMEM media with phenol red, 5% Horse Serum, 5% Fetal Bovine Serum (FBS), 1% penicillin/streptomycin (*see* **Note 1**).

   For HeLa cells: MEM/DMEM media without phenol red, 10% FBS, 1% penicillin–streptomycin, 1% sodium pyruvate, and 1% L-GlutaMAX.
5. Coverslip cleaning solution: 30% ammonium hydroxide solution, 30% hydrogen peroxide solution, and ultrapure water.

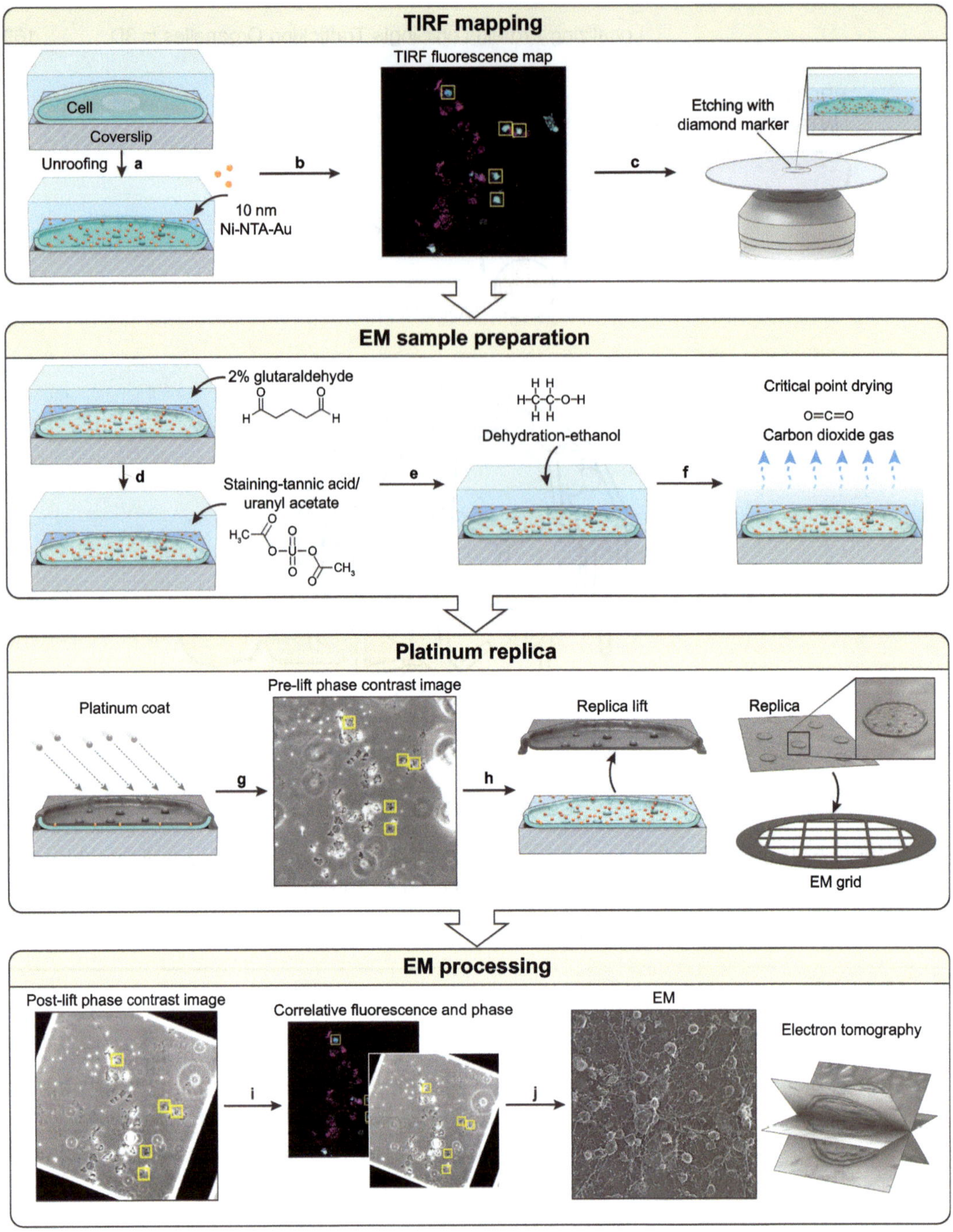

Fig. 2 Imaging workflow for Ni-NTA-Au labeled proteins. (**a**) Cells overexpressing histidine and GFP-tagged proteins of interest are unroofed and labeled with Ni-NTA-Au. (**b**) Cells are mapped for fluorescence to obtain a montage. (**c**) The mapped region is marked by etching the coverslip using a diamond objective marker. Sample is placed in 2% glutaraldehyde until ready for imaging. For EM sample preparation, (**d**) staining with tannic acid and uranyl acetate, (**e**) ethanol dehydration, and (**f**) critical point drying are done. Dried samples are then coated with platinum and carbon. (**g**) The region of interest on the coated coverslip is imaged with a 20 × phase-contrast objective. (**h**) The replica is lifted and transferred to a TEM grid. For EM imaging, first postlift phase-contrast image is obtained, (**i**) Phase-contrast and fluorescence images are correlated to find GFP exhibiting cells, and (**j**) 2D-TEM and electron tomogram of the selected cells are acquired and tomograms reconstructed [13]

6. 6-well tissue culture plates.
7. 25 mm #1.5 glass coverslips.
8. Poly-L-lysine.
9. Opti-MEM.
10. Lipofectamine 2000 (Invitrogen, 11668019).

2.2 Unroofing and Labeling

1. Intracellular stabilization buffer: 30 mM HEPES-KOH, pH 7.4, 70 mM KCl, 5 mM $MgCl_2$, and 3 mM EGTA.
2. 19-gauge needles and 10 mL syringes.
3. Primary fixative: 2% paraformaldehyde (in stabilization buffer) fresh from 16% stock.
4. Tweezers.
5. Dulbecco's phosphate buffered saline (PBS).
6. Blocking buffer: 3% bovine serum albumin (BSA) in PBS.
7. 10 nm Ni-NTA-Au in PBS (Nanoprobes 2084).
8. Ultrasonic bath cleaner.
9. Actin label: Alexa Fluor 647-Phalloidin (Life Technologies A22287).
10. Orbital shaker.

2.3 Cell Mapping with Fluorescence Imaging

1. Attofluor cell chamber (ThermoFisher Scientific).
2. Inverted fluorescence microscope with software that enables capturing large field of view by stitching montages of smaller images. We use a CFI SR HP Apochromat TIRF 100× 1.49NA oil objective on a Nikon Eclipse Ti2 microscope with Nikon Elements NSTORM module.
3. 488 nm and 647 nm lasers.
4. Andor iXon Ultra 897 emccd (Andor DU 897U).
5. Diamond objective marker (Leica, M25 #11505059).
6. EM grade 2% glutaraldehyde in PBS (freshly made from 8% stock).

2.4 EM Preparation

1. EM grade 0.1% (w/v) tannic acid in ultrapure water (freshly made).
2. EM grade 0.1% (v/v) uranyl acetate in ultrapure water.
3. Serial ethanol dehydration: 15%, 30%, 50%, 70%, 80%, 90%, and 100% ethanol solutions (EM grade).
4. Liquid carbon dioxide (CO_2) dehydration setup: custom made critical point drying (CPD) sample holder, wavy washers (Tousimis 8767-01), 100 mL beakers with stir bar as the CPD sample holder, and bent metal mesh as shelf for the holder [14].

5. Critical point dryer equipped with liquid CO_2 filter. We use Tousimis Samdri 795.
6. Liquid CO_2 (ARC3).
7. Freeze fracture (JEOL/RMC 9010 FF machine or Leica ACE900).
8. Light microscope with 20 × phase-contrast objective. We use Nikon Eclipse Ti2 microscope equipped with phase objective.
9. Glow discharge (PELCO easiGlow or similar).
10. EM grade 5% Hydrofluoric Acid (HF) from 50% HF diluted in ultrapure water.
11. Platinum 4–5 mm loop.
12. Diamond scriber.
13. Formvar/carbon-coated 75 mesh copper TEM grids (e.g., Ted Pella).
14. 60 mm cell culture dish.
15. Filter paper.
16. 12-well plate.
17. Polypropylene plastic bulb transfer pipettes (e.g., Fisher 13-711-7M).

2.5 EM Imaging and Tomography

1. Electron microscope: TEM, high tilt specimen retainer for tomogram, and Serial EM Freeware [15]. We use a JEOL 1400 TEM equipped with a specimen quartet holder for 2D imaging.
2. IMOD/3dmod freeware [16].

3 Methods

3.1 Cell Culture and Transfection

1. For cell culture, use clean coverslips stored in 100% ethanol after sterilizing in boiling solution of 3 parts water, 1 part ammonium hydroxide, and 1 part hydrogen peroxide (30%).
2. Pick the air-dried coverslips with tweezer and place in 6-well plate and incubate in ~500 μL of poly-L-lysine for 20 min before seeding cells at the density of 3–3.5 × 10^5 cells per well.
3. After 20 min, wash coverslips twice with ultrapure water and replace with culture medium until ready for seeding cells.
4. Allow cells to adhere to the coverslips and grow overnight in a 37 °C incubator supplied with 5% CO_2.
5. Transfect cells with plasmids of interest using transfection method of your choice. Our plasmids have both histidine and GFP tags (*see* **Note 2**). We transfect cells with Lipofectamine 2000. Place 3 μL of Lipofectamine and 1–2 μg of plasmids in

350 μL of Opti-MEM, separately, and incubate for 20 min. Add the mixture to the cells cultured on coverslip.

6. Incubate cells in the lipofectamine-plasmid solution for 4 h at 37 °C and then replace with fresh cell culture media.
7. Allow cells to grow overnight until ready for experiment the next day. Do not grow cells longer than overnight after transfection as the lipofectamine can affect cell viability.

3.2 Unroofing and Labeling

1. Secure 19-gauge needle on to a 10 mL syringe (*see* **Note 3**).
2. In 6-well plate, add 2 mL of stabilization buffer to the first two wells, 1.5 mL to the third well, and 3.5 mL to the fourth well.
3. Add 500 μL of 16% PFA to the third and fourth well, making 4% and 2% PFA solution, respectively (*see* **Note 4**).
4. Transfer coverslip with transfected cells to the first well for washing, then to the second well.
5. With the syringe, draw 2 mL of the 4% PFA solution from the third well, and then draw 2 mL of buffer from the second well containing coverslip, and quickly inject the so formed 2% PFA solution onto the coverslip (*see* **Note 3**).
6. Immediately transfer the coverslip with the now unroofed cells to the fourth well with 4 mL of 2% PFA solution. Incubate for 20 min at room temperature or in a 4 °C refrigerator.
7. Wash the coverslip four times by transferring it to wells containing PBS (*see* **Note 5**).
8. Incubate in BSA solution for 45 min-1 h in a 4 °C refrigerator (*see* **Note 6**).
9. Wash coverslip four times to remove any unbound BSA (*see* **Note 6**).
10. Leave the coverslip in PBS while preparing gold nanoprobe solution.
11. In a 50 mL scintillation vial, prepare 1:5 dilution of 10 nm gold nanoprobe in PBS (Fig. 3a) (*see* **Note 7**).
12. Sonicate the nanoprobe solution in bath sonicator for 5 min (Fig. 3b). This step ensures proper and uniform dispersion of gold particles.
13. Transfer unroofed cells to this sonicated nanoprobe solution in a 6-well plate. Place the plate on an orbital shaker with the motion knob set at 3 and shake for 15 min (Fig. 3c). After 15 min, turn off the shaker and continue incubation for another 45 min at room temperature (*see* **Note 8**).
14. During the last 15 min, add 16.5 pmol of Alexa Fluor 647–Phalloidin to the solution (*see* **Note 9**).
15. After 1 h of incubation, transfer the coverslip to a well with PBS solution (*see* **Note 10**).

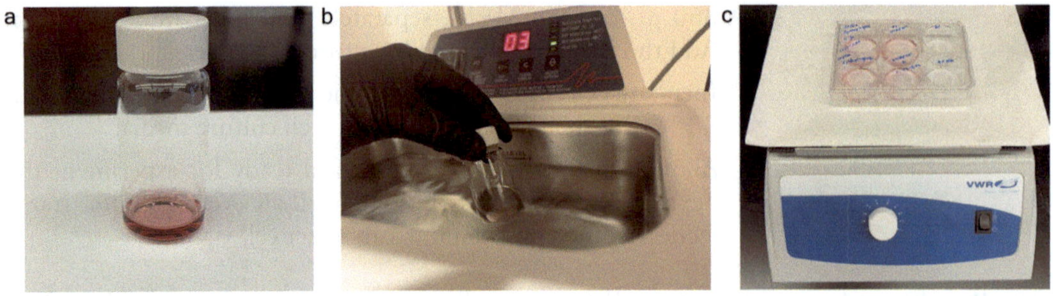

Fig. 3 Preparation for gold nanoprobe labeling. (**a**) 1:5 dilution of Ni-NTA-Au is made in PBS in a scintillation vial. (**b**) The nanoprobe solution is sonicated for 5 min in a bath sonicator. (**c**) Samples incubated with the nanoprobe solution are placed on a shaker for 15 min

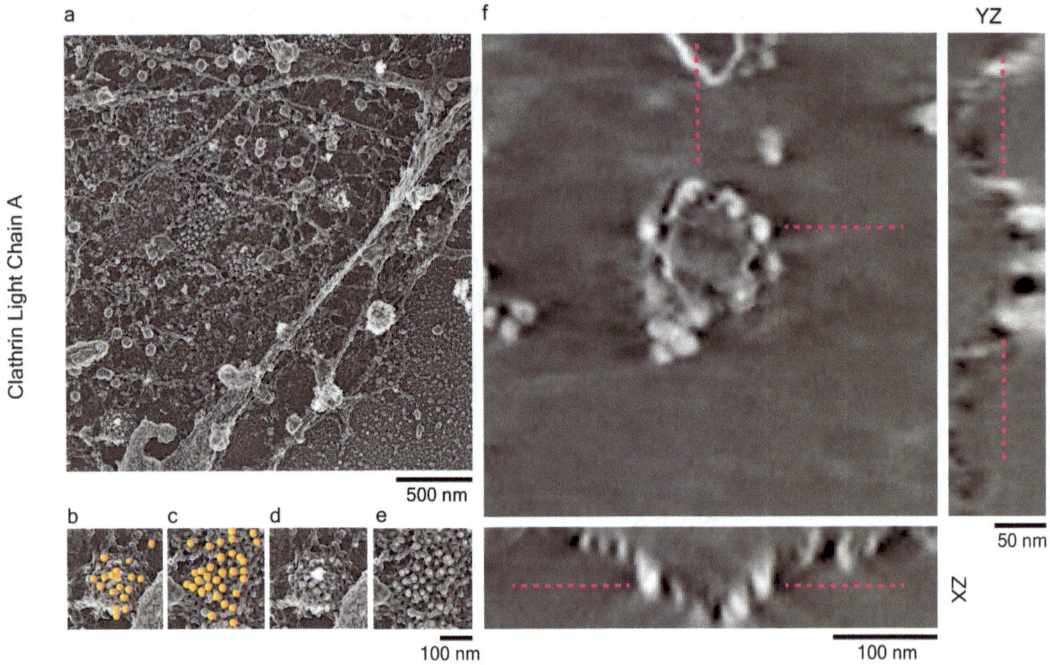

Fig. 4 Nanogold-based platinum replica protein labeling at the plasma membrane. (**a**) Platinum replica image of a HeLa cell (cropped from larger PREM image of a cell) expressing His-GFP-clathrin light chain A and treated with gold nanoprobe. Scale bar is 500 nm. (**b**, **c**) Magnified views of individual clathrin-coated vesicles from (**a**) with gold particles marked with orange circles on their surface. (**d**, **e**) Structures in **a** and **b** without orange markings on gold particles are shown. Scale bar is 100 nm. (**f**) Tomogram section (XY view, scale bar = 100 nm) of an individual clathrin structure labeled with gold nanoprobe for clathrin light chain A. The orthogonal views in XZ (scale bar = 100 nm) and YZ (scale bar = 50 nm) dimensions are also shown. Magenta dashed lines mark the gold particles seen in XY view of a z slice and denote their location in orthogonal views

3.3 Cell Mapping with Fluorescence Imaging

1. Mount the diamond objective marker on to the microscope turret before imaging (*see* **Note 11**). We use a custom machined adaptor to attach the M25 marker objective to the Nikon microscope objective turret.
2. Place coverslip in an Attofluor imaging chamber with PBS (*see* **Note 12**).
3. Use the 100× objective to create a fluorescent map of cells. After making sure the lasers to be used for mapping cells are collimated at epi-illumination, bring lasers at TIRF illumination and adjust TIRF angle such that shallow evanescent illumination is produced. We use laser lines 488 nm to excite GFP and 647 nm for phalloidin-647.
4. Search for an area with cells mostly expressing GFP (*see* **Note 13**).
5. Once you find the desired area, create a 20 × 20 (or 1.3–1.5 mm^2) fluorescent map of that region. If you intend to perform TIRF/CLEM, you can obtain individual images of cells at this time.
6. Copy and paste this map onto software such as Microsoft Paint to create a smaller file. Mark cells if imaged in the map copied in Microsoft Paint. This way the larger map file generated from NSTORM software need not be used for finding cells later on.
7. Position your stage at the center of the 20 × 20 map. Shift the turret to bring the diamond objective under the coverslip.
8. Raise the objective up close to the coverslip until you see the tip of the objective touch the coverslip, which is indicated by slight displacement of oil. Hold the chamber with one hand and under the stage using another hand turn the collar on the diamond marker to create a 4 mm sized marking on the coverslip. To ensure the diamond successfully etched the glass, raise the marking objective slightly and turn the collar once again to mark the glass.
9. For transferring coverslip from Attofluor chamber, seal the chamber such as that done during STORM experiment. For this, fill the chamber with PBS and cover it with a 25 mm coverslip. Remove excess PBS with vacuum suction. This creates a vacuum seal between coverslips securely sealing the chamber. The top coverslip can now function as base of the chamber, and the etched coverslip at the bottom part of the coverslip holder can be accessed without any spills. Proper cleaning of the oil used during imaging becomes much more convenient with this setup.
10. Once the oil residue is cleaned with 70% ethanol, transfer the coverslip to 2% glutaraldehyde solution (*see* **Note 4**). To ensure the ultrastructure is preserved, incubate samples in glutaraldehyde no more than a day.

3.4 EM Preparation

1. Transfer coverslip to 6-well plate with 0.1% (w/v) tannic acid. Incubate for 20 min in a 4 °C refrigerator (*see* **Note 14**).
2. Thoroughly wash the samples incubated in tannic acid before treating with uranyl acetate (*see* **Note 14**).
3. Incubate coverslips in uranyl acetate for 20 min in a 4 °C refrigerator (*see* **Note 15**).
4. Wash coverslips with water three times. Dispose used water in radioactive waste.
5. Transfer the coverslip into a CPD holder placed in a beaker with water. Separate more than one coverslips with wavy washers (*see* **Note 16**). Keep track of the position of your sample coverslips.
6. Dehydrate samples in series of 15%, 30%, 50%, 70%, 80%, 90%, and 100% (three times) ethanol solutions. For dehydration steps, move the sample CPD holder with tweezer from one beaker (lower concentration) to another beaker (higher concentration) with ethanol solution, and stir the solution for 4 min. Also, use new ethanol bottle for 100% wash to prevent introducing any moisture in the samples.
7. During 90% or first 100% ethanol dehydration step, turn on the CPD instrument to warm up and open the liquid CO_2 tank.
8. After the third 100% ethanol dehydration, transfer the sample holder to specimen chamber in CPD instrument. At high pressure, the purging of ethanol and filling of the chamber with liquid CO_2 occurs. Purge ethanol for 15–20 min. After heating step, once temperature reaches 35–40 °C at 1200–1300 psi, critical point is reached when liquid CO_2 coverts to gas. This step ensures complete drying while preserving ultrastructure of specimens.
9. Using diamond scriber, cut the etched region. For the purpose of identification, mark each sample with distinct marker around the etched region under the coverslip.
10. Mount the cut coverslip piece onto the stage of freeze fracture specimen holder using double sided tape. Four to six coverslip pieces can fit on a sample stage.
11. Coat the sample with 3 nm platinum at 17° angle and 5.5 nm carbon at 90° angle, consecutively. Coat while the specimen stage is rotating.
12. Image the coated coverslip piece with 20 × phase-contrast objective to obtain a phase map of the etched region. Compare this map with the fluorescent map to ensure that the region imaged with fluorescent microscope is present in the etched region.

13. Further cut the coverslip such that it is similar in size as the loop used for lifting the replica and place it on a filter paper on 60 mm cell-culture dish.
14. Before lifting the replica, glow discharge the formvar/carbon coated grid and place it on filter paper along with the coverslip piece.
15. To separate the metal replica from the coverslip, use 5% hydrofluoric acid. In 12-well plate, pipet 0.7–1 mL hydrofluoric acid and drop the marked coverslip piece. In a couple of minutes, the glass sinks to the bottom of the well after separating from the metal replica, which remains afloat.
16. To replace hydrofluoric acid with water, gently fill the well with water up to the rim, and gently pipet out the aqueous HF solution using the polypropylene plastic bulb transfer pipette. Perform this serial dilution and exchange with about 50 mL of water (*see* **Note 17**).
17. Using a loop, carefully pick the replica from under the water surface and place it on the grid on filter paper. The capillary action from filter paper helps deposit the replica onto the grid.
18. Take a phase-contrast map of the lifted replica and compare it with the prelift phase map.
19. Choose the GFP expressing cells present on the grid for EM imaging.

3.5 EM Imaging and Tomography and Image Processing

1. Image each region of interest in the electron microscope. We use SerialEM on a JEOL 1400 transmission electron microscope at 15,000× magnification to make montages according to the size of the unroofed cell regions [15].
2. With 2D EM, survey the gold tagged organelles of interest. Select the cell regions with appropriate labeling for electron tomogram.
3. Collect the single-axis tilt series (−60° to 60°, 1° increments) at 8000×. Stitch the montages and reconstruct the tilt series into tomograms using IMOD software [16].
4. Segment the reconstructed tomograms using 3dmod, model editing, and image display program. We use the 3dmod drawing tool to manually outline the sections of reconstructed vesicle tomograms with closed contours. This outline is a collection of coordinate points marking the locations of the vesicle membrane in the image. Scattered points are used as separate objects to mark the independent gold particles. Coordinates from these two objects can be obtained from more vesicles and spatial analysis can be performed [13].

4 Notes

1. Percentage of horse serum affects the morphology and adherence of PC12 cells on the substrate [17]. We suggest maintaining consistency in the concentration of serum used for reproducible results.
2. We need our cells to fluoresce to enable us to detect cells expressing histidine-tagged proteins, which are targets for Ni-NTA-Au. Since fluorescent proteins simply function as marker for histidine-tagged proteins, you may choose fluorescent proteins as per your experimental plan.
3. We perform unroofing manually. The syringe is positioned a few centimeters above the coverslip before injection. The force with which the fixative solution is injected determines the degree of unroofing. For PC12 cells, gentle injection (slow plunging of the syringe) is enough to obtain reasonable unroofing, while HeLa cells requires injection with greater force (faster plunging of the syringe pump). The type of needle used also affects the unroofing efficiency. Smaller diameter or higher gauge needles will produce higher force during unroofing. You may need to test to see which needle best suits the cell type you are unroofing. Make sure an appropriate substrate is used to seed the cells because the agitation caused during the unroofing may dislodge all cells from your coverslip.
4. To ensure better quality of ultrastructure, we always open fresh paraformaldehyde and glutaraldehyde stock. Paraformaldehyde for unroofing is made in stabilization buffer, while glutaraldehyde for storing imaged samples is made in PBS.
5. We have found BSA can bind with paraformaldehyde and cause nonspecific binding of the gold nanoprobe. To avoid this effect, after 20 min of fixation, we wash samples by transferring the coverslips to fresh PBS four times in 6-well plate to remove any unbound paraformaldehyde.
6. We try not to block samples longer than 1 h. In fact, 30–40 min of blocking also produces similar result as 1 h of blocking. We have found that BSA can cause aggregation of gold nanoparticles, so any unbound BSA is not preferable.
7. We have found nanoprobe solution made in PBS without BSA or other serums or additives is the most appropriate to obtain dispersed gold nanoprobe labeling with minimum clumping of the gold.
8. As we work with unroofed cells, 1 h incubation is more than enough time to obtain appropriate labeling. Longer incubation times may cause nonspecific and background labeling. You

many need to optimize your protocol based on the type of cells (intact or thin-sections) you are trying to label.

9. Phalloidin labels actin filaments and helps visualize the cell shape. If you are following the correlative protocol described in this method, actin labeling is essential for finding the cells on the replica. When comparing the fluorescent map with the phase-contrast map before and after replica lift, one cannot identify cells without the cells shape visualized by fluorescent actin filaments. If cell sorting with flow cytometry is done to ensure that all cell present on the coverslip express GFP and histidine-tagged proteins, then this step is irrelevant.
10. We skip the postfixation step in this protocol. We have found high background/nonspecificity when postfixed with paraformaldehyde. In Fig. 4, we show specifically labeled clathrin coated vesicles with minimum background gold labeling.
11. For diamond objective, we use a custom-made 14 mm spacer tapped with m25 to receive the marker and m25 threading to fit into the microscope turret [14].
12. Attofluor chambers allows us to secure the coverslip, which becomes necessary when etching the glass with a diamond marker. It also makes transferring the coverslip to glutaraldehyde after imaging convenient.
13. Alternatively, you can use flow cytometry to sort GFP-expressing cells. If you choose to sort cells, you can skip this correlative method to screen GFP expressing cells and move straight to EM preparation. In such case, Attofluor chambers can be replaced with other coverslip holders.
14. We use ultrapure water for all dilutions and washing during EM sample preparation. To prevent any potential precipitate formation due to interactions between tannic acid and uranyl acetate, samples in tannic acid must be washed properly before transferring to uranyl acetate solution.
15. Uranyl acetate is light sensitive and precipitates when exposed to light. Hence, the base of the stock bottle usually contains its precipitates. When preparing 0.1% dilution, ensure that the uranyl acetate particles are not used [18].
16. While transferring coverslips, make sure the time it stays in air is minimal to prevent drying the samples.
17. Transfer pipettes with larger diameters are preferable for pipetting water in and out of the well. Larger diameters generate less force when pipetting water, which prevents perturbations that can disintegrate the replica.

References

1. De Mey J, Moeremans M, Geuens G, Nuydens R, De Brabander M (1981) High resolution light and electron microscopic localization of tubulin with the IGS (immuno gold staining) method. Cell Biol Int Rep 5(9): 889–899
2. Stirling JW (1990) Immuno- and affinity probes for electron microscopy: a review of labeling and preparation techniques. J Histochem Cytochem 38(2):145–157
3. Dapson RW (2007) Macromolecular changes caused by formalin fixation and antigen retrieval. Biotech Histochem 82(3):133–140
4. Yamashita S (2010) The post-embedding method for Immunoelectron microscopy of mammalian tissues: a standardized procedure based on heat-induced antigen retrieval. In: Schwartzbach SD, Osafune T (eds) Immunoelectron microscopy: methods and protocols. Humana Press, Totowa, NJ, pp 237–248
5. Risco C, Sanmartín-Conesa E, Tzeng W-P, Frey TK, Seybold V, de Groot RJ (2012) Specific, sensitive, high-resolution detection of protein molecules in eukaryotic cells using metal-tagging transmission electron microscopy. Structure 20(5):759–766
6. Hainfeld JF, Liu W, Halsey CM, Freimuth P, Powell RD (1999) Ni-NTA-gold clusters target his-tagged proteins. J Struct Biol 127(2): 185–198
7. Li S, Raychaudhuri S, Lee SA, Brockmann MM, Wang J, Kusick G, Prater C, Syed S, Falahati H, Ramos R, Bartol TM, Hosy E, Watanabe S (2021) Asynchronous release sites align with NMDA receptors in mouse hippocampal synapses. Nat Commun 12(1):677
8. Heuser J (2000) The production of 'cell cortices' for light and electron microscopy. Traffic 1(7):545–552
9. Hochuli E, Döbeli H, Schacher A (1987) New metal chelate adsorbent selective for proteins and peptides containing neighbouring histidine residues. J Chromatogr 411:177–184
10. Kovtun O, Tillu VA, Ariotti N, Parton RG, Collins BM (2015) Cavin family proteins and the assembly of caveolae. J Cell Sci 128(7): 1269–1278
11. Pearse BM (1976) Clathrin: a unique protein associated with intracellular transfer of membrane by coated vesicles. Proc Natl Acad Sci U S A 73(4):1255–1259
12. Tebar F, Sorkina T, Sorkin A, Ericsson M, Kirchhausen T (1996) Eps15 is a component of Clathrin-coated pits and vesicles and is located at the rim of coated pits. J Biol Chem 271(46):28727–28730
13. Prasai B, Haber GJ, Strub M-P, Ahn R, Ciemniecki JA, Sochacki KA, Taraska JW (2021) The nanoscale molecular morphology of docked exocytic dense-core vesicles in neuroendocrine cells. Nat Commun 12(1):3970
14. Sochacki KA, Taraska JW (2017) Correlative fluorescence super-resolution localization microscopy and platinum replica EM on unroofed cells. In: Erfle H (ed) Super-resolution microscopy: methods and protocols. Springer, New York, NY, pp 219–230
15. Mastronarde DN (2005) Automated electron microscope tomography using robust prediction of specimen movements. J Struct Biol 152(1):36–51
16. Kremer JR, Mastronarde DN, McIntosh JR (1996) Computer visualization of three-dimensional image data using IMOD. J Struct Biol 116(1):71–76
17. Meth JL, Schoenfeld AR (2019) Higher percentage of horse serum in culture media blocks attachment of PC12 cells. BioTechniques 67(6):256–258
18. Chen S, Goode AE, Skepper JN, Thorley AJ, Seiffert JM, Chung KF, Tetley TD, Shaffer MSP, Ryan MP, Porter AE (2016) Avoiding artefacts during electron microscopy of silver nanomaterials exposed to biological environments. J Microsc 261(2):157–166

Chapter 14

Endocytosis Assays Using Cleavable Fluorescent Dyes

Shifeng Wang, Chun Wan, Galen T. Squiers, and Jingshi Shen

Abstract

Endocytosis mediates the entry of surface and extracellular cargoes into the cell. In this chapter, we describe assays to quantitively measure the endocytosis of both soluble and transmembrane cargo proteins, taking advantage of cleavable fluorescent dyes labeling cargo proteins or antibodies recognizing cargo proteins. After removing surface-bound fluorescent dye, internalized cargoes are measured using confocal imaging and flow cytometry. We also describe strategies to determine the role of clathrin-mediated endocytosis (CME) in the internalization of a cargo by using a small molecule inhibitor of CME and knockout (KO) of the *AAGAB* gene, which encodes an essential regulator of CME.

Key words Membrane trafficking, Endocytosis, Clathrin-mediated endocytosis, Cargo protein, Transferrin, Transferrin receptor, GLUT4, AAGAB

1 Introduction

Under normal conditions, most surface and extracellular proteins are internalized through clathrin-mediated endocytosis (CME) [1–3]. Moreover, CME is often exploited by pathogens and toxins to gain access to intracellular compartments [4]. Since the coat protein clathrin does not bind directly to cargoes, it relies on adaptors to recruit cargo proteins to endocytic sites [4–6]. A central clathrin adaptor in CME is the heterotetrameric AP2 adaptor complex [4, 5, 7]. Our previous research revealed that AP2 adaptor assembly is an ordered process controlled by alpha and gamma adaptin binding protein (AAGAB) [8]. Without the assistance of AAGAB, AP2 subunits fail to form the adaptor complex, resulting in CME defects [8].

To monitor endocytosis, soluble cargoes are usually labeled with fluorescent dyes. Transmembrane cargo proteins, on the other hand, are often labeled indirectly through fluorophore-conjugated antibodies recognizing the proteins. To accurately measure endocytosis, it is critical to distinguish between surface-exposed and internalized cargoes. A routinely used method

Jingshi Shen (ed.), *Membrane Trafficking: Methods and Protocols*,
Methods in Molecular Biology, vol. 2473, https://doi.org/10.1007/978-1-0716-2209-4_14,

to remove surface-exposed molecules is washing with an acidic buffer (e.g., a pH 3.0 acetic acid buffer), in which low pH loosens ligand-receptor or antibody–antigen interactions. The efficiency of acidic washing, however, varies widely across applications and often cannot completely dissociate surface-bound molecules. Moreover, prolonged exposure to low pH solutions is cytotoxic, likely due to disruption of essential biological complexes and acidification of intracellular compartments [9].

In this chapter, we describe endocytosis assays using ATTO-565-3-[(2-aminoethyl)dithio]propionic acid N-hydroxysuccinimide ester (ATTO-565-AEDP-NHS-ester), in which the ATTO-565 fluorescent dye (excitation/emission: 563 nm/589 nm) is linked to the NHS moiety through a disulfide bond (Fig. 1a). Surface-exposed ATTO-565 dyes can be removed by breaking the disulfide bond using membrane-impermeable reducing agents such as sodium 2-mercaptoethane sulfonate (MESNa) [10, 11], whereas endocytosed dyes are resistant to cleavage. ATTO-565-AEDP-NHS-ester is used to label transferrin (Tf), an iron-binding soluble protein internalized through binding to Tf receptors (Fig. 1b) [12–14], as well as antibodies recognizing the transmembrane protein GLUT4, a facilitative glucose transporter (Fig. 1b) [8, 15]. MESNa washing fully removes surface-exposed ATTO-565 dyes (Fig. 2). Internalized ATTO-565-conjugated Tf is quantified using confocal imaging and flow cytometry (Fig. 2). We also describe procedures to examine how Tf endocytosis is affected by *AAGAB* knockout (Fig. 3), or treatment with ikarugamycin (IKA) (Fig. 4), a CME inhibitor [16]. Endocytosis of GLUT4 is measured using ATTO-565-labeled anti-HA antibodies that recognize an HA-tagged GLUT4 reporter (Figs. 1b and 5). These methods can also be used to study other CME cargoes as well as clathrin-independent endocytosis.

2 Materials

2.1 Cell Culture

1. HeLa cells.
2. HeLa cells stably expressing the HA-GLUT4-GFP reporter [8, 17].
3. *AAGAB* KO HeLa cells, developed using CRISPR genome editing [8].
4. Cell culture media: Dulbecco's Modified Eagle Medium with high glucose and sodium pyruvate, supplemented with 10% FB Essence (VWR, Cat# 10803-034), 50 units/mL penicillin, and 50 μg/mL streptomycin. FB Essence can be replaced with an equal concentration of fetal bovine serum (FBS).
5. Trypsin.

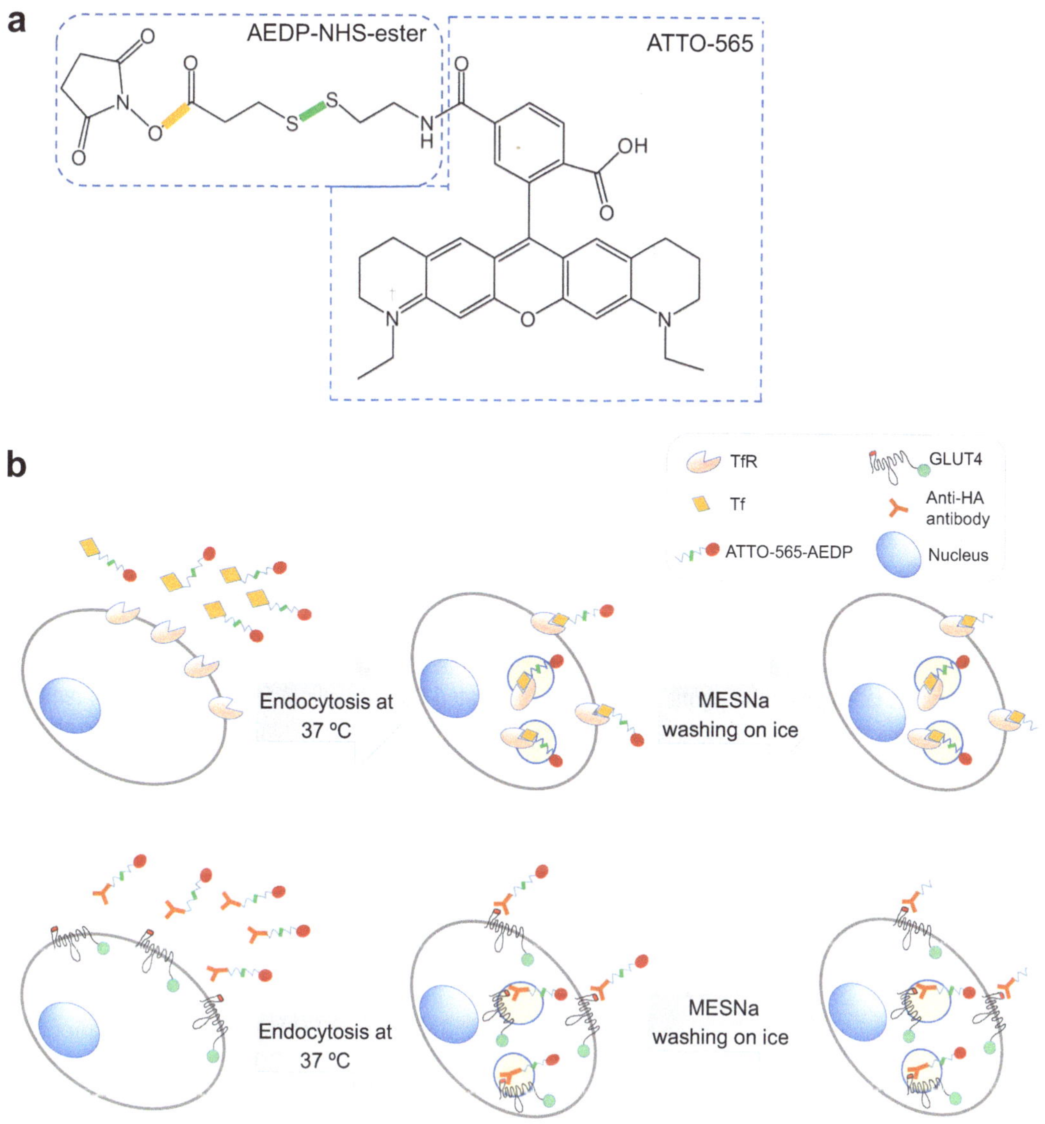

Fig. 1 Workflow of the endocytosis assay. (**a**) Chemical structure of ATTO-565-AEDP-NHS-ester, in which the disulfide bond is labeled in green and the bond reacting with a primary amine on a cargo protein is labeled in orange. The salt form of the chemical contains the anion ClO_4^-. (**b**) Diagrams illustrating the endocytosis assays of Tf (top) and HA-GLUT4-GFP (bottom). The membrane-impermeable reducing agent MESNa breaks disulfide bonds (labeled in green) between ATTO-565 and proteins, selectively removing ATTO-565 dyes from surface-exposed proteins

6. Accutase (Innovative Cell Technologies, Inc., Cat# AT104) (*see* **Note 1**).
7. Phosphate-buffered saline (PBS): 137 mM NaCl, 10 mM Na_2HPO_4, 2.7 mM KCl, 1.7 mM KH_2PO_4, pH 7.4. Store at 4 °C.
8. Cell culture incubator set at 37 °C with 5% CO_2.

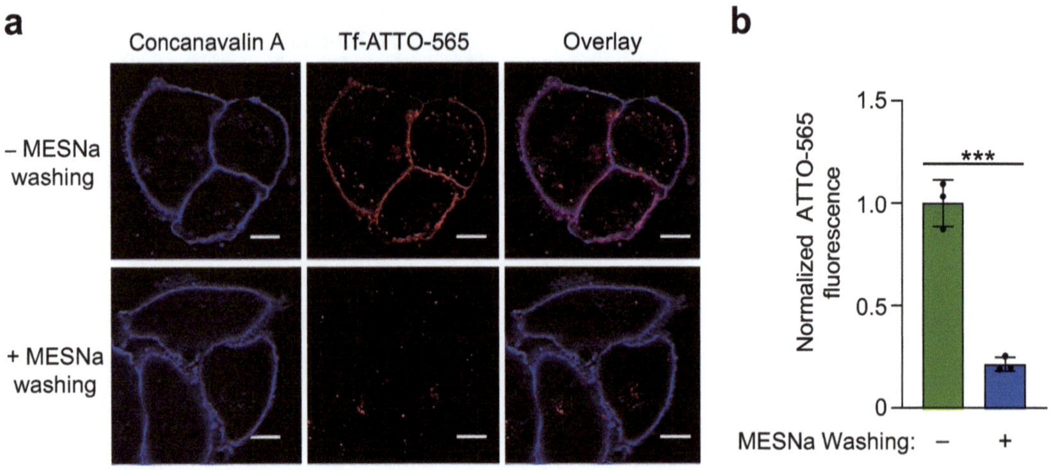

Fig. 2 MESNa fully removes surface-exposed ATTO-565 dyes. (**a**) Representative confocal images of Tf-ATTO-565 in *AAGAB* KO HeLa cells. *AAGAB* KO cells were used here because Tf and TfR are mainly localized to the surfaces of these mutant cells due to endocytosis defects (*see* Fig. 3), enabling us to better assess the efficiency of MESNa washing. The cells were incubated with 5 μg/mL Tf-ATTO-565 for 10 min on ice, which allowed Tf-ATTO-565 to bind to surface-localized Tf receptors without endocytosis. After washing with 150 mM MESNa three times on ice (5 min each time), the cells were fixed with 4% PFA and the cell membrane was labeled with 50 μg/mL of CF405M-conjugated Concanavalin A for 20 min at room temperature. The images were captured using a 100× oil immersion objective on a Nikon A1 laser-scanning confocal microscope. Scale bars: 10 μm. MESNa fully removed surface ATTO-565 dyes. The residual fluorescence inside the MESNa-washed cells was likely resulted from slight Tf-ATTO-565 internalization during the incubation on ice. (**b**) Flow cytometry analysis of Tf-ATTO-565 fluorescence in *AAGAB* KO HeLa cells with or without MESNa washing. The cells were labeled and washed in a similar way as in (**a**). The mean fluorescence of MESNa-washed cells was normalized to that of unwashed cells. Data are presented as mean ± SD, $n = 3$. *** $P < 0.001$. P values were calculated using Student's t-test

9. 24-well plates.
10. Ikarugamycin (IKA): dissolve in DMSO to a final concentration of 5 mM, aliquot and store at −20 °C.

2.2 Protein Labeling

1. ATTO-565-AEDP-NHS-ester: custom synthesis by ATTO-TEC (Siegen, Germany). Dissolve in DMSO to a final concentration of 10 mM, aliquot and store at −70 °C (*see* **Note 2**).
2. Human Apo-Tf. We use a product from Sigma-Aldrich (Cat# 616395). Dissolve in PBS with a pH of 8.3–8.5 (adjusted using NaOH) to a final concentration of 5 mg/mL (*see* **Note 3**). Prepare fresh for each labeling experiment.
3. Anti-HA monoclonal antibodies (BioLegend, Cat# 923502). Other anti-HA antibodies suitable for immunofluorescence and flow cytometry can also be used.
4. Coomassie protein assay reagent. We use a product from ThermoFisher (Cat# 1856209).
5. Cuvettes with two clear windows and a pathlength of 10 mm.

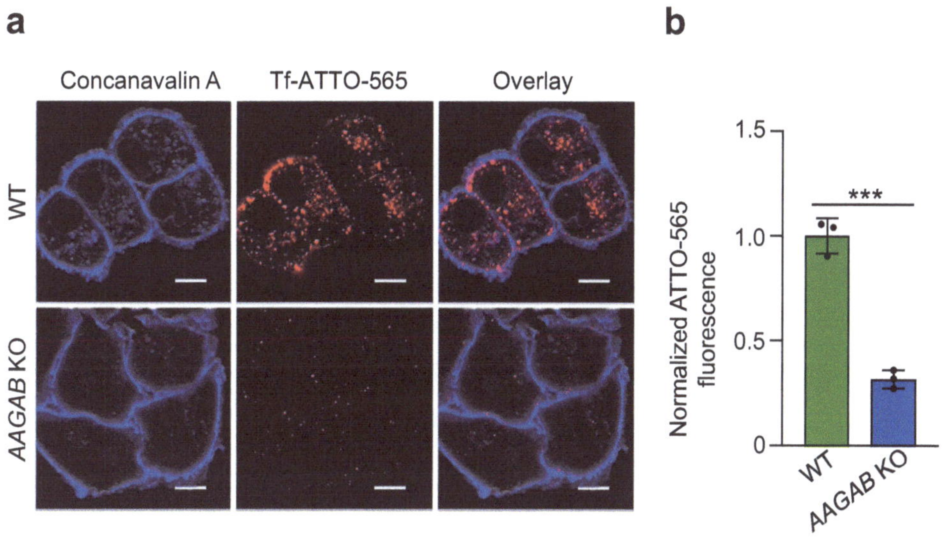

Fig. 3 Endocytosis of Tf-ATTO-565 in WT and *AAGAB* KO HeLa cells. (**a**) Representative confocal images of Tf-ATTO-565 endocytosis in WT and *AAGAB* KO HeLa cells. After incubation in the serum-free KRH buffer for 30 min, the cells were chilled on ice for 10 min. Subsequently, the cells were incubated with 5 μg/mL Tf-ATTO-565 for 5 min at 37 °C to allow endocytosis to occur. After washing with 150 mM MESNa three times on ice (5 min each time), the cells were fixed with 4% PFA and the cell membrane was labeled with 50 μg/mL CF405M-conjugated Concanavalin A for 20 min at RT. Scale bars: 10 μm. (**b**) Flow cytometry analysis of Tf-ATTO-565 fluorescence in WT and *AAGAB* KO HeLa cells. Endocytosis was carried out as in (**a**). The mean fluorescence of the KO cells was normalized to that of WT cells. Data are presented as mean ± SD, $n = 3$. *** $p < 0.001$. *P* values were calculated using Student's *t*-test

6. Dialysis tubing with a diameter of 14.6 mm and a molecular cutoff (MWCO) of 6–8 kDa.
7. Dialysis tubing closures (also known as clamps) with a width of at least 35 mm.
8. Kathon CG/ICP.

2.3 Immuno-fluorescence and Flow Cytometry

1. Round micro cover glasses with a diameter of 12 mm and a thickness of 0.17–0.25 mm. Sterilize using 70% ethanol before use.
2. Microscope slides.
3. KRH buffer: 12 mM HEPES-KOH, 121 mM NaCl, 4.9 mM KCl, 1.2 mM $MgSO_4$, and 0.33 mM $CaCl_2$, pH 7.4. Store at 4 °C (*see* **Note 4**).
4. CF405M-conjugated Concanavalin A (Biotium, Cat# 29074): dissolve in PBS to a final concentration of 2 mg/mL and store at 4 °C. Other fluorescent dyes can also be used to stain the plasma membrane.
5. MESNa washing buffer: 150 mM MESNa, 50 mM Tris–NaOH, 100 mM NaCl, 1 mM EDTA, 0.2% BSA, pH 8.6. Prepare fresh for each experiment and keep on ice.

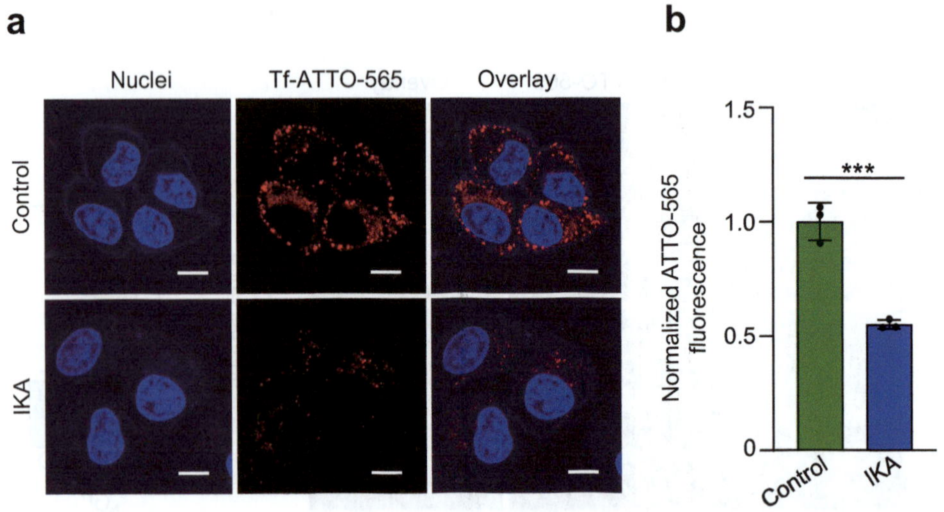

Fig. 4 IKA inhibits Tf-ATTO-565 endocytosis in HeLa cells. (**a**) Representative confocal images of Tf-ATTO-565 endocytosis in WT HeLa cells treated with or without the CME inhibitor IKA. Scale bars: 10 μm. The cells were treated with or without 5 μM IKA for 5 min, chilled on ice for 10 min, and incubated with 5 μg/mL Tf-ATTO-565 for 5 min at 37 °C in the presence or absence of IKA. After washing with 150 mM MESNa three times on ice (5 min each time), the cells were fixed with 4% PFA and stained with 10 μg/mL Hoechst 33342 for 15 min at RT. (**b**) Flow cytometry analysis of Tf-ATTO-565 fluorescence in HeLa cells treated with or without IKA. Endocytosis and IKA treatment were carried out as in (**a**). The mean fluorescence of IKA-treated cells was normalized to that of DMSO-treated control samples. Data are presented as mean ± SD, $n = 3$. *** $p < 0.001$. *P* values were calculated using Student's *t*-test

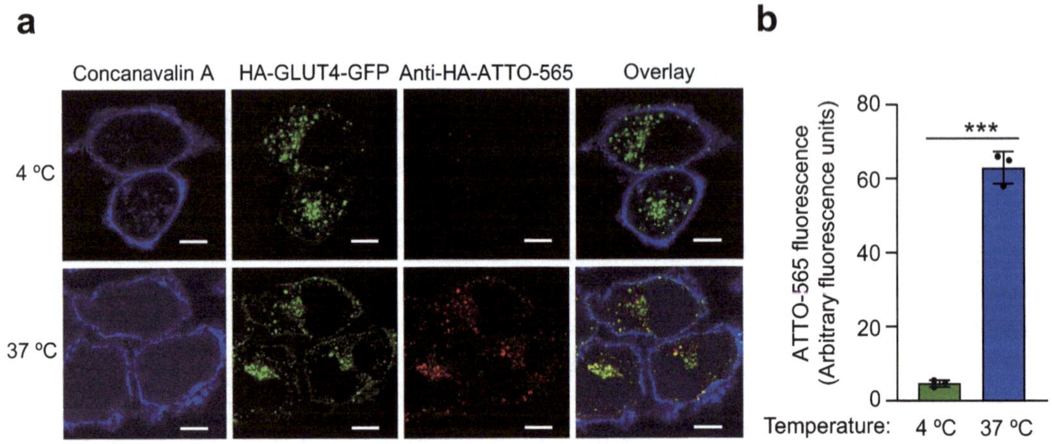

Fig. 5 Endocytosis of GLUT4 in HeLa cells. (**a**) Representative confocal images of HA-GLUT4-GFP reporters labeled with ATTO-565-conjugated anti-HA antibodies (anti-HA-ATTO-565). The cells were incubated with 1 μg/mL anti-HA-ATTO-565 at either 4 °C or 37 °C. After washing with 150 mM MESNa three times on ice (5 min each time), the cells were fixed with 4% PFA and the cell membrane was labeled with 50 μg/mL CF405M-conjugated Concanavalin A. Scale bars: 10 μm. (**b**) Flow cytometry analysis of anti-HA-ATTO-565 fluorescence after endocytosis was carried out as in (**a**). Data are presented as mean ± SD, $n = 3$. *** $p < 0.001$. *P* values were calculated using Student's *t*-test

6. Hoechst 33342, dilute to 10 μg/mL in PBS before usage.
7. Paraformaldehyde (PFA): the 16% PFA stock solution is kept at 4 °C. Immediately before an experiment, dilute the stock solution to 4% using PBS.
8. Mounting medium. We use the SlowFade™ Gold Antifade Mountant (Thermo Fisher, Cat# S36940), store at −20 °C.
9. Cover glass sealant such as nail polish.
10. Cell strainers with a pore size of 50 μm.
11. Flow cytometry buffer (PBS with 5% FBE): add 5 mL of FBE to 95 mL of ice-cold PBS, mix well, and store on ice. Prepare fresh for each experiment. FBE can be replaced with an equal concentration of FBS.
12. 5 mL round bottom polystyrene tubes.
13. Calibration beads for flow cytometry. We use the MACS-Quant® calibration beads (Miltenyi Biotec, Cat# 130-093-607).

2.4 Equipment

1. A laser-scanning confocal microscope. We use a Nikon A1 laser-scanning confocal microscope equipped with 405 nm, 488 nm and 565 nm lasers (Nikon Inc., Japan).
2. Flow cytometer. We use a MACSQuant Analyzer (Miltenyi Biotec Inc., Germany) equipped with 405 nm, 488 nm and 561 nm lasers.
3. Spectrophotometer. We use a SmartspecTM plus spectrophotometer (Bio-Rad Laboratories Inc., USA).

2.5 Software

1. Fiji (https://imagej.net/software/fiji/).
2. FlowJo (https://www.flowjo.com/).
3. GraphPad Prism (https://www.graphpad.com/).

3 Methods

3.1 Labeling of Tf Using ATTO-565-AEDP-NHS-Ester

1. Add 1 μL of 10 mM ATTO-565-AEDP-NHS-ester to 1 mL of 5 mg/mL Tf such that the molar ratio of the dye to Tf is approximately 10:1. Incubate the sample at room RT for 1 h. Continue the incubation for 18 h at 4 °C.
2. Centrifuge the labeled sample at 6769 × *g* for 30 s to remove precipitates, and collect the supernatant.
3. Cut 10 cm of dialysis tubing and immerse it in Milli-Q water for 10 min. Seal one end of the dialysis tubing using a closure, transfer the labeled protein sample into the dialysis tubing, and seal the other end of the tubing using another closure.

4. Dialyze the sample at 4 °C against 1 L of PBS. Change to fresh PBS three times during a period of 24 h. This dialysis step removes free dyes.
5. Transfer ATTO-565-labeled Tf (Tf-ATTO-565) to a 1.5 mL microtube and keep on ice.
6. To measure the protein concentration of the dialyzed sample, add 2 μL of Tf-ATTO-565 to 600 μL of the coomassie protein assay reagent. In parallel, prepare a series of BSA dilutions in the coomassie protein assay reagent to calculate the standard curve. Incubate the samples at RT for 5 min. Measure the absorbance of the samples at 595 nm using a spectrophotometer and calculate the protein concentration of Tf-ATTO-565 based on the BSA standard curve.
7. Add the preservative Kathon CG/ICP to the labeled sample to a final concentration of 22.5 ppm active ingredients. Store the labeled proteins at 4 °C protected from light.

3.2 Labeling of Anti-HA Antibodies Using ATTO-565-AEDP-NHS-Ester

1. Add 1 μL of 10 mM ATTO-565-AEDP-NHS-ester to 100 μL of anti-HA tag antibodies (1 mg/mL) such that the molar ratio of the dye to antibody is approximately 10:1. Mix well and incubate at RT for 1 h (*see* **Note 5**).
2. Centrifuge the sample at 8000 rpm for 30 s, collect the supernatant.
3. Dialyze the sample against PBS as described in **steps 3** and **4** of Subheading 3.1 using an appropriate length of dialysis tubing.
4. Transfer the labeled antibodies to a 1.5 mL microtube. Measure the concentration of the labeled antibodies using the coomassie protein assay reagent as described in **step 6** of Subheading 3.1.
5. Add the preservative Kathon CG/ICP to the labeled antibodies to a final concentration of 22.5 ppm active ingredients. Store the labeled antibodies at 4 °C protected from light.

3.3 Determining the Degree of Labeling (DOL)

To determine DOL, calculate the average dye-to-protein ratio of a sample using the following equation (c represents molar concentration):

$$\mathrm{DOL} = \frac{C(\mathrm{dye})}{C(\mathrm{protein})} \tag{1}$$

The protein concentration is determined in Subheadings 3.1 and 3.2 using the coomassie protein assay.

1. To determine the molar concentration of the dye, measure the absorbance of ATTO-565-conjugated proteins at 564 nm using a spectrophotometer.

2. Calculate the concentration of the dye in a labeled sample using the following equation:

$$C(\text{dye}) = {}^{A}\!/_{\varepsilon} \times d \tag{2}$$

In this equation, A represents absorption at 564 nm, ε represents the extinction coefficient, which is $1.2 \times 10^5\ \text{M}^{-1}\ \text{cm}^{-1}$ for the ATTO-565 dye, and d represents the path length (1 cm).

3. Determine the DOL of a sample using Eq. (1). In our assays, the DOL of Tf is usually 4–5, whereas the DOL of anti-HA antibodies is usually 1–2.

3.4 Endocytosis of Tf-ATTO-565

1. One day prior to endocytosis assays, disassociate HeLa cells using trypsin and seed them into a 24-well plate at a density of 50,000 cells/well. When plating cells for immunofluorescence, put sterile micro cover glasses into the wells of the 24-well plate before seeding cells.
2. Keep the plate in a 37 °C incubator with 5% CO_2 and culture overnight.
3. On the day of endocytosis assays, wash the cells with prewarmed KRH buffer three times, and keep the cells in the KRH buffer for 30 min in the 37 °C incubator.
4. Take out the plate and chill on ice for 10 min.
5. While the cells are on ice, dilute Tf-ATTO-565 in prewarmed or ice-cold KRH buffer to a final protein concentration of 5 μg/mL.
6. Remove the KRH buffer from the cells, add 250 μL of the diluted prewarmed Tf-ATTO-565 KRH solution, and incubate the cells in a 37 °C incubator for desired times. Efficient endocytosis of Tf occurs within 5 min of incubation at 37 °C (Figs. 3 and 4). In a negative control, add 250 μL of the diluted ice-cold Tf-ATTO-565 KRH solution and incubate on ice for the same durations.
7. To test the effects of the CME inhibitor IKA on endocytosis, add IKA to the KRH buffer to a final concentration of 5 μM and incubate at 37 °C for 5 min before endocytosis assays (*see* **Note 6**). The same concentration of IKA is also included during endocytosis. Use DMSO as a vehicle control.
8. After endocytosis, chill the plate on ice. Wash the cells with the ice-cold MESNa washing buffer three times, 5 min each.
9. Wash the cells once with ice-cold PBS and proceed to Subheadings 3.6 and 3.7.

3.5 Endocytosis of HA-GLUT4-GFP

1. One day prior to endocytosis assays, disassociate HeLa cells stably expressing the HA-GLUT4-GFP reporter using trypsin, and seed the cells into a 24-well plate at a density of 50,000 cells/well. When plating cells for immunofluorescence, put sterile micro cover glasses into the wells of the 24-well plate before seeding cells.
2. Keep the plate in a 37 °C incubator with 5% CO_2 and culture overnight.
3. On the day of endocytosis assays, wash the cells three times with prewarmed KRH buffer, and keep the cells in the KRH buffer for 30 min in the 37 °C incubator.
4. Take out the plate and chill on ice for 10 min.
5. While the cells are on ice, dilute anti-HA-ATTO-565 in prewarmed or ice-cold KRH buffer to a final concentration of 1 μg/mL.
6. Remove the KRH buffer from the cells, add 250 μL of the diluted prewarmed anti-HA-ATTO-565 KRH solution to the cells, and incubate in a 37 °C incubator for desired times. Efficient endocytosis of anti-HA antibodies in the reporter cells occurs within 10 min of 37 °C incubation (Fig. 5). In a negative control, add 250 μL of the diluted ice-cold anti-HA-ATTO-565 KRH solution and incubate on ice for the same durations.
7. After endocytosis, chill the plate on ice. Wash the cells with ice-cold MESNa washing buffer three times, 5 min each.
8. Wash once with ice-cold PBS and proceed to Subheadings 3.6 and 3.7.

3.6 Confocal Imaging of Endocytosis

1. To fix cells, add 500 μL of 4% PFA solution to each well in the 24-well plate and incubate at RT for 20 min.
2. Remove the PFA solution, wash three times with PBS.
3. To stain the cell membrane, dilute the CF405M-conjugated Concanavalin A solution in PBS to a final concentration of 50 μg/mL, and add 250 μL to each well. Incubate at RT for 20 min (*see* **Note 7**). To stain cell nuclei, dilute Hoechst 33342 in PBS to a final concentration of 10 μg/mL, and add 250 μL to each well. Incubate at RT for 15 min.
4. Remove the CF405M Concanavalin A or Hoechst 33342 solution, wash once with PBS.
5. Wash the cells with Milli-Q water once to prevent formation of salt precipitates. Add a small drop of mounting medium to the middle of a microscope slide, remove a micro cover glass from the plate and place it onto the mounting medium with cells facing down. Repeat for other micro cover glasses.

6. Keep the slides at RT protected from light for at least 1 h. Use sealants to seal the edges of the cover glasses.
7. Use a 100× oil objective on a confocal microscope to capture cell images. Use the 405 nm laser to detect CF405M-conjugated Concanavalin A (cell membrane staining) and Hoechst 33342 (nucleus staining), the 488 nm laser to detect the GFP fluorescence of HA-GLUT4-GFP, and the 561 nm laser to detect ATTO-565. Pay attention to laser power and exposure time (*see* **Note 8**).
8. Analyze fluorescence intensity of confocal images using the Fiji software. First, use "Polygon selections" to draw edges of selected cells. Second, add the selected cells to ROI Manager (*see* **Note 9**) by clicking "Analyze" > "Tools" > "ROI Manager" > "Add [t]". Third, set the lower and upper threshold levels by clicking "Image" > "Adjust" > "Threshold" > "Apply". The lower threshold is set according to the minimal intensity of single vesicles, and the upper threshold is set according to the maximum intensity of single vesicles.
9. Forth, analyze the total fluorescence intensity for each cell on the ROI list by clicking "Analyze Particles," type in the ranges of the size and circularity of the particles, and click "OK." The analyzed data will show up in a new window. "IntDen" indicates total fluorescence intensity per cell. Copy the data to a spreadsheet for statistical analysis.

3.7 Flow Cytometry Analysis of Endocytosis

1. Dissociate cells by adding 200 μL of prewarmed Accutase to each well of the 24-well plate. Keep the plate in a 37 °C incubator for 5–10 min until all cells detach.
2. Add 300 μL of ice-cold flow cytometry buffer to each well, pipette 3–5 times to resuspend the cells. Pass the cell suspension through a strainer to remove cell clumps.
3. Transfer the resuspended cells to 5 mL round bottom polysty rene tubes on ice.
4. Turn on the MACSQuant analyzer, wash the system with 5% bleach, followed by backflush.
5. To calibrate the system, scan the barcode of MACSQuant calibration beads, add one drop of the bead solution to a new 5 mL round bottom polystyrene tube, and click "run" to start calibration. When all the channels have passed calibration, proceed to run cell samples (*see* **Note 10**).
6. Measure the fluorescence of Tf-ATTO-565 and anti-HA-ATTO-565 using the Y1-PE channel (filter 586/15 nm) on the MACSQuant Analyzer. Analyze about 10,000 cells for each sample (*see* **Note 11**).

7. To analyze flow cytometry data, open the data file in FlowJo. Identify the cell population based on the forward scatter (FSC) versus side scatter (SSC) density plot and name this population "cells". Identify single cells from the "cells" population based on the forward scatter height (FSC-H) versus forward scatter area (FSC-A) density plot and name this population "single cells" (*see* **Note 12**). Add mean fluorescence intensities of the Y1-PE channel for the "single cells" population. Export the data to a spreadsheet for further analysis.

4 Notes

1. Accutase is gentler than trypsin for dissociating cells and thus better preserves surface antigens in flow cytometry assays.
2. Other cleavable dyes can also be used to label cargo proteins, depending on the lasers available in a user's microscope or flow cytometer as well as fluorescent molecules already present in cells. ATTO-TEC also offers custom synthesis of other cleavable dyes such as ATTO-488-AEDP-NHS-ester and ATTO-647-AEDP-NHS-ester.
3. The pH of the buffer to dissolve Tf is critical and the buffer should be free of amines.
4. KRH buffer is not recommended for washing cells before cell dissociation. Calcium ions in KRH buffer reduce the efficiency of Accutase and trypsin. Cells should be washed with PBS before adding Accutase or trypsin.
5. The labeling time and dye-protein ratio need to be optimized for each protein. We find that longer labeling times and higher dye concentrations cause anti-HA antibodies to precipitate. Tf, by contrast, tolerates wider ranges of labeling time and dye concentration.
6. In our assays, 5 μM IKA significantly inhibits Tf endocytosis. Higher concentrations of IKA lead to stronger inhibition but with increased cytotoxicity.
7. Concanavalin A labels the plasma membrane, which is useful for assessing the efficiency of ATTO-565 removal from the cell membrane by MESNa washing.
8. Appropriate laser power and exposure time are critical for confocal imaging. High laser power or long exposure times may lead to image saturation and photobleaching. Low laser power or short exposure times, on the other hand, may yield faint images. Laser power, exposure time, resolution, and other parameters should be kept the same for all samples.

9. To ensure analysis accuracy, select only whole cells in Fiji software and avoid partial cells on the edges of fields.
10. Perform bead calibration once a day on an MACSQuant flow cytometer before running cell samples. If calibration of one channel fails, this is often due to clogging or low concentration of calibration beads.
11. To ensure the quality of flow cytometry data, a sufficient number of cells should be analyzed. In addition, comparable numbers of cells should be analyzed for all samples.
12. While cell strainers remove large cell clumps, small cell clumps such as doublets may pass through cell strainers. The profiles of small cell clumps collected by the flow cytometer must be excluded from subsequent analysis to ensure that only single cells are used for data analysis. The FSC-H vs. FSC-A density plot is used to exclude cell clumps as well as cells undergoing division based on size.

Acknowledgments

We thank Yan Ouyang, Harrison Puscher, and Ximing Dou for technical assistance, and Dr. Stefan Höning for advice. This work was supported by National Institutes of Health (NIH) grants DK124431 (J.S.), GM126960 (J.S.), AG061829 (J.S.), an American Diabetes Association Basic Science Award (J.S.).

References

1. McMahon HT, Boucrot E (2011) Molecular mechanism and physiological functions of clathrin-mediated endocytosis. Nat Rev Mol Cell Biol 12(8):517–533. https://doi.org/10.1038/nrm3151
2. Haucke V (2006) Cargo takes control of endocytosis. Cell 127(1):35–37. https://doi.org/10.1016/j.cell.2006.09.012
3. Traub LM, Bonifacino JS (2013) Cargo recognition in clathrin-mediated endocytosis. Cold Spring Harb Perspect Biol 5(11):a016790. https://doi.org/10.1101/cshperspect.a016790
4. Mettlen M, Chen PH, Srinivasan S, Danuser G, Schmid SL (2018) Regulation of Clathrin-mediated endocytosis. Annu Rev Biochem 87:871–896. https://doi.org/10.1146/annurev-biochem-062917-012644
5. Kaksonen M, Roux A (2018) Mechanisms of clathrin-mediated endocytosis. Nat Rev Mol Cell Biol 19(5):313–326. https://doi.org/10.1038/nrm.2017.132
6. Wood LA, Larocque G, Clarke NI, Sarkar S, Royle SJ (2017) New tools for "hot-wiring" clathrin-mediated endocytosis with temporal and spatial precision. Int J Cell Biol 216(12):4351–4365. https://doi.org/10.1083/jcb.201702188
7. Kelly BT, McCoy AJ, Späte K, Miller SE, Evans PR, Höning S, Owen DJ (2008) A structural explanation for the binding of endocytic dileucine motifs by the AP2 complex. Nature 456(7224):976–979. https://doi.org/10.1038/nature07422
8. Gulbranson DR, Crisman L, Lee M, Ouyang Y, Menasche BL, Demmitt BA, Wan C, Nomura T, Ye Y, Yu H, Shen J (2019) AAGAB controls AP2 adaptor assembly in Clathrin-mediated endocytosis. Dev Cell 50(4):436–446 e435. https://doi.org/10.1016/j.devcel.2019.06.013
9. Kamei N, Yamamoto S, Hashimoto H, Nishii M, Miyaura M, Tomada K, Nakase I, Takeda-Morishita M (2019) Optimization of

the method for analyzing endocytosis of fluorescently tagged molecules: impact of incubation in the cell culture medium and cell surface wash with glycine-hydrochloric acid buffer. J Control Release 310:127–140. https://doi.org/10.1016/j.jconrel.2019.08.020

10. Bitsikas V, Corrêa IR Jr, Nichols BJ (2014) Clathrin-independent pathways do not contribute significantly to endocytic flux. eLife 3: e03970. https://doi.org/10.7554/eLife.03970
11. Tobys D, Kowalski LM, Cziudaj E, Muller S, Zentis P, Pach E, Zigrino P, Blaeske T, Honing S (2020) Inhibition of clathrin-mediated endocytosis by knockdown of AP-2 leads to alterations in the plasma membrane proteome. Traffic 22(1-2):6–22. https://doi.org/10.1111/tra.12770
12. Mayle KM, Le AM, Kamei DT (2012) The intracellular trafficking pathway of transferrin. Biochim Biophys Acta 1820:264–281. https://doi.org/10.1016/j.bbagen.2011.09.009
13. Ishikawa Y, Tacheva-Grigorova SK, Maeda M, Aguet F, Lee S-U, Ericsson M, Kirchhausen T, Maeda T (2013) Cellular mechanisms for transferrin receptor endocytosis vital for terminal erythroid differentiation. Blood 122(21): 3434. https://doi.org/10.1182/blood.V122.21.3434.3434
14. Harding C, Heuser J, Stahl P (1983) Receptor-mediated endocytosis of transferrin and recycling of the transferrin receptor in rat reticulocytes. Int J Cell Biol 97(2):329–339. https://doi.org/10.1083/jcb.97.2.329
15. Bryant NJ, Govers R, James DE (2002) Regulated transport of the glucose transporter GLUT4. Nat Rev Mol Cell Biol 3(4):267–277
16. Elkin SR, Oswald NW, Reed DK, Mettlen M, MacMillan JB, Schmid SL (2016) Ikarugamycin: a natural product inhibitor of Clathrin-mediated endocytosis. Traffic 17(10): 1139–1149. https://doi.org/10.1111/tra.12425
17. Gulbranson DR, Davis EM, Demmitt BA, Ouyang Y, Ye Y, Yu H, Shen J (2017) RABIF/MSS4 is a Rab-stabilizing holdase chaperone required for GLUT4 exocytosis. Proc Natl Acad Sci U S A 114(39): E8224–E8233. https://doi.org/10.1073/pnas.1712176114

Chapter 15

Reconstituting and Purifying Assembly Intermediates of Clathrin Adaptors AP1 and AP2

Bing Wang, Rui Yang, Yuan Tian, and Qian Yin

Abstract

Clathrin-coated vesicles mediate membrane cargo transportation from the plasma membrane, the trans-Golgi network, the endosome, and the lysosome. Heterotetrameric adaptor complexes 1 and 2 (AP1 and AP2) are bridges that link cargo-loaded membranes to clathrin coats. Assembly of AP2 was previously considered to be spontaneous; however, a recent study found AP2 assembly is a highly orchestrated process controlled by alpha and gamma adaptin binding protein (AAGAB). Evidence shows that AAGAB controls AP1 assembly in a similar way. Insights into the orchestrated assembly process and three-dimensional structures of assembly intermediates are only emerging. Here, we describe a protocol for reconstitution and purification of the complexes containing AAGAB and AP1 or AP2 subunits, known as AP1 and AP2 hemicomplexes. Our purification routinely yields milligrams of pure complexes suitable for structural analysis by X-ray crystallography and electron microscopy.

Key words Membrane trafficking, Endocytosis, Clathrin-mediated endocytosis, Golgi, Endosome, Adaptor complex, AP1, AP2, AAGAB, Protein purification

1 Introduction

Clathrin-coated vesicles (CCVs) mediate membrane cargo transportation from the plasma membrane, the trans-Golgi network, the endosome, and the lysosome [1–4]. Clathrin constitutes the outer scaffold of CCV, which assembles from three-legged triskelions and forms a polyhedral cage [5–7]. Multisubunit adaptor protein complexes (APs) are major bridges between cargo-loaded membranes and clathrin coats within CCV, as subunits of the adaptors bind membrane lipids, sorting signals on cargoes, and clathrin triskelions [1, 4]. AP1 and AP2 are two most studied clathrin adaptors. AP1

Bing Wang and Rui Yang contributed equally to this work.

Jingshi Shen (ed.), *Membrane Trafficking: Methods and Protocols*,
Methods in Molecular Biology, vol. 2473, https://doi.org/10.1007/978-1-0716-2209-4_15,

mediates membrane trafficking from the Golgi to the endosome, whereas AP2 facilitates endocytosis from the plasma membrane [8–10].

APs are heterotetrameric complexes composed of two ~100 kDa large subunits (β1 and γ for AP1, β2 and α for AP2), one medium subunit of ~50 kDa (μ, specifically μ1 for AP1, μ2 for AP2), and one small subunit of ~18 kDa (σ, specifically σ1 for AP1, σ2 for AP2) [11–13]. The two large subunits of AP1 and AP2 complex are composed of a "trunk" domain, an "appendage" domain, and a flexible connecting "hinge" region (Fig. 1a, left panel). The trunk domains of the two large subunits plus the entire medium (μ) and small (σ) subunits are structured as a "core" domain, which has been extensively studied by crystallography and cryogenic electron microscopy (cryo-EM) [14–25]. AP1 and AP2 core structures can be considered as composed of two hemicomplexes: γ/σ1 and β1/μ1 for AP1, and α/σ2 and β2/μ2 for AP2 [14, 17]. The hemicomplexes γ/σ1 and α/σ2 share significant structural similarity, and β1/μ1 and β2/μ2 closely resemble each other. The core domain of APs recognizes PI(4,5)P_2 or PI(4)P phospholipids of membranes and sorting motifs of cargo proteins, either tyrosine-based YXXΦ or dileucine-based [ED]XXXL[LI] (where Φ represents a bulky hydrophobic residue and X is any amino acid), whereas the hinge and appendage domains interact with clathrin and other accessory proteins.

High-resolution structures of AP1 and AP2 core domain in different conformational states, with or without ligands (sorting motifs), have been reported [14–23]. In the ligand-free conformation, the two cargo-binding sites on AP1 or AP2 are blocked by residues from the β1/β2 subunit [14, 15]. In the [ED]XXXL[LI]-bound AP2 core structure, the dileucine motif displaces the N terminus of β2, which relieves the blocking for [ED]XXXL[LI], whereas the YXXΦ-binding site on μ2 remains blocked [16]. In the YXXΦ and [ED]XXXL[LI] double-liganded AP2 structure, both cargo-binding sites are exposed and occupied by ligands, in which the μ2 C-terminal domain undergoes a significant shift, resulting in AP2's PI(4,5)P_2- and cargo-binding sites becoming coplanar, thus facilitating coordinated interactions with membrane-embedded cargoes [17]. On the basis of these different states, it is hypothesized that APs initially assemble in cytosol, adopting an inactive conformation that prevents premature engagement with cargoes and clathrin, and subsequently undergo conformational changes to an active state that orchestrates cargo selection and clathrin recruitment, which ultimately drives vesicle formation [2, 26].

It is still poorly understood how AP1 and AP2 assemble in vivo. A recent discovery showed that AP2 adaptor assembly is a highly orchestrated process in human controlled by alpha and gamma adaptin binding protein (AAGAB, also known as p34) [27]. Mutations in AAGAB are associated with human diseases such as

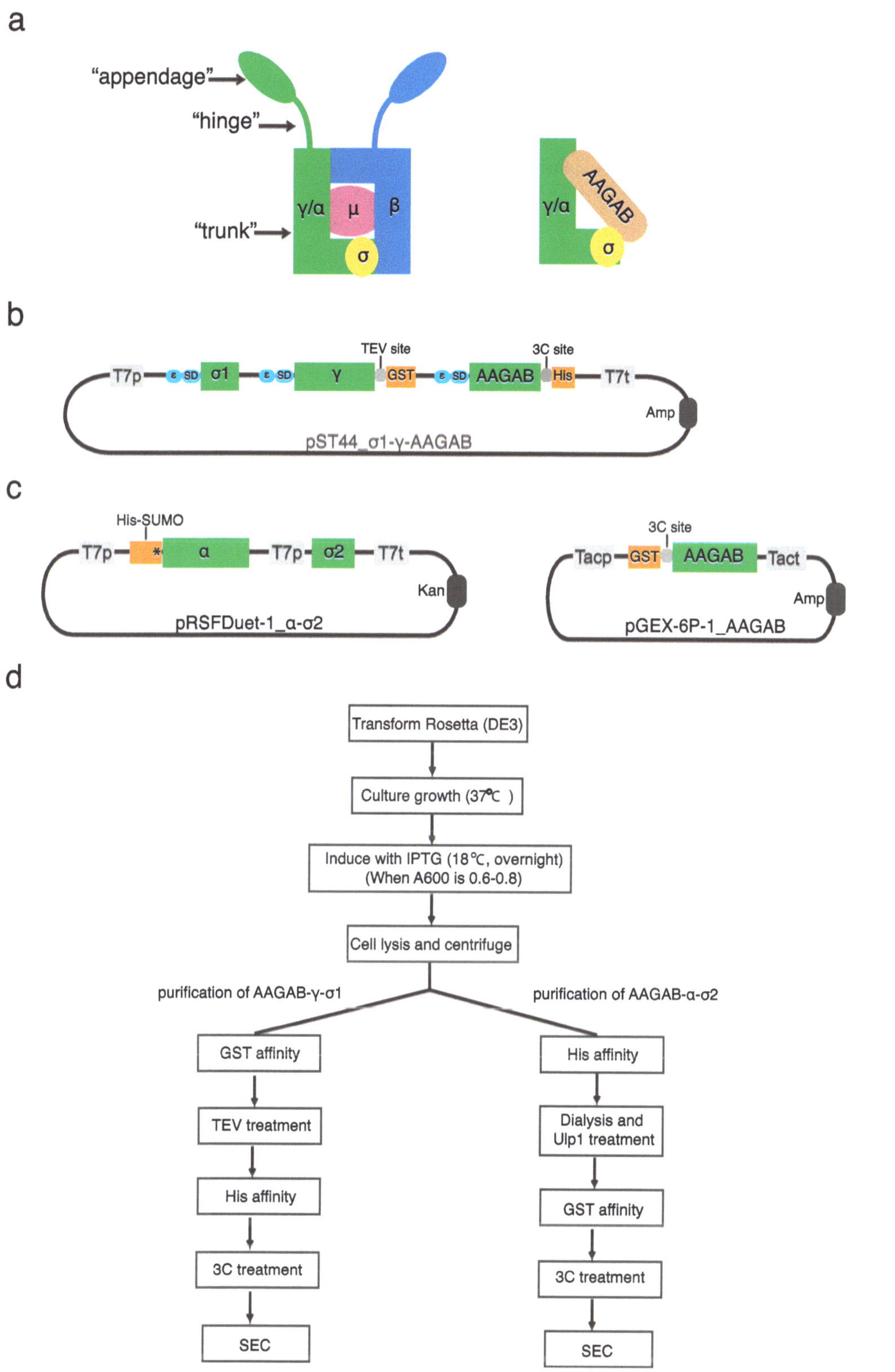

Fig. 1 Graphic outline of coexpression and purification of AAGAB-γ-σ1 and AAGAB-α-σ2 complexes. (**a**) Composition diagrams of AP1/2 complex (left) and AAGAB-AP1/2 assembly intermediates (right) described in

punctate palmoplantar keratoderma type 1 (PPKP1), a skin disorder characterized by lesions on palms and soles [28]. To elucidate the role of AAGAB in AP2 assembly, we started off by examining how AAGAB interacts with AP2 subunits. In our reconstituted system, we discovered that AAGAB has direct interactions with both α subunit (i.e., α trunk domain for all following) and σ2 subunit of AP2. Next, we proved that the two interactions are not mutually exclusive, as we were able to reconstitute a ternary complex of AAGAB-α-σ2 with our *E. coli* coexpression system. Similarly, we discovered direct interactions between AAGAB and γ subunit (i.e., γ trunk domain for all following) and σ1 subunit of AP1 in a ternary complex of AAGAB-γ-σ1, consistent with previous observation that AAGAB binds to and stabilizes γ subunit [27]. To unravel the molecular mechanisms of AAGAB-regulated AP1 and AP2 assembly, it is imperative to reconstitute and purify AAGAB-AP1 and AAGAB-AP2 complexes for biochemical characterizations and structural studies (X-ray crystallography and single-particle cryo-EM). Here, we provide a detailed protocol for expressing and purifying the two ternary complexes, AAGAB-γ-σ1 and AAGAB-α-σ2 (Fig. 1a, right panel). For expression of AAGAB-γ-σ1, we exploited a polycistronic coexpression strategy to insert all three genes into a single plasmid, which encodes untagged σ1, γ with a C-terminal GST tag removable using Tobacco Etch Virus protease (TEV), and AAGAB with a C-terminal hexahistidine tag (His_6) removable using human rhinovirus 14 3C protease (HRV14 3C, or 3C) (Fig. 1b). For expression of AAGAB-α-σ2, we cotransformed two plasmids with different selectable markers. One is the Duet plasmid that produces α with an N-terminal His_6-SUMO tag cleavable by ubiquitin-like-specific protease 1 (Ulp1) and untagged σ2, each under its own promoter. The other plasmid contains AAGAB with an N-terminal 3C-cleavable GST tag (Fig. 1c). Tagging subunits with His_6, His_6-SUMO, or GST increases protein solubility and facilitates affinity purification. This dual tagged system allows tandem purification to obtain stoichiometric ternary complexes. For purification, we followed a general procedure composed of first round of affinity purification,

Fig. 1 (continued) this chapter. (**b**) Plasmid diagram of coexpression strategy for AAGAB-γ-σ1. A single polycistronic coexpression plasmid was constructed on pST44 vector. Each transgene follows the same translation cassette that includes translational enhancer element (ε) and Shine–Dalgarno sequences (SD). σ1, γ and AAGAB were inserted sequentially. σ1 is without any tag, γ is followed by a C-terminal GST tag with TEV cleavage sequence in between, and AAGAB is followed by a C-terminal His tag with 3C cleavage sequence in between. *T7p* T7 promoter; *T7t* T7 terminator. (**c**) Plasmid diagram of coexpression strategy for AAGAB-α-σ2. Two plasmids were constructed in this strategy. N-terminal His-SUMO tagged α and tagless σ2 were cloned into the first and second multiple cloning sites of pRSFDuet vector respectively, whereas AAGAB was cloned into pGEX-6P-1 vector, in which the GST tag can be removed by 3C protease. *: Ulp1 cleavage site. (**d**) Outline of purification procedures, see text for details protease cleavage of first purification tag, second round of affinity purification, protease cleavage of second purification tag, and size-exclusion chromatography (Fig. 1d). In this chapter, we provide detailed procedures and highlight important points for ensuring reproduction.

2 Materials

Prepare all solutions using ultrapure water. Chemicals and reagents are of analytical grade unless otherwise stated. Filter all stock solutions through 0.22 μm membrane. If a buffer is made only from stock solutions that have been filtered and diluted with ultrapure water, further filtration is not necessary.

2.1 Bacterial Strains, Plasmids, Media, and Stock Solutions

1. Competent *E. coli* cells: We use Rosetta (DE3) (Novagen) cell for expression, but other strains suitable for expressing genes from T7 promoters can also be used.
2. pST44_σ1-γ-AAGAB: σ1 (encoded by human *AP1S3*), γ (encoded by human *AP1G1*) and AAGAB (encoded by human *AAGAB*) are cloned with polycistronic expression elements into pST44 vector (Addgene Cat # 64007). In this construct, the GST tag is fused to the C terminus of γ with a TEV cleavage site in between. A His_6 tag is fused to the C-terminus of AAGAB with a 3C cleavage site in between (Fig. 1b).
3. pRSFDuet-1_α-σ2: α (encoded by human *AP2A2*) is inserted into the first multiple cloning site of pRSFDuet-1 (Novagen Cat # 71341-3) with an N-terminal His_6-SUMO tag cleavable by Ulp1 protease. Untagged σ2 (encoded by human *AP2S1*) was cloned into the second multiple cloning site of pRSFDuet-1 (Fig. 1c).
4. pGEX-6P-1_AAGAB: The human *AAGAB* gene is inserted into the cloning site of pGEX-6P-1 (GE, now Cytiva, Cat # 28954648). A GST tag is fused to the N-terminus of AAGAB with a 3C cleavage site in between (Fig. 1c).
5. LB medium (1 L): dissolve 25 g of Luria Broth powder (10 g of tryptone, 10 g of sodium chloride, and 5 g of yeast extract) in 1 L of water. Sterilize by autoclave for 20 min at 121 °C. Depending on the plasmid(s) used, add antibiotics to the LB medium before usage.
6. SOC medium (1 L): dissolve 28 g of SOB broth powder (20 g of casein enzymic hydrolysate, 5 g of yeast extract, 0.5 g of NaCl, 2.4 g of $MgSO_4$, and 0.186 g of KCl) in 1 L of water. Sterilize by autoclave for 20 min at 121 °C. Allow the medium to cool to 60 °C, and add 20 mL of filter-sterilized 20% glucose solution.

7. LB-Agar plates (1 L): dissolve 35 g of LB agar powder (10 g of casein peptone, 5 g of sodium chloride, 5 g of yeast extract, and 15 g of agar) into 1 L of water. Sterilize by autoclave at 121 °C for 20 min, add 1 mL of required antibiotics when the solution temperature drops to 60 °C, mix evenly and prepare plates.
8. 10× PBS stock (1 L): dissolve 80 g of NaCl, 2 g of KCl, 14.4 g of Na_2HPO_4, and 2.4 g of KH_2PO_4 in about 800 mL of water, adjust the pH to 7.4 with HCl and then add water to 1 L.
9. 1 M Tris–HCl buffer stock (1 L): dissolve 121 g of Tris base in about 800 mL of water, adjust the pH to 8.0 with HCl and then add water to 1 L.
10. 1 M HEPES-Na buffer stock (1 L): dissolve 238.3 g of HEPES in about 800 mL of water, adjust pH to 7.6 with NaOH and then add water to 1 L.
11. 5 M imidazole buffer stock (1 L): dissolve 340.4 g of imidazole in about 800 mL of water, adjust pH to 8.0 with HCl and then add water to 1 L.
12. 5 M NaCl stock (1 L): dissolve 292 g of NaCl in about 900 mL of water, add water to 1 L.
13. 1 M dithiothreitol (DTT): dissolve 7.7 g of DTT into 50 mL of water, filtered with 0.22 μm membrane filter.
14. Antibiotic stocks (all 1000×):

    Ampicillin (Amp): 100 mg/mL in water.

    Kanamycin (Kan): 50 mg/mL in water.

    Chloramphenicol (Cm): 34 mg/mL in 70% ethanol.
15. 1 M Isopropyl β-D-1-thiogalactopyranoside (IPTG): dissolve 9.53 g of IPTG into 50 mL of water.

2.2 Equipment and Supplies

1. 37 °C incubator.
2. Shaker incubator with temperature control between 18 °C and 37 °C.
3. 2000 mL baffled flasks.
4. Centrifuge bottles and tubes: 1000 mL bottles for pelleting cells and 50 mL tubes for separating supernatant from cell debris after lysis.
5. Sonication device with appropriate tip for volumes of above 50 mL. We use Branson Sonifier 250.
6. Centrifuge with appropriate rotors. We use Beckman Coulter Avanti JXN-26 centrifuge together with JLA 8.1000 rotor for pelleting large volume of bacterial culture at 4000 × *g* and JA-25.50 rotor for separating supernatant from cell debris at 14,000 × *g*.
7. 0.22 μm and 0.45 μm cellulose acetate membrane filter.

8. Peristaltic pumps (optional if using gravity columns).
9. Ni-NTA columns. We use GE HisTrap HP 5 mL.
10. Glutathione columns. We use GE GSTrap HP 5 mL.
11. Gravity flow columns.
12. Fast protein liquid chromatography (FPLC) system with gradient mixer. We use GE ÄKTA pure.
13. Size-exclusion chromatography (SEC) column with proper injection loops. We use a GE HiLoad 16/600 Superdex 200 pg column with a bed volume of 120 mL.
14. Dialysis membranes and clamps with a molecular weight cutoff (MWCO) of 6–8 kDa. We use Spectra Por S/P 1 dialysis membrane (Spectrum Laboratories/Repligen).
15. Centrifugal filters for protein concentration. We use Amicon Ultra centrifugal filters with a MWCO of 50 kDa.
16. Magnetic stirrer and magnetic stir bars.
17. Device for measuring bacterial culture optical density and protein concentration. We use a Nanodrop ND-1000 spectrophotometer.
18. SDS-PAGE equipment.

2.3 Reagents and Buffers

1. Protease inhibitor cocktails. We use the Complete protease inhibitor cocktail tablets, EDTA-free, from Roche.
2. HRV 3C protease (homemade, modified from Addgene #78571 with a His_6 tag) [29]. Commercially known as PreScission Protease Protein (PSP).
3. TEV protease (homemade, with His_6 tag) [18].
4. Ulp1 protease (homemade) [30].
5. Nickel affinity resin. We use Takara His60 Ni Superflow resin or Qiagen Ni-NTA agarose resin.
6. Glutathione resin. We use GE glutathione agarose or Sepharose resin.
7. Lysis Buffer: 50 mM Tris–HCl pH 8.0, 150 mM NaCl, 5% glycerol (*see* **Note 1**), 10 mM imidazole.
8. Ni Wash Buffer: 50 mM Tris–HCl pH 8.0, 150 mM NaCl, 5% glycerol, 25 mM imidazole.
9. Ni Elution Buffer: 50 mM Tris–HCl pH 8.0, 150 mM NaCl, 5% glycerol, 500 mM imidazole.
10. Dialysis Buffer: 20 mM Tris–HCl pH 8.0, 150 mM NaCl, 5% glycerol, 5 mM β-mercaptoethanol (freshly added).
11. GST Wash Buffer: 15 mM HEPES-NaOH pH 7.6, 150 mM NaCl.

12. GST Elution Buffer: 15 mM HEPES-NaOH pH 7.6, 150 mM NaCl, 20 mM reduced glutathione (freshly prepared, adjust pH to 7.6 after adding glutathione).
13. SEC Buffer: 10 mM HEPES-NaOH pH 7.6, 150 mM NaCl, 5 mM β-mercaptoethanol (freshly added).

3 Methods

3.1 *E. coli* Transformation

For expression and purification of AAGAB-γ-σ1, transform the single plasmid pST44_σ1-γ-AAGAB into Rosetta (DE3) *E. coli* cell and select a positive bacterial colony from LB-Agar plate supplemented with ampicillin. For expression and purification of AAGAB-α-σ2, cotransform the two plasmids, pRSFDuet-1_α-σ2 and pGEX-6P-1_AAGAB to Rosetta (DE3) *E. coli* cell (*see* **Note 2**) and select a positive bacterial colony from LB-Agar plate supplemented with both kanamycin and ampicillin.

1. Thaw competent cells on ice and aliquot 50 μL into the autoclaved microcentrifuge tubes for each transformation.
2. Add 1 μL of plasmids (no more than 50 ng) to competent cells, mix well, and incubate on ice for 30 min.
3. Heat-shock the competent cells in a 42 °C water bath for 90 s and place on ice for 1 min.
4. Add 900 μL of SOC medium to each sample and incubate in a 37 °C shaker incubator for 1 h.
5. Plate 50 μL of the sample onto a LB-Agar plate supplemented with corresponding antibiotics. Incubate in a 37 °C incubator overnight. Keep the plates in the cold room for future use (*see* **Note 3**).

3.2 Protein Expression in *E. coli*

Before large-scale expression, we usually perform an expression trial with 50 mL of culture to evaluate the expression level for each construct (*see* **Note 4**). The procedures for large-scale expression are described below.

1. Pick a single bacterial colony from the transformation plate using a 10 μL sterile pipette tip and inoculate it into 50 mL of autoclaved LB medium (supplemented with corresponding antibiotics) in a 125 mL flask as a preliminary culture (*see* **Note 5**). Incubate the preliminary culture overnight in a 37 °C shaker incubator set at 200 rpm.
2. The following morning, transfer 6 mL of preliminary culture to each 2000 mL flask that contains 1000 mL of autoclaved LB medium supplemented with proper antibiotics (*see* **Note 6**). Incubate the flasks for around 5 h in a 37 °C shaker incubator set at 200 rpm. Lower the temperature of the shaker incubator

to 18 °C when the optical density at 600 nm (OD600) reaches 0.6–0.8. If required, measure the OD600 of cells several times during culturing using a spectrophotometer.

3. Add 400 μL of 1 M IPTG to each flask to reach a final IPTG concentration of 0.4 mM (*see* **Note 7**). Continue incubation at 18 °C in a shaker incubator set at 200 rpm overnight (no less than 16 h).
4. Harvest the cells by centrifugation at 4000 × *g* for 20 min using 1 L centrifugation bottles.
5. Resuspend bacterial pellets with 1× PBS buffer and transfer to 50 mL tubes. Centrifuge at 4000 × *g* for 1 h and discard the supernatant. At this point, cell pellets can be stored at −80 °C if not proceeding to the following purification steps immediately.

3.3 Cell Lysis

1. Prepare resuspension buffer by adding 5 mM β-mercaptoethanol and one tablet of EDTA-free protease inhibitor to 50 mL of Lysis Buffer (*see* **Note 8**). Add resuspension buffer of appropriate volume (*see* **Note 9**) and resuspend cells by vigorous vortex.
2. Transfer the cell suspension to a beaker precooled on ice. From this step, all operations should be carried out either on ice or in a cold room. Sonicate the cell suspension on ice at 40% output in pulse mode of 0.2 for 5 min. Be very careful not to overheat the sample. The sample will become transparent when cells are fully lysed.
3. Transfer the cell lysate to centrifuge tubes that can withstand high-speed centrifugation and centrifuge at 14,000 × *g* for 2 h at 4 °C. Decant the supernatant carefully into a precooled and clean container and avoid disturbing the fuzzy membrane sitting at the interface of supernatant and pellet.
4. Sonicate the supernatant again for 1 min to fragment nucleic acids. Filter the supernatant through a 0.45 μm membrane filter.

3.4 Affinity Purification (See Note 10)

3.4.1 GST-Affinity Purification

We use both prepacked GST-affinity column and glutathione beads packed into gravity columns to purify GST-tagged proteins. For prepacked GST-affinity column, *see* **steps 1–6**; for glutathione beads packed in gravity columns, *see* **steps 7–12**.

1. Equilibrate a 5 mL GST-affinity column with at least 5 column volumes (CV) of Lysis Buffer using a peristaltic pump or FPLC equipment.
2. Load filtered supernatant (or solution containing desired proteins from another step of purification) onto the column at a flow rate of 1 mL/min using a peristaltic pump.
3. Wash the column with at least 5 CV of GST Wash Buffer.

4. On an FPLC system equilibrate pump B with GST Elution Buffer, and equilibrate pump A with GST Wash Buffer.
5. Attach the protein-loaded GST-affinity column and elute bound proteins with a 50 mL gradient from pump A to pump B, collect 5 mL fractions.
6. Check the identity and purity of eluted proteins in the fractions by SDS-PAGE (Fig. 2a).

Alternatively, purification may be achieved using glutathione beads packed in gravity columns instead of prepacked GST-affinity columns. In this case, we adopt the following procedures.

7. Place 5 mL of glutathione resin in a gravity flow column. Drain all the storage solution.
8. Equilibrate the column with at least 5 bed volumes (i.e., 25 mL) of GST Wash Buffer.
9. Load filtered supernatant (or solution containing desired proteins from another step of purification) onto the equilibrated gravity column. Adjust the stopper so that the flow rate is about 1 mL/min.
10. Wash the column with at least 5 bed volumes of GST Wash Buffer.
11. Elute the bound proteins with 50 mL of GST Elution Buffer. Collect 5 mL fractions.
12. Check the identity and purity of eluted proteins in the fractions by SDS-PAGE (Fig. 3c).

3.4.2 Ni-Affinity Purification

We used both prepacked Ni-affinity column and Ni-affinity beads packed in gravity columns to purify His_6-tagged proteins. For prepacked Ni-affinity column, *see* **steps 1–6**; for Ni-affinity beads packed in gravity columns, *see* **steps 7–12**.

1. Equilibrate a 5 mL Ni-NTA column with at least 5 column volumes (CV) Ni Wash Buffer using a peristaltic pump or FPLC equipment.
2. Load filtered supernatant (or solution containing desired proteins from another step of purification) onto the column at a flow rate of 1 mL/min using a peristaltic pump.
3. Wash the column with at least 5 CV of Ni Wash Buffer (supplemented with 5 mM β-mercaptoethanol) (*see* **Note 11**).
4. On an FPLC system equilibrate pump B with Ni Elution Buffer, then equilibrate pump A with Lysis Buffer.
5. Attach the protein-loaded Ni-NTA column and elute the bound protein with a 50 mL gradient from pump A to pump B. Collect 5 mL fractions.
6. Check the identity and purity of eluted proteins in the fractions by SDS-PAGE (Fig. 2c).

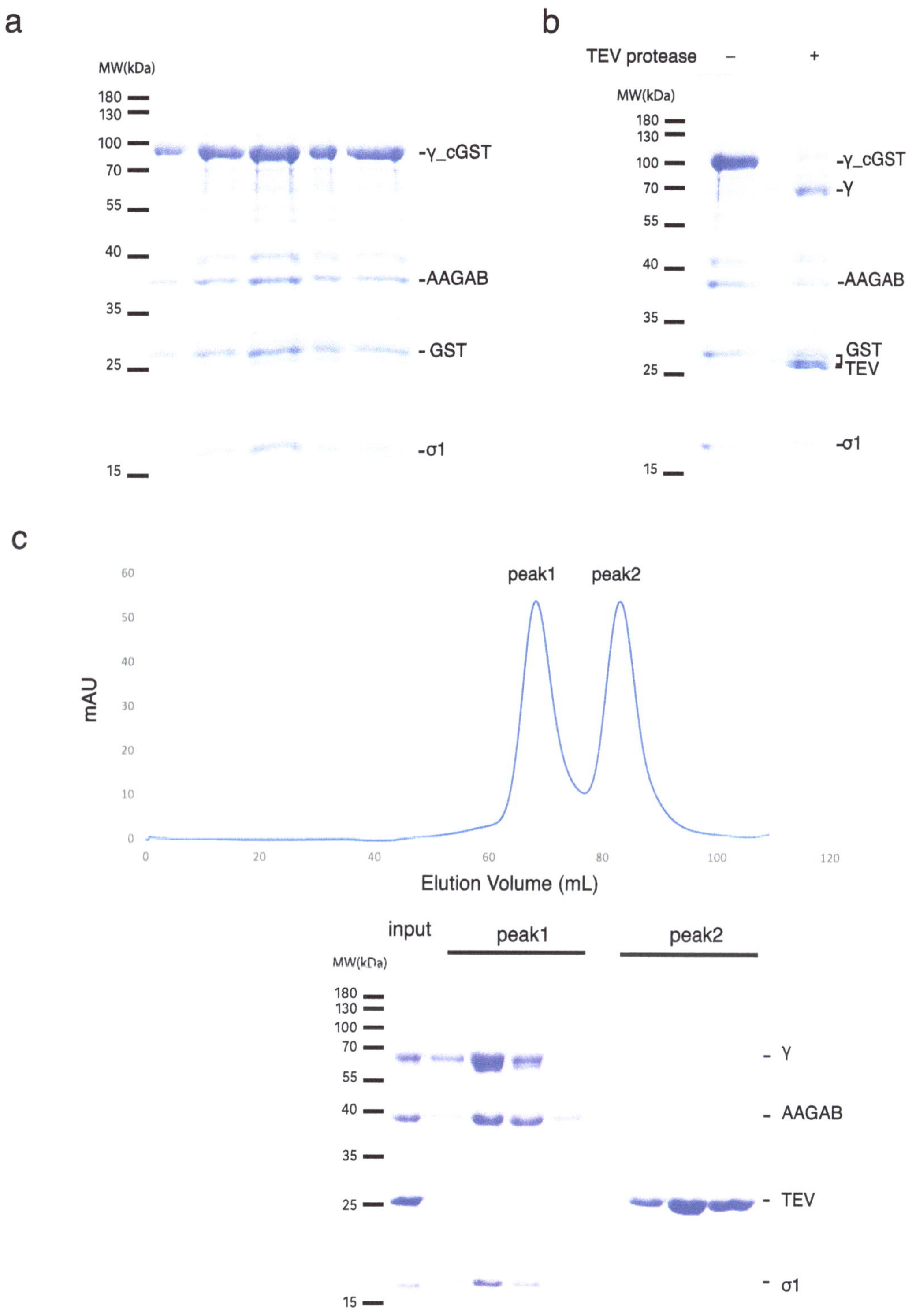

Fig. 2 Coomassie-stained SDS-PAGE gels and size-exclusion chromatography (SEC) profile of AAGAB-γ-σ1 purification. (**a**) SDS-PAGE gel of elution fractions from the first round GST affinity purification. In addition to

Alternatively, purification may be achieved using Ni-affinity beads packed in gravity columns instead of prepacked Ni-NTA columns. In this case, we adopt the following procedures:

7. Place 5 mL of Ni-affinity resin in a gravity flow column. Drain all the storage solution (usually 70% ethanol).
8. Equilibrate the column with at least 5 bed volumes (i.e., 25 mL) of Lysis Buffer.
9. Load filtered supernatant (or solution containing desired proteins from another step of purification) onto the equilibrated gravity column. Adjust the stopper so that the flow rate is about 1 mL/min.
10. Wash the column with at least 5 bed volumes of Ni Wash Buffer.
11. Elute bound proteins with 50 mL of Ni Elution Buffer. Collect 5 mL fractions.
12. Check the identity and purity of eluted proteins in the fractions by SDS-PAGE (Fig. 3a).

3.5 Buffer Exchange and Cleavage of the First Affinity Tag

1. Prepare 2 L of Dialysis Buffer in a beaker and cool it down to 4 °C.
2. Add protease (TEV protease for AP1 complex, 50 μg of TEV protease per mL of elution; Ulpl protease for AP2 complex, 1 μg of Ulpl protease per mL of elution) to the protein to cut off the C-terminal GST tag from the AP1 γ subunit (Fig. 2b) or the N-terminal His_6-SUMO tag from the AP2 α subunit (Fig. 3b).
3. Cut one piece of dialysis membrane to a size appropriate for the volume of the elution volume.
4. Wash the dialysis membrane briefly in Dialysis Buffer.
5. Seal one end of the membrane with a clamp, fill the membrane with the elution solution containing proteases, and close the open end with another clamp.
6. Put the filled dialysis membrane into the beaker containing Dialysis Buffer. Stir gently overnight in the cold room to allow buffer exchange and complete cleavage of the affinity tags.

Fig. 2 (continued) the three subunits in this complex, a standalone GST tag band is visible, likely due to proteolysis at the C-terminus of GST-γ. (**b**) SDS-PAGE gel of GST elution fractions before and after TEV treatment. (**c**) SEC profile and SDS-PAGE gel for 3C treated AAGAB-γ-σ1 ternary complex. The input for SEC is from the ternary complex after 3C protease treatment to remove the C-terminal His tag. His tagged TEV protease initially added to cleave GST tag is still present but can be separated from the ternary complex during SEC

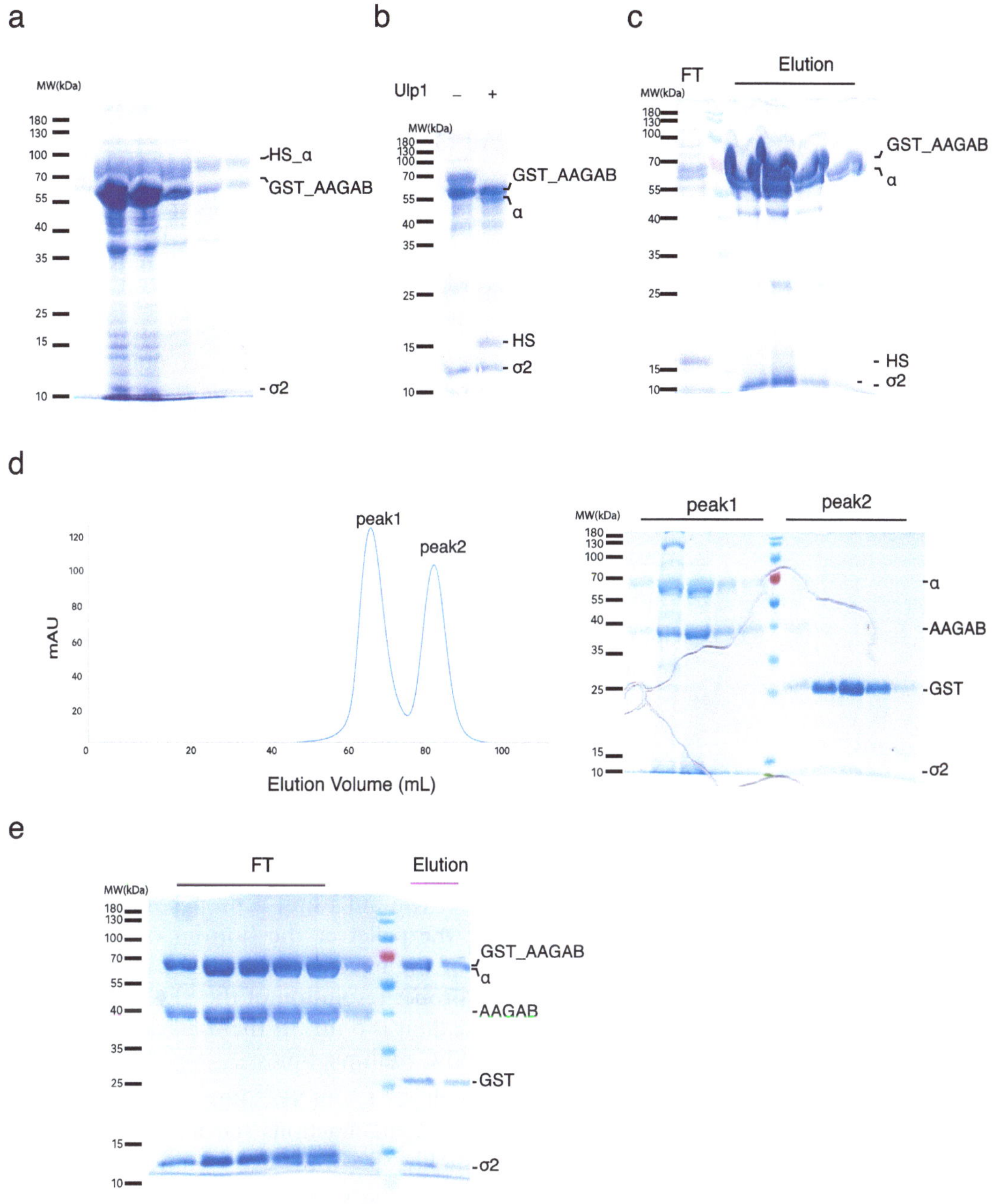

Fig. 3 Coomassie-stained SDS-PAGE gels and size-exclusion chromatography (SEC) profile of AAGAB-α-σ2 purification. (**a**) SDS-PAGE gel of elution fractions from the first round Ni affinity purification. HS: His-SUMO. (**b**) SDS-PAGE gel of Ni elution fraction before and after Ulp1 treatment. (**c**) SDS-PAGE gel of elution fractions from the second round GST affinity purification. (**d**) SEC profile and SDS-PAGE gel of 3C treated AAGAB-α-σ2 ternary complex. AAGAB-α-σ2 ternary complex eluted in peak 1, well separated from peak 2 containing cleaved GST. (**e**) SDS-PAGE gel of a "reverse" round of GST affinity purification if AAGAB-α-σ2 ternary complex is contaminated by GST in SEC. Note that now AAGAB-α-σ2 is in flow-through (FT) fractions instead of elution fractions

3.6 Cleavage of the Second Affinity Tag

After cleavage of the first affinity tag, the protein solution is subjected to a second round of affinity purification using the other tag, following steps described in Subheadings 3.4.1 and 3.4.2. In our case, it is Ni-affinity purification for AAGAB-γ-σ1 and GST affinity purification for AAGAB-α-σ2.

1. Pool the eluted AAGAB-γ-σ1 or AAGAB-α-σ2 complex in a 50 mL conical tube.
2. Add 3C protease to a final concentration of 1 μg of 3C protease per mL of elution.
3. Place the tube at 4 °C overnight to remove the C-terminal His_6 tag from AAGAB in AP1-containing complex or the N-terminal GST tag from AAGAB in the AP2-containing complex (*see* **Note 12**).

3.7 Size-Exclusion Chromatography (SEC)

1. Connect a HiLoad® 16/600 Superdex® 200 pg SEC column to an FPLC instrument.
2. Wash pump A with SEC Buffer.
3. Equilibrate the column with 1.2 CV of SEC Buffer.
4. Spin the conical tube containing the protein complex after second tag cleavage at 3000 × *g* for 10 min to remove insoluble particles.
5. Prepare an SDS-PAGE sample with the protein complex before SEC—"SEC input." Alternatively, prepare SDS-PAGE samples from second affinity purification peak elution fractions (Figs. 2c and 3c).
6. Inject the protein-containing solution after the second tag cleavage onto a SEC column using a proper injection loop without disturbing the pellet at the bottom of the conical tube. The volume of the injected sample should not exceed the upper limit for proper resolution of the SEC column. For HiLoad 16/600 the limit is 6 mL. If the total sample volume exceeds 6 mL, run SEC multiple times.
7. Elute the proteins with 1.2 CV of SEC Buffer at a flow rate of 0.8 mL/min. Collect 4 mL fractions starting from the void volume, which is roughly 35% of the bed volume. For HiLoad 16/600 column we start collection at 40 mL.
8. Check the identity and purity of the eluted proteins in the fractions by SDS-PAGE (*see* **Note 13**) (Figs. 2c and 3d, e).
9. Pool and concentrate the fractions containing desired ternary complexes using a new centrifugal filter to a final volume of less than 500 μL.
10. Determine the concentration of the ternary complexes using a UV spectrometer.

11. Aliquot the proteins in microcentrifuge tubes. Snap freeze protein samples in liquid nitrogen and store at −80 °C for future usage.

4 Notes

1. 5% glycerol is optional in Lysis Buffer and other buffers. Presence of glycerol increases system pressure if FPLC is used for purification. However, we find glycerol in general stabilizes proteins, especially during the freezing–thawing cycle. One exception is SEC buffer. We DO NOT include glycerol in SEC buffer because it complicates downstream structural characterization.
2. Cotransformation with multiple plasmids usually requires high efficiency of competent cells. We prepare ultracompetent *E. coli* cells for chemical transformation based on Inoue et al. (1990), Gene, 96:23–28 [31], with modifications, which is normally 100 to 1000 times more efficient than traditional calcium method.
3. Plates with transformed *E. coli* colonies can be stored in cold room for a few months; however, in some cases the expression level decreases for old transformations. As a rule, we use colonies transformed within 2 weeks for expression.
4. In our preliminary test, we noticed that AAGAB-α-σ2 has a better yield than AAGAB-γ-σ1. We tried to optimize the constructs by using the same cotransformation and same tag as AAGAB-α-σ2, but the yield for AAGAB-γ-σ1 remained low. To gain milligrams of proteins, we increased the total culture volume for AAGAB-γ-σ1 by eight times in comparison to AAGAB-α-σ2, thus 24 liters for AAGAB-γ-σ1 in contrast to 3 L for AAGAB-α-σ2.
5. We use Rosetta DE3 cells, which contain a plasmid for expressing *E. coli* rare-codon tRNAs to enhance expression of eukaryotic proteins. The plasmid is chloramphenicol resistant. To avoid plasmid loss during culture, we keep the selection pressure by adding chloramphenicol to a final concentration of 34 μg/mL in all cultures.
6. For AAGAB-γ-σ1 culture up to 24 L, we split the large-scale culture to 3 equal batches to comply with incubator space.
7. For some proteins, it is crucial to strictly follow the low temperature rule for induction. We typically wait for 20 min after lowering shaker incubator temperature to make sure the temperature inside the chamber has dropped to 18 °C at the time of IPTG addition.

8. Depending on protein properties and successive purification steps, Lysis Buffer could be made in high (>300 mM) or low (100–150 mM) salt (NaCl) concentrations. Low salt buffers usually facilitate the subsequent step of ion exchange chromatography if required. If a protein binds ion exchange resin only weakly, the protein sample in a low salt buffer can be directly loaded onto the column without dialysis or dilution treatment. High salt buffer may be helpful to reduce nonspecific binding to HisTrap column but may interfere with subsequent ion exchange chromatographic step. Some proteins are only soluble in high salt buffers, while some complexes are only stable under low salt conditions. Test various salt concentrations to identify the most suitable one. In our case, we skip the ion exchange after first affinity purification, but add a second affinity purification instead. In our pretests, 150 mM salt works well for our samples.
9. The volume of Lysis Buffer is about 0.6–1.5 times of total volume of cell pellet. For 3 L of bacterial culture, the total pellet volume is about 30 mL, requiring 18–45 mL of Lysis Buffer.
10. We introduced two affinity tags for tandem purification, GST tag and His_6/His_6-SUMO tag. In contrast to His_6 tag, GST tag has higher specificity but lower affinity. We choose the sequence of two affinity purifications based on the yield, purity, protein properties and convenience. In our cases, the expression level of AAGAB is much higher than all other proteins. To avoid accumulation of excessive free AAGAB during purification, we always purify the other tagged subunit in the first round of affinity purification and tagged AAGAB in the second round. Thus, in the case of AAGAB-γ-σ1 purification, in which γ subunit is GST tagged and AAGAB is C-terminally His_6 tagged, we performed GST affinity purification first. Similarly, in the case of AAGAB-α-σ2 purification, in which GST tag is at the N-terminus of AAGAB while α subunit bears an N-terminal His_6-SUMO tag, we performed His affinity purification first (Fig. 1d).
11. β-mercaptoethanol is recommended as the reducing agent for Ni-NTA affinity purification. Ni-NTA beads can well tolerate 5–10 mM β-mercaptoethanol. DTT is a stronger reducing reagent and 1 mM DTT may reduce Ni-NTA beads and turn them into a dark brown color.
12. We found buffer exchange unnecessary for cleavage of the second affinity tag. 3C protease is active in both Ni Elution Buffer and GST Elution Buffer. However, if 3C protease-mediated cleavage is not complete, consider buffer exchange to reduce imidazole concentration in addition to increasing 3C concentration.

13. In SEC, AAGAB-γ-σ1 and AAGAB-α-σ2 complexes are usually well separated from TEV protease (and trace amount of 3C protease) or cleaved GST tag, respectively (Figs. 2c and 3d). If TEV protease is not well separated from AAGAB-γ-σ1 ternary complex, consider using tagless TEV in cleavage of GST tag in Subheading 3.5. Tagless TEV stays in the flow-through fractions in the second round of Ni affinity purification. If a substantial amount of GST tag comigrate with the AAGAB-σ2-α ternary complex in the final SEC, run an additional "reverse" round of GST affinity purification to remove the GST tag following the procedures described in Subheading 3.4.1. The only difference is that since the desired ternary complex is in the flow-through, collect flow-through in 5 mL or smaller fractions (Fig. 3e).

Acknowledgments

This work was supported by NIH grant 1R01GM138685 to Q.Y.

References

1. Kaksonen M, Roux A (2018) Mechanisms of clathrin-mediated endocytosis. Nat Rev Mol Cell Biol 19(5):313–326. https://doi.org/10.1038/nrm.2017.132
2. Mettlen M, Chen PH, Srinivasan S, Danuser G, Schmid SL (2018) Regulation of Clathrin-mediated endocytosis. Annu Rev Biochem 87: 871–896. https://doi.org/10.1146/annurev-biochem-062917-012644
3. Kim K, Gadila SKG (2016) Cargo trafficking from the trans-Golgi network towards the endosome. Biol Cell 108(8):205–218. https://doi.org/10.1111/boc.201600001
4. Traub LM, Bonifacino JS (2013) Cargo recognition in clathrin-mediated endocytosis. Cold Spring Harb Perspect Biol 5(11):a016790. https://doi.org/10.1101/cshperspect.a016790
5. Fotin A, Cheng Y, Sliz P, Grigorieff N, Harrison SC, Kirchhausen T, Walz T (2004) Molecular model for a complete clathrin lattice from electron cryomicroscopy. Nature 432(7017): 573–579. https://doi.org/10.1038/nature03079
6. Paraan M, Mendez J, Sharum S, Kurtin D, He H, Stagg SM (2020) The structures of natively assembled clathrin-coated vesicles. Sci Adv 6(30):eaba8397. https://doi.org/10.1126/sciadv.aba8397
7. Kovtun O, Dickson VK, Kelly BT, Owen DJ, Briggs JAG (2020) Architecture of the AP2/clathrin coat on the membranes of clathrin-coated vesicles. Sci Adv 6(30): eaba8381. https://doi.org/10.1126/sciadv.aba8381
8. Park SY, Guo X (2014) Adaptor protein complexes and intracellular transport. Biosci Rep 34(4):e00123. https://doi.org/10.1042/BSR20140069
9. Ghosh P, Dahms NM, Kornfeld S (2003) Mannose 6-phosphate receptors: new twists in the tale. Nat Rev Mol Cell Biol 4(3):202–212. https://doi.org/10.1038/nrm1050
10. Gan Y, McGraw TE, Rodriguez-Boulan E (2002) The epithelial-specific adaptor AP1B mediates post-endocytic recycling to the basolateral membrane. Nat Cell Biol 4(8):605–609. https://doi.org/10.1038/ncb827
11. Kirchhausen T (1999) Adaptors for clathrin-mediated traffic. Annu Rev Cell Dev Biol 15: 705–732. https://doi.org/10.1146/annurev.cellbio.15.1.705
12. Traub LM (1997) Clathrin-associated adaptor proteins - putting it all together. Trends Cell Biol 7(2):43–46. https://doi.org/10.1016/S0962-8924(96)20042-X
13. Kirchhausen T, Owen D, Harrison SC (2014) Molecular structure, function, and dynamics of clathrin-mediated membrane traffic. Cold Spring Harb Perspect Biol 6(5):a016725. https://doi.org/10.1101/cshperspect.a016725

14. Collins BM, McCoy AJ, Kent HM, Evans PR, Owen DJ (2002) Molecular architecture and functional model of the endocytic AP2 complex. Cell 109(4):523–535. https://doi.org/10.1016/s0092-8674(02)00735-3
15. Heldwein EE, Macia E, Wang J, Yin HL, Kirchhausen T, Harrison SC (2004) Crystal structure of the clathrin adaptor protein 1 core. Proc Natl Acad Sci U S A 101(39):14108–14113. https://doi.org/10.1073/pnas.0406102101
16. Kelly BT, McCoy AJ, Spate K, Miller SE, Evans PR, Honing S, Owen DJ (2008) A structural explanation for the binding of endocytic dileucine motifs by the AP2 complex. Nature 456(7224):976–979. https://doi.org/10.1038/nature07422
17. Jackson LP, Kelly BT, McCoy AJ, Gaffry T, James LC, Collins BM, Honing S, Evans PR, Owen DJ (2010) A large-scale conformational change couples membrane recruitment to cargo binding in the AP2 clathrin adaptor complex. Cell 141(7):1220–1229. https://doi.org/10.1016/j.cell.2010.05.006
18. Ren X, Farias GG, Canagarajah BJ, Bonifacino JS, Hurley JH (2013) Structural basis for recruitment and activation of the AP-1 clathrin adaptor complex by Arf1. Cell 152(4):755–767. https://doi.org/10.1016/j.cell.2012.12.042
19. Jia X, Weber E, Tokarev A, Lewinski M, Rizk M, Suarez M, Guatelli J, Xiong Y (2014) Structural basis of HIV-1 Vpu-mediated BST2 antagonism via hijacking of the clathrin adaptor protein complex 1. eLife 3:e02362. https://doi.org/10.7554/eLife.02362
20. Kelly BT, Graham SC, Liska N, Dannhauser PN, Honing S, Ungewickell EJ, Owen DJ (2014) Clathrin adaptors. AP2 controls clathrin polymerization with a membrane-activated switch. Science 345(6195):459–463. https://doi.org/10.1126/science.1254836
21. Shen QT, Ren X, Zhang R, Lee IH, Hurley JH (2015) HIV-1 Nef hijacks clathrin coats by stabilizing AP-1:Arf1 polygons. Science 350(6259):aac5137. https://doi.org/10.1126/science.aac5137
22. Morris KL, Buffalo CZ, Sturzel CM, Heusinger E, Kirchhoff F, Ren X, Hurley JH (2018) HIV-1 Nefs are cargo-sensitive AP-1 Trimerization switches in Tetherin downregulation. Cell 174(3):659–671. e614. https://doi.org/10.1016/j.cell.2018.07.004
23. Buffalo CZ, Sturzel CM, Heusinger E, Kmiec D, Kirchhoff F, Hurley JH, Ren X (2019) Structural basis for Tetherin antagonism as a barrier to zoonotic lentiviral transmission. Cell Host Microbe 26(3):359–368. e358. https://doi.org/10.1016/j.chom.2019.08.002
24. Partlow EA, Baker RW, Beacham GM, Chappie JS, Leschziner AE, Hollopeter G (2019) A structural mechanism for phosphorylation-dependent inactivation of the AP2 complex. eLife 8:e50003. https://doi.org/10.7554/eLife.50003
25. Wrobel AG, Kadlecova Z, Kamenicky J, Yang JC, Herrmann T, Kelly BT, McCoy AJ, Evans PR, Martin S, Muller S, Sroubek F, Neuhaus D, Honing S, Owen DJ (2019) Temporal ordering in endocytic Clathrin-coated vesicle formation via AP2 phosphorylation. Dev Cell 50(4):494–508. e411. https://doi.org/10.1016/j.devcel.2019.07.017
26. Beacham GM, Partlow EA, Hollopeter G (2019) Conformational regulation of AP1 and AP2 clathrin adaptor complexes. Traffic 20(10):741–751. https://doi.org/10.1111/tra.12677
27. Gulbranson DR, Crisman L, Lee M, Ouyang Y, Menasche BL, Demmitt BA, Wan C, Nomura T, Ye Y, Yu H, Shen J (2019) AAGAB controls AP2 adaptor assembly in Clathrin-mediated endocytosis. Dev Cell 50(4):436–446. e435. https://doi.org/10.1016/j.devcel.2019.06.013
28. Pohler E, Mamai O, Hirst J, Zamiri M, Horn H, Nomura T, Irvine AD, Moran B, Wilson NJ, Smith FJ, Goh CS, Sandilands A, Cole C, Barton GJ, Evans AT, Shimizu H, Akiyama M, Suehiro M, Konohana I, Shboul M, Teissier S, Boussofara L, Denguezli M, Saad A, Gribaa M, Dopping-Hepenstal PJ, McGrath JA, Brown SJ, Goudie DR, Reversade B, Munro CS, McLean WH (2012) Haploinsufficiency for AAGAB causes clinically heterogeneous forms of punctate palmoplantar keratoderma. Nat Genet 44(11):1272–1276. https://doi.org/10.1038/ng.2444
29. Wang B, Ye K (2017) Nop9 binds the central pseudoknot region of 18S rRNA. Nucleic Acids Res 45(6):3559–3567. https://doi.org/10.1093/nar/gkw1323
30. Mossessova E, Lima CD (2000) Ulp1-SUMO crystal structure and genetic analysis reveal conserved interactions and a regulatory element essential for cell growth in yeast. Mol Cell 5(5):865–876. https://doi.org/10.1016/s1097-2765(00)80326-3
31. Inoue H, Nojima H, Okayama H (1990) High efficiency transformation of *Escherichia coli* with plasmids. Gene 96(1):23–28. https://doi.org/10.1016/0378-1119(90)90336-p

Chapter 16

Monitoring Endosomal Cargo Retrieval to the Trans-Golgi Network by Microscopic and Biochemical Approaches

Amulya Priya and Sunando Datta

Abstract

The endosomal recycling pathway plays a crucial role in diverse physiologically important biological processes such as cell-to-cell signaling, nutrient uptake, immune response, and autophagy. A selective subset of these recycling cargoes, mostly transmembrane proteins, is retrieved from endosomes to the trans-Golgi network (TGN) by a retrograde transport process. Endosome-to-TGN retrograde trafficking is crucial for maintaining cellular homeostasis and signaling by preventing proteins and lipids from degradation in the lysosome. Many of the membrane sorting machinery, such as the retromer complex and sorting nexins (SNXs) are involved in endosomal retrieval and recycling of various transmembrane proteins. Recent technological advances in the resolution of light microscopy and unbiased analytical approaches in quantitative image analysis enable us to explore and understand the regulation of membrane trafficking pathways in greater detail. In this chapter, we describe quantitative imaging-based methods for analyzing the roles of proteins involved in the retrograde trafficking in retromer dependent or independent fashion, using cation-independent mannose-6-phosphate receptor (CIM6PR) as an example.

Key words Golgi, Trans-Golgi network, Endosome, Retrograde trafficking, CIM6PR, Immunofluorescence, Membrane recruitment, Retromer

1 Introduction

Together, the endosomal and biosynthetic pathways maintain an active pool of proteins and lipids required for sorting and retrieval of the resident proteins [1, 2]. The equilibrium between the endocytic uptake and recycling controls the composition of the plasma membrane, thereby regulating various cellular processes within the cell [3–5]. Following endocytosis, the cargo is delivered to the sorting endosome, from where it can be recycled directly to the plasma membrane or via the endocytic recycling compartment (ERC) [6–8]. Further, the transport of proteins via TGN-to-endosomal pathway is balanced by independent routes that divert proteins from either endosomes to the TGN or from the Golgi to the endoplasmic reticulum [9, 10]. Many of the retrograde transport

Jingshi Shen (ed.), *Membrane Trafficking: Methods and Protocols*,
Methods in Molecular Biology, vol. 2473, https://doi.org/10.1007/978-1-0716-2209-4_16,

cargoes such as SNARE, sorting receptors and others are continuously retrieved from the endosomal system back to the biosynthetic pathway [11–13]. Several transmembrane proteins, such as CIM6PR, SNAREs, peptidases, and protein toxins undergo retrograde trafficking from endosomes to TGN [14]. This endosome-to-TGN retrograde trafficking route prevents proteins and lipids from degradation in endo-lysosomes [9, 10]. The dysregulation or defects in endosomal sorting and recycling process results in the lysosomal degradation of the receptors or their mislocalization to the cell surface, thereby resulting in a range of pathological situations [4, 15, 16]. Following internalization from the cell surface, most receptors undergo one or more of the following three potential fates: (1) ubiquitinated receptors are transported to multivesicular bodies and finally undergo degradation in lysosomes [17, 18] (2) receptors can be recycled from endosomal compartments back to the plasma membrane [3, 7, 19, 20] (3) receptors undergo trafficking from endosomes to the Golgi in a retrograde fashion [9, 10, 21]. The most widely studied cargo in the retrograde trafficking pathway is the cation independent mannose-6-phosphate receptor (CIM6PR) [22]. They are involved in delivery of lysosomal acid-hydrolases from the TGN to endosomes. Due to the acidic environment of the endosomes, acid hydrolases-CIM6PR complex dissociates and receptors (CIM6PR) are recycled back to TGN whereas acid hydrolases are transported to lysosomes where they catalyze the degradation of variety of macromolecules [23]. Following delivery of the cargo in the endosomes, CIM6PRs are retrieved back to the Golgi by several retrograde transport pathways in order for the receptor to be involved in another round of recycling [24–26]. Since the last 20 years, many studies have identified the role of the retromer complex in retrograde trafficking of CIM6PR [15, 24, 25]. However, recently this theory has been challenged showing the SNX-BAR dimer as a principle regulator of CIM6PR transport independent of retromer [27, 28].

Here, we describe methods that can be used to assay the role of different proteins in trafficking the retrograde cargo, CIM6PR [29, 30]. Many labs have established an array of methodologies for studying retrograde trafficking of CIM6PR [24, 25, 27, 30, 31]. Although this protocol is specifically written to monitor the state of CIM6PR in HeLa cell line, the assays could also be used to examine other cargoes that are trafficked in a retrograde fashion in variety of cell lines. Also, these assays could be used to study endosome-to-Golgi trafficking in retromer-dependent or -independent ways in various mammalian cell lines. The protocols described here can be used either with overexpression or loss-of-function approach to investigate the role(s) of proteins in retrograde trafficking.

The methods described here are:

1.1 Microscopy Based Methods

1.1.1 Steady-State CIM6PR Distribution

The steady state distribution of CIM6PR can be monitored using immunofluorescence (IF) and visualized by fluorescence microscopy. This assay relies on the fact that under normal circumstances, most of CIM6PR molecules are localized at the TGN but a small population is also detected at the plasma membrane and in endocytic compartments. RNAi-mediated knockdown or protein overexpression of the molecules of interest might cause redistribution of CIM6PR from the TGN, causing more CIM6PR to localize to the endosomal and lysosomal compartments. The association of CIM6PR with the Golgi (*see* **Note 1**) marked by endogenous antibodies against either TGN46, GM130 or Syntaxin 6 (Table 1) or GFP-Golph3 construct and endosomes marked by endogenous antibodies against either Rab5, Rab7, EEA1, CD63, or Lamp1 (Table 1) can be quantified using image-based quantitative approaches [21].

1.1.2 Pulse–Chase CD8 Antibody Uptake Assay Describing Endosome-to-Golgi Retrieval

A change in the steady-state distribution of CIM6PR only suggests a possible defect in its endosome-to-Golgi trafficking without specifying any step at which the process falters. To decipher the same, one would need to design an assay to reveal the trafficking of the cargo in real time under physiological conditions to understand the steps during endocytosis at which the cargo is mis-sorted. A small population of CIM6PRs were detected on the plasma membrane and their endocytosis enabled us to study the retrograde trafficking pathway to the TGN. The population of CIM6PR present at the plasma membrane could be detected by antibodies that recognize an extracellular domain of the protein. The antibody bound to the transmembrane receptor (such as CIM6PR) at the plasma membrane and was then internalized by the cell, followed by trafficking via endosomal compartments (Fig. 1). Finally, cells are fixed at different time points followed by immunostaining, and the amount of antibodies that reached the TGN or endosomal compartments were quantified (Fig. 1). For this assay, we will be using HeLa cells stably expressing GFP-Golph3 and CD8-CIM6PR reporter construct [24].

1.1.3 Membrane Recruitment of Core Retromer Vps Components to Endosomes

To decipher the role of the proteins of interest in endosome-to-Golgi trafficking mediated by retromer, here we describe immunofluorescence based assays. We quantitatively measure the association of the core retromer components Vps26A and Vps35 co-localizing with endosomal compartments upon knockdown of the proteins of interest or upon their overexpression (Fig. 2) [21]. The endosomes can be labeled by endogenous antibodies against either Rab5, EEA1, Rab7 or CD63 (Table 1). This in addition to the above assays would be indicative of their role in retrograde trafficking of CIM6PR in a retromer-dependent fashion.

Table 1
List of all antibodies used

Stock concentration (SC), Working concentration (WC)

S.No	Antibody	Reference	Concentration/Dilution	Description
1.	CIM6PR	AbD Serotec, MCA2048	SC: 1 mg/ml, WC: 2 μg/ml	transmembrane protein
2.	CD8	Ancell, 153-020	SC: 1 mg/ml, WC: 2 μg/ml	transmembrane protein
3.	TGN46	AbD Serotec, AHP 1586	1:150 dilution from stock	Golgi
4.	GM130	BD Biosciences, 610822	SC: 250 μg/ml WC: 1.66 μg/ml	Golgi
5.	Syntaxin6	Cell Signaling Technology, C34B2	1:50 dilution from stock	Golgi
6.	Rab5	BD Biosciences, 610724	SC: 250 μg/ml, WC: 2.5 μg/ml)	endosomes
7.	Rab7	Cell Signaling Technology, D95F2	1:100 dilution from stock	endosomes
8.	EEA1	Marino Zerial Lab	1:500 dilution from stock	endosomes
9.	CD63	DSHB, H5C6	1:200 dilution from stock	endosomes
10.	Lamp1	BD Biosciences, 555798	SC: 0.5 mg/ml, WC: 5 μg/ml	endosomes
11.	Vps26	Abcam, ab23892	SC: 0.7 mg/ml, WC: 0.7 μg/ml	retromer
12.	Vps35	Santa Cruz, sc-374372	SC: 200 μg/ml, WC: 2 μg/ml	retromer
13.	Transferrin Receptor	Invitrogen, 136800	SC: 0.5 mg/ml, WC: 0.5 μg/ml	transmembrane protein
14.	Tubulin	Sigma, T6557	SC: 1.3 mg/ml, WC: 0.43 μg/ml	cytosolic protein
15.	Vinculin	Sigma, V9131	SC: 9.3 mg/ml, WC: 9.3 μg/ml	cytosolic protein
16.	Anti-rabbit IgG Alexa Fluor 647	Invitrogen, A21244	SC: 2mg/ml, WC: 4μg/ml	secondary antibody (IF)
17.	Anti-mouse IgG Alexa Fluor 647	Invitrogen, A21235	SC: 2mg/ml, WC: 4μg/ml	secondary antibody (IF)
18.	Anti-mouse IgG Alexa Fluor 488	Invitrogen, A11029	SC: 2mg/ml, WC: 4μg/ml	secondary antibody (IF)
19.	Anti-rabbit IgG Alexa Fluor 488	Invitrogen, A11034	SC: 2mg/ml, WC: 4μg/ml	secondary antibody (IF)
20.	Anti-mouse IgG Alexa Fluor 568	Invitrogen, A11031	SC: 2mg/ml, WC: 4μg/ml	secondary antibody (IF)
21.	Anti-rabbit IgG Alexa Fluor 568	Invitrogen, A11036	SC: 2mg/ml, WC: 4μg/ml	secondary antibody (IF)
22.	HRP-conjugated anti-rabbit IgG	Sigma, A0545	SC: 0.25 mg/ml, WC: 0.25μg/ml	secondary antibody (IB)
23.	HRP-conjugated anti-mouse IgG	BD Biosciences, 554002	SC: 0.25 mg/ml, WC: 0.25μg/ml	secondary antibody (IB)

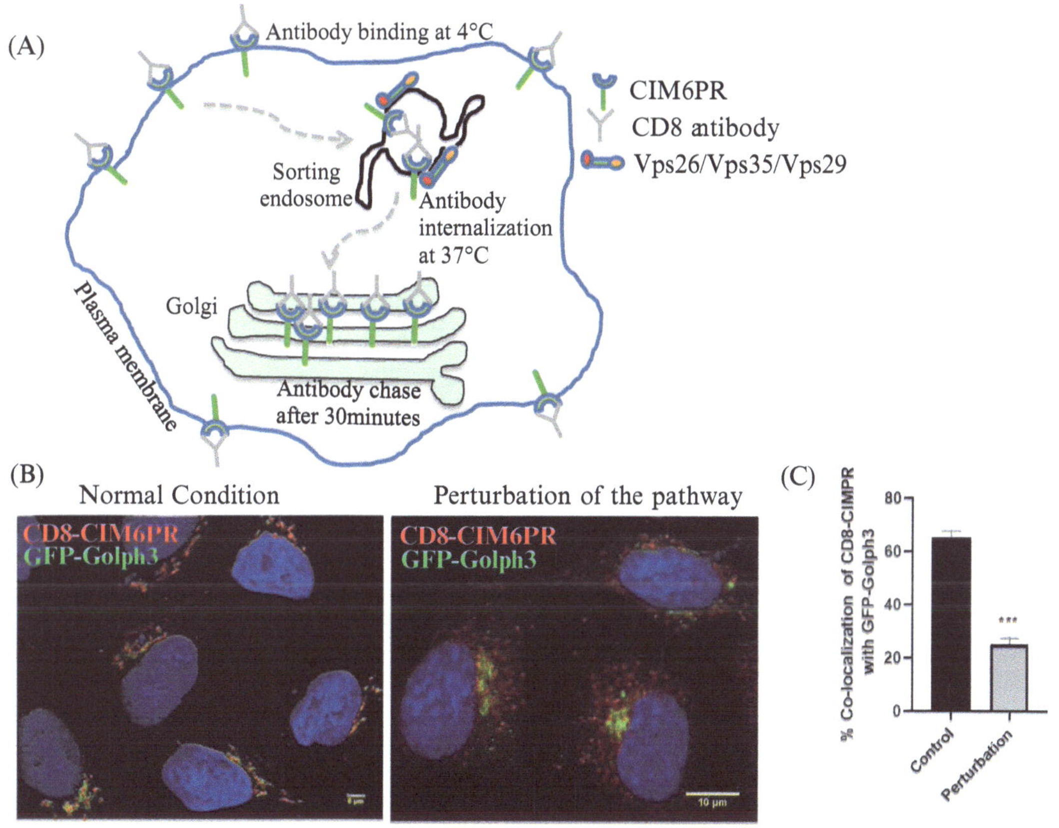

Fig. 1 Endosome-to-Golgi retrieval of CD8-CIM6PR. (**a**) Schematic model showing pulse–chase kinetics of the uptake of CD8-CIM6PR from plasma membrane en route to the Golgi as described in Subheading 1. (**b**). Representative confocal immunofluorescence images showing the majority of CD8-CIM6PR localizing in the Golgi in normal conditions whereas upon perturbation of the pathway they are mis-localized in endosomes. Scale bar 5-10 μM. (**c**) Quantitative colocalization analysis representing CD8-CIM6PR retrieval to the Golgi under normal conditions and upon perturbation of the pathway as analyzed by MotionTracking software

1.2 Biochemical Based Methods

As an additional readout, a biochemical fractionation experiment could also be performed to measure the amount of membrane associated Vps26 or Vps35 to investigate the role of proteins of interest in recruitment of the retromer core complex to endosomal membranes [21, 30]. The advantage of this assay is that the results are based on assaying millions of cells which is unbiased whereas microscopy-based approaches are limited to 200–500 cells maximum. By snap-freezing to permeabilize the membranes, the pellet and supernatant fractions of the cells are separated and the amount of membrane associated Vps26A or CIM6PR can be quantified by immunoblot (IB) analysis (measuring the intensities of protein bands with ImageJ software) for the proteins of interest upon their siRNA-mediated knockdown or overexpression (Fig. 3).

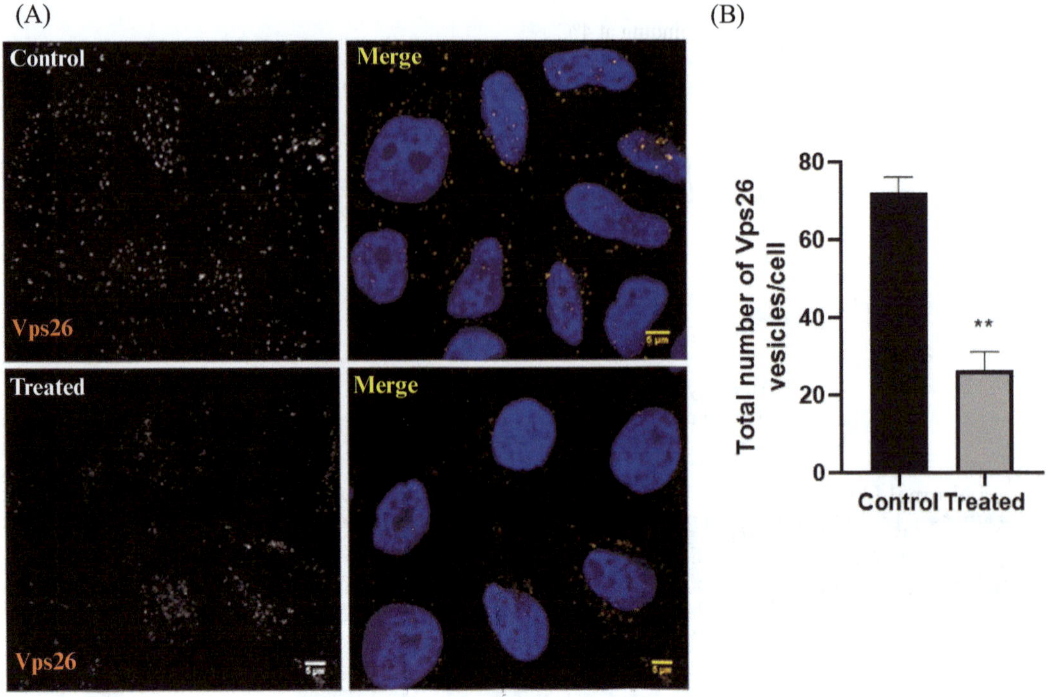

Fig. 2 Membrane recruitment of core retromer complex (**a**) Representative confocal images showing immunostaining of endogenous Vps26 antibody in control and treated HeLa cells. Scale bar 5 μM. The images represent that after treatment there is drastic reduction in the number of Vps26 endosomes. (**b**) The number of Vps26 endosomes as quantified using MotionTracking software. The quantitative representation of the total number of Vps26 vesicles per cell

2 Materials

Prepare all solutions with analytical grade chemical reagents using ultrapure deionized water (dH_2O). Store reagents at room temperature or as indicated.

2.1 Cell Culture Reagents

1. Plasticware: 10-cm round dish, T-25 flasks with filter cap, Filter tips: 0.2–1 mL, sterilized serological pipettes: 2–25 mL, sterilized microcentrifuge tubes, 35 mm dish, 6-well and 24-well cell culture plates. Sterilized glass coverslip.
2. HeLa cells.
3. Dulbecco's Modified Eagle's Medium (DMEM) with high glucose and pyruvate. Store at 4 °C.
4. Heat-inactivated fetal bovine serum (FBS). Store at −80 °C.
5. Penicillin–Streptomycin (PenStrep) (10,000 U/mL). Store at −20 °C.
6. Puromycin dihydrochloride (10 mg/mL). Store at −20 °C.

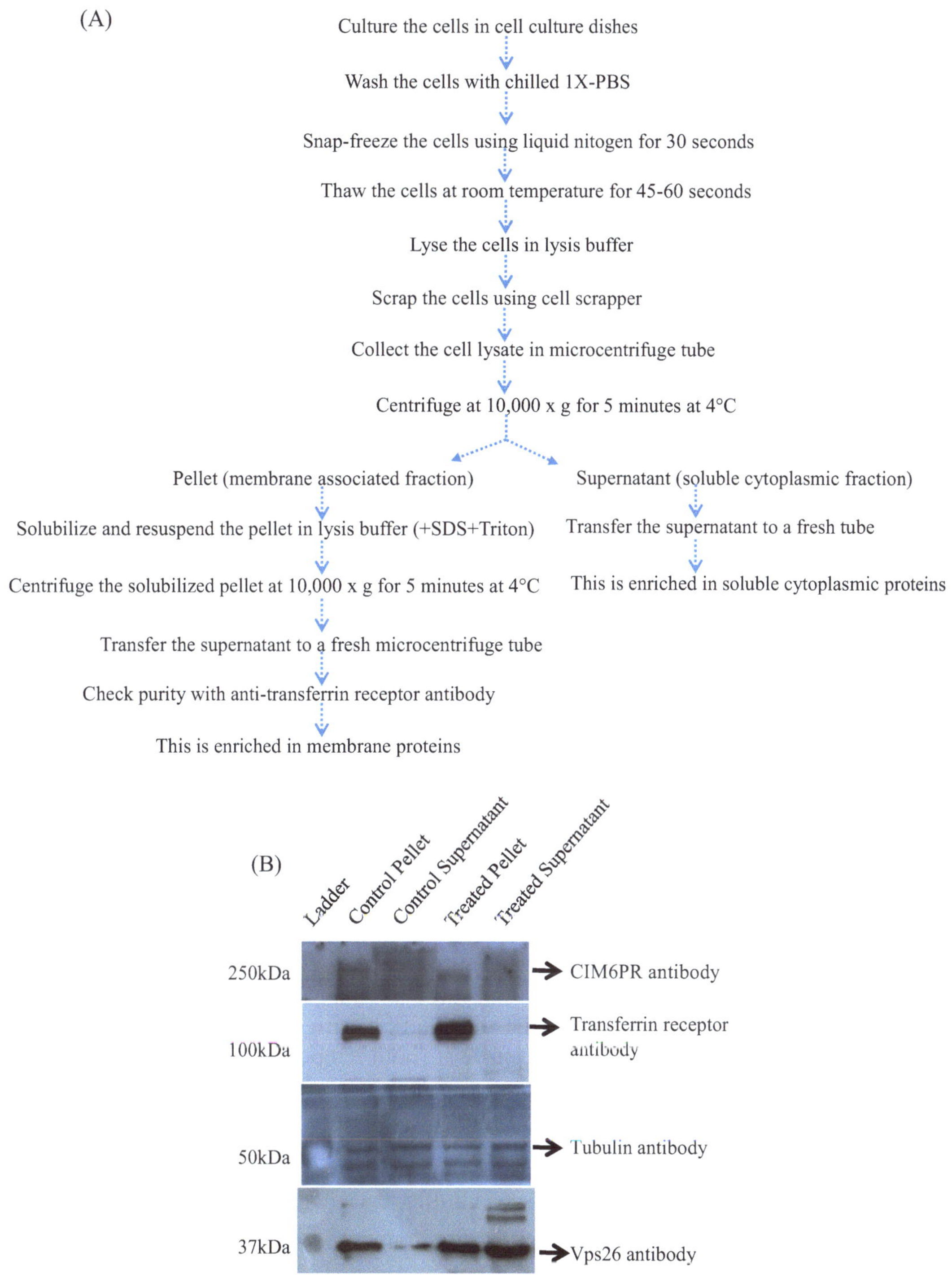

Fig. 3 Snap freezing method to measure the total amount of proteins in membrane and cytosolic fractions. (**a**) Representative flowchart depicting step by step process of snap freezing methodology to separate membrane associated vs soluble fractions of the proteins, for details refer to Subheading 3.9. (**b**) Representative immunoblot showing pellet and supernatant fractions of control and treated HeLa cells which were snap frozen for permeabilization to separate pellet (membrane proteins) and supernatant (cytoplasmic) fractions. The purified membrane fractions were probed with transferrin receptor antibody to check their purity. Immunoblot analysis showing the amount of Vps26 and CIM6PR in membrane vs soluble fractions

7. Geneticin selective antibiotic G418 sulfate (50 mg/mL). Store at 4 °C.
8. Culturing medium for HeLa cells: DMEM with High glucose and pyruvate, 10% fetal bovine serum, 2 mM L-glutamine, and PenStrep (10,000 U/mL). Store at 4 °C.
9. HeLa cells stably expressing GFP-Golph3 and CD8-CIM6PR construct.
10. Culturing medium for stably expressing cell line: DMEM with high glucose and pyruvate, supplemented with 10% FBS, Pen-Strep (10,000 U/mL), 2 mM L-glutamine, 400 μg/mL Geneticin, and 1 μg/mL puromycin. Store at 4 °C.
11. 0.25% Trypsin–EDTA. Store at −20 °C, short-term at 4 °C.
12. Lipofectamine LTX reagent with Plus reagent (Thermo Fisher Cat. No. 15338100).
13. DharmaFECT siRNA transfection reagent (GE Healthcare Cat. No. T-2001-01).
14. Opti-MEM Reduced Serum.
15. Nontargeting siRNA control pool and other smartpools for interested genes (GE Healthcare Dharmacon).
16. Humidified cell culture incubator set at 37 °C with 5% CO_2.
17. Immunofluorescence staining chamber.
18. Microscope glass slides.

2.2 Reagents and Solutions

Use analytical grade chemical reagents to prepare solutions. Prepare all solutions in distilled water and store at room temperature, unless mentioned otherwise.

1. 10× phosphate-buffered saline (PBS): 1.37 M NaCl, 27 mM KCl 100 mM Na_2HPO_4, 18 mM KH_2PO_4. Adjust pH to 7.4. Filter-sterilize the solution with 0.2 μm filter. Autoclave the solution and store at room temperature or 4 °C.
2. 1× PBS: Dilute 10×-PBS with dH_2O.
3. 4% Paraformaldehyde (PFA) in 1× PBS. 4% PFA, 0.1 mM $CaCl_2$, 0.1 mM $MgCl_2$ in 1× PBS. Store at −20 °C.
4. Permeabilization buffer: 0.1% Triton X-100 (v/v) in 1 × PBS.
5. Blocking buffer for immunofluorescence (IF) labeling: 10% FBS in 1× PBS.
6. Blocking buffer for antibody uptake assay: 3% BSA (w/v) in 1× PBS.
7. Mounting media: Mowiol 4–88, 0.12 M Tris–HCl, pH 8.5, and 20% glycerol.
8. 10% sodium dodecyl sulfate (SDS).

9. Protease inhibitor cocktail (PIC): aprotinin (2 mg/mL), leupeptin (10 mM), and pepstatin A (1 mM). Store at −20 °C.
10. 0.2 M phenylmethylsulfonyl fluoride (PMSF): Store at −20 °C.
11. Cell lysis buffer: 0.1 M MES–NaOH, pH 6.5, 0.25 M sucrose, 1 mM magnesium acetate, 200 μM sodium vanadate, 1× PIC, and 200 μM PMSF.
12. Blocking buffer for western blot analysis: 5% BSA in 1× TBST (Tris-buffer saline with 0.1% Tween 20). Alternatively, 1% skimmed milk in 1× TBST can be used.
13. Primary antibodies: The source and concentrations of all antibodies used are listed in Table 1.
14. Secondary antibodies for IF: anti-rabbit IgG Alexa Fluor 488, 546, and 633 and anti-mouse IgG Alexa Fluor 488, 546, and 633 antibodies (working concentration: 4 μg/mL). Refer to Table 1.
15. Secondary antibodies for immunoblot (IB): HRP-conjugated anti-rabbit and anti-mouse IgG antibodies (working concentration: 0.25 μg/mL). Refer to Table 1.

2.3 Microscopy and Image Analysis

1. Zeiss LSM 780 laser-scanning confocal microscope with Zen 2010 software.
2. Image analysis software: MotionTracking (http://motiontracking.mpi-cbg.de).
3. ImageJ 1.52p (https://imagej.nih.gov/ij/download.html).

3 Methods

3.1 Cell Plating

Perform all the following steps in a Biosafety-level-II laminar air flow cabinet that has been pretreated with UV-light for at least 30 min. Sterilize all materials to be used inside the cabinets with 70% ethanol.

1. One day prior to the experiment, place 12 mm sterile glass coverslips in a 24-well cell culture plate using sterile forceps. For every experimental condition, make sure to use at least two wells.
2. Trypsinize and split HeLa cells (70–80% confluency) grown in 10-cm cell culture dishes.
3. Plate 40,000 HeLa cells per well on sterile coverslips in 500 μL complete culturing media in 24-well cell culture plate. The cells will be ready for experiments the next day (*see* **Note 2**).
4. Place the 24-well cell culture plate in an incubator for 12–16 h for the adherent cells to attach and spread onto coverslips.

3.2 siRNA-Mediated Knockdown of Proteins of Interest

1. Seed HeLa cells ($n = 20{,}000$) on coverslips in 24-well cell culture plate 1 day prior to siRNA transfection, as described in Subheading 3.1 (*see* **Note 3**).
2. For transfection, prepare 5 μM siRNA solution in 1× siRNA buffer from the 20 μM stock solution. Do not vortex the reaction tubes.
3. For a single siRNA transfection in a 24-well cell culture dish: In tube 1, pipette 47.5 μL of Opti-MEM and to this add 2.5 μL of 5 μM siRNA such that the final concentration of siRNA per well is maintained at 25 nM (in 500 μL medium per well). In tube 2, pipette 48 μL of Opti-MEM and to this add 2 μL of DharmaFECT (per each reaction per well). Pipette and mix both tubes well. Do not vortex the reaction tubes (*see* **Notes 4** and **5**).
4. Incubate the two reaction mixtures, that is, tube 1 and tube 2, each for 5 min separately at room temperature.
5. Mix and transfer the solution of tube 2 to tube 1. Mix the solutions (total volume 100 μL) and incubate for another 20 min at room temperature.
6. After 20 min, take the 24-well cell culture plate from the incubator. Remove the culturing media from the wells of the 24-well cell culture plate and add 400 μL of warm culturing medium per well. Add 100 μL of the transfection mixture dropwise to each well of the plate. Gently swirl the plate.
7. Incubate the cells in 37 °C incubator for 72 h. Change the media after 24 h with complete media. After 72 h posttransfection, proceed for further analysis.

3.3 Transient Overexpression of Proteins of Interest

1. For transient transfection of HeLa cells, seed 40,000 cells on coverslips in a 24-well cell culture plate 1 day prior to experiment.
2. Add DNA (500 ng) in 100 μL of Opti-MEM in a microcentrifuge tube. This reaction is for one well of a 24-well cell culture plate but can be scaled up or down.
3. Add 1 μL of Lipofectamine PLUS reagent to the above solution, mix and incubate for 5 min at room temperature.
4. Add 1 μL of Lipofectamine LTX to the above mixture and gently mix without introducing air bubbles. Incubate for 30 min.
5. During the incubation time, gently aspirate culture media from the wells of the 24-well cell culture plate. Add 400 μL of DMEM complete media to each well. Place the 24-well plate back in the cell culture incubator.

6. Add DNA-LTX complex dropwise to the wells of the 24-well cell culture plate. Gently swirl the plate and put it back into the cell culture incubator.
7. Change the media after 14–16 h.
8. Observe fluorescence under a microscope with a 10× or 20× objective. The transfection time can be optimized between 20 and 36 h. Analyze the cells based on the experimental requirements.

3.4 Immunofluorescence Labeling to Monitor Steady State Distribution of CIM6PR

1. Aspirate media gently from the corner of the wells by suction pipette without disturbing the coverslips.
2. Wash the cells once with sterile 1× PBS (500 μL/well) and add 4% paraformaldehyde in 1× PBS (500 μL/well). Incubate for 15 min at room temperature.
3. Wash the cells twice with 1× PBS (500 μL/well). At this stage, the plate can be stored at 4 °C or proceed to the permeabilization step as described below.
4. Add 0.1% Triton-X-100 in 1× PBS (500 μL/well). Incubate at room temperature for 15 min. Wash the coverslips once with 1× PBS.
5. Add 500 μL of blocking buffer in each well of 24-well cell culture plate and incubate for 1 h at room temperature.
6. To perform primary and secondary antibody labeling, prepare a dark humidified chamber. We use an immunofluorescence staining chamber and cover it with moist tissue. Spray with 70% ethanol and place a layer of Parafilm on the moist tissue paper.
7. To probe the endogenous state of CIM6PR, the cells can be co-stained with endogenous CIM6PR antibody and TGN46 antibody as major population of CIM6PR would be localized at TGN at the steady state (Table 1). Alternatively, cells can be co-stained with EEA1 antibody to label early endosomes where CIM6PR might localize upon perturbation of its trafficking (*see* **Note 6**).
8. Prepare the dilution of antibodies using the blocking buffer (*see* Fig. 1 and Table 1). Pipette 500 μL of blocking buffer in 1.5 mL microcentrifuge tube. Add 1 μL of CIM6PR antibody (working concentration: 2 μg/mL) and 3.4 μL of TGN46 antibody (dilution: 1:150) into each tube. In another microcentrifuge tube, add 1 μL of CIM6PR antibody (working concentration: 2 μg/mL) and 1 μL of EEA1 antibody (dilution: 1:500) for labeling on different coverslips (Table 1). Mix well by vortexing or pipetting (*see* **Note 7**).

9. Add 50 μL of diluted antibodies to Parafilm placed in an immunofluorescence staining chamber. Gently hold a coverslip from the sides with the help of forceps and place the coverslip onto the primary antibody solution (cells facing down). Repeat this step for other coverslips. Incubate at room temperature for 1 h.
10. After 1 h, transfer the coverslips (with cells facing up) back to the 24-well cell culture plate with 1× PBS in each well. Wash the coverslips twice with the blocking buffer.
11. Prepare the dilution of anti-rabbit IgG Alexa Fluor 488 and anti-mouse IgG Alexa Fluor 546 antibodies (working concentration: 4 μg/mL). Add 500 μL of blocking buffer to a 1.5 mL microcentrifuge tube. Add 1 μL of anti-rabbit IgG Alexa Fluor 488 and 1 μL of anti-mouse IgG Alexa Fluor 546 antibodies. Mix well.
12. Transfer 50 μL of diluted secondary antibodies onto Parafilm and place a coverslip on the top of this solution (cells facing down). Repeat this for the other coverslips. Incubate for 1 h at room temperature.
13. After 1 h, transfer the coverslips carefully back to the plate with cells facing up.
14. Wash the coverslips twice with 1× PBS by incubating at room temperature for 5 min each.
15. Mount the coverslips on glass slides using the mounting medium (4–5 μL) with cells facing the mounting medium. Cover the glass slides with the lid of immunostaining chamber and let them air dry at room temperature in the dark.
16. Examine the samples under a confocal microscope with appropriate filters.

3.5 CD8 Antibody Uptake Assay Describing Endosome-to-Golgi Retrieval

This procedure is adapted from [21, 24, 31].

1. Seed HeLa cells stably expressing GFP-Golph3 and CD8-CIM6PR reporter construct on sterile coverslips in a 24-well cell culture plate 1 day prior to the experiment as described in Subheading 3.1.
2. Next day, transfect the cells with siRNA of specific proteins of interest as described in Subheading 3.2 (**steps 2**–7). After 72 h posttransfection, proceed with the steps below.
3. Take out the 24-well cell culture plate from the incubator. Aspirate the media of the wells gently and add 1 mL of ice-cold 1× PBS on the wall of the 24-well plate in each well. Keep the plate at 4 °C (preferably in cold room) covered with aluminum foil for 10 min to stop membrane trafficking inside the cells.

4. During the 10 min incubation period, prepare a dark humidified chamber as described in **step 6** of Subheading 3.4.
5. Prepare the dilution of CD8 antibody (working concentration: 2 μg/mL) in complete DMEM culturing media. 50 μL of diluted antibodies per coverslip is required (*see* **Note 9**).
6. In the cold room, aliquot 50 μL of diluted CD8 antibodies onto the required number of spots onto the Parafilm in the humidified chamber.
7. After 10 min of incubation (**step 4** above), wash all the wells with cold 1× PBS and take the coverslips out of the 24-well cell culture plate (cell side up) with the help of sterile forceps. Hold the coverslip and tap it onto a paper towel for a fraction of second and invert the coverslip onto the dispensed antibody solution placed on the Parafilm (cell facing down). Incubate at 4 °C for 20 min. During this time, CD8 antibody will bind to the receptors on the plasma membrane and will remain bound without being internalized.
8. After 20 min, quickly transfer the coverslips (cell side up) to a 24-well cell culture plate with 1 mL 1× PBS added to each well. Wash the coverslips with 1× PBS twice to remove unbound CD8 antibodies.
9. After washing, add 1 mL of prewarmed complete culturing media to each well. Immediately incubate this plate at 37 °C water bath with circulating motion (*see* **Note 8**).
10. Incubate the plate at 37 °C for the desired time (2 min, 5 min, 10 min, 15 min, 20 min, or 30 min) to allow the uptake of CD8 antibody to occur.
11. At the desired time points, wash the coverslips with 1× PBS once and fix the cells with 4% paraformaldehyde (500 μL/well) for 15 min. This step can be performed in the water bath at 37 °C.
12. After 30 min, fix all conditions with 4% PFA for 15 min.
13. Proceed with washing and permeabilization of cells as described in Subheading 3.4 (**steps 2–4**).
14. Block the cells in blocking buffer (3% BSA in 1× PBS) for at least for 1 h at room temperature.
15. Prepare the dilution of endogenous EEA1 antibody (1:500 dilution) in IF blocking buffer (3% BSA in 1× PBS). Aliquot 50 μL of the diluted antibodies on Parafilm and invert the coverslips (with the cell side down) on top of this antibody. Incubate for 1 h in the dark (*see* **Note 9**).
16. Follow the steps for primary and secondary antibody incubation as detailed described in Subheading 3.4 (**steps 9** and **10**).

17. Prepare the dilution of secondary antibodies, anti-mouse IgG Alexa Fluor 546 and anti-rabbit IgG Alexa Fluor 647 (working concentration: 4 μg/mL) in blocking buffer (3% BSA in 1× PBS).
18. Monitor the samples for fluorescence probing specific antibody labeling under confocal microscope with appropriate filter selection.

3.6 Membrane Recruitment of Vps26A or Vps35 on Early or Late Endosomes by Immunofluorescence

1. Seed HeLa cells on coverslips and proceed with siRNA to knockdown the proteins of interest as described in Subheading 3.2 (**steps 2–7**).
2. Proceed with fixation, permeabilization, and blocking steps as described in Subheading 3.4 (**steps 2–5**).
3. Next, probe the endogenous levels of retromer proteins, that is, Vps26 and Vps35—this will give an estimate that upon knockdown of the proteins of interest whether Vps26 or Vps35 are cytosolic or endosomal membrane bound localizing with Rab5, EEA1 positive early endosomes or Rab7, CD63, or Lamp1 positive late endosomes.
4. Co-stain the cells with Vps26 (working concentration: 0.7 μg/mL) and Rab5 (working concentration: 2.5 μg/mL) antibodies. In parallel costain on different coverslips for Vps35 (2 μg/mL) and Rab7 (1:100 dilution) antibodies.
5. Follow **steps 9–16** as described in Subheading 3.4.
6. As shown in Fig. 2, there is significant reduction in the number of Vps26 positive endosomes in control versus treated condition. This implies that the amount of Vps26 localizing on endosomes is drastically reduced upon this treatment as compared to control conditions.

3.7 Confocal Microscopy and Image Acquisition

1. Image acquisition can be done using Zeiss LSM 780 Laser Scanning Confocal Microscope.
2. For a given experimental setup, imaging is done randomly scanning most areas on the coverslips. Also, for each condition, it is advisable to acquire 10 images from each coverslip. At least acquire 20–30 images for each condition (at least 200 cells) in a given experimental setup [21, 26].
3. Process all the images together for quantification using the same object identification parameters.
4. In order to avoid the image distortion by z-shift of different channels and improve the image quality, use a maximum projection of four optical sections with thickness 1.0 μm and step 0.8 μm.

5. Analyze the data from three independent experiments by the automated image analysis program, MotionTracking (http://motiontracking.mpi-cbg.de) [32, 33].

3.8 Image Analysis and Quantification

Perform automated image analysis using MotionTracking software as explained below (Fig. 4) [21, 26].

3.8.1 Colocalization Analysis

1. Import acquired images: Use the command **File > Import > Import microscope images > Import Zeiss LSM file**. Click "OK" to project.
2. Save project: Once all the images are imported, use **File > Save Project** to create the Project File for the future analysis.
3. Identify objects: The software performs object based analysis by identifying the position of objects in the acquired images, such as intracellular vesicles. Each object represents a group of pixels. The number of pixels combined with the scale will lead to "Object Size" and "Area."
4. To run the Object Search procedure use command **Object Search > Find Objects Parameters**. The opened dialog box is enlisted with the list of parameters to be adjusted according to the quality of the acquired image frames. Basic parameters to be adjusted are Background window size, Resolution limit, Minimum Object Area, Threshold Scale, Object Combination Coefficient, Cut-off Intensity Threshold, and Binning (Fig. 4).
5. Once the Object Search Parameters are set, search for the Objects through all of your acquired images by running the command: **Object > Find Objects in Batch Mode**. Save the project once objects are identified in all the frames.
6. Once all objects are identified, to compute various associated parameters run statistical analysis by running command: **Statistics > Batch statistics > Open batch statistic window.**
7. Measure the intensity of vesicles identified as objects: **Statistics > Distributions > Object Characteristics Distributions > Mean Intensity**.
8. Similarly, the total number of objects like number of Vps26 vesicles per cell is quantified as shown in Fig. 2.
9. Once intensity of objects is calculated proceed for colocalization analysis using command: **Image set analysis > Colocalization > Colocalization**.
10. Further parameters for calculating colocalization will be adjusted in a new window. A base channel will be selected, usually selected as the channel with overexpressed molecule or the ligand (in case of uptake assays) (Fig. 4). **Colocalization > Select Base channel > Number of objects > Execute**.

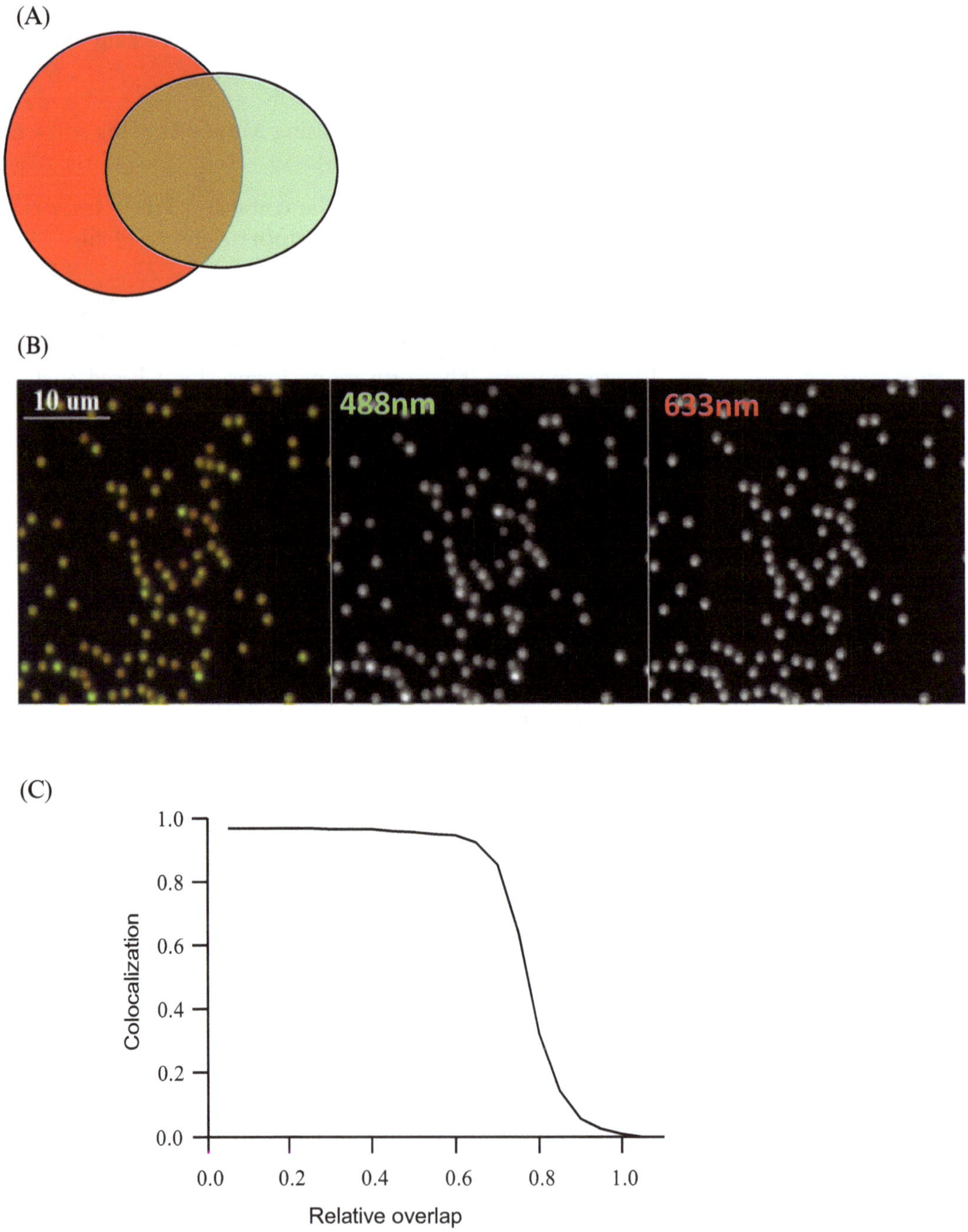

Fig. 4 Colocalization analysis by MotionTracking software. (**a**). The green vesicle is co-localized with red if more than x% of its area is covered by red vesicle. For example, if $x = 35\%$, then green is colocalized with red, but red may still not be co-localized with green [26]. It means that colocalization is asymmetric. (**b**) Example of colocalization of images of 1 μm multi-spec beads with excitation by 488 nm and 633 nm lasers. (**c**) Colocalization of multi-spec beads (ratio of number of co-localized images of beads to the total number of detected beads) as function of relative area overlap threshold [26]. (**d–f**). Screenshots of the MotionTracking software (**d**) List of different parameters to identify the objects as vesicles such as minimum object size, cut off intensity, threshold etc. (**e**) Detailed list of options in Image set Analysis so as to quantify the number of vesicular objects, mean intensities, or to measure colocalization. (**f**) Description of colocalization window to

(D)

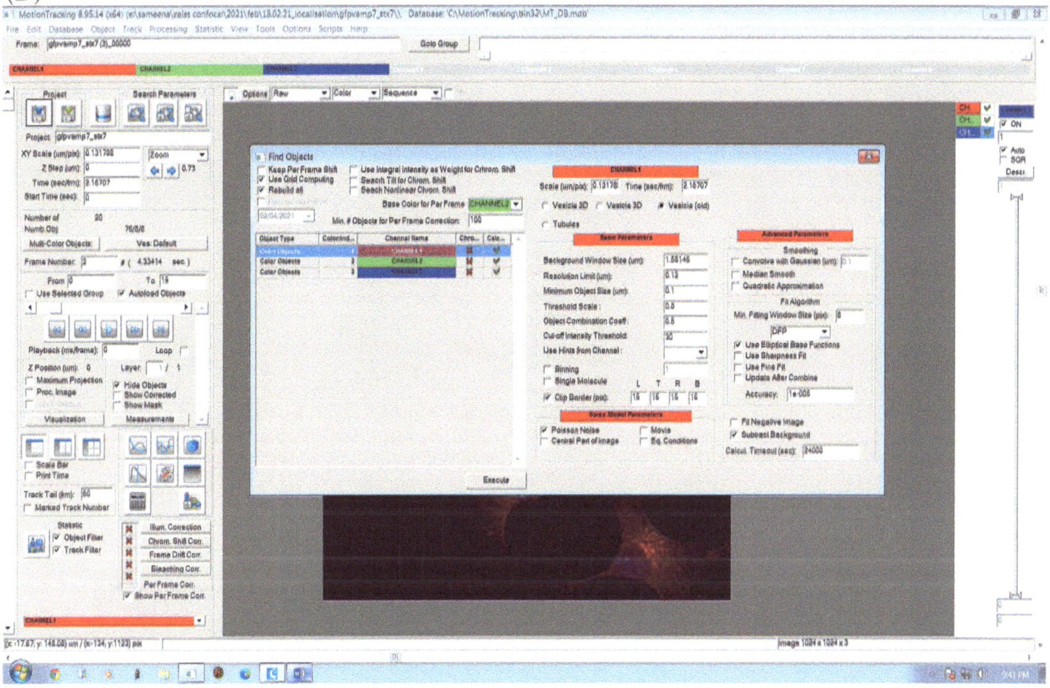

(E)

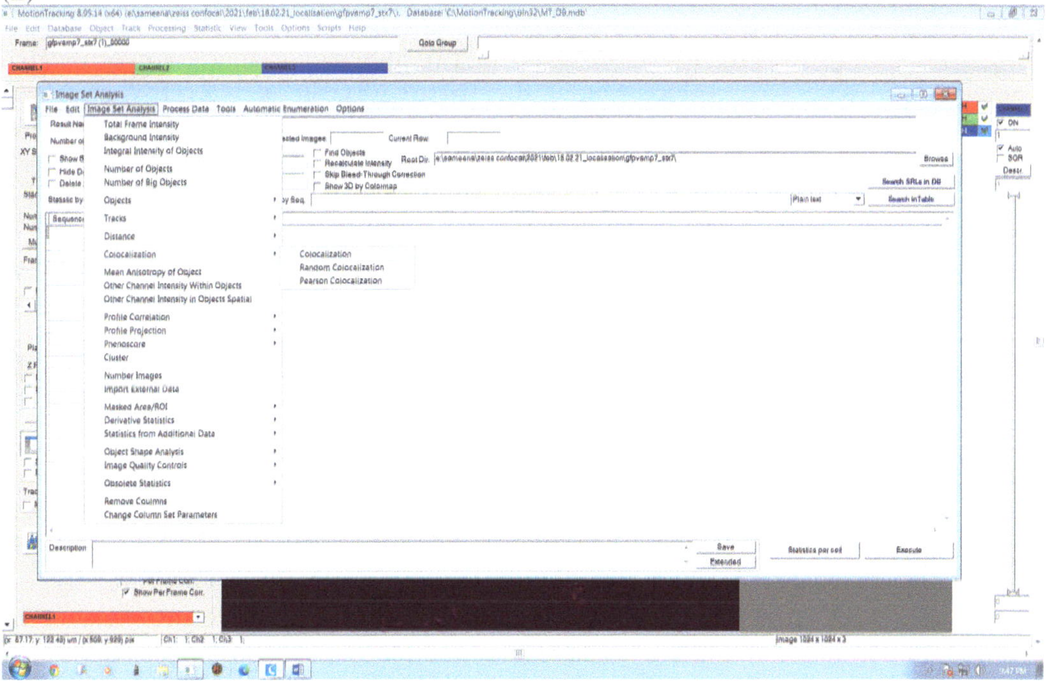

Fig. 4 (continued)

Fig. 4 (continued) select base channel colocalizing to other channels based on experimental requirements. The calculation of colocalization is based on integral intensity of objects taking into account subtraction of random colocalization as described in Subheading 3.8 (**steps 9–12**)

(F)

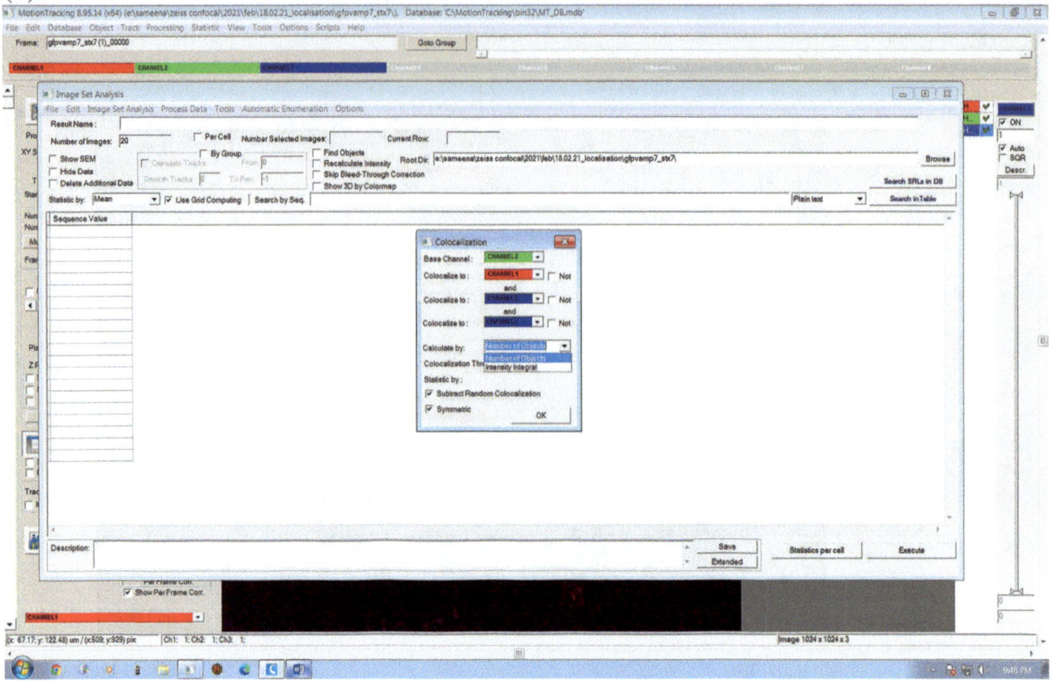

Fig. 4 (continued)

11. Objects detected in two different channels are considered colocalizing if the relative overlap of respective areas was above 35%. The apparent colocalization value is calculated as a ratio of integral intensities of co-localized objects to the integral intensities of all objects carrying the given marker and varies from 0.0 to 1.0 (Fig. 4) [21, 26].
12. The colocalization-by-chance (random colocalization) is estimated by random permutation of object localization in different channels. The apparent colocalization was corrected for random colocalization [21, 26].
13. Import the values for colocalization on an excel sheet. Plot the graph for colocalization with GraphPad Prism software between CIM6PR/TGN46 or CIM6PR/EEA1 or CIM6PR/Rab7 to estimate the amount of CIM6PR residing in Golgi and compare with the amount of CIM6PR residing on endosomes [21].
14. Export Data: **Image set analysis > File > Export**. Export the data in an excel file and plot the graph. Graph can also be plotted in Motion Tracking using command: **Image set analysis > Tools > Graph**.

3.9 Cell Fractionation Assay by Snap Freezing Method

This procedure is adapted from [24, 30].

1. Seed HeLa cells onto 6-well cell culture plate or 60 mm cell culture dishes (250,000 cells per well) 1 day prior to the experiment and proceed for experimental setup (siRNA mediated knockdown or overexpression of proteins of interest) as described in Subheadings 3.2 and 3.3.
2. On the day of the experiment, aspirate the medium of the cell culture dishes and wash the cells in 2 mL of cold 1× PBS. Gently release 1× PBS on side wall of the dishes in order to not detach the cells. Place the dishes or 6-well cell culture plate on ice.
3. Take liquid nitrogen in an ice bucket (in which 6-well dish or 35 mm plate can be placed inside). Take sufficient liquid nitrogen in the bucket and keep it covered with the lid of the bucket.
4. Aspirate 1× PBS solution from the cell culture dishes by Pasteur pipette.
5. With the help of forceps or tweezers hold the 6-well cell culture plate or 35 mm dishes from their side walls, and place them on liquid nitrogen. Let them float in liquid nitrogen for 30 s while still holding the plate with forceps or tweezers. Keep the dishes or the plate at an angle taking precautions not to let it fall and keeping your hands and eyes safe (*see* **Note 10**).
6. Take the dishes from liquid nitrogen and place them on ice. Proceed for the other plate or dishes.
7. Take the dishes from ice and rapidly thaw the cells at room temperature for 20–30 s.
8. Add 120–150 μL of cold lysis buffer in each well of the 6-well cell culture plate or 35 mm dishes (*see* **Note 11**).
9. Scrape the cells gently from the dishes or 6-well cell culture plate using a cell scraper.
10. Gently pipette and collect the cells in lysis buffer using a P200 pipette and transfer the lysate to 1.5 mL microcentrifuge tube on ice (*see* **Note 12**).
11. Centrifuge the cells in a refrigerated centrifuge at 4 °C for 10 min at 10,000 × *g* to separate the membrane-associated fraction in the pellet from the soluble cytoplasmic fraction in the supernatant.
12. Collect the supernatant (cytoplasmic fraction) in a fresh 1.5 mL microcentrifuge tube on ice. Label the tubes and keep the fractions on ice. This can be checked by western blot as described below by probing with tubulin or vinculin antibodies.
13. Gently resuspend the pellet in lysis buffer with 1% triton-X-100 and 0.1% SDS. Add 60–75 μL of lysis buffer to the pellet and

solubilize the pellet by gently pipetting up and down on ice with a P200 without introducing air bubbles. Do not vortex the tube (*see* **Note 11**).

14. Resuspend the pellet by gently petting up and down and allow the pellet to solubilize on ice for 5 min.
15. Centrifuge the resuspended solubilized fraction in a refrigerated centrifuge at 4 °C at 10,000 × *g* for 10 min. Transfer the supernatant to a fresh tube. This supernatant fraction, referred to as the "pellet fraction," is enriched in membrane-associated proteins. The purity of these fractions can be checked by western blot as described below by probing with transferrin receptor antibody or Lamp1 antibodies.
16. The amount of protein in pellet (membrane) and the supernatant (cytoplasmic) fractions, can be quantified by Bradford or BCA reagent. This is important to ensure equal loading of all the samples of different experimental conditions.
17. Prepare the samples by diluting them in 1× PBS and using equal quantity of protein for every experimental condition. Add 1× SDS sample dye to each of these samples and boil at 100 °C in a thermomixer for 5 min.
18. Load the samples preferably on 4–12% gradient polyacrylamide gel.
19. After separation on SDS-PAGE, transfer the proteins to a PVDF (Polyvinylidene fluoride) membrane in the western blot transfer set-up, overnight at 4 °C.
20. Carefully remove the membrane from the transfer setup. Incubate the membrane in blocking buffer (5% BSA in 1× TBST) or 1% skimmed milk in 1× TBST.
21. Prepare primary antibody solution of Vps26 (0.7 μg/mL) in 5% BSA in 1× TBST. Similarly, prepare the antibody solution of transferrin receptor (0.5 μg/mL), tubulin (0.43 μg/mL) and CIM6PR (2 μg/mL). Cut the membranes at correct molecular weight and incubate them with primary antibody solutions in small western boxes at room temperature for 1 h or overnight at 4 °C (*see* **Notes 13** and **14**).
22. After the incubation time, put the primary antibody back in a 15 mL tube. This can be stored at −20 °C for reuse.
23. Wash the membrane with 1× TBST for 10 min on a rocker to get rid of nonspecific antibody binding. Repeat this step three times.
24. Prepare a dilution of HRP-conjugated anti-mouse and anti-rabbit IgG antibodies separately (working concentration: 0.25 μg/mL). Incubate the membrane with secondary antibody for 1 h at room temperature.

25. Remove the secondary antibody. Wash the membrane with 1× TBST for 10 min on a rocker. Repeat this step twice.
26. To develop the membranes, use ECL detection reagent and incubate the membrane in ECL solution in the dark for 2–5 min. Scan the membrane in LICOR scanner or a Chemi-Doc setup.
27. Refer to Fig. 3 for a step-by step flowchart and immunoblot showing membrane and pellet fractions in different experimental setups.

4 Notes

1. One should be careful in choosing a neutral Golgi marker so that the distribution of the chosen marker is not dependent on the other pathway being studied. Golgi glycosyltransferases are membrane proteins involved in maintaining the architecture of the Golgi and are considered neutral Golgi markers.
2. Here, the procedure is described for a 24-well cell culture plate but could be scaled up or down for 96, 12 or 6-well plate. It is recommended to have two wells or duplicate coverslips for each condition in an experimental setup.
3. If the cell plating is done for siRNA, plate 20,000 cells per well as these would grow for 72 h so less cells are seeded to avoid overconfluency. Also, for other cell lines, the number of cells to be seeded might differ depending on their growth and doubling time.
4. Here, the final concentration of siRNA used is 25 nM, but the concentrations can be optimized (25–100 nM) depending on the knockdown efficiency of the genes of interest.
5. This is for a single reaction so multiply the volumes according to the number of transfections. As a control, include nontargeting control siRNA. Be careful while pipetting to avoid introducing air bubbles.
6. For labeling endosomes, any of the classical endosomal markers can be used. These can be marked by endogenous antibodies against either Rab5, Rab7, EEA1, CD63, or Lamp1. Similarly, for labeling Golgi one can use endogenous antibodies against either TGN46, GM130 or Syntaxin6. Golgi marker should be chosen carefully (*see* **Note 1**).
7. Here, the dilution is made in 500 μL of blocking buffer but it can be calculated according to the number of coverslips and the conditions of the experiment.
8. It is advisable to use the water bath with constant circulation in order to maintain accurate temperature. Take precautions to

keep the 24-well cell culture plate constant and stable without letting it fall.

9. To enhance the signal for GFP-Golph3, GFP primary antibody can also be used. Here we could detect good signal without using GFP antibody.
10. This step is crucial in this procedure. Be careful that the liquid nitrogen does not enter the plate. After 30 s, remove the dish or the plate from the liquid nitrogen and place them on ice. Proceed with the next dish or plate. It is advisable to do this procedure for only 4–5 dishes at a time to avoid delays in the subsequent steps.
11. The volume of the lysis buffer to be added can be optimized depending on the surface area of the plate and can vary between different cell lines.
12. It is important to keep all cell lysates on ice in order to prevent them from degradation. Microcentrifuge tubes can be labeled and precooled on ice to maintain the temperature.
13. For probing these antibodies in the pellet and cytosolic fractions, the antibody dilution should be well worked out in normal lysates of specific cell lines. This assay can also be performed in CD8-CIM6PR expressing cell lines.
14. To cut the membrane, hold it gently with forceps and place it on Parafilm. Cut the membrane with a scalpel at the correct molecular weight ensuring all the lanes are intact. Be careful not to place any fingerprints or touch any part of membrane.

Acknowledgments

We sincerely thank Prof. M.N.J Seaman for providing us HeLa cells stably expressing GFP-Golph3 and CD8-CIM6PR reporter construct. We also thank Prof. Yannis Kalaidzidis for his tremendous help in image quantifications with MotionTracking software, and Angelika Giner for providing us with endogenous EEA1 antibody and the protocol for preparing the Mowiol mounting media. We also thank Dr. Priyanka Sharma for her help with the image quantification write-up.

References

1. Blom T, Somerharju P, Ikonen E (2011) Synthesis and biosynthetic trafficking of membrane lipids. Cold Spring Harb Perspect Biol 3: a004713. https://doi.org/10.1101/cshperspect.a004713
2. Lippincott-Schwartz J, Phair RD (2010) Lipids and cholesterol as regulators of traffic in the endomembrane system. Annu Rev Biophys 39:559–578. https://doi.org/10.1146/annurev.biophys.093008.131357
3. Seaman MNJ (2021) A dimmer switch for endosome-to–cell surface recycling. J Cell Biol 220:e202102130. https://doi.org/10.1083/jcb.202102130

4. O'Sullivan MJ, Lindsay AJ (2020) The endosomal recycling pathway—at the crossroads of the cell. Int J Mol Sci 21:6074. https://doi.org/10.3390/ijms21176074
5. Grant BD, Donaldson JG (2009) Pathways and mechanisms of endocytic recycling. Nat Rev Mol Cell Biol 10:597–608. https://doi.org/10.1038/nrm2755
6. Hsu VW, Bai M, Li J (2012) Getting active: protein sorting in endocytic recycling. Nat Rev Mol Cell Biol 13:323–328. https://doi.org/10.1038/nrm3332
7. Sheff DR, Daro EA, Hull M, Mellman I (1999) The receptor recycling pathway contains two distinct populations of early endosomes with different sorting functions. J Cell Biol 145: 123–139. https://doi.org/10.1083/jcb.145.1.123
8. MacDonald E, Savage B, Zech T (2020) Connecting the dots: combined control of endocytic recycling and degradation. Biochem Soc Trans 48:2377–2386. https://doi.org/10.1042/BST20180255
9. Tu Y, Zhao L, Billadeau DD, Jia D (2020) Endosome-to-TGN trafficking: organelle-vesicle and organelle-organelle interactions. Front Cell Dev Biol 8:163. https://doi.org/10.3389/fcell.2020.00163
10. Bonifacino JS, Rojas R (2006) Retrograde transport from endosomes to the trans-Golgi network. Nat Rev Mol Cell Biol 7:568–579. https://doi.org/10.1038/nrm1985
11. Xie S, Bahl K, Reinecke JB, Hammond GRV, Naslavsky N, Caplan S (2016) The endocytic recycling compartment maintains cargo segregation acquired upon exit from the sorting endosome. Mol Biol Cell 27:108–126. https://doi.org/10.1091/mbc.E15-07-0514
12. Yong X, Zhao L, Deng W, Sun H, Zhou X, Mao L, Hu W, Shen X, Sun Q, Billadeau DD, Xue Y, Jia D (2020) Mechanism of cargo recognition by retromerlinked SNX-BAR proteins. PLoS Biol 18:e3000631. https://doi.org/10.1371/journal.pbio.3000631
13. Progida C, Bakke O (2016) Bidirectional traffic between the Golgi and the endosomes – machineries and regulation. J Cell Sci 129(21): 3971–3982. https://doi.org/10.1242/jcs.185702
14. Burd C, Cullen PJ (2014) Retromer: a master conductor of endosome sorting. Cold Spring Harb Perspect Biol 6:a016774. https://doi.org/10.1101/cshperspect.a016774
15. Cui Y, Carosi JM, Yang Z, Ariotti N, Kerr MC, Parton RG, Sargeant TJ, Teasdale RD (2019) Retromer has a selective function in cargo sorting via endosome transport carriers. J Cell Biol 218:615–631. https://doi.org/10.1083/jcb.201806153
16. McNally KE, Cullen PJ (2018) Endosomal retrieval of cargo: Retromer is not alone. Trends Cell Biol 28:807–822. https://doi.org/10.1016/j.tcb.2018.06.005
17. Piper RC, Dikic I, Lukacs GL (2014) Ubiquitin-dependent sorting in endocytosis. Cold Spring Harb Perspect Biol 6:a016808. https://doi.org/10.1101/cshperspect.a016808
18. Piper RC, Lehner PJ (2011) Endosomal transport via ubiquitination. Trends Cell Biol 21: 647–655. https://doi.org/10.1016/j.tcb.2011.08.007
19. Sharma P, Parveen S, Shah LV, Mukherjee M, Kalaidzidis Y, Kozielski AJ, Rosato R, Chang JC, Datta S (2020) SNX27–retromer assembly recycles MT1-MMP to invadopodia and promotes breast cancer metastasis. J Cell Biol 219: e201812098. https://doi.org/10.1083/jcb.201812098
20. Mao L, Liao C, Qin J, Gong Y, Zhou Y, Li S, Liu Z, Deng H, Deng W, Sun Q, Mo X, Xue Y, Billadeau DD, Dai L, Li G, Jia D (2021) Phosphorylation of SNX27 by MAPK11/14 links cellular stress–signaling pathways with endocytic recycling. J Cell Biol 220:e202010048. https://doi.org/10.1083/JCB.202010048
21. Priya A, Sugatha J, Parveen S, Lacas-Gervais S, Raj P, Gilleron J, Datta S (2017) Essential and selective role of SNX12 in transport of endocytic and retrograde cargo. J Cell Sci 130: 2707–2721. https://doi.org/10.1242/jcs.201905
22. Lin SX, Mallet WG, Huang AY, Maxfield FR (2004) Endocytosed cation-independent mannose 6-phosphate receptor traffics via the endocytic recycling compartment en route to the trans golgi network and a subpopulation of late endosomes. Mol Biol Cell 15(2):721–733. https://doi.org/10.1091/mbc.e03-07-0497
23. Ghosh P, Dahms NM, Kornfeld S (2003) Mannose 6-phosphate receptors: new twists in the tale. Nat Rev Mol Cell Biol 4:202–212. https://doi.org/10.1038/nrm1050
24. Seaman MNJ (2004) Cargo-selective endosomal sorting for retrieval to the Golgi requires retromer. J Cell Biol 165:111–122. https://doi.org/10.1083/jcb.200312034
25. Arighi CN, Harmell LM, Aguilar RC, Haft CR, Bonifacino JS (2004) Role of the mammalian retromer in sorting of the cation-independent mannose 6-phosphate receptor. J Cell Biol 165:123–133. https://doi.org/10.1083/jcb.200312055

26. Priya A, Kalaidzidis IV, Kalaidzidis Y, Lambright D, Datta S (2015) Molecular insights into Rab7-mediated endosomal recruitment of Core Retromer: deciphering the role of Vps26 and Vps35. Traffic 16:68–84. https://doi.org/10.1111/tra.12237
27. Simonetti B, Danson CM, Heesom KJ, Cullen PJ (2017) Sequence-dependent cargo recognition by SNX-BARs mediates retromer-independent transport of CI-MPR. J Cell Biol 216:3695–3712. https://doi.org/10.1083/jcb.201703015
28. Kvainickas A, Jimenez-Orgaz A, Nägele H, Hu Z, Dengjel J, Steinberg F (2017) Cargo-selective SNX-BAR proteins mediate retromer trimer independent retrograde transport. J Cell Biol 216:3677–3693. https://doi.org/10.1083/jcb.201702137
29. Osborne DG, Phillips-Krawczak CA, Billadeau DD (2015) Monitoring receptor trafficking following retromer and WASH deregulation. Methods Cell Biol 130:199–213. https://doi.org/10.1016/bs.mcb.2015.03.018
30. Breusegem SY, Seaman MNJ (2014) Image-based and biochemical assays to investigate endosomal protein sorting. Methods Enzymol 534:155–178. https://doi.org/10.1016/B978-0-12-397926-1.00009-3
31. Wassmer T, Attar N, Bujny MV, Oakley J, Traer CJ, Cullen PJ (2006) A loss-of-function screen reveals SNX5 and SNX6 as potential components of the mammalian retromer. J Cell Sci 120:45–54. https://doi.org/10.1242/jcs.03302
32. Collinet C, Stöter M, Bradshaw CR, Samusik N, Rink JC, Kenski D, Habermann B, Buchholz F, Henschel R, Mueller MS, Nagel WE, Fava E, Kalaidzidis Y, Zerial M (2010) Systems survey of endocytosis by multiparametric image analysis. Nature 464(7286):243–249. https://doi.org/10.1038/nature08779
33. Rink J, Ghigo E, Kalaidzidis Y, Zerial M (2005) Rab conversion as a mechanism of progression from early to late endosomes. Cell 122:735–749. https://doi.org/10.1016/j.cell.2005.06.043

Chapter 17

Measuring Plasma Membrane Recycling Using Microscopic and Biochemical Approaches

Priyanka Sharma, Sameena Parveen, and Sunando Datta

Abstract

The endocytic pathway has an intricate network of vesicular compartments carrying a variety of proteins referred to as cargoes. Endosomal trafficking is exclusively required to transport these cargoes through various intracellular routes for their delivery to the site of action. Among these, recycling of cargoes to the plasma membrane is a crucial pathway for the efficient functioning of the cell. Hence, endosomal cargo recycling assays are crucial to gain insight into the molecular mechanism governing recycling of the cargoes and in turn to understand their key role in maintaining cellular physiology. These assays have been efficiently utilized to study the recycling of adhesion molecules, transporters, channels, receptors, and so on to the plasma membrane. The basic methodology involves labelling of the cargo at the surface, allowing its internalization followed by direct or indirect measurement of the amount of the cargo recycled back to the plasma membrane. These microscopy-based and biochemical methods can be used as a tool to study the role of various trafficking or signaling molecules on the cell surface involved with the recycling of the membrane proteins, by altering their expression either by silencing or overexpressing the gene.

Key words TIRF microscopy, Antibody uptake, Cell surface recycling, Biotinylation, Immunofluorescence, Immunoblotting, Recycling kinetics

1 Introduction

Living systems have a unique set of mechanisms to regulate their functions for an optimized response to a dynamic external milieu and maintenance of cellular homeostasis [1, 2]. One such mechanism is endocytic trafficking that regulates the timely delivery of proteins to the site of action, that is, either intracellular organelles or the plasma membrane [3, 4]. Proteins that are internalized from the plasma membrane are destined for either lysosomal degradation or recycled back to the cell surface [5–7]. To specifically study recycling to the cell surface, membrane proteins such as Transferrin receptor, β1 integrin, or Membrane-type matrix metalloproteinase (MT1-MMP) are well-established cargoes [8–10] that are extensively used for gaining insight into the underlying molecular

Jingshi Shen (ed.), *Membrane Trafficking: Methods and Protocols*,
Methods in Molecular Biology, vol. 2473, https://doi.org/10.1007/978-1-0716-2209-4_17,

mechanisms [3]. In a recent study, we found that MT1-MMP is sorted and recycled to the cell surface by retromer and SNX27, crucial cargo sorting proteins. We used multiple approaches to label MT1-MMP at the cell surface to investigate the effect of either retromer or SNX27 perturbation on the amount of the protease at the cell surface [11]. Using quantitative microscopy and biochemical assays, surface population as well as recycling of MT1-MMP from endosome to the plasma membrane was measured. These methods are elaborately discussed in this chapter.

Two types of methods detailed are:

1.1 Microscopy Based Methods

1. **Antibody uptake assay**: One of the most promising probes is the monoclonal antibody (directly conjugated with the fluorophore or bound to fluorophore conjugated secondary antibody) [12, 13]. Labelling is done to specifically tag cell surface proteins, and then the antigen–protein complex is allowed to internalize inside the cells. Cells are subjected to fixation and analyzed by automated imaging based microscopy to minimize user introduced bias at the image acquisition step. Further, automated image analysis is done to quantify the fluorescence signal of the antibody internalized by the cells.
2. **Measure MT1-MMP vesicles at the plasma membrane**: Working with antibodies can be problematic, since not all antibodies are equally efficient or sometimes not available for a novel molecule. Additionally, one might be interested in studying the dynamics of protein recycling in live cell conditions. To address these concerns, live cell imaging can be used where the cargoes tagged with fluorescent molecules like GFP, mCherry, RFP etc. are expressed in cells. Cells are then analysed on the TIRF microscope where the tagged proteins are monitored selectively near the plasma membrane with a resolution of ~100 nm [14, 15]. A mutant GFP such as pHluorin has emerged as a preferable tag in such studies [16–18]. The fluorescence of pHluorin is quenched in the acidic environment so labeled proteins that enter late endosomal compartments do not emit any signal. However, proteins in recycling endosomes or vesicles with low acidic pH emit GFP fluorescence. Hence background noise can be minimized by using cargo tagged with pHluorin.

1.2 Biochemical Assay Based Methods

1. **To measure surface population:** Although microscopy-based methods unveil plenty of cell-based information, they are still limited by scanty population size. To complement microscopic approaches one may use biochemical assays to study protein recycling in millions of cells where a biotin label is used as a probe for surface proteins. Similar to the antibody, biotin labelling can be used to measure the surface population of

protein at a steady-state [19, 20]. We used Sulfo-NHS-Biotin, a commercially available modified form of biotin that has a highly reactive ester group *N*-Hydroxysuccinimide (NHS). NHS can be covalently attached to any polypeptide that has a free terminal amine or exposed amino group of its lysine residue present at the surface. Cells are lysed and the lysate is incubated with Neutravidin beads that specifically bind to biotin followed by immunoblotting to measure the amount of membrane protein.

2. **To measure recycling kinetics**: To monitor recycling kinetics, the protein is labelled with NHS-SS-Biotin (cleavable biotin) which has a disulfide bond that can be cleaved by a reducing agent. The biotin-labelled protein is allowed to internalize and recycle to the cell surface, at different time points, in live cells. At every recycling time point, biotin is cleaved from the labeled protein that reenters to the plasma membrane by using MeSNa reagent. Cells are lysed followed by immunoblotting and the amount of recycled protein is quantified.

Although the methods described above are optimized for studying MT1-MMP recycling in MDA-MB-231 cells, these can be easily adapted to measure recycling of any other membrane proteins from endosomes to the cell surface in other cell lines.

2 Materials

2.1 Cell Culture Reagents

1. Plastic ware: T-25 flasks with filter cap, Filter tips: 0.2–1 ml, sterilized serological pipettes: 2–25 ml, sterilized microfuge tubes, 35 mm dish, 6-well, 24-well. Sterilized glass coverslip.
2. MDA-MB-231 cell line.
3. Leibovitz's L-15 medium. Store at 4 °C.
4. Heat-inactivated fetal bovine serum (FBS). Store at −80 °C (*see* **Note 1**).
5. Penicillin–Streptomycin (PenStrep) (10,000 U/ml). Store at −20 °C.
6. 0.25% Trypsin–EDTA. Store at −20 °C, short-term at 4 °C.
7. Complete cell culture medium: L-15 medium supplemented with 10% HI FBS and PenStrep (10,000 U/ml). Store at 4 °C.
8. DNA of gene of interest, prepared using silica columns.
9. Lipofectamine LTX reagent with Plus reagent.
10. DharmaFECT siRNA transfection reagent.
11. Nontargeting siRNA control pool, siRNA SMARTpool targeting gene of interest (GE Healthcare Dharmacon).

2.2 Immunofluorescence Reagents

Use analytical grade chemical reagents to prepare solutions. Prepare all solutions in distilled water and store at room temperature, unless mentioned otherwise.

1. 1× PBS solution: Prepare from 10× PBS stock in distilled water.

   10× Phosphate Buffer Saline (PBS) stock: 1.37 M NaCl, 27 mM KCl 100 mM Na_2HPO_4, 18 mM KH_2PO_4.
2. 4% Paraformaldehyde (PFA) in 1× PBS: 4% PFA, 0.1 mM $CaCl_2$, 0.1 mM $MgCl_2$ (*see* **Note 2**).
3. 0.1% Triton X-100 IN 1× PBS: 0.1% Triton X-100 (v/v) in distilled water.
4. Blocking buffer: 5% fetal bovine serum in 1× PBS.
5. Mounting media: Mowiol 4–88, 0.1 M Tris–Cl, pH 8.5, 20% Glycerol.
6. Antibody used: MT1-MMP (Millipore, Cat. No. MAB3328) for immunofluorescence (IF) at working concentration of 2 μg/ml and immunoblotting (IB) 1 μg/ml (*see* **Note 3**).
7. Alexa 488–conjugated secondary antibody, working concentration 1 μg/ml.

2.3 Biotinylation Reagents

1. EZ-link Sulfo-NHS-Biotin (Invitrogen Cat. No. 21217), EZ-Link Sulfo-NHS-SS-Biotin (Invitrogen Cat. No. 21331).
2. Buffer I: 3.3 mM NaH_2PO_4, 6.7 mM Na_2HPO_4, 140 mM NaCl.
3. Biotin solution: 0.2 mg/ml biotin in Buffer I. (Always prepare fresh).
4. Quenching solution: 50 mM Tris–HCl, pH 8.0. (Store the buffer at 4 °C).
5. Buffer II: 100 mM MeSNa, 50 mM Tris–HCl, pH 8.6, 100 mM NaCl, 1 mM EDTA, 0.2 w/vol bovine serum albumin (BSA) (*see* **Note 4**).
6. Wash Buffer: 0.5% Triton, 0.1% SDS, 50 mM Tris–HCl, pH 7.5, 150 mM NaCl, 5 mM EDTA.
7. Protease inhibitor cocktail (PIC) 1000×: 2 mg/ml aprotinin, 10 mM leupeptin, 1 mM pepstatin A.
8. Lysis Buffer: 50 mM Tris–HCl, pH 7.4, 150 mM NaCl, 2 mM DTT, 1% Triton X-100, 1 mM EDTA, 1× PIC, 200 μM PMSF.

2.4 Microscopy

1. Nikon Instruments Eclipse Ti inverted Microscope equipped with TIRF setup, a temperature controller, an Agilent laser combiner MCL400 (Agilent Technologies) with an electron multiplying charge-coupled device camera (Photometrics Inc.), Live cell imaging chamber with glass bottom dish holder, Micro-Manager software version 1.4.23.

2. Zeiss LSM 780 laser-scanning confocal microscope with the Zen 2010 software.

2.5 Image/Western Blot Analysis Software

1. Motion Tracking (http://motiontracking.mpi-cbg.de).
2. ImageJ 1.52p (https://imagej.nih.gov/ij/download.html).

3 Methods

All the experiments should be carried out mandatorily by following the research guidelines of the respective institute.

3.1 Mammalian Cell Culture

Perform all the following steps strictly in the Biosafety level-II laminar airflow hood.

1. Subculturing of cells: Remove the conditioned media using an aspirator. Wash the cells by adding prewarmed PBS. Remove the PBS and add prewarmed trypsin to the flask and incubate it at 37 °C for 3 min.
2. After trypsinization, add 5 ml of prewarmed complete cell culture media to the flask, this will inactivate the trypsin. Gently transfer the cell suspension to the 15 ml falcon tube. Take out small amount of suspension for cell counting, rest of the suspension in falcon is subjected to centrifugation at 300 × *g* for 3 min at room temperature.
3. Discard the supernatant and resuspend the cells in 5 ml of fresh complete media.
4. Prepare a dilution of the cells based on the cell counting for the planned experiment and add rest of cells to a new T-25 flask. Immediately place the flask with subcultured cells in a 37 °C humidified incubator.

3.2 Transient Knockdown of the Gene of Interest Via siRNA

All steps are performed inside the laminar flow hood.

1. Trypsinize and count the cells as discussed in **steps 1**, **2** of Subheading 3.1.
2. For siRNA transfection, grow 25,000 cells per well in a 24-well plate for 12–16 h (*see* **Note 5**).
3. Prepare 100 μl of 5 μM siRNA from 20 μM stock solution in 1× siRNA buffer. Use only sterilized RNase-free low adhesion 1.5 ml microfuge tubes.
4. Prepare a working concentration of siRNA 25 nM per well diluted in 47.5 μl of serum free L-15 media. Always prepare a siRNA mixture of the same concentration for non-targeting siRNA which acts as an appropriate control in the experiment.

5. Add 1 μl of DharmaFECT reagent to 49 μl of the serum-free medium in a separate tube. Mix by gently tapping the tube or pipetting up and down.
6. Incubate the two tubes from **steps 4** and **5** for 5 min at room temperature.
7. Using a pipette gently transfer the solution of tube 1 (**step 4**) to tube 2 (**step 5**). Mix gently by pipetting up and down. Incubate the siRNA and DharmaFECT solution for an additional 20 min at room temperature.
8. Add 400 μl of L-15 complete media to the solution of **step** 7. Aspirate culture medium from the 24-well tissue culture plate. Using a pipette, add 500 μl of the solution.
9. After 24 h of transfection, replace the transfection medium with 500 μl of complete media in each well. Allow cells to grow for an additional 72 h to proceed with the experiment.

3.3 Antibody Uptake Assay and Immunostaining

The sequence of the steps is summarized in the flow chart in Fig. 1.

1. Subculture cells as described in Subheading 3.1. Seed 20,000 cells per well by carefully adding cell suspension on top of the coverslip. Place the plate in 37 °C humidified incubator.
2. After 14–16 h, transfect the cells with control siRNA and siRNA targeting the gene of interest as described in Subheading 3.2. Proceed with the experiment post 72 h of the transfection.
3. Prepare MT1-MMP antibody dilution just before the experiment: working concentration 10 μg/ml in L-15 complete media. Perform all steps on the workbench at room temperature, unless mentioned. Wash cells with 1× PBS to remove traces of media, repeat this step twice.
4. Antibody binding at the cell surface: Remove PBS and add 400 μl of the media with antibody dilution in each well. Place 24-well plate on ice or in a cold room to ensure that endocytosis does not take place during the process of antibody binding. Incubate for 30 min.
5. Gently remove the media from each well. Wash cells by adding PBS and incubate for 5 min. Wash twice.
6. Internalization of the antibody bound to the cell surface MT1-MMP: Remove 24-well plate from the ice/cold room and add 0.5 ml of prewarmed L-15 media to each well, gently along the sidewall. Incubate cells in 37 °C water bath for 15 min (*see* **Note 6**). Remove media from the wells and wash with PBS twice as done in **step 5**.
7. Proceed for immunofluorescence staining. All steps are done at room temperature unless mentioned. Fix the cells by adding

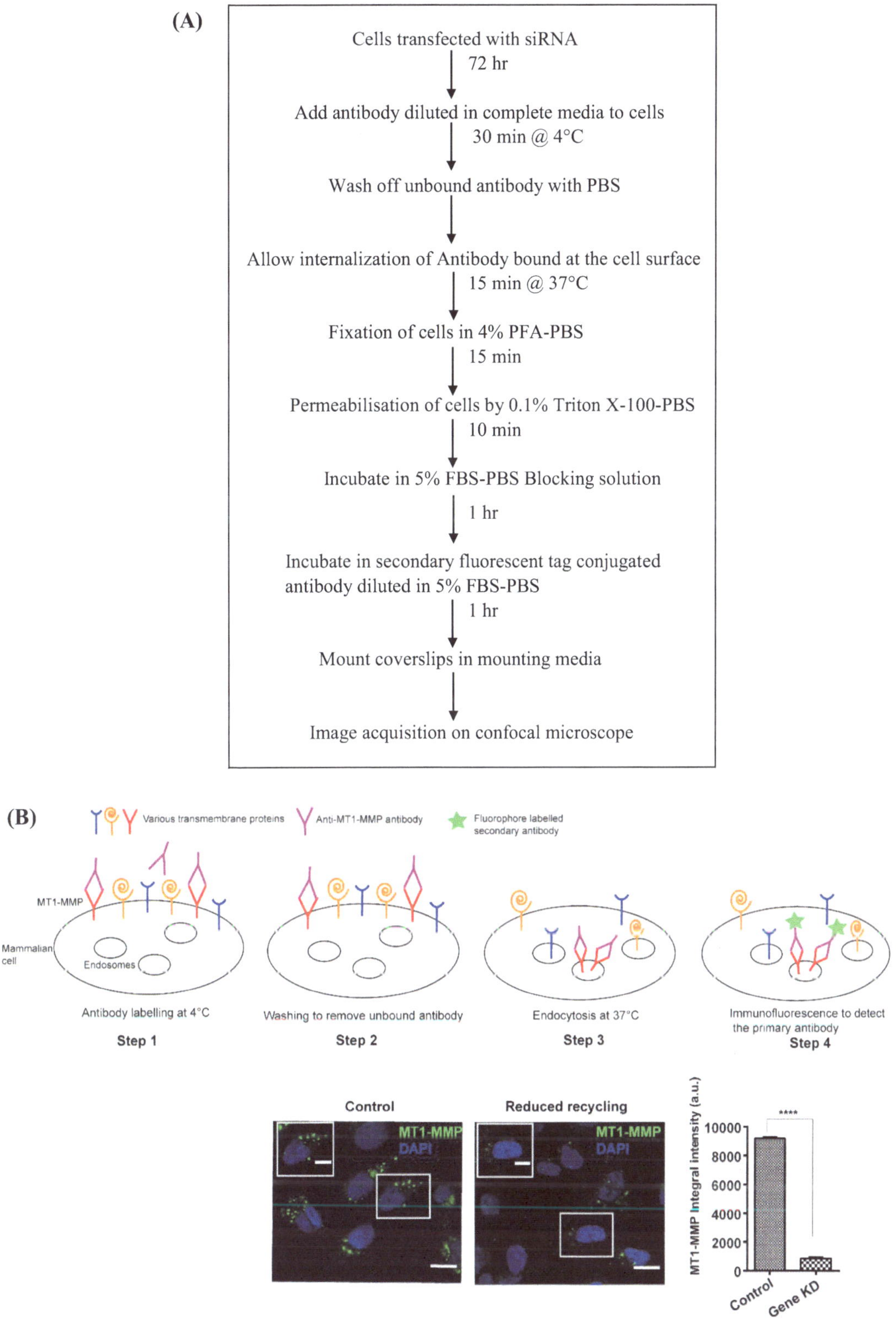

Fig. 1 Antibody uptake assay to measure membrane protein at the cell surface. (**a**) Flowchart of the steps to be followed while performing the assay, for details refer to Subheading 3.3. (**b**) Model showing various steps of

0.5 ml of 4% PFA to each well. Incubate for 15 min followed by PBS washing.

8. Permeabilize cells by adding 0.1% Triton X-100 solution in each well. Incubate for 10 min followed by PBS washing.
9. Remove PBS and add blocking buffer to each well. Incubate for 1 h.
10. Meanwhile prepare the humidified dark chamber where antibody staining will be performed. Clean the chamber by spraying 70% alcohol. Place tissue paper in the chamber and spray distilled water on it just to keep the tissue moist. Carefully place the Parafilm sheet on the wet tissue layer. Cover the entire chamber with aluminum foil.
11. Prepare secondary antibody dilution: Dilute fluorophore tagged antibody in blocking buffer. Using forceps, carefully lift the coverslip, and place it on the Parafilm layer. Gently add 50 μl of antibody dilution on the top of the coverslip. Cover the dark chamber and leave it undisturbed for 1 h at the room temperature.
12. Carefully lift the coverslip from the edge and drain off the antibody solution. Place it back into the 24-well plate followed by PBS washes and give final washing by MilliQ.
13. Clean the glass slide, add 4 μl of prewarmed Mowiol drop and gently place the coverslip (*see* **Note 7**) side where cells are attached, on the Mowiol (*see* **Note 8**). Leave the slide overnight for drying at room temperature in a dark box. Acquire images the next day.

3.4 Transient Overexpression of Gene of Interest

1. Dilute DNA (500 ng) in the 100 μl of serum-free L-15 media.
2. Add 1 μl of Lipofectamine PLUS reagent to the above media. Incubate for 10 min.
3. Add 1 μl of Lipofectamine LTX to the above mixture. Incubate for 30 min.
4. Gently remove culture media from the cells (plated in 24-well), add 400 μl of L-15 complete media to each well. Add the DNA-LTX complex dropwise to the entire well. Monitor cells after 16–18 h for fluorescence under fluorescence microscope.

3.5 Measure MT1-MMP Vesicles at the Plasma Membrane

The sequence of the steps is summarized in the flowchart in Fig. 2.

1. Trypsinize the cells and seed on glass-bottom dish. Incubate the cells for 4–5 h (*see* **Note 9**).

Fig. 1 (continued) the assay. Confocal micrograph of the control and gene knockdown (KD) cells, fixed after 15 min of endocytosis, showing puncta of MT1-MMP antibody. Graph showing the quantification of the integral intensity of MT1-MMP in control and KD cells. (Scale bar 5 μm, Inset 3 μm)

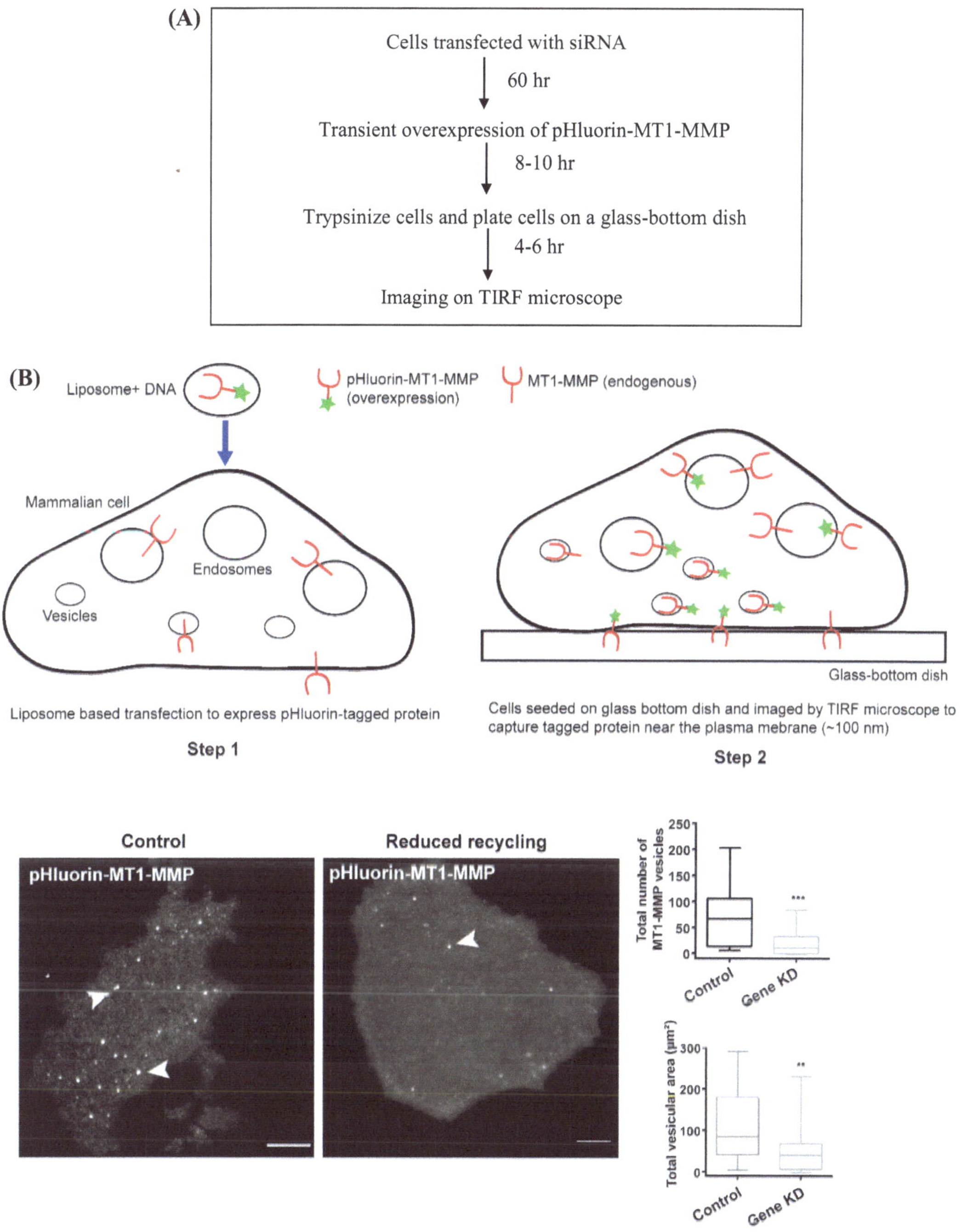

Fig. 2 TIRF microscopy-based assay to measure MT1-MMP at the cell surface. (**a**) Flowchart of the method to be followed, for details refer to Subheading 3.5. (**b**) Model representing lipid-mediated transfection of pHluorin-MT1-MMP in the cells. The endogenous MT1-MMP resides on endosomes, vesicles, and the ventral surface of the plasma membrane. Upon overexpression, the pHluorin-MT1-MMP localizes to the intracellular compartments and on the plasma membrane. Representative TIRF images of pHluorin-MT1-MMP in control cells and cells showing reduced recycling of MT1-MMP in MDA-MB-231 cell line. Arrows pointing to MT1-MMP puncta (Scale bar 10 μm). Graphs showing quantification of the total number of vesicles and total vesicular area in control and gene knockdown (KD) cells

2. Carry out TIRF microscopy on Nikon Eclipse Ti inverted microscope using a 488 nm laser to excite pHluorin tagged proteins. Choose the incident angle for the laser line suitable for creating the total internal reflection. The evanescent wave illuminates the ventral surface of the cells with penetration depth of ~100 nm.
3. Acquire images at 1 s interval for 5 min.

3.6 Biotinylation to Measure the Surface Population of MT1-MMP

The sequence of the steps is summarized in the flowchart in Fig. 3a. Perform all steps using repeater pipette to avoid handling error while adding solutions to multiple dishes.

1. Subculture MDA-MB-231 cells as described in Subheading 3.1. Seed 9×10^4 cells/ml in 35 mm plate.
2. After 24 h proceed with siRNA transfection as given in Subheading 3.2. Post 72 h of transfection prepare cells for the biotinylation assay.
3. Biotin binding: To label proteins at the cell surface, remove Buffer I from the dishes. Add 1 ml of freshly prepared non cleavable biotin solution (ice-cold). Dishes are kept on an orbital shaker moving at 90 rpm (low speed), at 4 °C for 45 min (*see* **Note 10**).
4. Aspirate out the biotin solution followed by washing with Buffer I. Repeat this step twice.
5. Quenching of the unbound/free biotin: Add quenching solution to each dish. Incubate for 15 min on the orbital shaker. Remove the solution and repeat this step (*see* **Note 11**).
6. Aspirate out the solution and wash cells with Buffer I. Remove the solution and repeat this step.
7. Solubilize the cells using 150 μl of lysis buffer to each dish. Incubate for 10 min at 4 °C. Collect the cell lysate from each dish and centrifuge at 19,000 × *g* for 15 min at 4 °C.
8. Equilibration of beads: Prepare the Neutravidin beads to capture the biotin labeled proteins from the cell lysate. Wash the beads twice with Buffer I followed by final wash with wash buffer using a centrifuge at 4 °C.
9. Carefully transfer an equal amount of protein (by quantifying the lysate) to the 1.5 ml microfuge tubes after equilibrating the Neutravidin beads. Place the microfuge tubes on the rotor at low speed for 45 min at room temperature.
10. Wash the protein bound beads with the wash buffer at 1000 g for 3 min at 4 °C. Repeat this step twice.
11. Add 2.5× of SDS dye to the beads and boil at 95 °C for 5 min. Load the samples on gradient polyacrylamide gel.

(A)

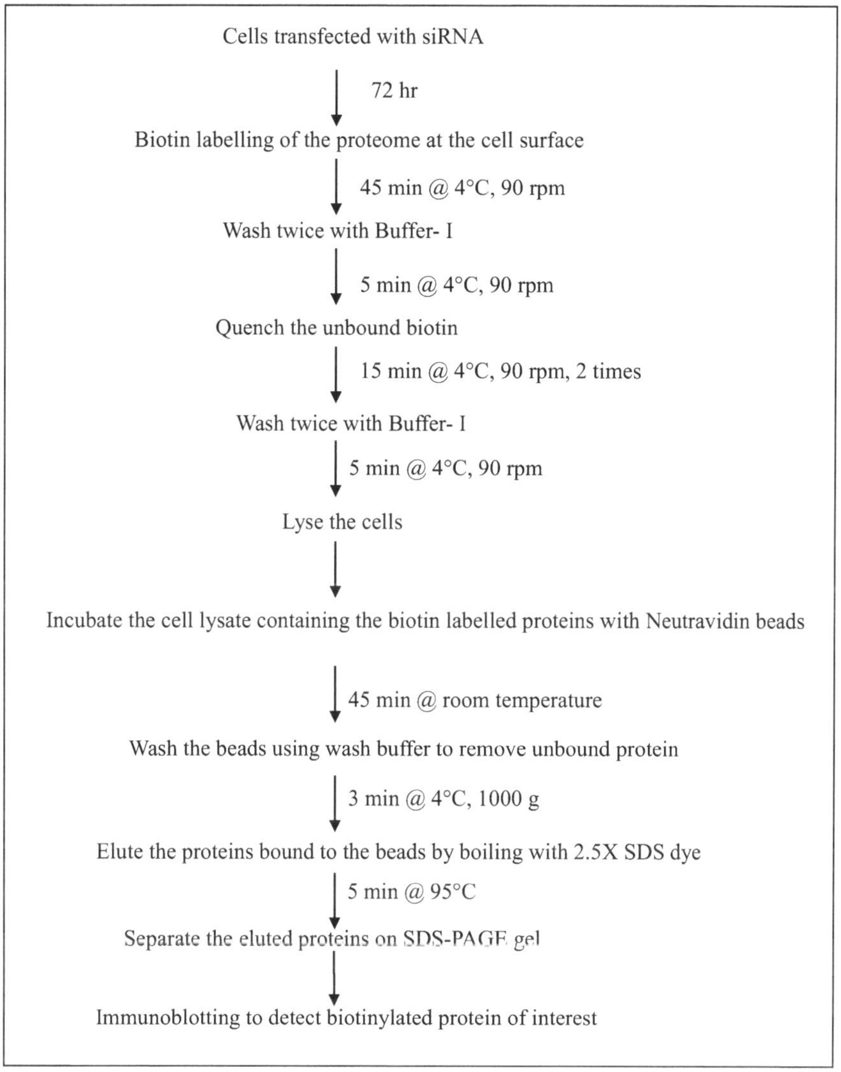

Fig. 3 Biotinylation assay to measure protein levels at the cell surface. (**a**) Flow chart of the method, for details refer to Subheading 3.6. (**b**) Model showing steps of the assay. Western blot of the lysate collected after biotin binding at 4 °C, followed by quenching for control cells and cells with reduced surface recycling of MT1-MMP. Graph showing the densiometric analysis to measure MT1-MMP intensity in control and gene depleted (KD) cells

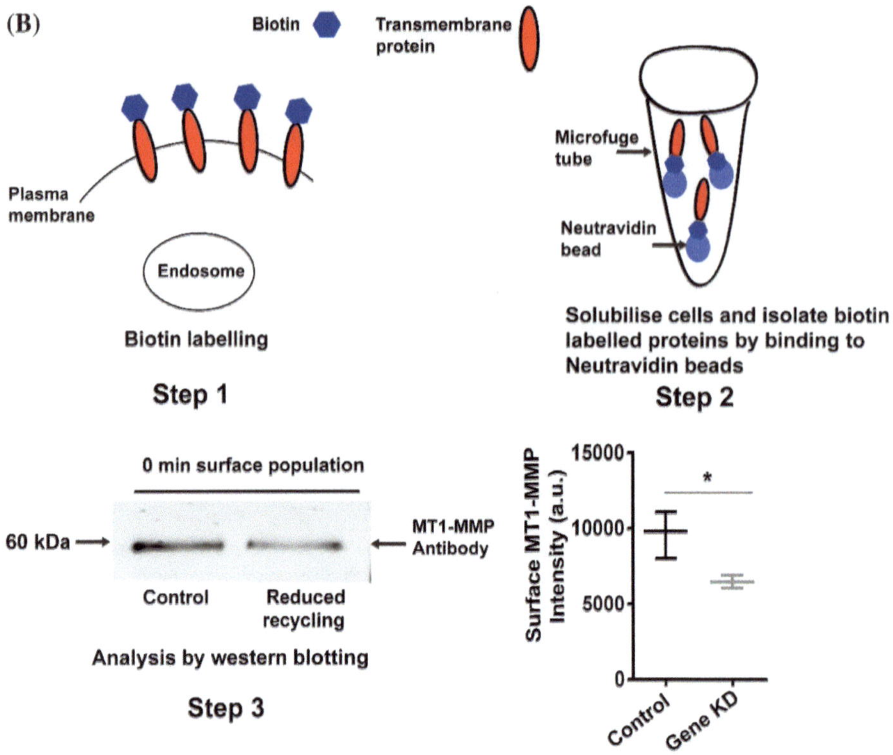

Fig. 3 (continued)

12. Transfer the proteins from polyacrylamide gel to PVDF (Polyvinylidene fluoride) membrane, using a western blot transfer setup. Run the transfer setup for 10–12 h at 4 °C.
13. Proceed with the immunoblotting using primary antibody solution of MT1-MMP (1:1000 dilution).

3.7 Biotinylation to Measure Recycling Kinetics of MT1-MMP to the Cell Surface

The sequence of the steps is summarized in the flowchart in Fig. 4. Perform all steps using repeater pipette to avoid handling error while adding solutions to multiple dishes.

1. Subculture MDA-MB-231 cells as described in Subheading 3.1. Seed 9 × 10^4 cells/ml in 35 mm plate.
2. After 24 h proceed with siRNA transfection as given in Subheading 3.2. Seventy two hours post transfection, prepare cells for the biotinylation assay.
3. Biotin binding: To label proteins at the cell surface, remove Buffer I from the dishes. Add 1 ml of freshly prepared cleavable biotin solution (ice-cold). Dishes are kept on an orbital shaker moving at 90 rpm (low speed), at 4 °C for 45 min.
4. Aspirate out the biotin solution followed by washing with Buffer I. Repeat this step twice.

(A)

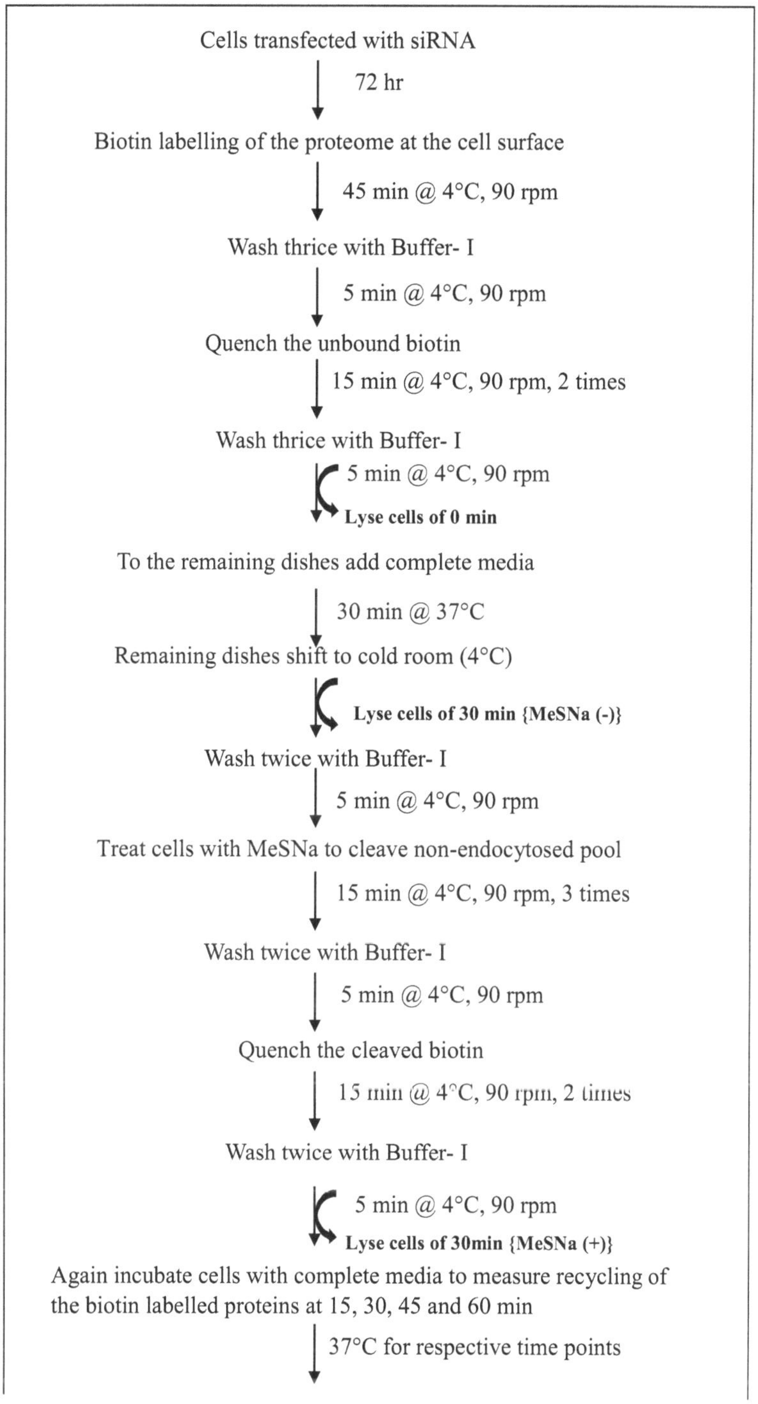

Fig. 4 Biotinylation assay to measure recycling kinetics of the membrane protein from endosomes to the plasma membrane. (**a**) Flow chart of the method, for details refer Subheading 3.7. (**b**) Model showing various steps of the assay. Immunoblot of the lysate collected after endocytosis and different recycling time points in control cells and cells showing reduced surface recycling of MT1-MMP. A gradual decrease in the protein signal represents loss of the biotinylated protein in the lysate because of its recycling to plasma membrane, whereas in cells with reduced recycling of protein, loss of the signal is less pronounced. Graph showing densiometric analysis of the percentage recycling of MT1-MMP in control and gene knockdown (KD) cells

At each time point give buffer- I wash followed by MeSNa buffer treatment to cleave recycled biotinylated protein at the cell surface.

↓ 15 min @ 4°C, 90 rpm, 3 times

Wash twice with Buffer- I

↓ 5 min, @ 4°C, 90 rpm

Quench the cleaved biotin

↓ 15 min @ 4°C, 90 rpm, 2times

Wash twice with Buffer- I

↓ 5 min @ 4°C, 90 rpm

Lyse the cells

↓

Incubate the cell lysate containing the biotin labelled proteins with Neutravidin beads

↓ 45 min @ room temperature

Wash the beads using wash buffer to remove unbound protein

↓ 3 min @ 4°C, 1000 g

Elute the proteins bound to the beads by boiling with 2.5X SDS dye

↓ 5 min @ 95°C

Separate the eluted proteins on SDS-PAGE gel

↓

Immunoblotting to detect biotinylated protein of interest

(B)

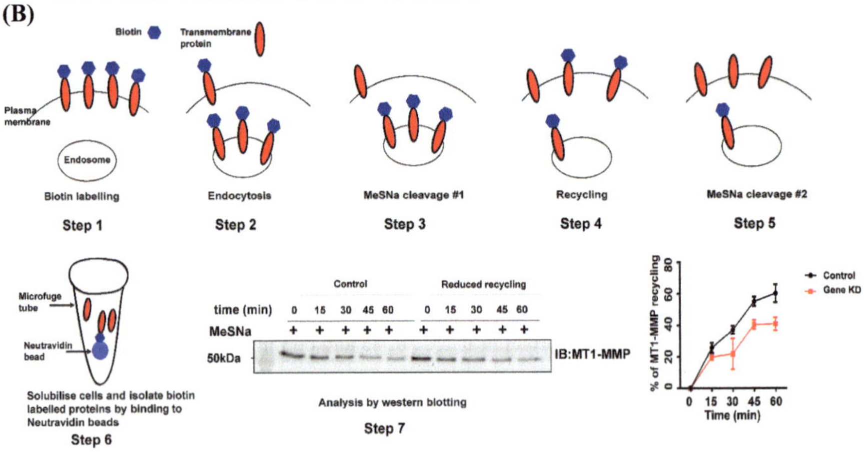

Fig. 4 (continued)

5. Quenching of the unbound/free biotin: Add quenching solution to each dish. Incubate for 15 min on the orbital shaker. Remove the solution and repeat this step.
6. Collect cell lysate for 0 min: Wash cells with Buffer I followed by solubilization using lysis buffer (*see* **Note 12**). Do not disturb dishes for which lysate will be collected at later time points.
7. Endocytosis of biotin labelled proteins: Add 1 ml of complete media to the rest of the dishes. Incubate cells at 37 °C for 30 min.
8. Collect cell lysate for 30 min (MeSNa-): Wash cells with Buffer I followed by solubilization using lysis buffer.
9. MeSNa treatment to remove biotin: Shift the dishes to 4 °C cold room and gently aspirate the media. Wash cells with Buffer I, add MeSNa buffer to each dish. Place dishes on the orbital shaker moving at low speed for 15 min. Repeat this step twice (*see* **Note 13**).
10. Quenching of the unbound/free biotin: Add quenching solution to each dish. Incubate for 15 min on the orbital shaker. Remove the solution and repeat this step.
11. Collect cell lysate for 30 min endocytosis: Wash cells with Buffer I followed by solubilization using lysis buffer (*see* **Note 14**).
12. Cell lysate collection at recycling time points: Add complete media and incubate cells at 37 °C. The endocytosed biotin labeled protein is allowed to recycle at different time points, that is, 15, 30, 45, and 60 min. At each time point move cells to 4 °C and follow **steps 9** and **10**. Prepare cell lysate as described in **step** 7 of Subheading 3.6.
13. Allow the biotin labelled proteins in the lysate to bind to equilibrated Neutravidin beads, elute, and analyze by immunoblotting as described in **steps 8–13** of Subheading 3.6.

4 Analysis

4.1 Measure MT1-MMP Vesicles at the Plasma Membrane

Automated image analysis was done using MotionTracking software as explained below.

1. Import acquired video: Use the command File Import, in the menu to import recorded video in .nd2 format, to read these files.
2. Save project: Once movie is imported, use **File** > **Save Project** (Fig. 5a).
3. Objects in MotionTracking: The basic purpose of Motion-Tracking is to find the position of objects in the acquired frames

(A)

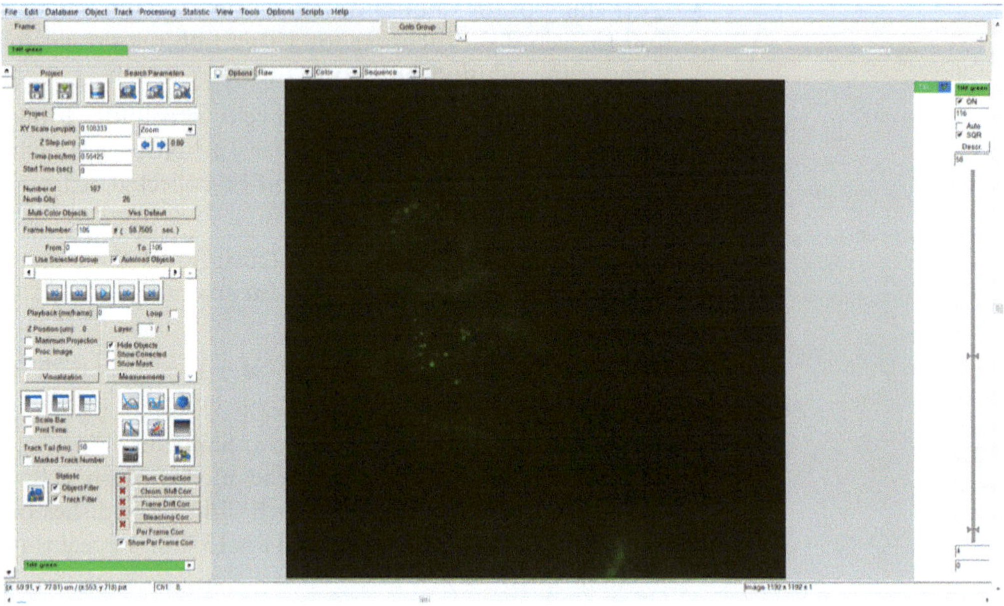

(B)

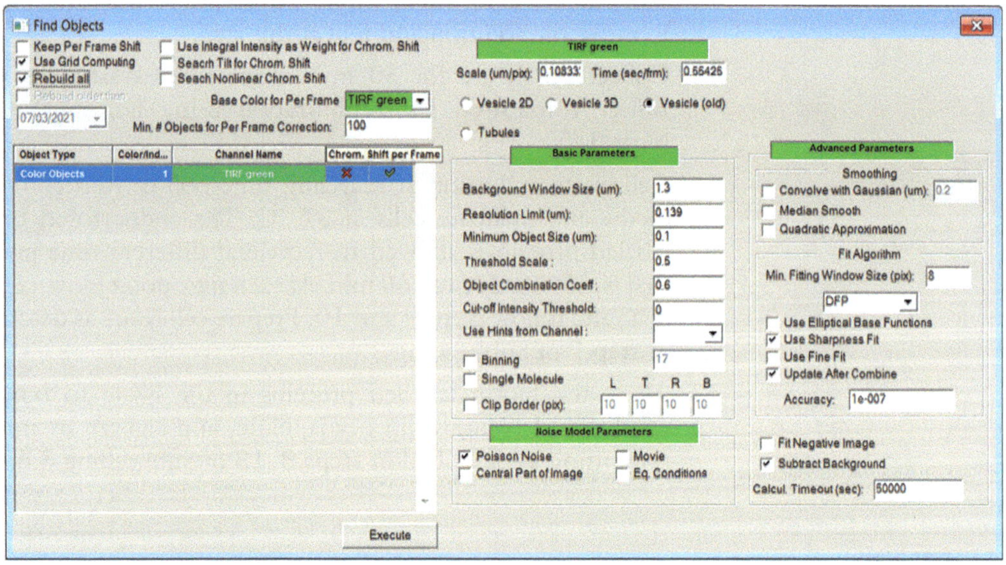

Fig. 5 Analysis of videos acquired on TIRF microscope to quantify the number and size of the MT1-MMP vesicles. (**a**) Video imported as a project in the Motion tracking software. (**b**) List of parameters to be adjusted so as to identify vesicles as objects. (**c**) Objects along with their contours after identified by the software. (**d**) Vesicles moving in *x*–*y* direction are detected as Tracks

(C)

(D)

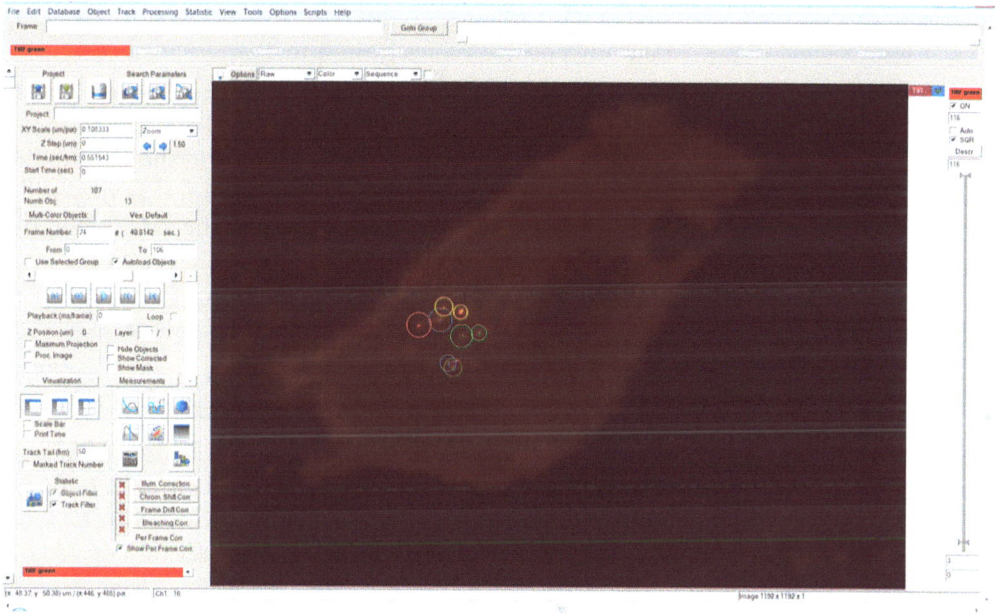

Fig. 5 (continued)

of the movie, such as intracellular vesicles. Once found, the software assigns an x-y position and quantifies the intensity profiles of the objects. Each object is a group of pixels. The number of pixels combined with the scale lead to different properties of the object such as “Size” and “Area.” Each pixel also has its own brightness called “Intensity.” The vesicles are identified as objects by setting various parameters.

4. Selecting objects: **Object Search** > **Find Objects Parameters** (Fig. 5b).
5. The opened dialog box is enlisted with the list of parameters to be adjusted according to the quality of the acquired image frames. Basic parameters to be adjusted are Background window size, Resolution limit, Minimum Object Area, Threshold Scale, Object Combination Coefficient, Cut-off Intensity Threshold, and Binning.
6. Once the parameters are assigned, the search for objects across the entire frames of the acquired movie is performed by following the command **Object search** > **Find Objects in Batch Mode** (Fig. 5c).
7. Identifying the Tracks: Once MotionTracking has successfully localized every object in the movie, the objects in consecutive frames can be linked into trajectories, or tracks. A track is the location of an object over time. Tracks have properties such as speed, direction, maximum displacement. Go to **Track Search** > **Track Objects Parameters**. Following parameters are adjusted to identify tracks: Direction, Position, Speed, Intensity, Area, Integral, unidirectional. The Relative Weights can take any positive values from 0 to 1.
8. Track break thresholds: Adjust the Track break thresholds, using the same dialogue. These are Total score threshold, Area score threshold, Intensity score threshold, and Integral score threshold. These thresholds are critical as they will command the program to select a range of the variation in properties of the object that it can tolerate while building a track.
9. Searching Tracks in entire movie: **Track search** > **Track finding procedure** (Fig. 5d).
10. Deselecting random Tracks: Delete tracks that do not move by using command Delete Non-moving Tracks.
11. Quantify number of vesicles near the cell surface: Analyse the number of tracks found in each frame by searching for the tracks with respect to the time of the movie. Select.

    Statistics > **Time Course** > **Number of Tracks course**.
12. Measure the intensity of vesicles: **Statistics** > **Distributions** > **Object Characteristics Distributions** > **Mean Intensity**.

4.2 Antibody Uptake Assay

Motion tracking software is used for automated image analysis.

1. Import acquired images in a single project using command **File** > **Import** > **Import. lsm.**
2. Save the project as in **step 2** of Subheading 4.1.
3. Identify the objects in the images of the project as detailed in **steps 4–6** of Subheading 4.1.

4. Measure the intensity of vesicles: **Statistics** > **Distributions** > **Object Characteristics Distributions** > **Mean Intensity**.
5. Import the values to an excel sheet. Plot the graph for intensity of the endosomes in control and gene knockdown images.

4.3 Surface Population of MT1-MMP by Biotinylation

Download and install ImageJ program from the link given in Subheading 2.5. Perform densiometric analysis to quantify intensity of the protein bands in the scanned western blot.

4.4 Recycling Kinetics of MT1-MMP

To calculate the recycling kinetics, percentage recycling of the protein will be obtained using the following formula.

$$\text{Amount recycled at time } t_1 = \left(\frac{A-B}{A}\right) \times 100$$

where time $t = 0$ {30 min of endocytosis (MeSNa +)}, Time $t = 0$, Intensity of the MT1-MMP band in the anti-MT1-MMP IB = A, $t = t1$, Intensity of the MT1-MMP band in the anti-MT1-MMP IB = B.

5 Notes

1. Heat inactivation can be done by keeping FBS in water bath (at 56 °C) for 30 min.
2. Wearing a mask and glasses are strongly recommended to avoid inhalation of the PFA fumes.
3. Alternatively, another antibody for MT1-MMP (MAB9181-SP-Human MMP-14/MT1-MMP Mab Clone 128527) from R&D systems can be used.
4. Alternatively, prepare the buffer by dissolving all reagents except the MeSNa and store at 4 °C. Just before the experiment add MeSNa, adjust pH at 4 °C, and then use the buffer.
5. Optimal cell density for siRNA transfection can vary among different cell lines. Overcrowding of the cells should be avoided as it leads to low efficiency of gene knockdown.
6. A water bath with a water circulation system is preferred for these experiments to maintain even temperature.
7. Do not place coverslips at the edge/corner of the glass slide. It will create difficulties while acquiring images. Leave some space (~0.5 cm) at the edges. Avoid using high amount of Mowiol, it will result in floating of the coverslips.
8. Avoid trapping air bubbles while mounting the coverslip. It hampers the image acquisition.
9. Most of the cell culture media have phenol red (pH indicator) as one of the components. The color of the indicator might

interfere with the image acquisition. Thus, it is recommended to use media without phenol red for image acquisition.

10. Maintaining temperature strictly at 4 °C will ensure that endocytosis is halted.
11. This step ensures the inactivation of the unreacted/free biotin present on the cell surface.
12. The lysate obtained here will represent the total amount of biotinylated population at the cell surface. This will be taken as a reference in the final analysis.
13. MeSNa cleaves non endocytosed biotin bound to the proteins.
14. The lysate collected at 30 min endocytosis will represent the amount of biotin labelled protein internalized by cells (considered as the 100% pool). It is this fraction of protein which will be tracked for recycling kinetics in the further steps of the assay.

Acknowledgments

We thank Prof. Yannis Kalaidzidis for helping with our TIRF image analysis using Motion Tracking software and Angelika Giner for providing us the protocol for preparing Mowiol. We also thank Dr. Amulya Priya for her critical suggestions while writing the chapter.

This work was supported by Department of Science and Technology (EMR/2016/00340 and SR/FST/LSI-643/2015 (C)), the Council of Scientific and Industrial Research 27(0326/17/EMR-II7), and Department of Biotechnology (102/IFD/SAN/1941/2018-2019).

References

1. Apaja PM, Lukacs GL (2014) Protein homeostasis at the plasma membrane. Physiology 29: 265–277. https://doi.org/10.1152/physiol.00058.2013
2. Okiyoneda T, Apaja PM, Lukacs GL (2011) Protein quality control at the plasma membrane. Curr Opin Cell Biol 23:483–491. https://doi.org/10.1016/j.ceb.2011.04.012
3. Evans PR, Owen DJ (2002) Endocytosis and vesicle trafficking. Curr Opin Struct Biol 12(6): 814–821. https://doi.org/10.1016/s0959-440x(02)00395-0
4. Elkin SR, Lakoduk AM, Schmid SL (2016) Endocytic pathways and endosomal trafficking: a primer. Wien Med Wochenschr 166(7–8): 196–204. https://doi.org/10.1007/s10354-016-0432-7
5. Wandinger-Ness A, Zerial M (2014) Rab proteins and the compartmentalization of the endosomal system. Cold Spring Harb Perspect Biol 6(11):a022616. https://doi.org/10.1101/cshperspect.a016162
6. Poteryaev D, Datta S, Ackema K, Zerial M, Spang A (2010) Identification of the switch in early-to-late endosome transition. Cell 141(3): 497–508. https://doi.org/10.1016/j.cell.2010.03.011
7. Klumperman J, Raposo G (2014) The complex ultrastructure of the endolysosomal system. Cold Spring Harb Perspect Biol 6:a016857. https://doi.org/10.1101/cshperspect.a016857
8. Kajiho H, Kajiho Y, Frittoli E, Confalonieri S, Bertalot G, Viale G, Di Fiore PP, Oldani A,

Garre M, Beznoussenko GV, Palamidessi A, Vecchi M, Chavrier P, Perez F, Scita G (2016) RAB2A controls MT1-MMP endocytic and E-cadherin polarized Golgi trafficking to promote invasive breast cancer programs. EMBO Rep 17:1061–1080. https://doi.org/10.15252/embr.201642032
9. Di Fiore PP, von Zastrow M (2014) Endocytosis, signaling, and beyond. Cold Spring Harb Perspect Biol 6:a016865. https://doi.org/10.1101/cshperspect.a016865
10. Dozynkiewicz MA, Jamieson NB, MacPherson I, Grindlay J, van den Berghe PVE, von Thun A, Morton JP, Gourley C, Timpson P, Nixon C, McKay CJ, Carter R, Strachan D, Anderson K, Sansom OJ, Caswell PT, Norman JC (2012) Rab25 and CLIC3 collaborate to promote integrin recycling from late endosomes/lysosomes and drive cancer progression. Dev Cell 22:131–145. https://doi.org/10.1016/j.devcel.2011.11.008
11. Sharma P, Parveen S, Shah LV, Mukherjee M, Kalaidzidis Y, Kozielski AJ, Rosato R, Chang JC, Datta S (2020) SNX27–retromer assembly recycles MT1-MMP to invadopodia and promotes breast cancer metastasis. J Cell Biol 219: e201812098. https://doi.org/10.1083/jcb.201812098
12. Teuchert M, Berghö S, Klenk H-D, Garten W (1999) Recycling of Furin from the plasma membrane functional importance of the cytoplasmic tail sorting signals and interaction with the AP-2 adaptor medium chain subunit. J Biol Chem 274(51):36781–36789. https://doi.org/10.1074/jbc.274.51.36781
13. Arancibia-Cárcamo IL, Fairfax BP, Moss SJ, Kittler JT (2006) Studying the localization, surface stability and endocytosis of neurotransmitter receptors by antibody labeling and biotinylation approaches. In: The dynamic synapse: molecular methods in ionotropic receptor biology. CRC Press, Boca Raton, Florida, pp 91–118
14. Morton PE, Parsons M (2011) Dissecting cell adhesion architecture using advanced imaging techniques. Cell Adhes Migr 5:351–359. https://doi.org/10.4161/cam.5.4.16915
15. Yildiz A, Vale RD (2015) Total internal reflection fluorescence microscopy. Cold Spring Harb Protoc 2015:801–810. https://doi.org/10.1101/pdb.top086348
16. Miesenböck G (2012) Synapto-pHluorins: genetically encoded reporters of synaptic transmission. Cold Spring Harb Protoc 7:213–217. https://doi.org/10.1101/pdb.ip067827
17. Miesenböck G, De Angelis DA, Rothman JE (1998) Visualizing secretion and synaptic transmission with pH-sensitive green fluorescent proteins. Nature 394:192–195. https://doi.org/10.1038/28190
18. Monteiro P, Rossé C, Castro-Castro A, Irondelle M, Lagoutte E, Paul-Gilloteaux P, Desnos C, Formstecher E, Darchen F, Perrais D, Gautreau A, Hertzog M, Chavrier P (2013) Endosomal WASH and exocyst complexes control exocytosis of MT1-MMP at invadopodia. J Cell Biol 203:1063–1079. https://doi.org/10.1083/jcb.201306162
19. Turvy DN, Blum JS (2000) Biotin labeling and quantitation of cell-surface proteins. Curr Protoc Immunol 36. https://doi.org/10.1002/0471142735.im1807s36
20. Elia G (2012) Cell surface protein biotinylation for SDS-PAGE analysis. Methods Mol Biol 869:361–372. https://doi.org/10.1007/978-1-61779-821-4_29

Chapter 18

Light Scattering Techniques to Assess Self-Assembly and Hydrodynamics of Membrane Trafficking Proteins

Marijn G. J. Ford and Rajesh Ramachandran

Abstract

Light scattering methods permit the determination of molar mass and hydrodynamic radius for a protein from first principles. They are, therefore, particularly useful for the biophysical characterization of any protein. Molar mass and hydrodynamic radius determinations may be used to demonstrate that the protein of interest multimerizes. In the endomembrane system, reversible and regulated assembly and multimerization of proteins is critical for building coats required for vesicle budding, for the function of membrane remodeling machines, for fission and fusion and for assembling and disassembling trafficking intermediates. Light scattering methods have therefore significantly contributed to the understanding of the underlying trafficking processes. Herein, we describe methods to express and purify the recombinant fungal SNX-BAR Mvp1, a membrane remodeling protein required for retrograde trafficking at the endosome. Using Mvp1 as an example, we provide protocols for determining its molar mass and hydrodynamic radius by multiangle static light scattering and dynamic light scattering, respectively. These methods can be applied directly to the study of other membrane trafficking proteins, yielding a wealth of biophysical and biochemical information.

Key words Light scattering, SLS, DLS, SEC-MALS, Membrane remodeling, Mvp1, Multimerization

1 Introduction

Accurate determination of a protein's molar mass, as opposed to the molecular mass predicted by its amino-acid sequence, is critical and can yield various insights, including on its capacity to homo-oligomerize or participate in complexes with other proteins. The former is highly relevant as self-assembling systems are widespread in biology, particularly amongst membrane trafficking proteins [1, 2].

Several methods exist for determining the molar mass of a protein, with varying levels of sophistication. While in silico predictions from the given amino acid sequence may be acceptable for some purposes, this approach will miss any solution multimerization equilibria as well as the effect of posttranslational modifications

Jingshi Shen (ed.), *Membrane Trafficking: Methods and Protocols*,
Methods in Molecular Biology, vol. 2473, https://doi.org/10.1007/978-1-0716-2209-4_18,

on protein self-association. If there is no requirement for a protein to be in its native, folded, state, an approximate estimate can be made using denaturing gel electrophoresis. The sizing is done by comparison to a range of standards. This approach will usually destroy transient or obligate multimerization that may be required for biological activity and highly charged proteins may run anomalously. While some non–denaturing gel-mediated separation approaches exist that can shed light on multimerization and complex formation, anomalous migration may result in difficulties in assesing size. This is particularly the case where complexes are stabilized with the use of extraneous cross-linkers, which shifts systems in equilibrium to assembled species and may therefore be misleading.

Analytical size exclusion chromatography (SEC) is often used to obtain an estimate of the molar mass of a protein of interest. Here, a standard curve is typically constructed relating elution volume to molar mass for several well-characterized proteins and this is used to assign a mass to the protein under study. A variation involves first constructing a standard curve of hydrodynamic radius versus elution volume for a range of standards that permits estimation of the hydrodynamic radius of a protein of interest, based on its elution volume from the same column. A second step then uses an additional standard curve to connect hydrodynamic radius to molar mass. While such approaches may give an indication of the multimeric state of the protein, they rely on assumptions that may be wrong. Chief among these is that SEC separates proteins on the basis of hydrodynamic radius and/or shape and sizing an unknown protein using standard curves constructed using hydrodynamic radii assumes that the standards and the protein of interest share the same shape.

Electrospray ionization mass spectrometry, or native mass spectrometry, may be used to determine molecular mass with exquisite precision with a size limit in the million Dalton range. Recent advances in protocols and instrumentation now permit routine determination of protein molecular mass, oligomerization and subunit stoichiometry, with small amounts of sample [3], though it requires maintenance of the native or oligomeric state in the gas phase during the ionization process and significant technical expertise [4].

Analytical ultracentrifugation (AUC) is a powerful and versatile set of techniques that may be used to extract molar mass, oligomeric state, hydrodynamic radius and assembly association constants for a wide range of proteins under an equally wide range of concentrations [5]. However, run times can be long and protein aggregation can cause difficulties in analysis and data interpretation. While molar mass estimations from sedimentation equilibrium measurements can be accurate if determined from well-fitted data, those from sedimentation velocity experiments can be misleading as sedimentation coefficients are again influenced by protein shape.

Therefore, an absolute measure of a protein's molar mass that is not influenced by its shape and is not reliant on its elution position from a chromatography column or comparison to disparate external protein standards is critical. Light scattering techniques, as elaborated below, offer the ideal and the most precise solutions.

Light scattering, in simple terms, refers to the deflection of light by matter. The application of light scattering in molecular biology entails a set of techniques that detect and measure the spatiotemporal distribution of scattered incident monochromatic light by a solution (typically protein) of interest, which can provide valuable information on molar mass, size and shape. Two overarching approaches exist: static light scattering (SLS) and dynamic light scattering (DLS).

In SLS, time-averaged intensities of scattered light are measured, often at many angles simultaneously. The intensity of scattered light may be used to directly determine the molar mass (*see* **Note 1**) of the scattering molecules in the solution and, in some cases, also their mass distributions (their radii of gyration, or R_g). In DLS, by contrast, fast fluctuations (ns-μs timescale) in the intensity of scattered light are recorded and are autocorrelated as a function of time. The residence time, τ_D, of these fluctuations reflects the average time taken for the scattering molecule to diffuse through the observation volume (also known as characteristic diffusion time) and yields the translational diffusion coefficient (D) for the molecule of interest, from which its molecular hydrodynamic radius (r_h) can be determined.

The main advantage in using SLS for mass determination is that molar mass is directly related to scattering intensity, so can be directly determined from the experiment [6]. In the rather special case of an ideal solution of a molecule or protein with dimensions much smaller than the wavelength of the light used for illumination:

$$R \propto CM$$

R, the Rayleigh ratio, is the quantity directly measured in the light scattering experiment. C is the weight concentration of the molecule (*see* **Note 2**) and M is the weight-averaged molar mass. In principle, if the weight concentration and a constant of proportionality, termed K, are both known, M may be directly determined. In a SLS experiment, the weight concentration is typically determined using a differential refractometer and the constant of proportionality may be calculated, estimated, or determined using a known standard.

As molar mass is directly calculated from scattered intensity, SLS is independent of any assumptions about molecular shape, conformation or structure. Consequently, it may be used regardless of whether a protein is natively structured or unstructured. SLS, like AUC, has a tremendous dynamic range and can be used to determine molar masses ranging from a few hundred Daltons up to 1 billion Daltons. In addition, like AUC, size assessments (in terms of radii of gyration) can be made for proteins or protein complexes with sizes larger than ~10 nm.

Furthermore, with hardware and software advances, analysis of proteins by multiangle SLS, particularly when coupled with size exclusion chromatography, is quick and straightforward, to the extent that a standard run takes only a few hours. It can therefore be readily incorporated into routine purification and quality assessment protocols in the lab, as well as being quick enough to thoroughly characterize a system and its response to perturbations such as changing the biochemical environment or the addition of ligands.

SLS has a rich history in the study of the proteins that function in membrane remodeling, particularly at the endosome, in endocytosis and at the mitochondrion. Members of the dynamin superfamily protein (DSP) family catalyze membrane remodeling, fusion and fission events throughout the cell [7]. Dynamin is the endocytic DSP that catalyzes the membrane scission event that releases an endocytic vesicle into the cytosol. Close dynamin relatives catalyze mitochondrial fission (Drp1 or Dnm1 in mammalian cells/yeast respectively) and at the fungal endosomal membrane (Vps1). All are complex, multidomain proteins that self-assemble into helical structures and, in response to nucleotide binding and hydrolysis, undergo complex conformational changes that result in disassembly and membrane remodeling. More distant family members catalyze outer and inner mitochondrial membrane fusion or homotypic endoplasmic reticulum (ER) tubule fusion.

Close dynamin relatives self-assemble via several interaction interfaces within their Stalk domains, helical bundles that connect their GTPase domains to elements that interact with their respective target membranes. In addition, their GTPase domains all undergo nucleotide-mediated dimerization, critical for interrung connections in assembled helices, and hydrolysis-dependent conformational changes, where a large reorientation of a helical bundle that flanks the GTPase domains occurs.

SLS studies contributed to elucidating all these interactions. SEC-MALS was used to demonstrate that dynamin (including the unstructured proline-rich domain or PRD at its C-terminus) is a stable tetramer in its resting state [8]. Purified Drp1 exists in a dimer-oligomer equilibrium and different Drp1 splice variants have differing oligomerization propensities [9, 10]. Right-angle light scattering (RALS) was used to probe the consequences of mutagenesis of the Stalk assembly interfaces in both Drp1 and in MxA, a DSP and innate immunity factor that is protective against influenza viruses [11, 12]. SEC-MALS analysis of minimal GTPase-helical bundle constructs from Vps1 were used to demonstrate the nucleotide-dependence of GTPase domain dimerization [13, 14]. RALS and MALS were both used to probe the nucleotide and ion dependence of dimerization of minimal GTPase-based constructs of Mfn1 and atlastin, required for mitochondrial outer membrane and ER homotypic fusion, respectively [15–18].

As DLS results in the direct determination of the hydrodynamic radius of a protein sample, it is an important component in the validation pipeline for any structural project. This is particularly true for crystallographic projects, where monodisperse samples are critical. Furthermore, changes in hydrodynamic radius on ligand binding can provide insight into conformational changes.

Here, we present a brief summary of some of the principles of SLS and DLS, addressed in the form of responses to queries. We then illustrate SLS and DLS data gathering and analysis using Mvp1, a *Saccharomyces cerevisiae* SNX-BAR protein required for endosomal membrane remodeling and trafficking, which we have studied extensively using light scattering techniques [19]. SNX-BAR proteins minimally consist of a typically lipid-binding PX domain in close association with a membrane curvature-sensing or generating BAR domain and are involved in several trafficking and autophagic processes [20–23]. As BAR domains invariably dimerize, all SNX-BARs were thought to be dimeric. Indeed, crystal structures of some minimal PX-BAR modules from SNX-BAR proteins supported this [24, 25]. However, light scattering approaches were used to show that full-length Mvp1, which contains a low-complexity intrinsically disordered region (IDR) of ~120 residues before its PX-BAR module, is tetrameric in solution [19]. Structure determination by cryogenic electron microscopy (cryo-EM) demonstrated that the tetramer is a dimer of dimers. By contrast, a minimal Mvp1 PX-BAR module is dimeric in solution, as are all other PX-BAR modules studied to date. Mvp1 lacking its IDR has trafficking defects in vivo. As several other fungal and mammalian SNX-BARs have N-terminal IDRs, regulation of PX-BAR module function by their IDR may be a recurring motif throughout the family—indeed, in the case of the mammalian endocytic SNX-BAR SNX9, the IDR has been shown to regulate membrane association of the PX-BAR module [26]. Several outstanding publications offer a more detailed discussion of the physics of light scattering and of the mathematics underpinning SLS and DLS [6, 27, 28].

1.1 Why Couple SLS with Size Exclusion Chromatography?

While SLS can be used in batch mode, it is often used downstream of size exclusion chromatography (online or in-line mode), an approach known as SEC-MALS. After separation on the basis of hydrodynamic radius through a size exclusion column, sample is passed through tandemly arranged, in-line MALS and refractive index (RI) (or, alternatively, UV) detectors for scattering measurement and concentration determination, respectively. This is convenient as it permits a sequential and separate analysis of all the components eluting from the column. Often this permits insight into protein oligomerization, as well as a qualitative insight into the dynamics of equilibration between the multimerizing species. The output of SLS is the weight-averaged molar mass of the protein of interest. As this is highly sensitive to the presence of even small concentrations of aggregates, an additional advantage of coupling SLS with size exclusion is therefore sample quality improvement. The most common use, however, is to couple SLS-mediated molar mass determination with theoretical masses determined from primary sequence analysis, such that insight into protein oligomerization and native multimeric states can readily be obtained.

1.2 Why Is Scattering Related to Mass?

A scattering molecule, or protein, is an assembly of individual point scatterers. The scattering of the molecule may then be understood by considering first the scattering from a single point scatterer and then considering what effects additional scatterers may have on the measured scattering.

For a single point scatterer, with dimensions much smaller than the wavelength of light used, scattering from an incoming polarized light source in the *xy* plane is isotropic, or equally intense in any azimuthal direction (*see* **Note 3**).

If the molecule consists of two point scatterers, a complication arises. If the first scatterer is situated at the origin of a hypothetical coordinate system, incoming light may have to travel a different distance to reach the second scatterer and, in any particular direction, a different distance to reach the detector. In other words, the light scattered from a second scatterer may have a different total path length compared to the light scattered from the first scatterer at the origin. Light from the second scatterer may therefore experience a relative phase shift when that light reaches the detector (*see* **Note 4**). This may constructively or destructively interfere with light scattered from the origin.

The same considerations can be extended to cover all the scatterers that constitute the molecule. The total scattering will therefore depend on the phase relationships of every scatterer (of which there will be n) with respect to all others. This can be captured in an expression with a double summation with a total of n^2 terms, where n is the number of scatterers in the molecule:

$$I_{\text{molecule}} = I_{\text{scatterer}} \sum_{i=1}^{n} \sum_{j=1}^{n} \cos\left(\phi_i - \phi_j\right)$$

$I_{\text{scatterer}}$ and I_{molecule} are the scattering intensities from a point scatterer and the whole molecule, respectively. ϕ_i and ϕ_j are the phases of scattering from the ith and j^{th} point scatterers with respect to a scatterer at the origin.

If the overall size of the molecule is much smaller than the wavelength of the light used, these phase differences, $\phi_i - \phi_j$, will be more or less zero. Hence, the term $\cos(\phi_i - \phi_j) \approx 1$. Therefore:

$$I_{\text{molecule}} \approx I_{\text{scatterer}} \tilde{n} n^2$$

n is clearly proportional to the molecular mass of the molecule. Total scattering from the molecule is, therefore, proportional to the square of the molecular mass of the molecule:

$$I_{\text{molecule}} \propto I_{\text{scatterer}} \tilde{n} M^2$$

1.3 Why Is Scattering Related to Concentration?

A solution consists of many copies of the molecule, which have no long-range order between them. (Different approaches, such as diffraction experiments, could be used for analysis, molecular mass and even structure determination if there were a long-range order!) Compared to a single molecule at the origin of a coordinate system, the total scattering from any other molecule may have a phase difference with respect to the molecule at the origin, such that, again, constructive and destructive interference may occur between light scattered from these molecules. However, in this case, because of the distances between the molecules relative to the wavelength of light used, phase differences are significant. Total scattering from the cohort of molecules, $I_{\text{total scattering}}$, is then:

$$I_{\text{total scattering}} = I_{\text{molecule}} \sum_{i=1}^{N} \sum_{j=1}^{N} \cos\left(\phi_i - \phi_j\right)$$

N is the total number of molecules in the cohort. Total scattering is simply obtained by adding up over all possible pairs of scattering molecules, here labeled i and j, each of which will have different phase relationships (captured in the term $\cos(\phi_i - \phi_j)$). The molecules are all moving independently so the phase relationships will constantly change. In SLS, measurements are therefore made of *time-averaged* scattering intensities, indicated with the angle brackets $\langle \ldots \rangle$:

$$\left\langle I_{\text{total scattering}} \right\rangle = I_{\text{molecule}} \sum_{i=1}^{N} \sum_{j=1}^{N} \left\langle \cos\left(\phi_i - \phi_j\right) \right\rangle$$

ϕ_i and ϕ_j can take any value between 0 and 2π. Hence the average value is of $\cos(\phi_i - \phi_j)$ is 0. The only exception is when $\phi_i = \phi_j$, where $\cos(\phi_i - \phi_j) = 1$. This occurs N times. So

$$\langle I_{\text{total scattering}} \rangle = I_{\text{molecule}} \tilde{n} N$$

N is directly related to concentration and can be written as $N = cVN_A$, where V is the volume doing the scattering, N_A is Avogadro's constant, and c is the molar concentration (*see* **Note 5**). Therefore, scattering also depends directly on the concentration of the molecules doing the scattering.

1.4 What Is the Rayleigh Ratio R?

The Rayleigh ratio is the parameter that is directly recorded in the SLS experiment. It is the normalized scattered intensity at some angle θ (where $\theta = 0$ is scattering in the same direction as the incident beam). The normalization accounts for the intensity of the incident light, I_0, the distance to the detector and the volume of the sample doing the scattering:

$$R(\theta) = \frac{I_{\text{total scattering}}(\theta) r^2}{I_0 V}$$

For ease of mathematical manipulation, it is more common to see R described in terms of the *scattering vector q*. The definition of q also shows its angle-dependence:

$$q = \frac{4\pi}{\lambda} \sin\left(\frac{\theta}{2}\right)$$

This means R will be referred to as $R(q)$.

1.5 What is Rayleigh Scattering Anyway?

Rayleigh scattering is the elastic scattering that arises from molecules that are very much smaller than the wavelength of light used (*see* **Note 6**). The electric field of the incident light induces an oscillating dipole in the molecule which then itself becomes a source for radiation of the same frequency. This may be reradiated (or scattered) in any direction and recorded. The intensity of the scattered light in any direction is directly proportional to the intensity of the incident light and inversely proportional to the fourth power of the wavelength of incident light used and the square of the distance from the molecule to the detector. If the incident light is polarized and the detected light is in the same plane, the intensity of the scattered light is:

$$I_{\text{scattered}} = \frac{I_0 \pi^2 \alpha^2}{\lambda^4 r^2}$$

α is the polarizability of the molecule, or how efficiently the incident light can induce an oscillating dipole.

1.6 What Is the Constant K?

K is the *optical contrast constant* and is the constant of proportionality in the equation relating light scattering to molar mass:

$$R = KCM$$

It is defined as:

$$K = \frac{4\pi^2 \left(\frac{dn}{dC}\right)^2 n_0^2}{N_A \lambda^4}$$

$\frac{dn}{dC}$ is the refractive index increment and quantifies the change of solution refractive index with protein weight concentration. This can be determined with a differential refractometer or, for proteins in standard buffers, is known precisely. n_0 is the refractive index of the buffer being used, $N_A$is Avogadro's constant and λ is the wavelength. An overall value for K can be calculated, using known $\frac{dn}{dC}$ values for proteins.

1.7 Are More Angles Better: Why Do Multiangle Light Scattering?

In certain cases, accurate molar mass information can be obtained by measuring scattering at only one angle. Why might it then be preferable to record scattering at more than one angle and under what circumstances must one do so to obtain accurate mass information? Does recording at more angles results in additional insights or better data?

Two separate cases need to be considered: (1), when a scattering molecule (or an oligomeric assembly of a molecule) is, as a whole, significantly smaller than the wavelength of the laser light used for the scattering and (2), when the dimensions of the molecule are not significantly smaller than the wavelength of light used.

If the molecule is appropriately small, the phase differences of light scattered by the constituent scatterers in the molecule are ≈ 0. Hence the whole molecule scatters in phase in any given direction and the scattering of polarized, incident light is isotropic, or the same in all directions in the plane of incident light. In this case, the molar mass may be determined from the scattered intensity at any angle and only one angle would in principle suffice. Right angle light scattering (RALS) employs this approach. However, if the molecule under investigation is too large, significant errors in the mass determination arise because the phase differences of the light scattered by the constituent scatterers comprising the molecule are not ≈ 0. Hence, there is an angle-dependence to the scattering, with the most efficient scattering occurring at a scattering angle of 0, where all components of the molecule scatter in phase. The falloff in scattering efficiency with respect to scattering angle θ is described by the form factor, $P(\theta)$, and is defined as the ratio of the scattering at angle θ compared to the scattering at angle 0:

$$P(\theta) = \frac{I_{\text{total scattering}}(\theta)}{I_{\text{total scattering}}(0)}$$

Because of this, the relation describing the dependence of scattering on the molecular mass must be modified to include the form factor, which is more conveniently expressed in terms of the scattering vector q:

$$R(q) = KMCP(q)$$

If the scattering molecule is appropriately small, $P(q)$ is essentially 1 for all scattering angles and disappears from the equation. For larger particles, the form factor depends on the scattering angle, but falls away from 1 at a scattering angle of 0. The approach is therefore to measure the Rayleigh ratio $R(q)$ at several scattering angles such that it is possible, by analytical means, to extrapolate the value of the form factor to a scattering angle of 0 (which cannot be measured), where $P(q) = 1$. For this reason, scattering at as many angles as possible needs to be assessed—for example, the DAWN® HELEOS® II multiangle light scattering detector from Wyatt Technology records scattering at 18 angles. A number of approaches exist for the extrapolation.

1.8 Is Other Information Available from an SLS Experiment?

Yes, provided that the value of the form factor has an angle dependence. In this case, the root mean square (rms) radius, also known as the radius of gyration, R_g, can also be determined. R_g describes the mass distribution of the constituent components of the protein around its center of mass. This shows that $P(q)$ also captures aspects of the shape of the scattering molecule. $P(q)$ and R_g are related by:

$$P(q) \approx e^{-\frac{1}{3}q^2 R_g^2}$$

1.9 Dynamic Light Scattering

Static light scattering derives the molar mass and, for large molecules, the radius of gyration, from average scattering intensities recorded at one or more angles. If fluctuations in scattering from a small region of sample are measured over a period of time, other information, particularly the *translational diffusion coefficient* (D) and the hydrodynamic radius (r_h) may be extracted. The hydrodynamic radius (also known as the *Stokes* radius) is the radius of a sphere that diffuses at the same rate as the molecule under study. This approach is termed *dynamic light scattering*, or DLS.

1.10 How Does DLS Work?

The translational diffusion coefficient is determined by monitoring and recording Brownian motion in the solution. The fundamental idea is that if the molecules of interest are small, they will move farther in a period of time than if they are large. This movement can be recorded and analyzed to yield the translational diffusion coefficient.

In SLS, the time average of the total scattering is measured. It was noted that, as the molecules that generate the total scattering are moving, the relative phase differences $\phi_i - \phi_j$ of all possible pairs of molecules themselves constantly vary. It is this that results in the intensity fluctuations. In DLS, it is the time course of this variation that is of interest and is recorded with a fine time temporal sampling, with a minimum interval in the nanosecond to microsecond range.

1.11 How Is the Sequence of Scattering Fluctuations Analyzed?

For all possible time intervals τ, an autocorrelation, G_2, is calculated:

$$G_2(\tau) = \langle I_{\text{total scattering}}(t)\tilde{n}I_{\text{total scattering}}(t+\tau)\rangle$$

In other words, for every possible τ increment, the average value of the product of $I_{\text{total scattering}}$ at all times t and $t + \tau$ is found. If τ is very small, $I_{\text{total scattering}}$ at t and $t + \tau$ are not very different so the expected value of G_2 is asymptotically:

$$G_2(0) = \left\langle I^2_{\text{total scattering}}\right\rangle$$

when τ gets large, the phases will have changed due to Brownian motion. How fast this happens will depend on the size of the molecules. Ultimately, $I_{\text{total scattering}}$ at t and $t + \tau$ will bear no relation to each other. At this point, the expected value of G_2 is:

$$G_2(\text{long time}) = \langle I_{\text{total scattering}}\rangle^2$$

Assuming that the solution consists of copies of only one molecule, a plot of G_2 versus τ will show an exponential decay from $\left\langle I^2_{\text{total scattering}}\right\rangle$ to $\langle I_{\text{total scattering}}\rangle^2$.

1.12 How Is the Translational Diffusion Coefficient Extracted?

G_2 and its normalized version, g_2, are autocorrelations of the scattered and fluctuating *intensity* and can be directly calculated from experimental measurements (*see* **Note 7**). Similar considerations can be applied to autocorrelations of the electric *field strengths* and give rise to G_1 and its normalized version g_1. These latter two are related mathematically to the translational diffusion coefficient but cannot be measured directly. Fortunately, g_2 and g_1 are related to each other via an equation known as the *Siegert* relationship:

$$g_2(\tau) = 1 + \beta|g_1(\tau)|^2$$

β is the coherence factor that depends on both the instrument and the scattering.

Assuming the solution is monodisperse, $g_1(\tau)$ is an exponential decay with a constant Dq^2:

$$g_1(\tau) = e^{-Dq^2}$$

We can substitute this into the Siegert relationship:

$$g_2(\tau) = 1 + \beta e^{-2Dq^2\tau}$$

This illustrates how recording intensity fluctuations can give the translational diffusion coefficient.

1.13 How Can the Hydrodynamic Radius Be Obtained?

Once D has been determined, use the Stokes–Einstein relation to calculate r_h:

$$r_h = \frac{kT}{6\pi\eta D}$$

here, k is Boltzmann's constant, T is the absolute temperature, and η is the solution viscosity. This shows why particular care must be taken to ensure that the temperature of the experiment (which has significant effects on viscosity) remains constant.

2 Materials

2.1 Transformation and Growth of *E. coli* BL21-CodonPlus (DE3)-RIPL

1. *E. coli* BL21-CodonPlus (DE3)-RIPL (Agilent product no. 230280).
2. LB media and LB-agar: 10 g Bacto tryptone, 5 g yeast extract, 10 g NaCl per liter, autoclaved. For plates add 15 g/l Bacto agar per liter prior to autoclaving.
3. 2xTY media: 16 g Bacto tryptone, 10 g yeast extract, 5 g NaCl per liter, autoclaved.
4. Isopropyl-β-D-thiogalactopyranoside: make a stock of 35 mg/ml in water and filter-sterilize.
5. Ampicillin: make a 100 mg/ml stock in water and filter-sterilize.
6. Chloramphenicol: make a 35 mg/ml stock in ethanol.
7. Orbital shaker; we use a New Brunswick™ Innova® 44R shaker (Eppendorf) but any temperature-controlled shaker can be substituted.
8. Glassware appropriate for growth of cultures; we use 125 ml and 2000 ml PYREX™ flasks.

2.2 Reagents for Purification of His_6-Mvp1 from *E. coli*

1. A homogenizer; we use the Avestin Emulsiflex C3 but other systems can be substituted.
2. TN150: 20 mM Tris–HCl pH 7.4, 150 mM NaCl, 1.93 mM β-mercaptoethanol.
3. TN250: 20 mM Tris–HCl pH 7.4, 250 mM NaCl, 1.93 mM β-mercaptoethanol.
4. Imidazole.
5. Ni-IDA resin.
6. Centrifugal filter units with a molecular weight cutoff of 30,000 Daltons.

7. Dialysis Cassettes, 30 ml, with a molecular weight cutoff of 20,000 Daltons.
8. Empty columns appropriate for low-pressure liquid or gravity chromatography; we use a C 10/10 column (Cytiva), but any other can be substituted.
9. FPLC system; we use an NGC Quest 10 Discover system equipped with a single wavelength detector, but any comparable system can be substituted.
10. Q Sepharose Fast Flow (Cytiva).
11. Superdex S200 16/60 gel filtration column (Cytiva).

2.3 Size-Exclusion Chromatography-Coupled Multiangle Light Scattering (SEC-MALS)

1. FPLC system at room temperature; we use ÄKTAexplorer™ or ÄKTApure (Cytiva) equipped with a UV detector.
2. MALS and refractive index (RI) equipment; we use in-line DAWN® HELEOS® II 18-angle MALS and Optilab® T-rEX™ RI detectors (Wyatt Technology) set at 25 °C in the FPLC→MALS→RI→Waste/Fraction collector flowpath configuration, and the ASTRA® software package (Wyatt Technology) for data acquisition and analysis.
3. Appropriate size-exclusion chromatography column and a 500 μl sample loop; for Mvp1, we use a Superdex 200 Increase 10/300GL column (Cytiva).
4. TN150 (filtered and degassed) maintained at room temperature with β-mercaptoethanol added to a final concentration of 1.93 mM immediately prior to use.
5. Protein stock diluted in TN150 at room temperature; for Mvp1, we use a final concentration of 20 μM in an injection volume of 500 μl (protein concentration and volume to be used as empirically determined; **Notes 8** and **9**).

2.4 Dynamic Light Scattering (DLS)

1. DLS equipment; we use the DynaPro® Nanostar® cuvette-based (batch mode) DLS instrument set at 25 °C (Wyatt Technology) with the DYNAMICS software package (Wyatt Technology) for data acquisition and analysis.
2. UV-transparent cuvettes compatible with the DLS instrument; we use UVette® 220–1600 nm disposable single-sealed cuvettes, 50–2000 μl volume.
3. Protein stock diluted in TN150 at room temperature; for Mvp1, we use a final concentration of 20 μM in a volume of 100 μl (as for SEC-MALS, the protein concentration to be used is empirically determined for optimal signal and S/N ratio; **Note 8**). Samples must be filtered using spin columns to remove aggregates, as for SEC-MALS.

3 Methods

3.1 Purification of His$_6$-Mvp1 from E. coli

1. Transform *Escherichia coli* strain BL21-CodonPlus (DE3)-RIPL with the construct encoding *Saccharomyces cerevisiae* His$_6$-Mvp1 (pET-15b *S cer. MVP1*) (*see* **Note 10**). Plate the transformed cells onto LB-agar supplemented with 100 μg/ml ampicillin and 35 μg/ml chloramphenicol. Incubate overnight at 37 °C.
2. Inoculate 2 to 4 50 ml volumes of LB supplemented with 100 μg/ml ampicillin and 35 μg/ml chloramphenicol, in appropriate glassware, with fresh colonies of *E. coli* BL21-CodonPlus (DE3)-RIPL containing pET-15b *S cer. MVP1.*
3. Grow the cultures at 37 °C on an orbital shaker (180 rpm) until the culture is saturated (approximately 8 h).
4. Transfer each entire 50 ml culture into 1 l 2xTY, in a 2000 ml flask, to which ampicillin (final concentration 100 μg/ml) and chloramphenicol (35 μg/ml) have been added.
5. Grow at 37 °C on an orbital shaker (180 rpm) until the culture reaches an optical density at 600 nm (OD_{600}) of ~0.6.
6. Add IPTG to a final concentration of 12.25 μg/ml (~51.4 μM) to induce expression (*see* **Note 11**).
7. Reduce the ambient temperature of the incubator to 21 °C and grow the cells overnight (~14 additional hours) with shaking (180 rpm). The final OD_{600} should be ~4.0.
8. Harvest the cells by transferring each liter of culture to a 1 l centrifuge bottle and centrifuging for 15 min at 3800 × *g* at 4 °C.
9. Decant the supernatant and drain all excess growth medium from the pellet.
10. Resuspend the pellet in ~20 ml TN150 per pellet generated from each liter of culture.
11. Snap-freeze in liquid nitrogen and store at −80 °C or proceed directly to cell lysis.

3.2 Lysis Using an Avestin Emulsiflex C3 Homogenizer

1. Thaw the cell pellet on ice. Prepare ice-cold TN150.
2. Prechill the homogenizer sample chamber on ice. Turn on the heat exchanger or place the sample loop in a wet ice water bath (*see* **Note 12**).
3. Switch on the homogenizer and turn on the air supply. Input air pressure needs to be ~100 psi (6.8 atm—**Note 13**).
4. Sequentially pass ice-cold tap water and TN150 through the homogenizer with no pressure.
5. Refill the sample chamber with ice-cold TN150. With the machine running, slowly increase the input air pressure using

the pressure regulator until the homogenizing pressure gauge begins pulsing (input pressure ~40 psi).

6. Allow the remaining buffer to flow through the homogenizer.
7. Pour the fully thawed cell pellet into the sample chamber (*see* **Note 14**).
8. With the homogenizer running, turn the pressure regulator until it is pulsing at ~10,000 psi (*see* **Note 15**).
9. Collect the lysate in a chilled beaker on ice.
10. Pass the lysate through the homogenizer an additional two times.
11. Flow an additional ~40 ml of TN150 through the homogenizer 10,000 psi to collect lysate remaining in the homogenization chamber.
12. Clarify the lysate by centrifugation at 142,000 × *g* for 45 min at 4 °C.

3.3 Purification of His$_6$-Mvp1 from the Clarified Lysate

1. To bind the recombinant protein, add an appropriate volume of Ni-IDA beads rewashed with TN150 directly to the clarified lysate (*see* **Note 16**).
2. Incubate the beads with the lysate with gentle mixing at 4 °C for 1 h (*see* **Note 17**).
3. Sediment the beads by centrifugation (1000 × *g*, 4 °C, 5 min) in a refrigerated benchtop centrifuge.
4. Wash the pellet thoroughly and several times with ice-cold TN150, pelleting the beads by centrifugation after each wash (*see* **Note 18**).
5. Pack the beads into a column of appropriate bed volume (*see* **Note 19**).
6. Using an FPLC system, wash the beads extensively with TN150 until no more protein washes off the beads, as assessed by monitoring absorbance at 280 nm (*see* **Note 20**).
7. Elute the protein with a shallow imidazole gradient from 0 to 250 mM in TN150.
8. Analyze peak fractions by running an SDS-polyacrylamide gel followed by staining with Coomassie Brilliant Blue, retaining those showing abundant, clean recombinant protein.
9. Pool the peak fractions and, using dialysis cassettes, dialyze overnight at 4 °C into 2 l TN150 to remove the imidazole.
10. Replace the TN150 and dialyze an additional 2 h.
11. Pass the dialyzed protein onto a Q Sepharose anion exchange column in TN150. His$_6$-Mvp1 will be retained on the column and may be eluted with a 150–500 mM NaCl gradient, with the elution peak at ~250 mM NaCl.

12. Analyze the eluted protein-containing fractions by SDS-polyacrylamide gel electrophoresis.
13. Concentrate the peak fractions by centrifugation at 4500 × *g* at 4 °C using a centrifugal filter unit.
14. Pass the concentrated protein over a Superdex S200 16/60 size exclusion column at 1 ml/min, using thoroughly degassed TN250 as the mobile phase.
15. Analyze peak fractions by SDS–polyacrylamide gel electrophoresis followed by Coomassie Brilliant Blue staining.
16. Concentrate the peak fractions as above using a centrifugal filter unit.
17. Determine the concentration of the protein by assessing absorption at 280 nm or using an appropriate colorimetric assay.
18. Aliquot and snap-freeze the proteins in liquid nitrogen and store at −80 °C until needed (*see* **Note 21**).

3.4 SEC-MALS Analysis of Protein Oligomerization State(s)

The following methods apply for the Wyatt instruments we use. For instruments from other manufacturers, the reader is referred to the relevant user guides; the basic methodologies are comparable, but the individual steps and other details may vary.

1. Pre-equilibrate the SEC column with 1.5 column volumes (24 ml bed volume for our Superose column × 1.5 = 36 ml) of TN150 using the FPLC system and a pre-programmed protocol (using UNICORN™ software in case of ÄKTA systems). For SEC-MALS experiments, we recommend a flow rate of 0.3 ml/min (not to exceed 0.5 ml/min when the column is in-line with the MALS and RI detectors). Adhere to manufacturer's recommendations for backpressure limits (*see* **Note 22**). Likewise, rinse the sample loop and injection valves with TN150 (2 ml).
2. Turn on and warm up the MALS and RI detectors for about an hour prior to the experiment and let the TN150 from the equilibrated FPLC column flow through their flow cells (*see* **Note 23**). Using the PURGE tab on the RI instrument, replace storage solution in the reference flow cell (pure H_2O or 20% ethanol or buffer from previous use) with TN150. Run until the RI signal is flat (absence of a slope) and baselined (readjust to zero if needed) before turning off PURGE (*see* **Note 24**).
3. Prepare a sample of 20 μM Mvp1 in a total volume of 550 μl in TN150 at room temperature and equilibrate for about 5–10 min (smaller sample loops and lower volumes may also be used).

4. Using a 1 ml syringe, carefully load the sample into the 500 μl sample loop via the dedicated injection port. Take care to avoid introducing air bubbles at the front and back ends of the injected solution.
5. We use the ASTRA software package that accompanies the Wyatt instrument. Open the ASTRA software (we currently use version 7) and ensure that the MALS and RI detectors are both recognized by the software in the "Instruments" tab. The FPLC, MALS, and RI systems must be daisy-chained, and the UV data from the FPLC system should be programmed to be transferred to the ASTRA software (please refer to the ASTRA user guide from Wyatt Technology). We use RI, and not UV, as the concentration source for proteins because of the various inherent limitations of UV spectroscopy (*see* **Note 25**).
6. Open a saved standard "Methods" file (specific for the particular instrument) that contains predetermined information on instrument parameters (calibration constants), solvent and sample properties, normalization coefficients for LS signal at each detector, interdetector delay volumes (for peak alignment), band broadening correction to account for sample dilution between detectors (please refer to the ASTRA user guide available from Wyatt Technology). Instrument parameters are usually determined using a monodisperse, isotropic scatterer such as bovine serum albumin (BSA: Sigma product no. A1900, monomeric fraction of BSA) previously, saved as a "Methods" file, and are recalled and applied for every subsequent sample run (*see* **Note 26**). Hit the "Run" tab in the ASTRA software for it to await sample autoinjection signal from the FPLC and trigger LS and RI data collection.
7. Run the SEC protocol from the FPLC (ideally programmed for 1.5 column volumes at a flow rate of 0.3 ml/min) and ensure that all signals (UV, LS and dRI) are acquired by the ASTRA software as shown in Fig. 1a.
8. After data acquisition is complete, set baselines for the LS and RI traces in the "Baseline" tab under the current Experiment, as shown in Fig. 1a. Typically signal from the LS11 detector (90° detector), or from the concentration source (in our case, RI), is used to define the start and end points of the baseline. The "Set All" option will apply these setting to signals from all detectors. Check the baselines of all detectors, especially that of the RI (or UV) detector. Zoom in on the traces and ensure that the baselines are flat (horizontal) and the ends straddle the peak (s). This is a critical step to ensure measurement accuracy.
9. In the "Peaks" tab under the current Experiment, zoom in on the relatively scaled traces and define the start and end of the peak(s) as shown in Fig. 1b. For a monodisperse sample, the

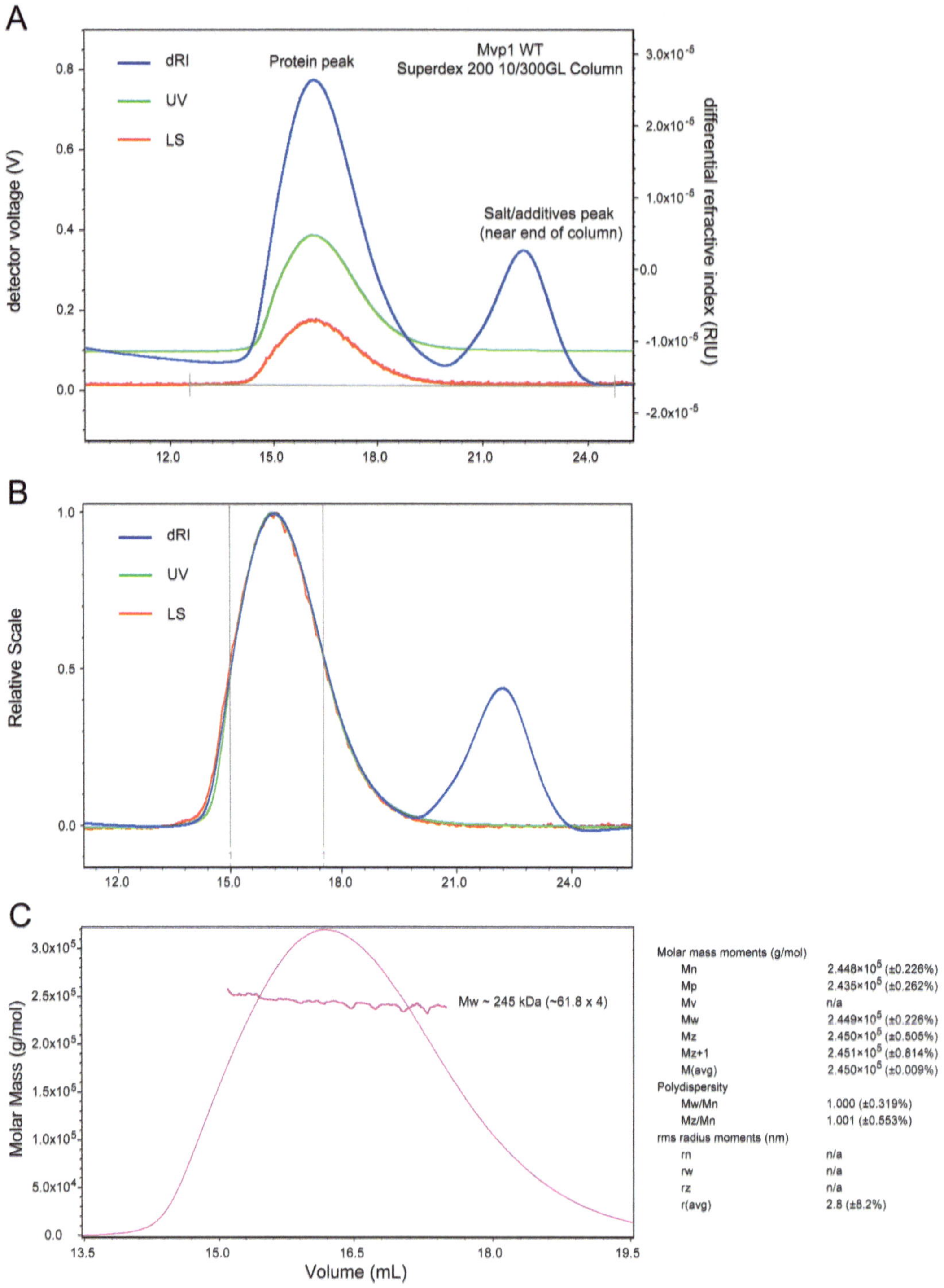

Fig. 1 SEC-MALS data acquisition and analysis. (**a**) Basic collection of dRI, UV and LS data traces in an SEC-MALS experiment, here using Mvp1 as an example. A large RI peak may appear near the end of the column (total inclusion volume) with no corresponding LS and UV signals as seen here. This contains dissolved gases, salts, and other molecules (additives, e.g., glycerol) that may be present (in different amounts) in the sample compared to the running buffer (mobile phase). Small differences in the ratio of oxidized vs. reduced

derived molar mass will not be sensitive to the peak region selected, whereas for a polydisperse sample in fast dynamic equilibrium [9], it is expected to be. Cover as much area of the peak (generally greater than full-width half-maximum or FWHM) keeping in mind the presence of a good LS signal (above background) underneath the RI peak.

10. Results are computed and (1) displayed in the "EASI graph" tab with the calculated molar mass (Y-axis) drawn across the selected peak region as shown in Fig. 1c, *left panel*, and (2) tabulated in the "Results (detailed)" tab indicating the various molar mass moments in g/mol as shown in Fig. 1c, *right panel*. The Zimm formalism is generally best for determining the mass/oligomerization state(s) of small to mid-sized molecules (rms radii up to 50 nm) and no shape (conformation) information is assumed during the calculation. The molar mass profile is generally flat across the peak for monodisperse protein species, whereas it typically slopes or curves down from left to right for polydisperse species with limited column resolution. The weight-averaged molar mass (M_w) does not depend on column resolution and is generally reported for monodisperse species (as for Mvp1; Fig. 1c, *left panel*), whereas a combination of other moments may need to be reported for polydisperse samples. A polydispersity index value of "1" as shown in Fig. 1c, *right panel* indicates a monodisperse species within the selected peak. The "Results (detailed)" tab also indicates the calculated rms radius or R_g as "r(avg)", but the lower limit for accuracy is ~10 nm. For monodisperse Mvp1, the molar mass obtained by SEC-MALS (~245 kDa) is approximately four-times the calculated theoretical mass of a monomer. Therefore, Mvp1 was determined to be a monodisperse tetramer, consistent with its cryo-EM structure [19].

Fig. 1 (continued) reducing agents such as DTT and β-mercaptoethanol between the injected sample and the running buffer could also give rise to this peak. The position of this peak should be consistent from one injection to another. Similarly, a ghost LS peak (not shown here) may appear in the included (void) volume without any corresponding RI or UV signal. This is caused by a very minor fraction of very large particles, which may include protein macroaggregates and/or shed resin. The start and end of the baseline drawn for the LS trace from the 90° detector is shown and is applied to all detectors. Ensure that the baseline straddles all peaks and is relatively flat. (**b**) Selection of peak area from the normalized dRI, UV, and LS data traces (defined by the two vertical lines labeled "1") for molar mass determination. (**c**) *left*, EASI graph representation of molar mass profile within the selected peak region. The weight-averaged molar mass (M_w) of Mvp1 (~245 kDa) is approximately four times the calculated theoretical mass of a monomer (~61.8 kDa). Therefore, Mvp1 exists as a monodisperse tetramer in solution. *Right*, Tabulated results (detailed) showing the different molar mass moments, the polydispersity index (with a value of 1 indicating a monodisperse species), and the rms radius (not reliable below ~10 nm)

3.5 DLS Determination of Diffusion Characteristics and Protein Hydrodynamics

The following methods apply for the Wyatt instrument. The reader is referred to the relevant user guides for instruments from other manufacturers as the individual steps may differ.

1. Turn on the instrument and let it warm up and equilibrate at set detector temperature (25 °C) for approximately 30 min before data acquisition. Open the DYNAMICS software and connect to instrument using the "connect to instrument" tab.
2. Under "Instrument" settings, set DLS acquisition time to 5 s, read interval to 1 s, and the number of acquisitions to 10. Autoattenuation should be "ON" ("YES") and autoattenuation time limit set to 60 s.
3. Dilute Mvp1 to a final concentration of 20 μM in TN150 maintained at room temperature and load 100 μl in the disposable Eppendorf cuvette. We recommend volumes above the lowest allowed volume of 50 μl to avoid interference from meniscus and/or potential air bubbles.
4. Prior to hitting "START" in the DYNAMICS software, check for stable signal intensity counts (counts/s) and laser power using the "Instrument control panel" tab. Once the signal appears relatively stable (fluctuations within 10% when monitored for approximately 30 s), close this window. An unstable signal (large sporadic deviations in counts/s) indicates the presence of large aggregates that will result in data spikes and preclude DLS measurements.
5. Hit "START" and acquire data. Repeat this set of 10 measurements three more times.
6. For each measurement, check the goodness-of-fit of the "autocorrelation graph" as shown in Fig. 2a. For a monodisperse species, the data will be best fit to a single exponential as shown in Fig. 2a. For a polydisperse sample in protein self-assembly or conformation, the fit will deviate systematically from the raw data, indicating a poor fit. The *cumulant fit* "assumes" a distribution of diffusion rates and determines their mean and width to calculate the average value of r_h, whereas the *regularization fit* fits to a set of exponentials to obtain the "real" underlying distribution of r_h, represented as a histogram in Fig. 2b. The average r_h is indicated above. Check whether these data are comparable across measurements and reject any outliers owing to data inconsistencies (spikes). Remarkably, for the Mvp1 tetramer, the DLS-derived r_h (6.3 nm) is consistent with the computed r_h from its cryo-EM structure using the HullRad server (5.3 nm) [29] (Fig. 3). The difference may be due to the IDRs, which are mostly disordered in the structure.

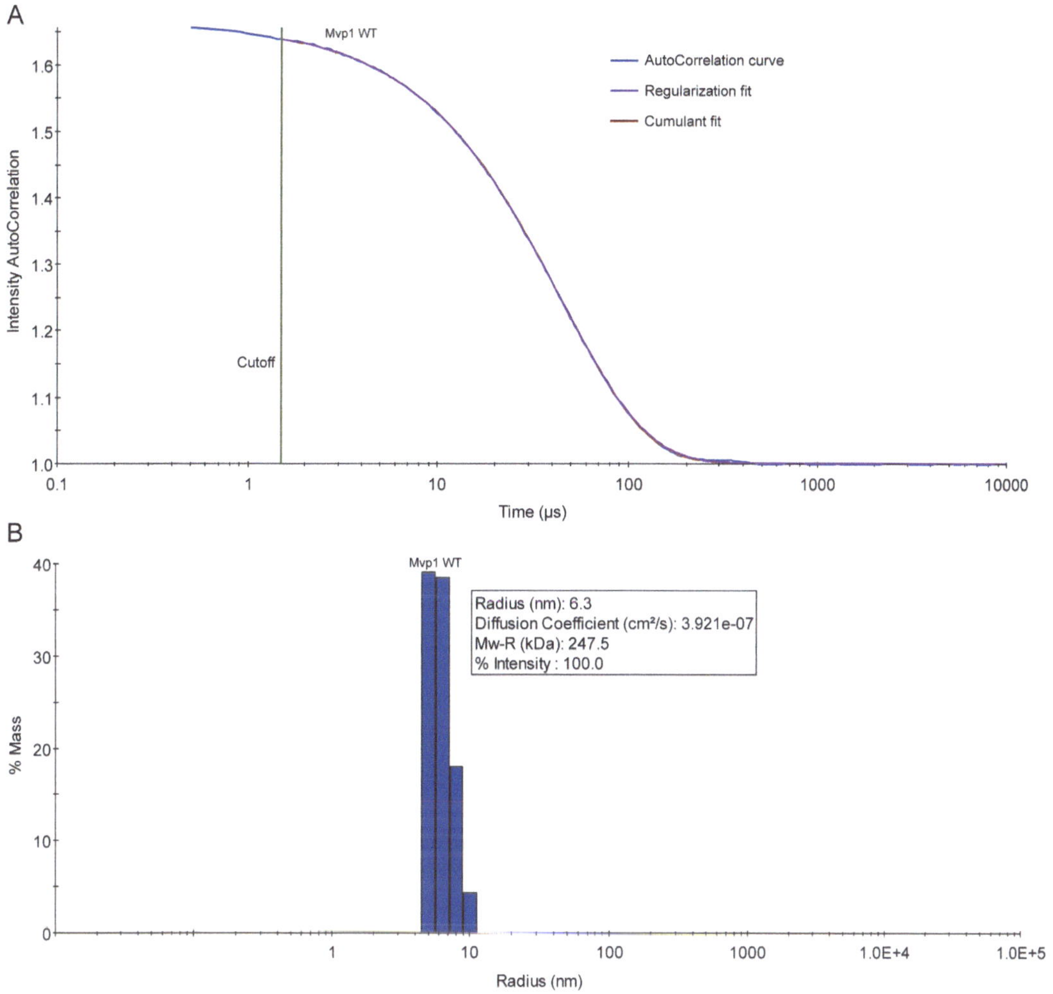

Fig. 2 DLS data analysis. (**a**) Representative DLS autocorrelation trace for Mvp1 (blue) best fit to a single exponential using *cumulant* and *regularization* fits. These indicate a monodisperse sample. (**b**) Histogram of hydrodynamic radii distribution with the average radius (r_h), diffusion coefficient (D), molecular mass calculated using r_h (labelled "Mw-R"), and the contribution of this monodisperse species to the scattering intensity indicated in the box

4 Notes

1. Note the distinction between molar and molecular masses of a protein: molar mass is the mass of 1 mole of the protein (reported in g/mol) whereas molecular mass is the mass of one molecule (reported in Daltons). If the sample is entirely homogeneous, the magnitudes of the molar and molecular masses are the same.

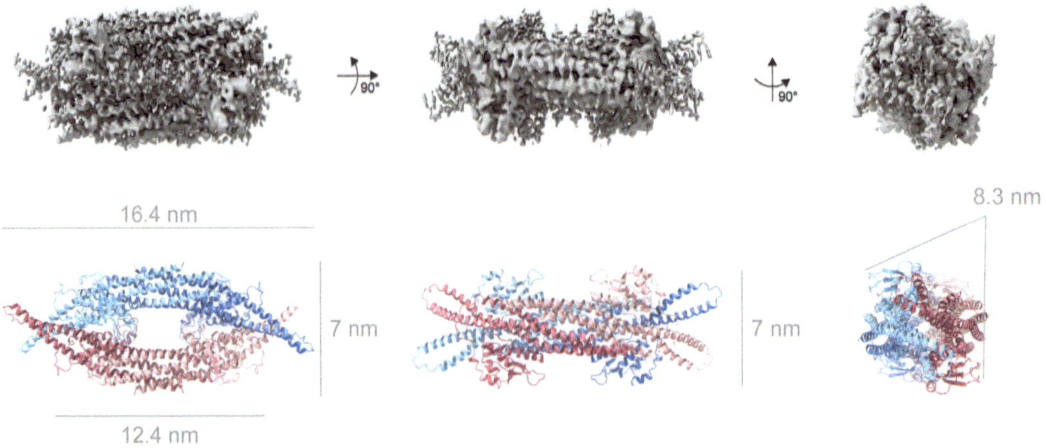

Fig. 3 Molecular dimensions of the Mvp1 tetramer. Top: Three views of the D2-symmetrized, sharpened ($B = -252$ Å^2) cryo-EM structure of the Mvp1 tetramer (EMDB entry number 20555), contoured at 4 σ. Bottom: Ribbon diagrams of the final real-space refined models of the Mvp1 tetramer (PDB ID 6QOX) shown in the same orientations. One Mvp1 dimer is shown in pinks and the other in blues. Several dimensions are annotated

2. The weight concentration C, in g/l, is related to the molar concentration c by multiplication by the molar mass.

$$C = cM$$

3. Scattering does depend on altitude from the xy plane, but, as the detectors in light scattering instruments typically lie in the same plane as the incoming light, scattering in directions other than the xy plane can be disregarded.
4. Path difference Δx is related to phase difference $\Delta\phi$ via the following relation.

$$\Delta x = \frac{\lambda}{2\pi}\Delta\phi$$

5. The ratio of N to N_A is the mole fraction.
6. Practically, a molecule very much smaller than the wavelength used means $\sim\frac{1}{20}\lambda$. The lasers used to generate the light in Wyatt instruments have a wavelength of ~660 nm. Hence, a molecule very much smaller than the wavelength used will have a longest dimension of ~33 nm.
7. The subscript 2 in G_2 and its normalized form g_2 indicates that the autocorrelation is second-order, as intensity is being autocorrelated to intensity. This contrasts with first-order G_1 and g_1, where field strengths are being autocorrelated.
8. A golden rule of thumb is that "the smaller the protein, the greater the concentration (at injection)" that would be

required to generate a light scattering signal at the LS detector. This must be empirically determined for each protein, and especially for polydisperse samples.

9. For self-assembling protein systems prone to concentration-dependent aggregation, we highly recommend clarifying frozen-and-thawed protein stocks using centrifuge tube filters (e.g., Costar® Spin-X 0.22 μm centrifuge tube filters from Corning or similar) and low-speed centrifugation at 4 °C according to the manufacturer's protocol before dilution.
10. To generate the Mvp1 expression construct, the coding sequence for *Saccharomyces cerevisiae MVP1* was cloned into the NdeI and BamHI sites of pET-15b (Novagen) by Gibson assembly. The final constructs contained only a histidine residue between the pET-15b thrombin cleavage site and the Mvp1 coding sequence.
11. Take samples of cells (0.25 OD_{600} units/50 μl) before induction and prior to harvesting and analyze by SDS–polyacrylamide gel electrophoresis and Coomassie Brilliant Blue staining to assess expression.
12. The Emulsiflex C3 homogenizes by pushing the sample through a small orifice. Pressure changes result in lysis by cavitation and shear forces. This can generate significant amounts of heat in the sample. Heat mitigation is therefore essential.
13. If the input air pressure is inadequate, use an inline air pressure booster. We use the VBA10A-T02-Z (SMC Pneumatics) that is compatible with the Avestin Emulsiflex C3.
14. As cell clumps may obstruct the emulsifier, we recommend passing the resuspended cell pellet through a narrow-lumen syringe needle to help dissipate remaining clumps.
15. Do not allow the sample chamber to run dry. Though sample retrieval is possible, mechanical losses will be severe. We stop the homogenizer for sample chamber reloading when there is ~1 pulse of buffer/sample remaining.
16. The theoretical binding capacity of the Ni-IDA beads is 20 mg/g resin and 10 mg/ml bed volume once hydrated. As a guide, we obtain 10–15 mg purified protein per l culture.
17. We use a rotating wheel with an appropriate number of 50 ml conical tubes. We ensure to fill these completely to minimize protein denaturation at air-liquid interfaces.
18. We use sufficient TN150 to completely fill the 50 ml conical tubes that contain the Ni-IDA beads.
19. For the Ni-IDA bead volumes we typically use, a C 10/10 column (Cytiva) or equivalent as appropriate.

20. We use a flow rate of 2 ml/min, when using a C 10/10 column.
21. We obtain 10–15 mg purified protein per l culture. This may be concentrated to >30 mg/ml.
22. Always use a dedicated, well-maintained, seasoned column with a spotless record for SEC-MALS. Very old columns as well as relatively new ones shed resin, which will create scatter artifacts (spikes) during data acquisition. As a rule, do not use columns that have been used previously for protein purification from cell lysates. Periodic backpressure fluctuations in the FPLC system (pumps + column) will be reflected as minor periodic deviations in the LS signal (Fig. 1b), as well as in the molar mass profile (Fig. 1c). Therefore, a well-functioning FPLC system and column are essential for obtaining noise-free, clean-looking data.
23. We recommend cleaning the MALS flow cell for 30 min either immediately before or after an SEC-MALS experiment using the COMET feature, which uses gentle ultrasonic vibrations to remove any biological matter that may remain adhered to the flow cell wall and may generate background (please also refer to the ASTRA user guide from Wyatt Technology for details).
24. Accurate dRI measurements require an exact buffer match between the reference and sample flow cells.
25. UV as a concentration source necessitates that the molar extinction (absorptivity) coefficient of the protein is accurately determined. Moreover, many proteins do not contain aromatic residues (especially tryptophan) that contribute to absorbance at 280 nm. The response factor (AU/V) and the cell path length (L) of the detector must also be known. On the other hand, the specific refractive index increment ($\frac{dn}{dC}$) for proteins in aqueous solutions is a constant (0.1850 ml/g). dRI, therefore, is preferred as a concentration source.
26. Monomeric BSA is typically injected at 2 mg/ml during instrument calibration. We recommend running BSA (~66 kDa monomer) as a validation standard during a sequence of sample runs to ascertain molar mass results. Note that the "Methods" file (calibration) from one instrument cannot be used to analyze data from a different instrument as this is instrument-specific. The "Methods" file needs to be replaced if changes are made to the hardware setup (e.g., interdetector tubing length).

Acknowledgments

We thank Bryan A. Tornabene for assistance with protein purification and Kayla K. Troutman for comments on the manuscript. This work was supported by the National Institutes of Health grants GM139546 (M. G. J. Ford) and GM121583 (R. Ramachandran).

References

1. Goodsell DS, Olson AJ (2000) Structural symmetry and protein function. Annu Rev Biophys Biomol Struct 29:105–153. https://doi.org/10.1146/annurev.biophys.29.1.105
2. Matthews JM, Sunde M (2012) Dimers, oligomers, everywhere. Adv Exp Med Biol 747: 1–18. https://doi.org/10.1007/978-1-4614-3229-6_1
3. Lorenzen K, van Duijn E (2010) Native mass spectrometry as a tool in structural biology. Curr Protoc Protein Sci. Chapter 17: Unit17.12. https://doi.org/10.1002/0471140864.ps1712s62
4. Leney AC, Heck AJR (2017) Native mass spectrometry: what is in the name? J Am Soc Mass Spectrom 28(1):5–13. https://doi.org/10.1021/jasms.8b05378
5. Cole JL, Lary JW, Moody TP, Laue TM (2008) Analytical ultracentrifugation: sedimentation velocity and sedimentation equilibrium. Methods Cell Biol 84:143–179. https://doi.org/10.1016/S0091-679X(07)84006-4
6. Wyatt PJ (1993) Light scattering and the absolute characterization of macromolecules. Anal Chim Acta 272(1):1–40. https://doi.org/10.1016/0003-2670(93)80373-S
7. Ford MGJ, Chappie JS (2019) The structural biology of the dynamin-related proteins: New insights into a diverse, multitalented family. Traffic 20(10):717–740. https://doi.org/10.1111/tra.12676
8. Mehrotra N, Nichols J, Ramachandran R (2014) Alternate pleckstrin homology domain orientations regulate dynamin-catalyzed membrane fission. Mol Biol Cell 25(6):879–890. https://doi.org/10.1091/mbc.E13-09-0548
9. Macdonald PJ, Stepanyants N, Mehrotra N, Mears JA, Qi X, Sesaki H, Ramachandran R (2014) A dimeric equilibrium intermediate nucleates Drp1 reassembly on mitochondrial membranes for fission. Mol Biol Cell 25(12): 1905–1915. https://doi.org/10.1091/mbc.E14-02-0728
10. Macdonald PJ, Francy CA, Stepanyants N, Lehman L, Baglio A, Mears JA, Qi X, Ramachandran R (2016) Distinct splice variants of dynamin-related protein 1 differentially utilize mitochondrial fission factor as an effector of cooperative GTPase Activity. J Biol Chem 291(1):493–507. https://doi.org/10.1074/jbc.M115.680181
11. Frohlich C, Grabiger S, Schwefel D, Faelber K, Rosenbaum E, Mears J, Rocks O, Daumke O (2013) Structural insights into oligomerization and mitochondrial remodelling of dynamin 1-like protein. EMBO J 32(9):1280–1292. https://doi.org/10.1038/emboj.2013.74
12. Gao S, von der Malsburg A, Dick A, Faelber K, Schroder GF, Haller O, Kochs G, Daumke O (2011) Structure of myxovirus resistance protein a reveals intra- and intermolecular domain interactions required for the antiviral function. Immunity 35(4):514–525. https://doi.org/10.1016/j.immuni.2011.07.012
13. Varlakhanova NV, Alvarez FJD, Brady TM, Tornabene BA, Hosford CJ, Chappie JS, Zhang P, Ford MGJ (2018) Structures of the fungal dynamin-related protein Vps1 reveal a unique, open helical architecture. J Cell Biol 217(10):3608–3624. https://doi.org/10.1083/jcb.201712021
14. Tornabene BA, Varlakhanova NV, Hosford CJ, Chappie JS, Ford MGJ (2020) Structural and functional characterization of the dominant negative P-loop lysine mutation in the dynamin superfamily protein Vps1. Protein Sci 29(6): 1416–1428. https://doi.org/10.1002/pro.3830
15. Cao YL, Meng S, Chen Y, Feng JX, Gu DD, Yu B, Li YJ, Yang JY, Liao S, Chan DC, Gao S (2017) MFN1 structures reveal nucleotide-triggered dimerization critical for mitochondrial fusion. Nature 542(7641):372–376. https://doi.org/10.1038/nature21077
16. Yan L, Qi Y, Huang X, Yu C, Lan L, Guo X, Rao Z, Hu J, Lou Z (2018) Structural basis for GTP hydrolysis and conformational change of MFN1 in mediating membrane fusion. Nat Struct Mol Biol 25(3):233–243. https://doi.org/10.1038/s41594-018-0034-8
17. Byrnes LJ, Sondermann H (2011) Structural basis for the nucleotide-dependent dimerization of the large G protein atlastin-1/SPG3A.

Proc Natl Acad Sci U S A 108(6):2216–2221. https://doi.org/10.1073/pnas.1012792108

18. Byrnes LJ, Singh A, Szeto K, Benvin NM, O'Donnell JP, Zipfel WR, Sondermann H (2013) Structural basis for conformational switching and GTP loading of the large G protein atlastin. EMBO J 32(3):369–384. https://doi.org/10.1038/emboj.2012.353
19. Sun D, Varlakhanova NV, Tornabene BA, Ramachandran R, Zhang P, Ford MGJ (2020) The cryo-EM structure of the SNX-BAR Mvp1 tetramer. Nat Commun 11(1):1506. https://doi.org/10.1038/s41467-020-15110-5
20. Carman PJ, Dominguez R (2018) BAR domain proteins-a linkage between cellular membranes, signaling pathways, and the actin cytoskeleton. Biophys Rev 10(6):1587–1604. https://doi.org/10.1007/s12551-018-0467-7
21. Cullen PJ (2008) Endosomal sorting and signalling: an emerging role for sorting nexins. Nat Rev Mol Cell Biol 9(7):574–582. https://doi.org/10.1038/nrm2427
22. Ma M, Burd CG (2020) Retrograde trafficking and plasma membrane recycling pathways of the budding yeast Saccharomyces cerevisiae. Traffic 21(1):45–59. https://doi.org/10.1111/tra.12693
23. van Weering JR, Verkade P, Cullen PJ (2010) SNX-BAR proteins in phosphoinositide-mediated, tubular-based endosomal sorting. Semin Cell Dev Biol 21(4):371–380. https://doi.org/10.1016/j.semcdb.2009.11.009
24. Wang Q, Kaan HY, Hooda RN, Goh SL, Sondermann H (2008) Structure and plasticity of Endophilin and Sorting Nexin 9. Structure 16(10):1574–1587. https://doi.org/10.1016/j.str.2008.07.016
25. Pylypenko O, Lundmark R, Rasmuson E, Carlsson SR, Rak A (2007) The PX-BAR membrane-remodeling unit of sorting nexin 9. EMBO J 26(22):4788–4800. https://doi.org/10.1038/sj.emboj.7601889
26. Lo WT, Vujicic Zagar A, Gerth F, Lehmann M, Puchkov D, Krylova O, Freund C, Scapozza L, Vadas O, Haucke V (2017) A coincidence detection mechanism controls px-bar domain-mediated endocytic membrane remodeling via an allosteric structural switch. Dev Cell 43(4): 522–529.e524. https://doi.org/10.1016/j.devcel.2017.10.019
27. Stetefeld J, McKenna SA, Patel TR (2016) Dynamic light scattering: a practical guide and applications in biomedical sciences. Biophys Rev 8(4):409–427. https://doi.org/10.1007/s12551-016-0218-6
28. Øgendal L (2019) Light scattering demystified. Theory and practice
29. Fleming PJ, Fleming KG (2018) HullRad: fast calculations of folded and disordered protein and nucleic acid hydrodynamic properties. Biophys J 114(4):856–869. https://doi.org/10.1016/j.bpj.2018.01.002

Chapter 19

Measurement of Lysosome Positioning by Shell Analysis and Line Scan

Chad D. Williamson, Carlos M. Guardia, Raffaella De Pace, Juan S. Bonifacino, and Amra Saric

Abstract

Lysosomes are membrane-bound organelles that degrade diverse biomolecules and regulate a multitude of other essential processes including cell growth and metabolism, signaling, plasma membrane repair and infection. Such diverse functions of lysosomes are highly coordinated in space and time and are therefore tightly coupled to the directional transport of the organelles within the cytoplasm. Thus, robust quantitative assessments of lysosome positioning within the cell provide a valuable tool for researchers interested in understanding these multifunctional organelles. Here, we present point-by-point methodology to measure lysosome positioning by two straight forward and widely used techniques: shell analysis and line scan.

Key words Organelle positioning, ImageJ analysis, Shell analysis, Line scan, Lysosome positioning measurements, Lysosome distribution, Lysosome transport

1 Introduction

Lysosomes are dynamic organelles that catabolize macromolecules delivered to them by endocytosis, phagocytosis, autophagy, and the biosynthetic pathway. They are central to such diverse processes as plasma membrane repair [1], pathogen killing [2], antigen presentation [2, 3], cell migration [4, 5], and autophagy [6]. Lysosomal functions are influenced by their retrograde and anterograde transport along microtubules by the motor proteins dynein and kinesin, respectively [3, 7–11] and perturbations in their transport and positioning are linked to numerous diseases including neurodegeneration and cancer [12, 13]. Thus, the ability to measure lysosome

Supplementary Information: The online version of this chapter (https://doi.org/10.1007/978-1-0716-2209-4_19) contains supplementary material, which is available to authorized users.

Jingshi Shen (ed.), *Membrane Trafficking: Methods and Protocols*,
Methods in Molecular Biology, vol. 2473, https://doi.org/10.1007/978-1-0716-2209-4_19,

positioning is a valuable approach if we are to understand the complete physiological (and pathophysiological) functions of these essential organelles.

Lysosome positioning has been measured using a variety of techniques [4, 14–23], typically applied to microscopy images of adherent cells in which lysosomes and other cellular structures (such as the nucleus or plasma membrane) have been fluorescently labeled. Some analyses include radial profile plots [22, 24, 25], radial kymographs [26, 27] and gross automated analyses suitable for high-content screening [28]. Two widely used and conceptually simple methods are shell analysis and line scan.

Shell analysis and line scan, like other described methods, are applied to 2D images of cells in which the xy plane is quantified, precluding analysis of the *z*-axis due to resolution limits. The shell analysis entails applying a series of concentric regions of interest (ROIs) to a cell, which may be circular or in the shape of the cell [14, 20, 29, 30], to analyze the total lysosomal population. Caveats to this analysis include cell shape restrictions that limit is application to cells that spread in an equibiaxial manner with a roughly centered nucleus. Alternatively, the line scan is based on signal intensity measurements along a straight line from the nucleus to a cell vertex and its speed and simplicity are its major advantages. Limitations behind this approach are its one-dimensional sampling of the lysosomal signal and that the axes are determined by the user. The researcher is encouraged to weigh the strengths and caveats of each method against their own image data to determine which method should be applied. Here, we describe how to conduct lysosome positioning measurements in ImageJ using shell analysis and line scan and we provide macroinstructions or "macros" for semiautomation and batch-processing.

2 Materials

1. ImageJ (https://imagej.nih.gov/ij/download.html) or Fiji (https://imagej.net/Fiji/Downloads), a version of ImageJ that bundles numerous plugins.
2. Microsoft Excel or other software for graphing and statistical analyses (i.e., Prism, R, or Python).
3. Micrographs with lysosomes, nuclei and cell edges labeled. Here, we utilize maximum intensity projections of confocal micrograph z-stacks acquired at 0.33 μm increments, typically collecting the entire cell volume in 10–13 z-slices.

3 Methods

Prepare cells for lysosome positioning analysis by labeling lysosomes, cell edges and the nucleus. Here, HeLa cells (Figs. 1, 2, and 3) were cultured on #1.5 German cover glass, transfected with plasmid vectors expressing GFP, the dynein adaptor GFP-RILP or the kinesin GFP-KIF1A, fixed in 4% PFA after 24 h, immunostained for LAMP1 and mounted on glass slides with DAPI-containing mounting medium. However, we note that the methods described here can be applied to images of either fixed or live cells labeled for the aforementioned cellular features in a variety of possible ways (*see* **Note 1**).

These methods can be applied to epifluorescence images or confocal images of either single focal planes or multiple *z*-slices which will be flattened for the analysis. Image acquisition settings should be optimized by the user depending on their microscope (i.e., saturated pixels should be avoided) and 16-bit images should be acquired, which have a substantially higher dynamic range in pixel values than 8-bit. For example, the micrographs in Figs. 1, 2, and 3 are flattened confocal *z*-stacks acquired on a Zeiss LSM880 microscope using a Plan-Apochromat ×63 1.4 NA objective lens with 0.33 μm spacing between *z*-slices (optimal for this particular system). If acquiring *z*-stacks as we have, the user should know the optimal *z*-step distance to set in order to prevent over- or undersampling. Ensure that acquisition settings are kept identical for an entire experimental cohort. Multicell analysis will depend on micrographs having matching pixel scaling and comparable lysosome marker intensities. Importantly, because these methods are applied to 2D images of cells, they are most suitable for the analysis of cultured adherent cells that are flat and wide, rather than tissue samples, organoid cultures, or columnar shaped cells that can have substantially higher *z* depths.

3.1 Shell Analysis

This analysis describes the generation of an ROI in the shape of a cell and iterative shrinking of that ROI by a fixed gap width multiple times, to create a series of shells within the cell. Since the image stack to be analyzed is first projected in the *z* axis, lysosomes throughout the entire cell volume are meant to be accounted for. Thus, the outermost shell is the extreme periphery, while the innermost shell is the center and usually includes all or part of the area occupied by the nucleus. Then, the lysosomal area or intensity is measured within each shell and can be represented as a fraction of the total. Since the intensity of the lysosome signal may vary, the shell analysis method described here uses the area occupied by fluorescent lysosomes, as opposed to averaged or summed fluorescence intensity.

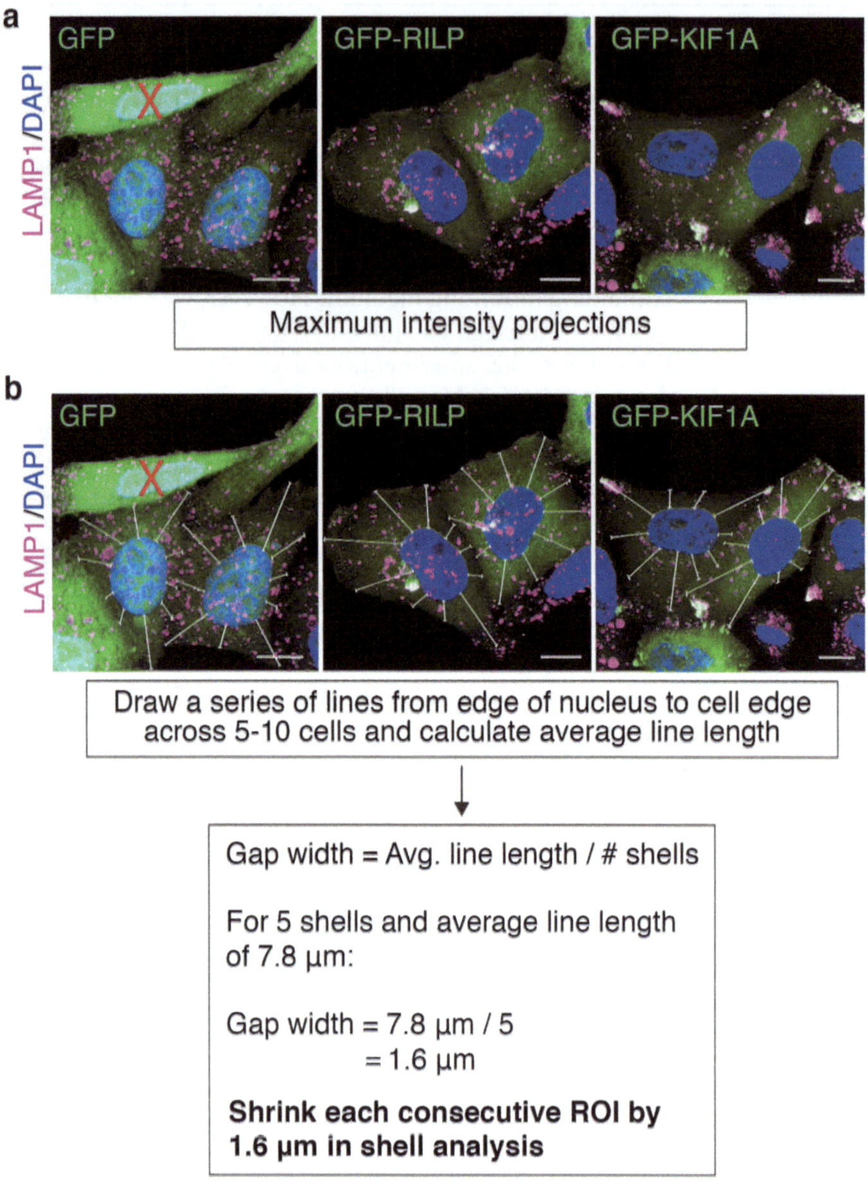

Fig. 1 Determining the shell gap width for shell analysis. (**a**) Maximum intensity projections of z-stack confocal fluorescence micrographs of HeLa cells transfected with plasmid vectors expressing GFP, GFP-RILP (a lysosomal dynein adaptor that causes lysosomal clustering in the juxtanuclear area) or GFP-KIF1A (a kinesin that moves lysosomes towards the cell periphery). The cells were fixed and immunostained for the lysosomal marker LAMP1 and nuclei were labeled with DAPI. (**b**) A series of manually drawn lines extending from the edge of the nuclei to cell edges were applied to the images in **a**. The average line length was determined to be 7.8 μm and used to compute the gap width of 1.6 μm for downstream shell analysis (bottom). Red "X" in **a** and **b** indicates a narrow cell that cannot be accurately measured by shell analysis and is excluded. Scale bars = 10 μm

Determining the gap width between consecutive shells and the number of shells analyzed will depend on cell size and desired granularity of measurements: The available cytosolic space between the nucleus and cell edges should be used as a determining factor. In order to set this value, it is recommended that the user first conduct a short analysis to determine the gap width (Subheading 3.1.1), then apply this value to all cells using the shell analysis (Subheading 3.1.2) (*see* **Note 2**).

3.1.1 Determining the Shell Gap Width

1. Conduct labeling of cells as described (*see* **Note 1**) and acquire 16-bit, *z*-stack, confocal fluorescence micrographs of cells or epifluorescence images of cells that meet the predefined criteria for shell analysis (*see* **Note 3**) (Fig. 1a, "X" marks a cell that cannot be measured accurately by shell analysis, given its elongated shape, and is therefore excluded). Save the files with embedded metadata.
2. Open an image file in ImageJ. If it is a *z*-stack file, generate a maximum intensity projection (Fig. 1a). Since the image file should have multiple channels (for lysosomes, nuclei, etc.), go to Image>Color>Make Composite and using the line selection tool draw several lines connecting the edge of the nucleus to the edge of the cell, per cell, for those cells that meet predefined criteria. Repeat this action across 5–10 cells and calculate the average line length (Fig. 1b) (*see* **Note 4**).
3. Divide this average value by the number of desired shells, a feature that is user-defined. For the purposes of this method, we describe a 5-shell analysis. Thus, for an average line length of 7.8 μm, the gap width between consecutive shells in the analysis should be set to 1.6 μm (7.8 μm/5 = 1.6 μm) (Fig. 1b).

3.1.2 Shell Analysis

1. Conduct **step 1** from Subheading 3.1.1.
2. Open an image file in ImageJ. If it is a *z*-stack file, generate a maximum intensity projection (*see* **Note 5**). Open the ROI Manager tool in Analyze>Tools>ROI Manager (*see* also Subheading 3.3.2 on using the ROI Manager). Go to Image>Color>Make Composite and using the freehand selection tool trace the edges of a cell and add this ROI to the ROI Manager (Fig. 2a).
3. Repeat this process for all cells in the field that meet the predefined criteria for shell analysis (Fig. 2a) (*see* **Note 6**).
4. Separate the individual channels by selecting Image>Color>Split Channels and close all except the lysosomal channel.
5. Set a threshold to eliminate background in Image>Adjust>Threshold (Fig. 2b) (*see* **Note 7**).

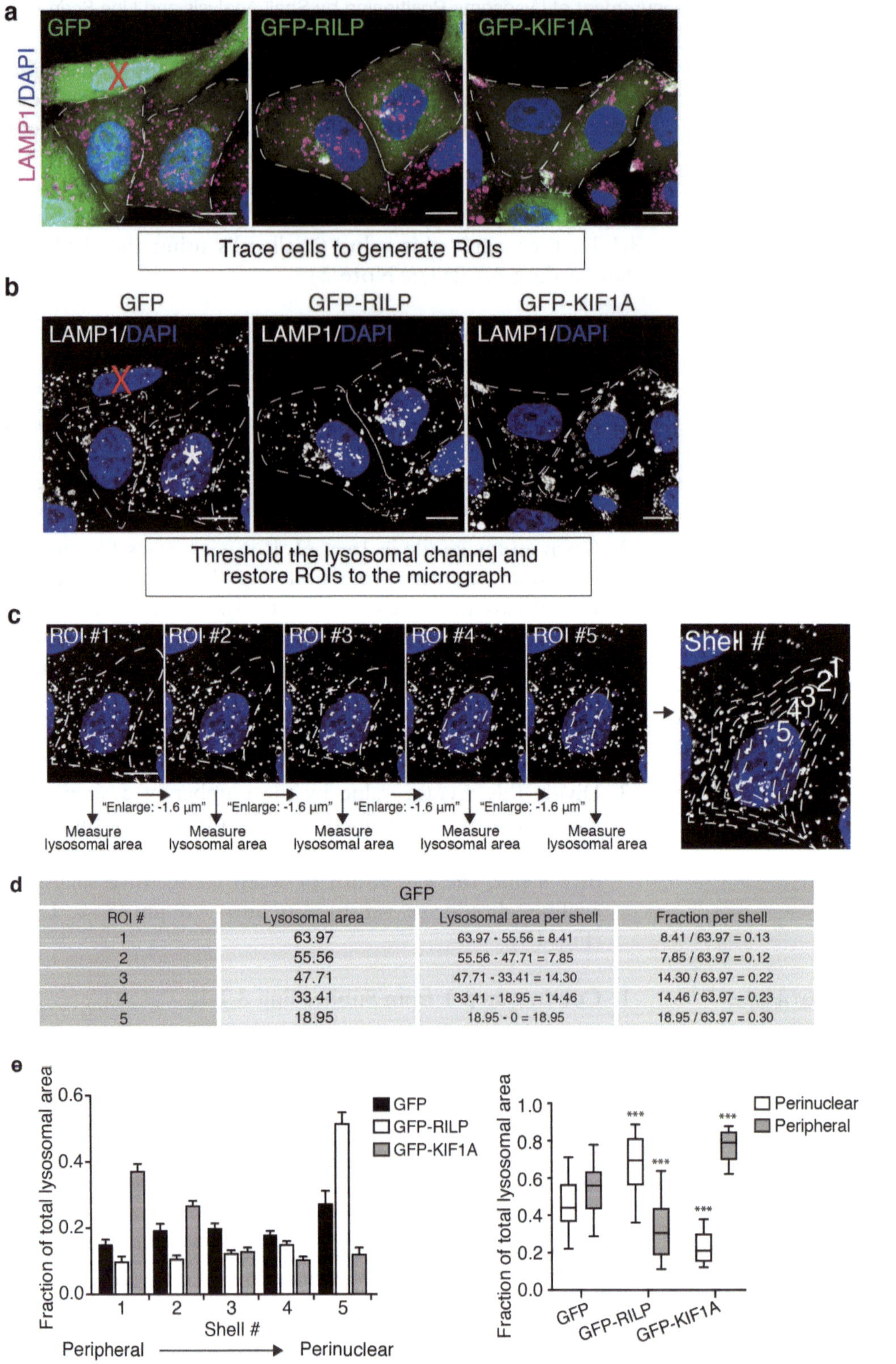

Fig. 2 Shell analysis workflow. (**a**) Maximum intensity projections of z-stack confocal fluorescence micrographs of HeLa cells from Fig. 1a, with manually traced cell edges (ROIs) in white dashed lines. (**b**) Cells and their ROIs from **a** showing only the lysosomal channel, to which a threshold has been applied to eliminate

6. Under Analyze>Set Measurements, select Area and Limit to Threshold.
7. From the ROI Manager, select the first ROI and select Analyze>Measure. In ImageJ, a new Results window will open that adds data to it each time a measurement is made. Keep this window open.
8. Under Edit>Selection, open the Enlarge option and input "-1.6" in order to shrink the ROI by the predetermined gap width of 1.6 μm. When applied, the ROI will shrink. Select Analyze>Measure (Fig. 2c).
9. Repeat **step 8** until a total of 5 measurements have been made for that ROI (Fig. 2c).
10. Repeat **steps 7–9** for each of the remaining ROIs in the ROI Manager (*see* **Note 8**).
11. Clear the ROI manager.
12. Repeat **steps 1–11** for each image file in a given condition (i.e., control).
13. When finished with one condition, copy the data from the Results window and paste it into a spreadsheet program such as Microsoft Excel.
14. Close the Results window in ImageJ. Results do not have to be saved in ImageJ as the Excel spreadsheet containing all data can be saved.
15. Repeat **steps 1–14** for the next condition (i.e., experimental treatment) and continue until all conditions are measured.
16. Within Excel, annotate the column headings (Fig. 2d) (*see* **Note 9**).
17. To determine the amount of lysosomal signal within each individual shell, subtract the lysosomal Area of each

Fig. 2 (continued) background. Asterisk denotes a cell that is used as an example in **c**. Red "X" in **a** and **b** indicates a narrow cell that cannot be accurately measured by shell analysis and is excluded. (**c**) Example cell showing iterative concentric shrinking of an ROI and measurement of lysosomal area within each ROI. To shrink each ROI by the gap width of 1.6 μm determined in Fig. 1, the ImageJ command "Enlarge" is used to enter "-1.6 μm". Lysosomal area is measured, and the steps are repeated 5 times to generate measurements of the area covered by lysosomal signal within 5 shells (right). (**d**) Example quantification of the area of lysosomal signal within each ROI such as in **c**, to determine the lysosomal area per shell and the fraction of total lysosomal area per shell. For the purposes of shell analysis, Area units can be arbitrary as long as all images are acquired with the same scaling. Note that due to the embedded metadata in our images, the Area unit here is square microns. In the absence of metadata, ImageJ reports square pixels for Area. (**e**) Data as in **d** collected across 14 cells per condition for each of the three conditions indicated, are plotted in 2 different ways. The graph on the left shows data for each shell as means ± SEM. The box and whiskers plot on the right shows a line at the median and the minimum and maximum values. One-way ANOVA with Dunnett's multiple comparisons test was applied to determine significance, $^{***}p < 0.001$. Scale bars = 10 μm

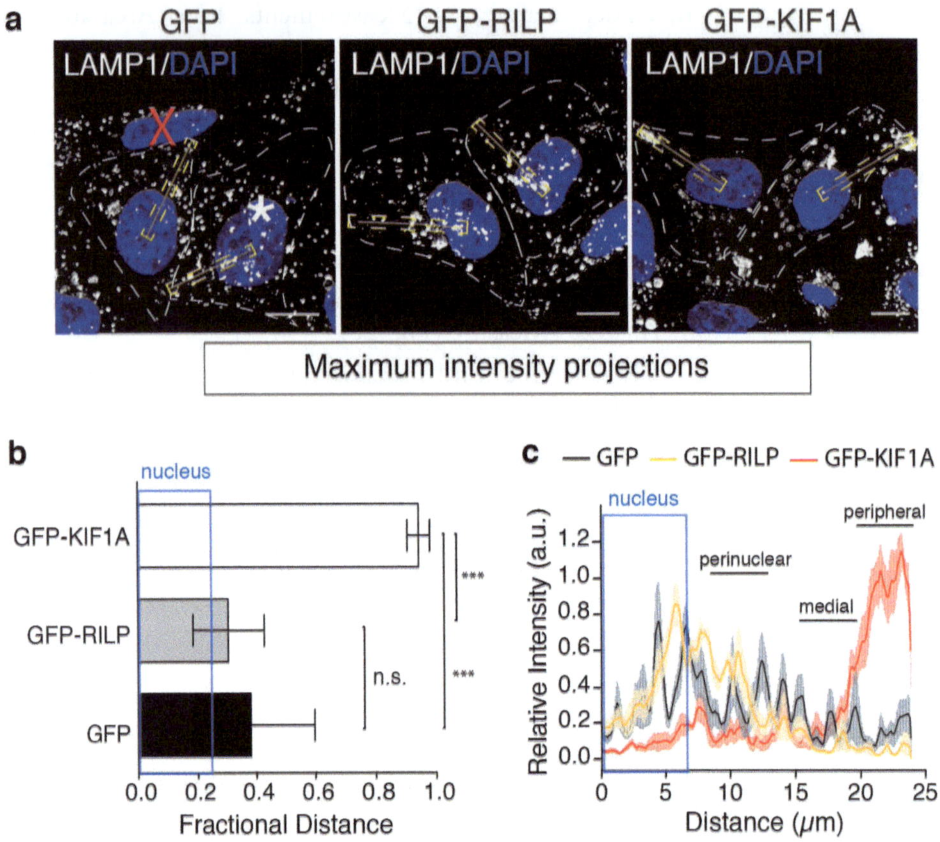

Fig. 3 Line scan analysis. (**a**) Maximum intensity projections of *z*-stack confocal fluorescence micrographs of HeLa cells from Fig. 1a, with manually traced cell edges (ROIs) in white dashed lines. Line segments (yellow lines), 30 pixels wide (yellow dashed lines), were drawn starting at the nucleus and ending at the cell periphery. Scale bars = 10 μm. (**b**) Positional information corresponding to the maximum intensity of the lysosome (LAMP1) channel was recorded as a fractional distance along the line segment (distance reading at the intensity maximum divided by total line segment distance) for 14 cells per condition. The mean fractional distance ± SD of lysosomes is plotted for each condition. Statistical significance was calculated using One-way ANOVA with Tukey's multiple comparisons test ($n = 14$, *n.s.* means not significant, ***$p < 0.001$). Blue box highlights average position along the line scan occupied by the nucleus. (**c**) Average lysosome position plotted as line graphs. Relative intensity (lysosomal intensity at each pixel divided by the sum of all intensity values for the line segment) along 24 ± 0.2 μm line segments from six cells per condition were recorded. The mean relative intensity (lines) ± SEM (vertical bars) of lysosome signal was plotted against the line distance (in μm). Blue box highlights average position along the line scan occupied by the nucleus

consecutive ROI from the ROI before it. For example, the Area of ROI#2 is subtracted from ROI#1, resulting in the lysosomal Area within the outermost Shell #1 (Fig. 2c, d, right).

18. To determine the fraction of total lysosomes within each shell, divide the Area per shell by the total lysosomal Area (ROI #1) (Fig. 2d).
19. There are multiple ways to represent data from the shell analysis and two examples are provided (Fig. 2e) (*see* **Note 10**).

3.2 Line Scan Analysis

A line scan is one of the most straightforward types of lysosome positioning measurements and the approach is easy to perform for users less familiar with ImageJ. It is accomplished by drawing a line segment from the nucleus extending to the cell border and using the built-in Plot Profile ImageJ analysis tool to output fluorescent signal intensities at each pixel along the length of the line. This generates an intuitive graphic with lysosome clustering appearing as signal peaks at measurable distances from each boundary. One drawback of this type of analysis is that a single line scan significantly under-samples the total lysosomal population, which can lead to high cell-to-cell or even line-to-line variability in the quantification.

1. Conduct labeling of cells with appropriate markers for lysosome positioning analysis (*see* **Note 1**).
2. Acquire 16-bit, *z*-stack confocal fluorescence micrographs, single focal planes or epifluorescence images of cells (*see* **Note 11**) that meet the predefined line scan criteria (*see* **Note 12**) with matched settings, and save files with embedded metadata (Fig. 3a, "X" marks a cell that varies considerably in shape from the majority of cells and is therefore excluded).
3. Open an image file in ImageJ. If it is a multichannel *z*-stack file, generate a maximum intensity projection.
4. Open the ROI Manager tool by selecting Analyze>Tools>ROI Manager. In "Composite" view, using the line selection tool, draw one line segment per cell (or more, if necessary; *see* **Note 13**), and add line segments to the ROI Manager by clicking Add. It is imperative that the directionality of the line segments drawn remain consistent during an analysis (i.e., draw all line segments starting at the nucleus and ending at the cell periphery).
5. Move the channel slider within the image window to the channel corresponding to the lysosomal marker. Highlight a single ROI by clicking on it in the ROI Manager and select Analyze>Plot Profile. Lysosome signal intensity (*y*-axis) is reported at each pixel along the length of the line segment (*x*-axis, distance). In this way one can analyze each line selection, one at a time (*see* **Note 14**).
6. Record the distance (*x*-axis) at which the lysosomal fluorescence intensity is at its highest value, as well as the total line segment length. Note that the entire raw data output can be recorded by selecting Data>Copy all Data in the Profiles window and pasting it into an Excel file (*see* **Note 15**).
7. To measure multiple line segments at once within a single image, click Show All within the ROI Manager to select all lines, click More then choose MultiPlot in the ROI manager.

8. Divide the distance at maximum fluorescence intensity by the total line segment distance to get a fractional distance for each measurement. The average fractional distances can be plotted as a bar graph and statistical analyses applied. As an example, Fig. 3b shows such data collected from 14 cells per condition and analyzed by One-way ANOVA.
9. As an alternative visual, it is also possible to plot *xy* scatter graphs of lysosome position information for multiple cells, if desired (Fig. 3c) (*see* **Note 16**).

3.3 Accelerating Image Analysis for Complete Datasets

To help expedite either of the analyses described in Subheadings 3.1 and 3.2, we provide methods to automate all or parts of these analyses, according to user needs and levels of technical expertise (Fig. 4).

3.3.1 Using Shortcuts and Creating Custom Shortcuts

Most ImageJ functions can be performed using a keyboard shortcut (Fig. 4a). Some of these shortcuts are already defined and are listed to the right of the function name in the main dropdown menus (e.g., Image>Show Info can be performed by clicking CMD + I). For functions without a preset shortcut key, ImageJ allows users to create personalized shortcuts. This is an incredibly simple method to expedite analysis with routinely used functions. We recommend that users remember to set their custom shortcuts to the same keystrokes on all workstations to prevent confusion.

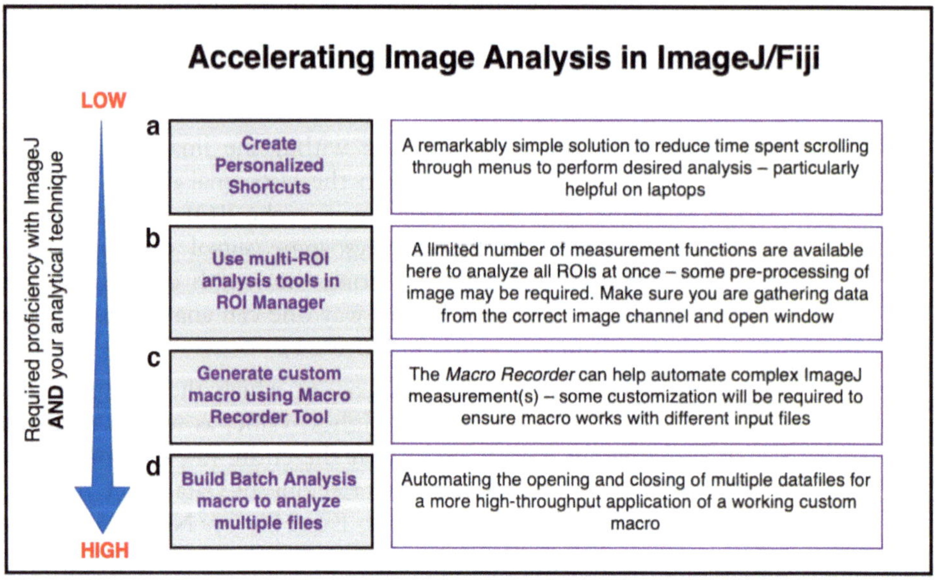

Fig. 4 Methods for automating image analysis in ImageJ/Fiji according to analysis complexity and user software proficiency

1. Both the shell and line scan analyses presented here require maximum intensity projections of confocal *z*-stack images. To set a custom shortcut for this function, go to Plugins>Shortcuts>Add Shortcut.
2. Two fields will appear, "Shortcut" and "Command". A limited number of letters, as well as F1 through F12, are available in the "Shortcut" field to be assigned a command. Choose one that is available.
3. In the "Command" field, choose the desired function name just as it appears in the main menu where you would normally go to select it (*see* **Note 17**). Click "OK".
4. Restart ImageJ to complete the shortcut key assignment.
5. If a letter was used in the "Shortcut" field, one can now call this function by pressing CMD + the assigned letter. If F1 through F12 were assigned, one can call these functions by pressing the appropriate F-key.
6. Repeat this procedure for other commands according to user needs. For example, the "Enlarge" command used in **step 8** of Subheading 3.1.2 could be given a custom shortcut. Also, note that Analyze>Plot Profile used in **step 5** of Subheading 3.2 already has a shortcut key, CMD + K.

3.3.2 Multi-ROI Measurements in the ROI Manager

When scaling up these procedures for many cells across various experimental conditions, much time can be invested in generating ROIs. The ROI Manager tool allows one to collect all ROIs into a compiled list. From the ROI Manager, one can save the list (to be recalled later in the event that an analysis must be repeated), or simply run the Measure or Plot Profile functions on all ROIs at once (Fig. 4b).

1. There are multiple ways to load cell outlines (shell analysis) or line segments (line scan) into the ROI Manager. One way to do this, presented in **step 2** of Subheading 3.1.2, is to open the ROI Manager first by selecting Analyze>Tools>ROI Manager..., and then clicking Add after each ROI is drawn (*see* **Note 18**).
2. After a list of ROIs has been generated, click More in the ROI Manager and then click Save. Label this file as "ROI_" with the exact name of the image file appended after the underscore and save it to the same folder where the raw image file is stored (*see* **Note 19**).
3. To measure all ROIs at once, move the channel slider in the image window to the desired channel (i.e., the lysosome signal). Make sure no individual ROIs are highlighted (toggle the Show All button on/off as described in **Note 18** if needed). Click the More button and then select either Multi Measure

(this would perform the Analyze>Measure function from **step 7** of Subheading 3.1.2 on all ROIs at once) or Multi Plot (this would perform the Analyze>Plot Profile function from **step 5** of Subheading 3.2). For Multi Measure, be sure to uncheck "measure all slices" and "one row per slice".

3.3.3 Using the Macro Recorder to Create a Custom Macro

ImageJ has a built-in feature that allows one to capture the command code for any process. Using this command recorder, the user can easily build custom macros to automate a single repetitive function, or series of functions, that a procedure requires. In practice, there are a few considerations to keep in mind when building a custom macro, which we explain here, to ensure that the macro will work in the automation procedures (Fig. 4c). As a reference for this section, and before considering how to create a custom macro, we begin by referring the reader to the supplemental "Shell analysis.ijm" file (Data 1) and describe it in **steps 1**–7. This is followed by instructions to build an individual macro (**steps 8–21**). In this section, as well as Subheading 3.3.4, we focus discussion on the Shell Analysis since this procedure is more complex than the line scan.

1. Open the supplemental "Shell analysis.ijm" file (Data 1) in ImageJ by dragging it into the ImageJ application (*see* **Note 20**).
2. The first command in this macro appears in line 6 and temporarily renames the file to "Image". This new file name will not be saved nor will any changes to the raw image file encoded within this macro. Raw data should always be stored as is. This renaming step should be used as the first command for any macro that is created.
3. The second command, line 9, is to create a maximum intensity projection of the confocal *z*-stack.
4. The max projection image is split into its individual acquisition channels (red, lysosome marker; green, transfected GFP construct; blue, nuclear stain). Notice the C3, C2, and C1 labeling in lines 16, 18, and 24, indicating that this macro expects to work on a three-channel image (*see* **Note 21**).
5. Line 27 sets an auto threshold for the channel containing the lysosome marker. This command marks the end of the image processing and is followed by image quantification steps.
6. Lines 32–48 define an iteration loop that allows one to make the same measurements for all ROIs loaded in the ROI Manager. Specifically, the sequence of Measure and Enlarge commands are performed, in order, from line 36 to 44 and then repeated for each ROI sequentially in the ROI Manager list (*see* **Note 22**).

7. For a custom macro that will work on multiple ROIs within a single open image, copy lines 32–35 (keeping the same indentation formatting) into the custom macro after any processing steps and replace lines 36–44 with the command or command sequence desired. Copy lines 47–52 (keeping the same indentation formatting) to ensure the ROI iteration loop ends correctly and the ROI list is cleared.
8. To build a macro, start with a well-defined procedure (*see* **Note 23**).
9. Open an image file in ImageJ and load corresponding ROI (s) into the ROI Manager tool.
10. Select Plugins>Macros>Record... to open the command recorder.
11. Select Image>Rename from the main menu and rename the image as "Image" (*see* **Note 24**).
12. Go through the image analysis procedure from start to finish for a single ROI manually, closing any extra windows that do not need to be open. If prompted, do not save any changes to your image file(s). Do not close any measurement windows that contain data.
13. When done, click Create in the command recorder. This will copy the analysis procedure that was performed in command code format for a custom macro. Notice the line numbers to the left of each line of code. Should a macro fail, an error message with a line number will be given, indicating where the macro encountered the error.
14. With the macro window selected (in the foreground), a new main menu will be visible. From that menu, select Language>IJ1 Macro.
15. If the macro only needs to work on a single ROI, proceed to **step 16**. If the macro will work on multiple ROIs, copy and paste the ROI iteration loop commands from the sample shell analysis macro (Shell analysis.ijm (Data 1), **steps 6** and 7).
16. Close all open images/windows except for the window containing the new macro, without saving changes, and clear the ROI Manager. Do not close the ImageJ program.
17. Reopen the raw image file that was used and reopen its ROI(s), which will automatically open the ROI Manager.
18. In the custom macro window, click Run. The image analysis will be automatically performed and should generate the same output as when the analysis was performed manually. If an error message appears, one should be able to identify where the macro failed.

19. If windows remain open after the macro runs, the macro can be modified to force these windows to close automatically. Open the command recorder, then close those remaining windows. Copy and paste the command code for closing the windows into the bottom of the macro.
20. Once the macro can successfully generate the correct output from the original file used to create the macro, close the image file and clear ROIs as in **step 16**. Open a different image file and its corresponding ROIs and run the macro.
21. If the macro was able to successfully generate measurements from the second file, save the macro by selecting File>Save. This should create a .ijm file of the macro.

3.3.4 Batch Processing

The macros discussed in Subheading 3.3.3 will automate image analysis on multiple ROIs for a given image file. This will still require that the user manually opens the image files and corresponding ROI files. Once a macro is working well, however, these last two steps may also be automated for a batch of files (Fig. 4d).

1. On the workstation, create a folder labeled "Batch_Analysis" and within that folder create two new folders, labeled "Images" and "ROIs".
2. Move (or copy) raw image files of one experimental set (for example, images from the GFP condition in the case of the shell analysis described in Subheading 3.1) into the "Images" folder, and the corresponding ROI lists into the "ROIs" folder.
3. Open the supplemental "Batch analysis.ijm" file (Data 2) in ImageJ. Notice that lines 1–24 are new code that did not appear in the "Shell analysis.ijm" file. These lines tell the macro to look in the folders that were just created. Each user will need to edit the path information (the pink text) in lines 6 and 7 to indicate where on their workstation these folders reside. One can use path bars at the bottom of the Finder window, or "view entire file path" menu commands to find the correct file path. Ensure that the final "/" after the folder name is included.
4. From the "Images" folder, which is given the designation "directory1" (line 6), the macro will get a list of the number of files contained in the folder (line10), and will open the files one at a time, sequentially, for processing. From the "ROIs" folder, which is given the designation "directory2" (line 7), ROI lists for each image are loaded into the ROI Manager. Files in both folders are matched by their listed order within the folder, not by comparing text in file names. It is imperative that the order of the image files and the ROI files match. If the user has adopted the naming procedure discussed in **step 2** of Subheading 3.3.2, the files should match.

5. Lines 19–23 set up the iteration loop to move through the files sequentially and match them to their ROI list set. This loop is ended by a closing bracket in line 88. Notice the indentation within the iteration loop, as has been previously pointed out, an important feature that ensures the code functions properly.
6. Once image files and ROI sets are in their correct folders, clicking Run on the batch analysis macro should output quantified shell data for all ROIs in all files (*see* **Note 25**).

4 Notes

1. If imaging live, one can use fluorescent dextran or acidophilic dyes (i.e., LysoTracker) to label lysosomes. Ectopic expression of fluorescent protein-tagged lysosomal markers is discouraged (unless it is the experimental condition), as high expression levels of some lysosomal proteins are known to alter their morphology and distribution. If the conditions being analyzed involve the expression of any fluorescent protein that has cytosolic background (such as the GFP signal in Figs. 1 and 2a), this is usually sufficient to see cell edges. Otherwise, one can additionally stain cortical actin with phalloidin at the end of an immunofluorescence protocol or use plasma membrane-labeling dyes like CellMask deep red for live cell imaging. Alternatively, a brightfield image may suffice to identify cell edges. While we used DAPI to visualize the nucleus, Hoechst may be used for live cells. The location of the nucleus can also be approximated by immunostaining for the centrosome at the MTOC using antibodies to γ-tubulin.
2. The procedure to set the gap width between consecutive shells needs only to be done once per experiment, prior to starting, and usually not for each condition within the experiment unless a condition (i.e., drug treatment) causes obvious changes in cell size. In that case, Subheading 3.1.1 should be conducted per condition. Subheading 3.1.1 should also be done for each cell type if comparing multiple different cell types to ensure that shells are appropriately distanced for each cell size.
3. Shell analysis can only be applied to cells that meet a set of predefined criteria. Cells must be of relatively uniform shape in *x* and *y* planes, with the nucleus as centered as possible for this analysis. Narrow, elongated cells cannot be accurately measured by this method as there is often too little room in the narrowest part of the cell for positioning measurements to be made. Furthermore, a highly skewed nucleus (and usually MTOC) in the corner of a cell will render the analysis inaccurate. As an example, the cell with an “X” in Fig. 1a, is not measured because it is too narrow and positional information

would be limited or impossible to ascertain. Of course, there will be cell-to-cell variability regarding the exact position of the nucleus relative to the cell edges. Such variations are expected to average out when many cells across multiple independent experiments are analyzed.

4. Theoretically, one can do these measurements from the center of the nucleus, or from the MTOC if the MTOC is labeled. However, we find that this approach usually results in an extremely small innermost shell, data from which is not very informative. We thus prefer to extend lines from the edge of the nucleus to the edge of the cell, a technique that usually results in the innermost shell capturing a region that contains all or the bulk of the space occupied by the nucleus.

5. Note that converting *z*-stacks into maximum intensity projections causes loss of *z*-axis data and is a caveat of this method. Lysosome positioning in the *z*-axis cannot be accurately determined for flat adherent cells in culture by most confocal systems due to substantially lower axial resolution (compared to the *xy*-plane) that can be achieved, and is not possible by epifluorescence.

6. One may wish to save the ROIs directly from the ROI Manager tool, to be used later. If automating this analysis using macros or batch processing as described in Subheading 3.3, ROIs must be saved.

7. Determining a threshold value is somewhat arbitrary and will depend on the specific experiment (i.e., signal specificity and intensity). One may opt to use the Auto Threshold setting within ImageJ. Alternatively, a threshold can be set at the top 1% of the signal. If it appears some lysosomes are excluded, change the threshold to include more data. The same thresholds for certain channels can then be applied to all images in that experiment.

8. Due to the repetitive nature of this process, one may wish to automate this part of the analysis by creating shortcuts or generating a macro, *see* Subheadings 3.3.1 and 3.3.3. Subheading 3.3.3 additionally provides a macro of the full shell analysis method (Data 1), though depending on user needs (Fig. 4), one may wish to only automate segments of the analysis, as described here.

9. For clarity, the user may wish to insert a blank row between every 5 rows of data, as every 5 rows correspond to shell analysis of a single cell. Note that copying and pasting these results from ImageJ to Excel will also provide the minimum and maximum threshold values, which are not needed for the analysis and can be ignored (for clarity, these values were excluded in Fig. 2d).

10. One may choose to graph the fraction of total lysosomal signal across all shells in order to visualize the dispersion of the signal in the different cellular regions (Fig. 2e, left). Alternatively, one may plot the "peripheral" fractions and "perinuclear" fractions, by choosing to graph, for example, the sum of Shells 1–3 and 4–5, respectively (Fig. 2e, right). The user should define and clearly state their definition of "peripheral" and "perinuclear" when presenting data this way. Alternatively, one may wish to show only a simple bar graph of the "peripheral" or "perinuclear" fractions per condition. Note that in these examples, multiple cells were analyzed, which allows for graphing mean values with standard deviation or standard error of the mean. One can additionally perform statistical analyses to determine significance.
11. It should be noted that although we describe the application of this method to flattened *z*-stack confocal micrographs, the line scan can be applied to single focal planes acquired by confocal microscopy. If a single focal plane is imaged, the user should attempt to collect a similar *z* position between all cells for consistency.
12. As with the shell analysis, one must apply a set of criteria when identifying cells for line scan analysis. These criteria are predefined and therefore must be applied to all conditions analyzed: Cells must display a similar morphology and area from the nucleus to at least one cell tip and the line segment should cross through an area where the bulk of lysosomes appear. In particular, avoid cells where the MTOC or perinuclear lysosomal population is positioned too close to a cell edge or otherwise cannot be sampled by a line segment drawn from the nucleus to a cell tip. Avoid cells that display a large density of lysosomes directly above the center of the nucleus in *z* as it is not possible to accurately determine if these lysosomes are perinuclear or peripheral from a *z*-projection. Note that although in Fig. 3a the cell with an "X" is not quantified due to its high variation in shape compared to other cells of this cell line (HeLa), other cell types that have uniform elongated morphologies (i.e., fibroblasts) are quantifiable by line scan. In this case the line can be drawn in similar orientations between cells, such as from the nucleus to the longest cell edge.
13. To increase the sampling area within the cell for line scan analysis, one can either increase the pixel width of the line segment by double clicking the line selection tool and entering an appropriate pixel value (line widths in Fig. 3a are set to 30 pixels and dashed yellow lines indicate these boundaries) or draw multiple line segments per cell. These criteria should be decided upon before running the image analysis and should be consistently used for all cells across all conditions. In the example in Fig. 3a, only one line per cell was drawn.

14. It is suggested that all line segments are saved by clicking More in the ROI Manager and then clicking Save. Label this file with "ROI_" with the exact name of the image file appended after the underscore. Save this file to the same folder where the raw image file is stored.
15. In the Plot Profile output window, hovering the cursor over any part of the graph will display x and y values for that position. In this way, one can find the distance corresponding to the lysosomal intensity maximum. Alternatively, once the entire dataset has been copied to Excel, one can use the Max() function to identify the maximum intensity (y-value) and find its corresponding distance (x-value). Optional: Move the channel slider in the image window to the channel corresponding to the nuclear marker and repeat the line scan Plot Profile. In this way, DAPI signal intensities may be used to indicate the average position of the edge of the nucleus, as exemplified in Fig. 3b, c).
16. If all line segment lengths match to within ±0.2 microns, then the Plot Profile "Distance" output (in microns) for the x-axis can be compared directly between all samples. Line scan measurements can simply be plotted as replicate y-values along a defined and matching x-axis. As an example, Fig. 3c shows the mean lysosome positions as lines ± SEM as vertical bars, from 6 cells per condition with matching line segment lengths. However, due to normal variability in cell shape and size, most cells cannot be analyzed with line segments of equal length. When line segment lengths between cells differ more than ±0.2 microns lysosome position information cannot be directly compared. In this case, distances can be normalized to the total line length for each cell, resulting in a fractional distance that ranges from zero (at the start of the line segment) to one (at the end of the line segment). XY scatter plots of multiple cells or conditions can also be generated from these normalized data.
17. Normally one would go to Image>Stacks>Z Project... to perform a maximum intensity projection. This means that "Z Project..." will be the command needed to set the shortcut. Other plugins, like "Image5D Viewer", also have Z Project commands that might come up in the command options list. Be sure to select "Z Project..." as the desired command to assign a shortcut key.
18. One can alternatively start drawing ROIs or line segments in an open image and hit CMD + B after drawing each. This tells the software to set that selection into memory and allows one to create a new ROI or line. This is the same as going to Image>Overlay>Add Selection. Once all ROIs have been drawn,

they can be loaded together into the ROI Manager by selecting Image>Overlay>To ROI Manager. Making a custom shortcut for the "To ROI Manager" command and using it with the preset Add Selection (CMD + B) shortcut can help streamline the process.

19. Be aware that if an individual ROI in the list is highlighted, only that single ROI will be saved (the file type will be .roi). Go back to the ROI Manager and toggle on/off the Show All button until no ROIs in the list are highlighted. Now when saved, the file type should be .zip (this is the correct file type). Not only does saving ROIs allow one to consider automating an analysis (described later), it also saves the time of having to redraw ROIs if the analysis needs to be repeated or different types of measurements made with the same ROIs in the future. To load saved ROIs, simply drag the .zip file into ImageJ as you would an image file, or open the ROI Manager tool by selecting Analyze>Tools>ROI Manager…, click More, select Open, and find the .zip file in the Finder window.
20. This action will open a sample shell analysis macro written to perform the shell analysis described in Subheading 3.1.2 on a single image file with multiple ROIs. Notice the row numbers on the left-hand side of the macro. When referring to a specific portion of the macro code, corresponding line numbers will be given for disambiguation. Green text that is preceded by "//" is explanatory text, not code. This text is included before a section of code to help explain what the code is doing or to emphasize a particular point about the code.
21. Windows for the individual C3 and C2 channels are immedi ately closed (lines 16 to 17 and 18 to 19), which means that the lysosomal marker must correspond to channel C1. This number is defined by the order of laser illumination in the microscope acquisition setup and could very well be different between users, even if the data all contain 3 channels. When an image file is opened, the channel labels will appear at the top left corner of the image. If one wishes to apply this macro as is, the acquisition settings on the user's microscope must be set so that the lysosomal signal corresponds to channel C1. Otherwise, this macro may be edited to change the channel numbers such that the nonlysosomal channels are closed (lines 16 and 18) and the lysosomal channel is kept open (line 24).
22. If one wishes to apply this macro for their own shell analysis, note that the numerical values that appear after "Enlarge= -" in lines 37, 39, 41, and 43 must be changed by the user to their predetermined gap width (Subheading 3.1.1). One can also modify the number of "Enlarge"/"Measure" command pairs to change the number of shells in the analysis.

23. Building a macro is a step to be taken only after a procedural workflow for an analysis has been established. The user must know what processing steps will need to be done to prepare images for quantification. The user should be comfortable with running the analysis manually and be familiar with any steps in the process where mistakes are likely to occur.
24. Many ImageJ commands must specify the image window to act on. They often do this by using the filename or some appended version of the filename. For a macro to work correctly on multiple files, all the files must have the same filename. Since this is not a useful method of storing raw data files, the simplest way to get around this requirement is to build a renaming step into the macro. This renaming step should always be the first step of a macro.
25. To help organize the batch analysis output, line 77 in the macro specifies that data from each image file will be separated into individual result tables named with a file identifier, the number of ROIs counted, and a reminder that this analysis creates five measurements per ROI. In this case, the file identifier is the number assignment indicating the order in which the file appeared in directory1 (starting from zero). An example table title from the first image file processed might be "Image#0 ROIs=5; 5 Measurements/ROI".

Acknowledgments

We thank Drs. Tal Keren-Kaplan and Spencer A. Freeman for critical review of the manuscript. This work was supported by the Intramural Program of NICHD, NIH (project # ZIA HD001607).

References

1. Encarnação M, Espada L, Escrevente C, Mateus D, Ramalho J, Michelet X, Santarino I, Hsu VW, Brenner MB, Barral DC, Vieira OV (2016) A Rab3a-dependent complex essential for lysosome positioning and plasma membrane repair. J Cell Biol 213(6):631–640. https://doi.org/10.1083/jcb.201511093
2. Garg S, Sharma M, Ung C, Tuli A, Barral DC, Hava DL, Veerapen N, Besra GS, Hacohen N, Brenner MB (2011) Lysosomal trafficking, antigen presentation, and microbial killing are controlled by the Arf-like GTPase Arl8b. Immunity 35(2):182–193. https://doi.org/10.1016/j.immuni.2011.06.009
3. Vyas JM, Kim YM, Artavanis-Tsakonas K, Love JC, Van der Veen AG, Ploegh HL (2007) Tubulation of class II MHC compartments is microtubule dependent and involves multiple endolysosomal membrane proteins in primary dendritic cells. J Immunol 178(11):7199–7210. https://doi.org/10.4049/jimmunol.178.11.7199
4. Pu J, Schindler C, Jia R, Jarnik M, Backlund P, Bonifacino JS (2015) BORC, a multisubunit complex that regulates lysosome positioning. Dev Cell 33(2):176–188. https://doi.org/10.1016/j.devcel.2015.02.011
5. Bretou M, Saez PJ, Sanseau D, Maurin M, Lankar D, Chabaud M, Spampanato C, Malbec O, Barbier L, Muallem S, Maiuri P, Ballabio A, Helft J, Piel M, Vargas P, Lennon-Dumenil A (2017) Lysosome signaling controls the migration of dendritic cells. Sci

Immuno 2(16):1–11. https://doi.org/10.1126/sciimmunol.aak9573

6. Eskelinen EL, Saftig P (2009) Autophagy: a lysosomal degradation pathway with a central role in health and disease. Biochim Biophys Acta 1793(4):664–673. https://doi.org/10.1016/j.bbamcr.2008.07.014
7. Pu J, Guardia CM, Keren-Kaplan T, Bonifacino JS (2016) Mechanisms and functions of lysosome positioning. J Cell Sci 129(23):4329–4339. https://doi.org/10.1242/jcs.196287
8. Bonifacino JS, Neefjes J (2017) Moving and positioning the endolysosomal system. Curr Opin Cell Biol 47:1–8. https://doi.org/10.1016/j.ceb.2017.01.008
9. Korolchuk VI, Saiki S, Lichtenberg M, Siddiqi FH, Roberts EA, Imarisio S, Jahreiss L, Sarkar S, Futter M, Menzies FM, O'Kane CJ, Deretic V, Rubinsztein DC (2011) Lysosomal positioning coordinates cellular nutrient responses. Nat Cell Biol 13(4):453–460. https://doi.org/10.1038/ncb2204
10. Chow A, Toomre D, Garrett W, Mellman I (2002) Dendritic cell maturation triggers retrograde MHC class II transport from lysosomes to the plasma membrane. Nature 418:988–994. https://doi.org/10.1038/nature01004
11. Tuli A, Thiery J, James AM, Michelet X, Sharma M, Garg S, Sanborn KB, Orange JS, Lieberman J, Brenner MB (2013) Arf-like GTPase Arl8b regulates lytic granule polarization and natural killer cell-mediated cytotoxicity. Mol Biol Cell 24(23):3721–3735. https://doi.org/10.1091/mbc.E13-05-0259
12. Kroemer G, Jäättelä M (2005) Lysosomes and autophagy in cell death control. Nat Rev Cancer 5(11):886–897. https://doi.org/10.1038/nrc1738
13. Ballabio A, Bonifacino JS (2020) Lysosomes as dynamic regulators of cell and organismal homeostasis. Nat Rev Mol Cell Biol 21(2):101–118. https://doi.org/10.1038/s41580-019-0185-4
14. Johnson DE, Ostrowski P, Jaumouille V, Grinstein S (2016) The position of lysosomes within the cell determines their luminal pH. J Cell Biol 212(6):677–692. https://doi.org/10.1083/jcb.201507112
15. Li X, Rydzewski N, Hider A, Zhang X, Yang J, Wang W, Gao Q, Cheng X, Xu H (2016) A molecular mechanism to regulate lysosome motility for lysosome positioning and tubulation. Nat Cell Biol 18(4):404–417. https://doi.org/10.1038/ncb3324
16. Caviston JP, Zajac AL, Tokito M, Holzbaur EL (2011) Huntingtin coordinates the dynein-mediated dynamic positioning of endosomes and lysosomes. Mol Biol Cell 22(4):478–492. https://doi.org/10.1091/mbc.E10-03-0233
17. Willett R, Martina JA, Zewe JP, Wills R, Hammond GRV, Puertollano R (2017) TFEB regulates lysosomal positioning by modulating TMEM55B expression and JIP4 recruitment to lysosomes. Nat Commun 8(1):1580. https://doi.org/10.1038/s41467-017-01871-z
18. Jongsma ML, Berlin I, Wijdeven RH, Janssen L, Janssen GM, Garstka MA, Janssen H, Mensink M, van Veelen PA, Spaapen RM, Neefjes J (2016) An ER-associated pathway defines endosomal architecture for controlled cargo transport. Cell 166(1):152–166. https://doi.org/10.1016/j.cell.2016.05.078
19. Rosa-Ferreira C, Munro S (2011) Arl8 and SKIP act together to link lysosomes to kinesin-1. Dev Cell 21(6):1171–1178. https://doi.org/10.1016/j.devcel.2011.10.007
20. Saric A, Hipolito VE, Kay JG, Canton J, Antonescu CN, Botelho RJ (2016) mTOR controls lysosome tubulation and antigen presentation in macrophages and dendritic cells. Mol Biol Cell 27(2):321–333. https://doi.org/10.1091/mbc.E15-05-0272
21. Pu J, Keren-Kaplan T, Bonifacino JS (2017) A ragulator-BORC interaction controls lysosome positioning in response to amino acid availability. J Cell Biol 216(12):4183–4197. https://doi.org/10.1083/jcb.201703094
22. Guardia CM, Farias GG, Jia R, Pu J, Bonifacino JS (2016) BORC functions upstream of kinesins 1 and 3 to coordinate regional movement of lysosomes along different microtubule tracks. Cell Rep 17(8):1950–1961. https://doi.org/10.1016/j.celrep.2016.10.062
23. Aittaleb M, Chen PJ, Akaaboune M (2015) Failure of lysosome clustering and positioning in the juxtanuclear region in cells deficient in rapsyn. J Cell Sci 128(20):3744–3756. https://doi.org/10.1242/jcs.172536
24. Sengupta D, Truschel S, Bachert C, Linstedt AD (2009) Organelle tethering by a homotypic PDZ interaction underlies formation of the Golgi membrane network. J Cell Biol 186(1):41–55. https://doi.org/10.1083/jcb.200902110
25. Li K, Yang L, Zhang C, Niu Y, Li W, Liu JJ (2014) HPS6 interacts with dynactin p150Glued to mediate retrograde trafficking and maturation of lysosomes. J Cell Sci 127

(Pt 21):4574–4588. https://doi.org/10.1242/jcs.141978

26. Kapitein LC, Schlager MA, van der Zwan WA, Wulf PS, Keijzer N, Hoogenraad CC (2010) Probing intracellular motor protein activity using an inducible cargo trafficking assay. Biophys J 99(7):2143–2152. https://doi.org/10.1016/j.bpj.2010.07.055
27. Guardia CM, De Pace R, Sen A, Saric A, Jarnik M, Kolin DA, Kunwar A, Bonifacino JS (2019) Reversible association with motor proteins (RAMP): a streptavidin-based method to manipulate organelle positioning. PLoS Biol 17(5):e3000279. https://doi.org/10.1371/journal.pbio.3000279
28. Circu ML, Dykes SS, Carroll J, Kelly K, Galiano F, Greer A, Cardelli J, El-Osta H (2016) A novel high content imaging-based screen identifies the anti-helminthic niclosamide as an inhibitor of lysosome anterograde trafficking and prostate cancer cell invasion. PLoS One 11(1):e0146931. https://doi.org/10.1371/journal.pone.0146931
29. Saric A, Freeman SA, Williamson CD, Jarnik M, Guardia CM, Fernandopulle MS, Gershlick DC, Bonifacino JS (2021) SNX19 restricts endolysosome motility through contacts with the endoplasmic reticulum. Nat Commun 12(1):4552. https://doi.org/10.1038/s41467-021-24709-1
30. Keren-Kaplan T, Saric A, Ghosh S, Williamson CD, Jia R, Li Y, Bonifacino JS (2022) RUFY3 and RUFY4 are ARL8 effectors that promote coupling of endolysosomes to dynein-dynactin. Nat Commun 13(1):1506. https://doi.org/10.1038/s41467-021-24709-1

Part III

Emerging Topics in Membrane Trafficking

Chapter 20

Optogenetic Control of Membrane Trafficking Using Light-Activated Reversible Inhibition by Assembly Trap of Intracellular Membranes (IM-LARIAT)

Bobae Hyeon, Mai Khanh Nguyen, and Won Do Heo

Abstract

Intracellular membrane trafficking is a dynamic and complex cellular process. To study membrane trafficking with a high spatiotemporal resolution, we present an optogenetic method based on a blue-light inducible oligomerization of Rab GTPases, termed light-activated reversible inhibition by assembly trap of intracellular membranes (IM-LARIAT). In this chapter, we focus on the optical disruption of the dynamics and functions of previously studied intracellular membrane trafficking events, including transferrin recycling and growth cone regulation in relation to specific Rab GTPases. To aid general application, we provide a detailed description of transfection, imaging with a confocal microscope, and analysis of data.

Key words IM-LARIAT, Intracellular trafficking, Optogenetics, CRY2, PHR, CIB1, Photoactivation, Confocal microscope, Rab GTPase, Vesicle fusion, Organelle transport

1 Introduction

Intracellular membrane trafficking is a highly dynamic and complex cellular process, which underlies various cellular properties such as polarity, motility, and signaling [1]. Thus, perturbation of the trafficking process is important to gain insights into the causal relationship between the diverse trafficking pathways and functions. In this context, genetic perturbations such as deletion of target genes and overexpression of mutant proteins have been widely used to study altered intracellular membrane trafficking and consequences. However, this approach has suffered from intrinsic impediments, including a slow experimental onset latency and side effects resulting from the irreversible modifications to the proteins involved [2]. Another commonly used approach is taking advantage of small molecules targeted to proteins involved in the process, yet their actions often involve off-target effects with poor reversibility [3].

Jingshi Shen (ed.), *Membrane Trafficking: Methods and Protocols*,
Methods in Molecular Biology, vol. 2473, https://doi.org/10.1007/978-1-0716-2209-4_20,

Over recent years, advancements in optogenetics have expanded the available intervention toolkits in the context of intracellular membrane trafficking, with high target-specificity, reversibility, and precise spatiotemporal regulation [1, 4, 5]. The toolkits are photosensitive genetically encoded protein systems, designed to induce oligomerization of target proteins involved in the trafficking process, including regulatory proteins, adaptor proteins, and motor proteins [4, 6, 7]. The development and application of these tools have enabled us to address many previously unanswerable questions about the dynamics of the trafficking process and functions [1, 4, 6, 7]. However, as a vast pool of such tools has emerged, it has become increasingly important to choose the tool that best suits one's needs.

Rab GTPases constitute over 60 known members in humans [8]. They are key regulators of diverse trafficking pathways and contribute to the maintenance of membrane physiology [8–10]. Although not all their identities are well-defined, they are broadly considered markers for specific organelles in a trafficking network [9]. Given these facts, the Rab GTPase based light-inducible oligomerization system, termed light-activated reversible inhibition by assembly trap of intracellular membranes (IM-LARIAT) [7], provides a versatile platform for revealing undefined identities of Rab GTPases in relation to the highly dissected steps of the intracellular membrane trafficking process.

The photosensitivity of IM-LARIAT relies on the oligomerization properties of cryptochrome 2 (CRY2) [11], which can undergo CRY2-CIB1 (CRY2 interacting basic-helix-loop-helix) hetero-dimerization [12] and CRY2-CRY2 homo-oligomerization [13]. The IM-LARIAT system uses the N-terminal photolyase homology region (PHR) of CRY2 and N-terminal region of CIB1 (CIBN) fused to Rab GTPases (herein CRY2 and CIB1, respectively) [7]. Upon blue-light illumination, CRY2-CIB1 and CRY2-CRY2 interactions occur and the targeted organelles are trapped into clusters within seconds; the clusters disperse to their original state within minutes of light withdrawal (Fig. 1) [7]. This optical perturbation module has been shown to effectively alter the positioning of organelles or the rate of trafficking [7].

Here, we describe the IM-LARIAT system to perturb two different trafficking events, exemplified by Rab5b, Rab7, and Rab11a targeted modules. This protocol is subdivided into two parts: (1) optogenetic control of transferrin recycling and (2) optogenetic control of growth cones on neurites.

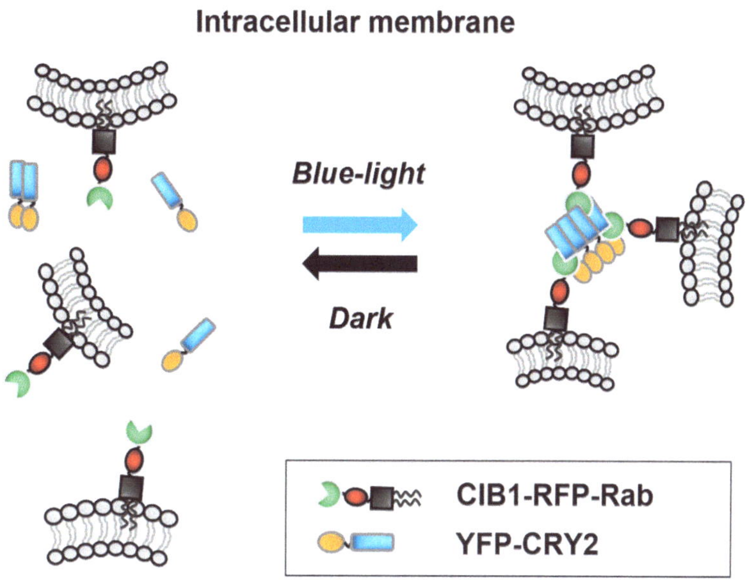

Fig. 1 Schematic of IM-LARIAT. Intracellular membrane aggregation is triggered by the light-induced hetero-dimerization between CRY2 and CIB1-Rab and homo-dimerization between CRY2-CRY2. They disperse back into the original state in minutes upon the withdrawal of blue-light. It is reproduced from [7] with permission from *Nature Chemical Biology*

2 Materials

2.1 Plasmids

1. YFP-CRY2 plasmid was constructed by cloning PCR-amplified CRY2 (amino acids 1–498) with mammalian codon optimization [11], flanked by AgeI/BsrGI sites, into the EGFP-C1 plasmid (Clontech). Original EGFP in the vector is replaced by mCitrine (YFP) at AgeI and NheI sites.
2. CIB1-RFP plasmid was constructed by cloning PCR-amplified sequences of CIB1 (amino acid 1–170, nuclear localization signal-deleted) [12] and FusionRed (RFP), flanked by NheI/AgeI and AgeI/BsrGI respectively, into EGFP-C1 plasmid in the replacement of the original EGFP.
3. CIB1-RFP-Rab small GTPases plasmids were constructed by cloning CIB1-RFP (2), flanked by NheI/BsrGI sites, into CFP-Rab plasmids obtained from the small GTPase library [14].
4. CIB1-CFP-Rab plasmids were constructed by cloning mCerulean (CFP), flanked by AgeI/BsrGI sites, into CIB1-RFP-Rab GTPases plasmids.
5. mCherry-Lifeact [15].

2.2 Cell and Transfection

Prepare all solutions using ultrapure water. Filter-sterilize all the solutions and media using 0.2 μm filters and store them at 4 °C unless otherwise indicated.

2.2.1 For Cell Handling of Both COS-7 Cell (ATCC) and Primary Cultured Neurons

1. A clean bench.
2. A water bath, temperature set at 37 °C.
3. A vortex.
4. A centrifuge.
5. An inverted light microscope.
6. A hemocytometer.
7. A razor blade.
8. A red-light desk lamp.
9. Plastic tape pads.
10. Aluminum foil.
11. Tubes: 1.5-mL microcentrifuge tubes, 15- and 50-mL conical tubes.
12. 0.25% Trypsin-EDTA solution.
13. Dulbecco's Phosphate Buffered Saline (DPBS).
14. Lipofectamine LTX with Plus Reagent (#15338-100, Invitrogen).
15. Opti-MEM (#31985-062, Invitrogen).

2.2.2 For COS-7 Cell (ATCC) Handling Only

1. A chemical safety hood.
2. Cell culture incubator: humidified, temperature set at 37 °C and CO_2 at 10%.
3. Plates: T75 flasks, 96-well plastic bottom plates.
4. Complete DMEM: Dulbecco's Modified Eagle's Medium (DMEM), supplemented with 10% fetal bovine serum (FBS).
5. Phosphate buffered saline (PBS).
6. Paraformaldehyde fixing solution (4%, v/v): 10 mL of 20% paraformaldehyde (v/v) and 30 mL of PBS (*see* **Note 1**).
7. Low-pH PBS: PBS pH at 5, brought with acetic acid.
8. Transferrin stock solution in PBS (5 mg/mL): 5 mg of Alexa Fluor 647-conjugated transferrin (#T23366, Invitrogen) in 1 mL of ultrapure water, containing 2 mM sodium azide.
9. Transferrin working solution (30 μg/mL): 6 μL of transferrin stock solution in PBS and 994 μL of the complete DMEM for COS-7 cells.

2.2.3 For Primary Cultured Neurons Handling Only

1. A dissecting microscope.
2. An alcohol lamp.
3. Forceps (#11295-10, Solco Biomedical).
4. Scissors (#004-0301, #004-1301, #005-1301 and #008-15-1, Solco Biomedical).
5. Cell strainers (#352350, BD Falcon) and glass Pasteur pipettes.
6. Animal: E18 pregnant Spraque-Dawley female rats (*see* **Note 2**).
7. Cell culture incubator: humidified, temperature set at 37 °C and CO_2 at 5%.
8. Plates: 24-well glass bottom plates, 6- and 10-cm dishes.
9. Complete neurobasal medium (CNB): Neurobasal Medium (#21103-049, Gibco), supplemented with 2 mM glutamine (GlutaMAX, #35050-061, Gibco) and 1% Penicillin-Streptomycin (P/S), (*see* **Note 3**).
10. Complete neuron culture medium (CNB/B27): CNB, supplemented with 2× B-27 (#17504-044, Gibco), (*see* **Note 3**).
11. Plating medium (NM10): CNB, supplemented with 10% horse serum, (*see* **Note 3**).
12. Dissection solution: 1× Hank's Balanced Salt Solution (HBSS), supplemented with 10 mM HEPES, 1 mM sodium pyruvate, 2 mM glutamine (GlutaMAX) and 1% P/S.
13. Triturating solution: 2.7 mL of dissection solution with 0.3 mL of 2.5% Trypsin solution (#15090-046, Gibco), (*see* **Note 4**).
14. 0.1 M Borate buffer (pH 8.5): 1.55 g of boric acid and 2.45 g of sodium tetraborate decahydrate in 500 mL of ultrapure water (*see* **Note 5**).
15. Poly-L-lysine coating solution (0.1 mg/mL): 25 mg of poly-L-lysine hydrobromide in 250 mL of Borate buffer (*see* **Note 6**).

2.3 Imaging and Photoactivation

1. A confocal microscope. We use the A1R (Nikon) mounted on an Eclipse Ti body (Nikon).
2. A stage chamber. We use the Chamlide TC system (Live Cell Instruments), equipped with a lens heater, a stage-top incubator with a cover, a humidifier, a temperature controller, and a gas flow rate controller.
3. Objective. We use the CFI Plan Apochromat VC 60× Oil objective with NA 1.4 (Nikon).
4. Immersion oil. We use Immersol-518F (#444962-0000-000, Carl Zeiss) (*see* **Note 7**).
5. Lens cleaning wiper.

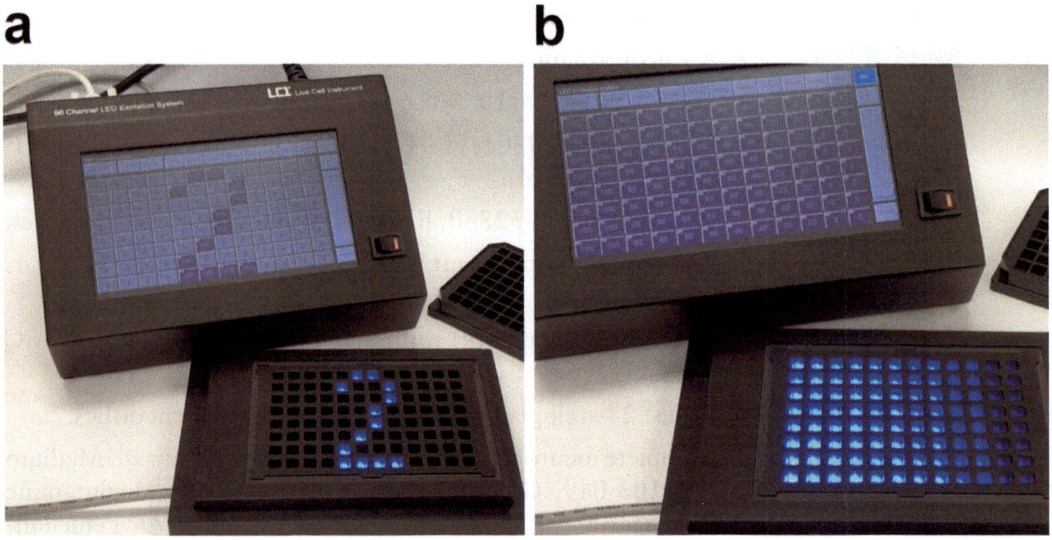

Fig. 2 A customized RGB LED array for global photoactivation of a 96-well plate. (**a**) Selective group excitation mode (**b**) Gradient excitation mode

6. Lasers. We use the multi-line Argon Laser (Nikon) for 457, 488, and 514 nm, Sapphire Solid Laser (Coherent) for 561 nm, and CUBE diode Laser (Coherent) for 640 nm. Local photoactivation through Galvano scanner with a hybrid confocal scan head containing a high-speed hyper selector (Nikon).
7. Optical power meter (#8230E, ADCMT): measurement of laser intensity on the focal plane.
8. A customized RGB LED array: a board with 96 RGB LED chips (Live Cell Instruments), designated for each well of a 96-well plate, connected to a controller for RGB selection, intensity and on/off time sequencing (Fig. 2).
9. Software: NIS-element AR (Nikon) and Microsoft Excel.

3 Methods

3.1 Optogenetic Control of Transferrin Recycling

3.1.1 Cell Handling

1. Bring 0.25% Trypsin-EDTA solution to the room temperature and pre-warm complete DMEM and DPBS at 37 °C.
2. Under a clean bench, place two 96-well plates and seal the wells with tape pads (*see* **Note 8**). For each plate, remove the tape above 3 wells with a sterile razor blade and sterilize the plates with UV light for 30 min. Label one as *Dark Plate* and the other as *Light Plate*.
3. Prepare the COS-7 cells at a confluence of ~80% in a T75 flask. Discard the medium and wash the flask with 10 mL of DPBS. Add 4 mL of 0.25% Trypsin-EDTA solution and place the flask

back into the incubator for 1–2 min (*see* **Note 9**). Meanwhile, prepare a new T75 flask for a sub-cultivation and label appropriately.

4. Add 6 mL of complete DMEM into the flask and harvest the cells into a 15 mL conical tube. Take 1 mL of the cells into the new flask and add 14 mL of complete DMEM. Put the flask into the incubator and centrifuge the tube for 1 min at 600 RCF.
5. Discard the supernatant and resuspend the cells with 5 mL of complete DMEM. Count the total cell numbers with a hemocytometer and then adjust the concentration of the cell suspension to 6×10^4 cells/mL.
6. Slowly add 200 μL of the cell suspension into each well of the sterilized 96-well plates (1.2×10^4 cells/well) (*see* **Note 10**) and do not move the plates for 5 min (*see* **Note 11**).
7. Place the plates into the incubator overnight (15–24 h).

3.1.2 Transfection

1. Aliquot 250 μL of opti-MEM in a microcentrifuge tube and pre-warm it at 37 °C.
2. Under a clean bench, prepare three microcentrifuge tubes and label them as *Control*, *Rab11b*, and *Rab7*. Add 150 ng of YFP-CRY2 into each tube and 200 ng of CIB1-RFP, CIB1-RFP-Rab11b, and CIB1-RFP-Rab7, respectively.
3. Add 1.25 μL of PLUS Reagent into the microcentrifuge tube containing 250 μL of the pre-warmed opti-MEM and mix well by pipetting 3–4 times.
4. Add 70 μL of the opti-MEM/PLUS mixture into each DNA-containing tube and mix well by pipetting 3–4 times. Seat the tubes at room temperature (RT) for 5 min.
5. Add 1.05 μL of Lipofectamine LTX into each tube and mix well by pipetting 3–4 times. Seat the tubes at RT for 25 min.
6. Place the plates under a clean bench and label the wells as *Control*, *Rab11b*, and *Rab7* on each plate. Add 20 μL of the reaction mixture drop-wisely into the corresponding wells (*see* **Note 12**).
7. Wrap each plate in aluminum foil completely to protect from light and label on the foil, accordingly (*see* **Note 13**).
8. Place the plates into the incubator overnight (15–24 h).

3.1.3 Preparation of Microscope and Imaging Settings

1. At least 15 min in advance, switch on the A1R confocal microscope, the multi-line argon laser, the Sapphire solid laser, the CUBE diode laser, and the Chamlide TC system at 37 °C with 10% CO_2 (*see* **Note 14**).

2. Select the 60× oil objective and clean the surface of the lens with the wipes dipped in 100% ethanol. Once it dries, clean the surface with another piece of lens wipe to remove any residual ethanol.
3. Using the Ti Remote Control pad, select the Optical Path to the A1 (*see* **Note 15**).
4. Run the NIS-Elements software (*see* **Note 16**).
5. In the A1plus Settings dialog box, open the Dye and Spectral Setting window. Click the Galvano button for the scan mode and DU4 bottom for the detector. Displace the transmitted light detector from the optical path by clicking the Out button. Select the Ch Series at the 4 → 1.
6. In the Dye and Spectral Setting window, click the Auto button for the setting mode. Uncheck the checkbox of the Ch1 and select the Ch2 for the YFP, the Ch3 for the RFP, and the Ch4 for the Alexa 647. Lasers will be selected automatically at 514, 561, and 640 nm, respectively. Click the Ok button to save and close the window.
7. Back in the A1plus Settings dialog box, click the Laser Power Monitor icon for an update and then click the Laser Inter-Locked button (*see* **Note 17**). Select the Scan Size at 512, the Scan Speed at 0.5 and the Line Average Count at None.
8. In the A1plus Settings dialog box, click the XYZ Time button to open the ND Acquisition dialog box.
9. In the ND Acquisition dialog box, check the checkbox of the Save to File. Select the Time tab and check a checkbox in the Phase column. Set the Interval at No Delay and the Loops at 1 (*see* **Note 18**).
10. Secure the imaging room according to the light clearance protocol.

3.1.4 Loading Transferrin into the Transfected Cells

1. Aliquot 994 μL of complete DMEM and pre-warm this and opti-MEM at 37 °C.
2. Mix the transferrin stock solution (5 mg/mL) using a vortex and centrifuge it briefly.
3. Add 6 μL of the supernatant in the transferrin stock solution to the pre-aliquoted complete DMEM to obtain transferrin working solution (30 μg/mL).
4. Secure the culture room according to the light clearance protocol.
5. Place the Dark plate under a clean bench. Gently remove all the media from each well and add 150 μL of the transferrin working solution to each well (*see* **Note 19**).

6. Place the plate back into the incubator for 10 min to allow for transferrin receptor internalization.
7. Place the plate on ice for 1 min to halt endocytosis.
8. Place the plates under a clean bench and gently replace the transferrin working solution with 150 μL of cold low-pH PBS to each well. Leave the plates under the clean bench for 1 min to strip surface-bound ligands.
9. Wash the wells with cold opti-MEM.
10. Add 200 μL of pre-warmed opti-MEM for live cell imaging.

3.1.5 Imaging and Photo-Stimulation

1. Drop an appropriate amount of immersion oil on the objective (*see* **Note 20**) and place the Dark plate on the microscope stage. Ensure that the plate is firmly fixed on the stage holder before covering with the stage cover (*see* **Note 21**).
2. In the A1plus Settings dialog box, uncheck the checkbox of the Ch2 (*see* **Note 22**) and then click the Find Mode button (*see* **Note 23**). Using the joystick, move the stage to locate the objective under the Control well. Then, move the z-coordinate up using the focus knob gradually and slowly until the bottom of the plate touches the immersion oil.
3. Turn on the Perfect Focus System Operation (PFS) on the microscope and locate the cells in the field of view. Adjust the *z*-coordinate finely with the PFS offset dial until you find an optimal focal plan (*see* **Note 24**).
4. In the A1plus Settings dialog box, adjust the laser powers or the pinhole size (*see* **Note 25**) and then click the Finding Mode button again to stop scanning.
5. In the ND Acquisition dialog box, select the XY tab and check the checkbox of the Include Z. Click the Finding Mode button and locate the desired cells using the joystick. Check the checkboxes in the Point Name column to register the position information.
6. Register five positions per well and save the multi-position information with the file name as *Dark plate* and click the XY button to acquire test images for the final check. Click the Run Now button on the ND Acquisition dialog box.
7. Once image acquisition is finished, press the Escape button on the microscope.
8. Displace the plate and clean the bottom of the plate.
9. Place the Dark plate back into the incubator for 1 h.
10. For the Light plate, repeat the transferrin loading in Subheading 3.1.4 (*see* **Note 26**).

11. Clean the surface of the lens and apply immersion oil. Place the Light plate on the microscope stage. Press the Refocus button and then turn on the PFS button on the microscope.
12. Clear the XY tab of the ND Acquisition dialog box and register new positions as before. Save the list with the file name as *Light plate* and run the ND Acquisition.
13. Once the image acquisition is finished, press the Escape button, displace the plate, and clean the bottom of the plate.
14. Place the Light plate on the customized LED array, which together should be placed in an incubator at 37 °C with 10% CO_2. Set the conditions of LED photoactivation at the blue channel with a time sequence of 10 s ON every 1 min for 1 h (*see* **Note 27**).
15. Go back to the microscope and clean the objective.
16. When the incubation time for the Dark plate is over, add 200 μL of the fixing solution (4% paraformaldehyde) into each well and incubate at RT for 20 min. Wash gently with PBS three times.
17. Meanwhile, acquire images from the Light plate immediately once the photoactivation protocol has finished. For the multi-position information, simply re-load the saved file, by clicking the Load button in the XY tab of the ND Acquisition dialog box.
18. Once image acquisition is finished, press the Escape button and clean the bottom of the plate. Place the Light plate back into the incubator for 1 h (*see* **Note 28**).
19. Meanwhile, acquire images from the fixed Dark plate. Before running the ND Acquisition, check the checkbox of the Ch2 and adjust the laser setting for the Ch2 (*see* **Note 29**).
20. Once the incubation time for the Light plate is over, fix the cells and then acquire images from the fixed Light plate.
21. Turn off all the imaging equipment and terminate the light clearance protocol.

3.1.6 Image Analysis

1. Run the Excel and the NIS-Elements AR software.
2. Open saved image files: for the Dark plate at 0 min and at 60 min; the Light plate at 0 min, 60 min, and at 120 min.
3. Using the Merge Opened Documents tab under the File tab in the ribbon bar, merge the files accordingly for the Dark and Light plates. Check if the Dark and Light plates have 2 and 3 time points (T frames), respectively.
4. On the merged Dark plate file, click the last T frame and the first M frame. On the image, select cells with both YFP and RFP signals. Press Ctrl + Shift + R on the keyboard and click

the Bezier line on the ROI toolbar to draw the region of interests (ROIs) on them. Be sure to cover nearly the whole cell area. Repeat this step for the all multi-positions (M frames) and press Esc when the final ROI has been drawn.

5. Press the Ctrl + Alt + R on the keyboard to open the ROI Statistics window. Select the Alexa 647 channel and click the ND icon in the window. In the Apply To pop-up window, click Selection with XY and then click the OK button to export the data to Excel.
6. In Excel, a spreadsheet with the ROI statistics will have appeared. Copy the data for the ROI Area and the ROI Mean Intensity. Paste the data into a new spreadsheet.
7. Select another T frame and check each ROI still covers your choice of the cells. If necessary, drag the ROIs to cover the cells (*see* **Note 30**). Export data, copy and paste ROI statistics for the current T frame.
8. Repeat **steps 4–7** for the Light plate.
9. Using the achieved data on the new Excel spreadsheet, calculate the intensity/μm^2 and plot a graph of the transferrin intensities (Fig. 3).

3.2 Optogenetic Control of Growth Cones on Neurites

3.2.1 Preparation of Primary Hippocampal Neurons

1. A day before neuron cultivation, place a 24-well plate (*see* **Note 31**) and seal the wells with tape pads (*see* **Note 8**). Remove the tape pads above 3 wells with a sterile razor blade and sterilize the plates with UV light for 30 min.
2. Add 0.5 mL of the coating solution into each well. Wrap the plate in aluminum foil to prevent evaporation and leave the plate at RT overnight (*see* **Note 32**).
3. On the day of neuron cultivation, aspirate the coating solution from each well and wash the wells with sterilized water for 5, 15, 30, and 60 min at RT.
4. Meanwhile, thaw a vial of FBS and a vial of 2.5% Trypsin solution at RT. Pre-warm NM10 at 37 °C.
5. After the last wash, add 0.5 mL of NM10 to each well. Place the plate in the culture incubator until further use.

3.2.2 Culturing Primary Hippocampal Neurons

1. Sterilize the working area and instruments, including forceps and scissors, with 70% ethanol.
2. On ice, prepare a 15-mL conical tube and 3 of 6-cm dishes, filled with the dissection solution and 2 of 10-cm dishes, filled with DPBS.
3. Euthanize an E18 pregnant Sprague-Dawley female rat with CO_2.

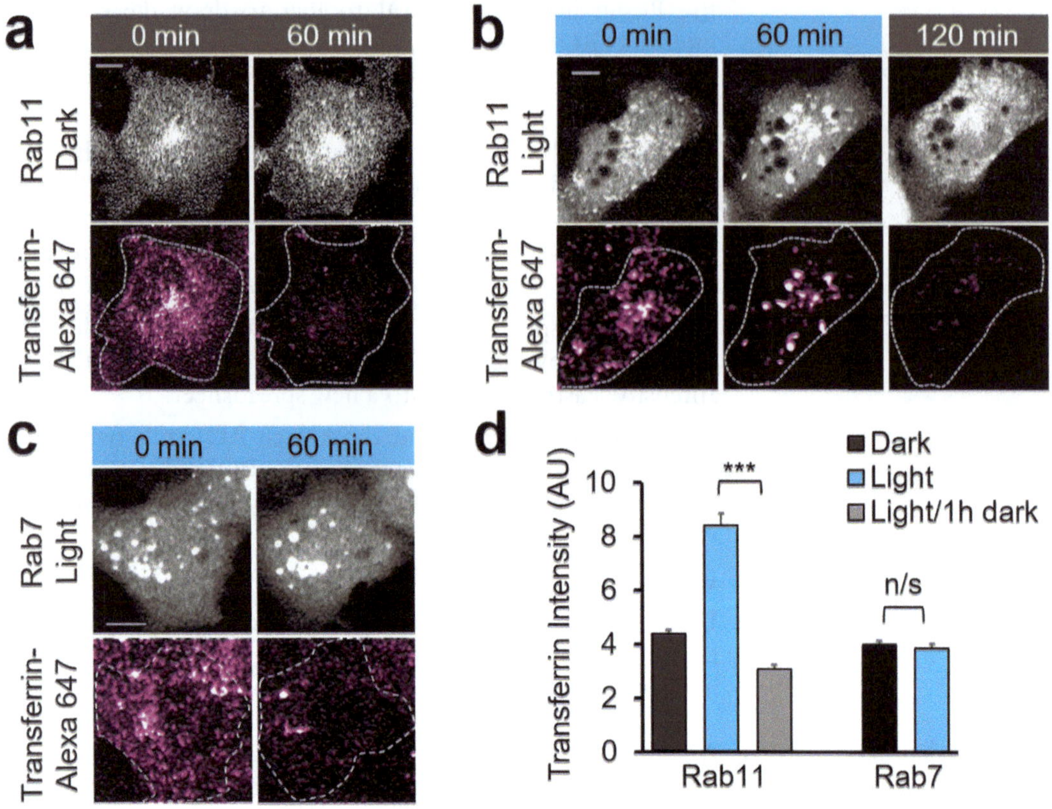

Fig. 3 Optogenetic perturbation of Rab11a specific endosomes in transferrin recycling. (**a**) Confocal images of Rab11a and transferrin (Alexa 647) in dark condition (0–60 min). (**b**) Confocal images of Rab11a and transferrin in light condition (0–60 min), followed by a recovery phase in dark condition for 1 h (120 min). (**c**) Confocal images of Rab7 and transferrin in light condition (0–60 min). (**d**) Bar graphs of the quantification data in **b** and **c**, showing the selective inhibitory effects on transferrin recycling by Rab11a-targeted IM-LARIAT but not Rab7-targeted: $P < 0.05$ (**) and $P > 0.5$ (n/s), respectively; in two-tailed Student's *t*-test. It is reproduced from [7] with permission from *Nature Chemical Biology*

4. Place the rat on its back and rinse its abdomen with 70% ethanol.
5. Using scissors and forceps, expose the abdomen and take the embryonic sacs out into the 10-cm dish. Rinse in the dish and transfer the sacs into the second 10-cm dish.
6. Using forceps, take two embryos out of the sacs and transfer them into the 6-cm dish. Cut heads off the embryos and transfer the heads into the second 6-cm dish. Cut off the skin and skull to take out the brains from the heads and transfer them into the third 6-cm dish.
7. Under a dissecting microscope, make a sagittal cut along the midline to separate the brain into two halves. Place the cortex facing down and cut off the midbrain from the cortex. Remove

the meninges from the cortex and dissect the hippocampus using forceps. Collect four pieces of the hippocampi in the 15-mL tube.

8. Under a clean bench, prepare the trituration solution by adding 2.7 mL of the dissection solution and 0.3 mL of 2.5% Trypsin solution in a 15-mL conical tube. Transfer the collected hippocampi into the 15-mL tube containing the trituration solution. Incubate this tube at 37 °C for 15 min while tapping the tube every 5 min.
9. Meanwhile, prepare a glass Pasteur pipette under a clean bench and turn the blower off. Flame the tip of the glass pipette with an alcohol lamp to obtain a smoothly polished narrow opening of the tip (*see* **Note 33**). Turn the blower on.
10. Prepare 3 of 15-mL conical tubes and label them as 10%, 5%, and 0%. Fill each tube with 9, 9.5, and 10 mL of the dissection solution and 1, 0.5, and 0 mL of FBS, respectively.
11. Once the incubation time is over, sterilize the tube containing hippocampi with 70% ethanol and place under a clean bench. Transfer the hippocampi into the 10% tube and set the tube on the bench until all the tissues have settled to the bottom of the tube. Remove 8 mL of the solution with a 10-mL pipette and transfer the tissues using a P1000 micropipette, to the 5% and then to the 0% tubes in the same manner (*see* **Note 34**).
12. Gently and slowly triturate the tissues with the remaining 2 mL of the dissection solution in the 0% tube; use the fire-polished glass pipette with a pipette-aid (*see* **Note 35**).
13. Prepare a 50-mL conical tube and place a cell strainer on the top. Pass the cell suspension through the strainer. Count cells with a hemocytometer and seed 4×10^4 cells per well (*see* **Note 36**). Incubate the cells in NM10 for 1–2 h. Meanwhile, pre-warm CNB/B27 at 37 °C.
14. Replace NM10 in each well with the pre-warmed CNB/B27 for maintenance and leave the cells to grow in the incubator until DIV2.

3.2.3 Transfection on DIV2 (See **Note 37**)

1. Aliquot 200 μL of opti-MEM, 1 mL of CNB, and 1 mL of CNB/B26 in each microcentrifuge tube and pre-warm them at 37 °C.
2. Under a clean bench, prepare three microcentrifuges and label them as Control, Rab5b, and Rab11a. Add 150 ng of YFP-CRY2 and 150 ng of mCherry-Lifeact into each tube. Add 200 ng of CIB1-CFP, CIB1-CFP-Rab5b, and CIB1-CFP-Rab11a, respectively.
3. Add 2 μL of PLUS Reagent to the tube containing 200 μL of pre-warmed opti-MEM and mix well by pipetting 3–4 times.

4. Add 50 μL of the mixture into each DNA-containing tube and mix well by pipetting 3–4 times. Seat the tubes at RT for 5 min.
5. Add 0.5 μL of Lipofectamine LTX into each tube and mix well by pipetting 3–4 times. Seat the tubes at RT for 25 min.
6. When 5 min of the incubation time remains, prepare three microcentrifuge tubes and label them as Control, Rab5b, and Rab11a. Add 100 μL of the pre-warmed fresh CNB/B27 into each tube. Place the plate containing DIV2 neurons under a clean bench and label the wells as the same as the tubes. Collect 200 μL of media from each well into the corresponding tubes. With a P1000 micropipette, gently remove all the remaining media and add 250 μL of CNB into each well.
7. When the incubation time is over, add the DNA-PLUS-LTX mixture into the corresponding wells dropwise. Incubate the plate and the tubes in the incubator for 45 min (*see* **Note 38**).
8. Place the plate and the tubes under a clean bench. Replace the whole media in each well with the stored media, correspondingly. Wrap the plate in aluminum foil completely and leave it in the incubator overnight.

3.2.4 Imaging and Photo-Simulation

1. At least 15 min in advance, switch on the A1R confocal microscope, the multi-line argon laser, the Sapphire solid laser, and the Chamlide TC system at 37 °C with 5% CO_2.
2. Repeat **steps 2–4** from Subheading 3.1.3.
3. In the Dye and Spectral Setting window, click the Manual button for the setting mode. Check the checkboxes of the Ch1, the Ch2, and the Ch3 and select the spectral setting for respective channels as follows: the names of the fluorescent proteins as the CFP, the YFP, and the RFP, respectively; the wavelengths as 457, 514, and 561 nm, respectively; and the filter ranges as 482/35, 540/30, and 595/50, respectively. Select the first Dichroic Mirror as 457/514/561. Click the OK button to save and close the window.
4. Back in the A1plus Settings dialog box, click the Laser Power Monitor and the Laser InterLocked button. Click the Photo-activation/Bleaching (Laser) tab. Check the checkbox of the All Stimulation Area Set to Same and select the Ch2 for a stimulating laser (*see* **Note 39**). Select the Stimulation Scan Speed at 0.25.
5. Click the Acquisition tab back. Uncheck the checkboxes of the Ch1 and the Ch2. Select the Scan Size at 512, the Scan Speed at 0.25, and the Line Average Count at None.
6. Repeat **steps 8–10** from Subheading 3.1.3.
7. Repeat **steps 1–5** from Subheading 3.1.5.

8. Find a growth cone using the Find Mode. Upon finding one, register the position and click the Find Mode button. Draw a square ROI (10 μm^2) on the tip of the growth cone in the Frozen image to designate the area of photoactivation. Right-click on the ROI and click on Use as Stimulation ROI: S1. Repeat twice to set two more stimulation ROI as S2 and S3. You now have S1, S2, and S3 on the Frozen image and 3 positions registered in the XY tab. Save the list.
9. Right-click in the empty space and click the Acquisition Controls and then the ND Sequence Acquisition. In the ND Sequence Acquisition dialog box, check the checkbox of the Time-lapse and set for Run Every at 1 min and Count at 30.
10. Click the + icon to add a new action and select action for the Run Command. Double-click on the description and enter the command as "CameraSet_ChannelSeriesMode(0);" (Fig. 4).
11. Add a new action and select action for the Run Command. Double-click on the description and enter the command as "StgMoveXYScan(0,0);".
12. Add a new action and select action for the Seq. Stimulation and click the Define button. In the pop-up window, check the checkbox in the Phase column and select the Stimulation in the Acq/Stim column. Select the S1 for the ROIs column and type 6 in the Loops column. Click the Apply Stimulation Settings button to allow that the current photoactivation

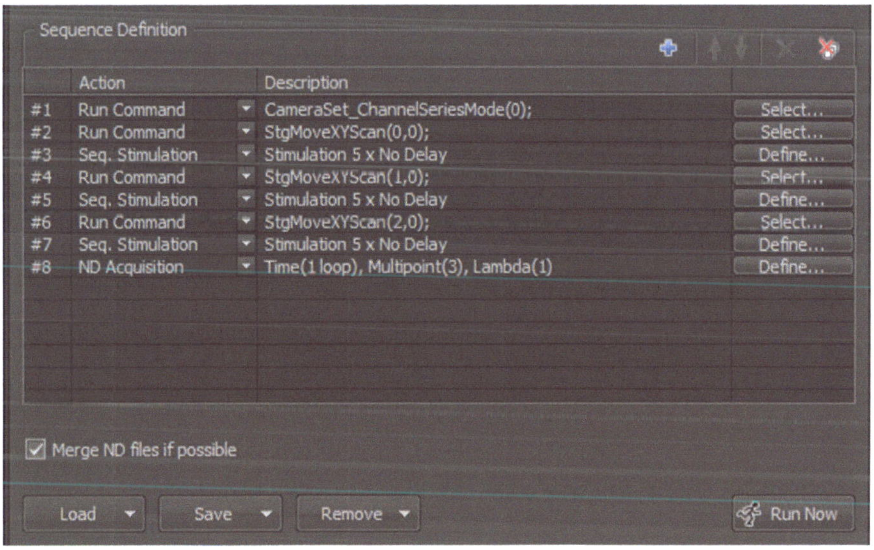

Fig. 4 Setting up of ND sequence acquisition for local photoactivation on multi-positions. The interface of NIS-element AR software is used to set up sequential imaging acquisition for applying photoactivation at three multi-positions

setting can be applied to the ROI positions and the duration for the photoactivation can be calculated (about 1.42 s for 10 μm^2). Click the OK button to close the pop-up window.

13. Click the Action #2 and click the + icon. When a copy of the Action #2 appears, double-click on the description and change the position from (0,0) to (1,0). In the same way, copy the Action #3 and click the define button. In the pop-up window, change the ROIs from the S1 to the S2.
14. Click the Action #2 and click the + icon. When a copy of the Action #2 appears, double-click on the description and change the position from (0,0) to (2,0). In the same way, copy the Action #3 and click the define button. In the pop-up window, change the ROIs from the S1 to the S3.
15. Add an action for the ND Acquisition and click the define button. For the Time tab, check a checkbox in the Phase column and set the Interval at No Delay and the Loops at 1. For the XY tab, load the saved list from **step 8**. For the λ tab, check a checkbox in the Optical Conf. column and click the Define New button to apply current optical configuration. Click the Ok button to close the pop-up window.
16. Check if all the ND Sequence Definition settings are correctly entered as shown in Fig. 4 (*see* **Note 40**).
17. Click the Run Now button in the ND Acquisition dialog box to acquire images during the *Dark phase I*.
18. Immediately after the process has finished, click the Run Now button in the ND Sequence Acquisition dialog box, to acquire images during the *Light Phase*.
19. Immediately after the process has finished, click the Run Now button in the ND Acquisition dialog box, to acquire images during *Dark Phase II*.
20. In the A1plus Settings dialog box, check the checkboxes of the Ch1 and the Ch2. Click the 1 Time Loop button in the ND Acquisition dialog box to check the expression levels of the constructs.
21. Uncheck the checkboxes of the Ch1 and the Ch2. Displace the plate from the stage and clean the lens.
22. Repeat **steps 1–3** from Subheading 3.1.5.
23. Repeat **steps 8–21** for the second well.
24. Repeat **steps 22** and **23** for the third well.

3.2.5 Image Analysis

1. Run the NIS-Elements AR software.
2. Open the saved image files for the Control, Rab5b, and Rab11a.
3. Merge files for each well.

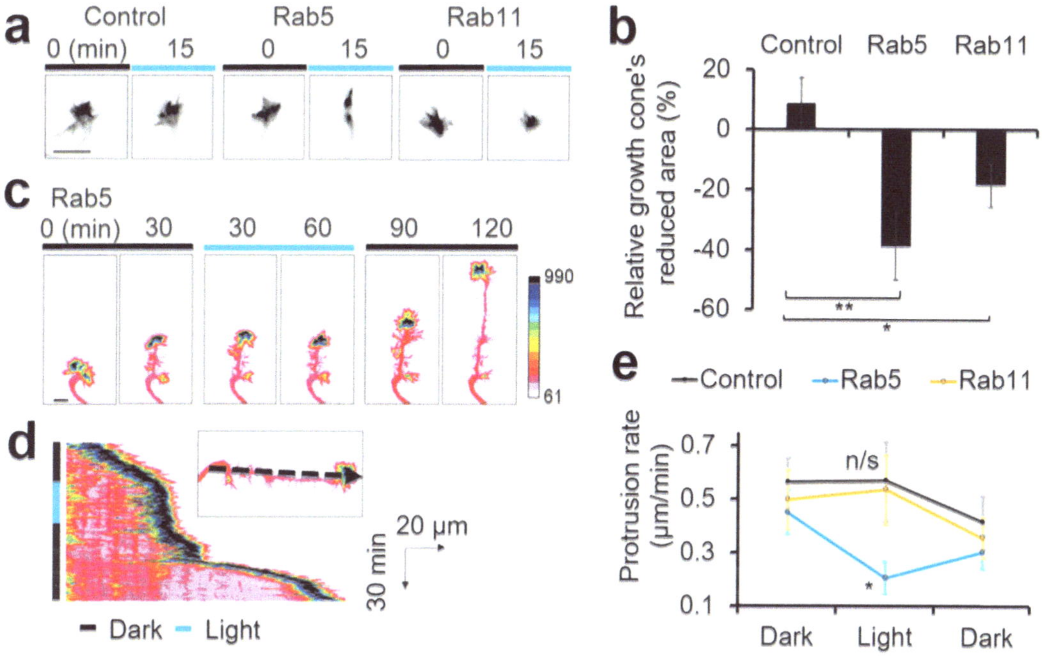

Fig. 5 Differential interferences of Rab5–Rab11-targeted IM-LARIAT in the growth cones of hippocampal neurons. (**a**) The changes in the signals of mChe-Lifeact in the growth cones subjected to the local stimulation with blue-light. (**b**) The area of growth cones with Rab5- and Rab11-targeted IM-LARIAT both significantly decreased: $P = 0.002193$ (**) and $P = 0.02683$ (*), respectively; two-tailed Student's *t*-test. (**c**) Representative images of blue-light induced perturbation in the growth of growth cones with Rab5-targed IM-LARIAT. (**d**) Kymograph of the growth cones in (**c**) along the growth cone's protrusion path, indicated with the arrow. (**e**) The protrusion rate of growth cones in the absence, in the present and in the absence of blue-light. In the present of blue-light, Rab5-targeted IM-LARIAT significantly decreased the protrusion rate ($P = 0.047$, *; two-tailed Student's *t*-test). Scare bars, 10 µm. Error bars, S.E.M. It is reproduced from [7] with permission from *Nature Chemical Biology*

4. To measure the growth area of the growth cones, click the last three T frames of each phase and draw ROIs on the fan-shape tip of the growth cones (Fig. 5b).
5. To generate a kymograph showing the spatiotemporal changes of the growth cones, click the last T frame and select the Image → ND Process → Create Kymograph by Line buttons in the ribbon box. Draw a line in the direction from the proximal to distal axon end, where the tip of a growth cone presents (Fig. 5d).
6. To measure the growth rate of the growth cones, mark the tip of the growth cone in the first and thirtieth T frames. Measure the distance between the two points with the Annotate and Measure tool in the toolbar (Fig. 5e).

4 Notes

1. This is a hazardous material. We recommend using paraformaldehyde under a chemical fume hood.
2. All experiments involving animals must be conducted in accordance with relevant institutional and governmental guidelines and regulations. All of our experimental procedures for handling animals were approved by the Animal Ethics Committee at the Korea Advanced Institute of Science and Technology (KAIST; Daejeon, Korea).
3. CNB can be stored at 4 °C for up to 3 months but it is not recommended to store CNB/B27 nor NM10 at 4 °C for more than 10 days. We recommend preparing them on the day of cultivation by adding B-27 and horse serum to CNB, respectively.
4. We recommend preparing the triturating solution by adding fresh trypsin solution to the dissection solution on the day of cultivation. We recommend storing the stock trypsin solution in aliquots at −20 °C for up to 3 months.
5. Dissolving the chemical powder takes few hours.
6. We recommend preparing a stock of poly-L-lysine solution at a concentration of 1 mg/mL by dissolving 25 mg of poly-L-lysine hydrobromide powder in 25 mL of Borate buffer—this stock solution can be stored in aliquots at −20 °C for up to 3 months. We recommend preparing the coating solution by diluting 2.5 mL of the poly-L-lysine stock in 22.5 mL of Borate buffer just before use.
7. We recommend using only the immersion oil that is compatible with the plate used for imaging. When using different types of plates or oil from ours, the product information should be checked for their compatibility, otherwise, there are risks to harming objectives and microscopes.
8. By using the tapes for covering wells, unused wells can be kept clean. We recommend using separate plates for dark and light samples to reduce risk of light exposure to the dark samples.
9. Additional time in the incubator helps in detaching the cells but the incubation time should not exceed 5 min. During the incubation, avoid agitating the flak, otherwise cells cluster together so that the cell count becomes inaccurate at the later step. As the residual complete DMEM in the flask can hinder trypsinization, the flask should be thoroughly washed with DPBS before adding 0.25% Trypsin-EGTA solution.
10. We recommend plating 1–1.5 $\times$ 10^4 cells/well for 96-well plates (growth area = 0.56 cm^2 per well). Higher densities should be avoided as transfection with Lipofectamine requires about 2 days from seeding to imaging.

11. We do not recommend handling the plates immediately after seeding as this can cause cells to congregate at the edges of the wells.
12. Following the procedures, each well will contain 43 ng of CRY2, 57 ng of CIB1, 0.1 μL of PLUS Reagent, and 0.3 μL of Lipofectamine LTX. We recommend the expression level of CRY2 to be lower than CIB1; we used the ratio of CRY2: CIB1 = 3:4. Alternatively, a stable cell line can be generated to stably express YFP-CRY2. For immortalized cell lines, including COS-7 here, we recommend 100–150 ng of total DNA to be transfected to 1–1.5 $\times$ 10^4 number of cells and the lipofection condition at a ratio of DNA:PLUS:LTX = 1:1:3.
13. Working with CRY2-based optogenetic tools requires strict restrictions of light to samples. Thus, we turn off the room light, cover or eliminate any light sources nearby, and work under a red-light ramp for the visual aid (wavelengths above 561 nm have no effects on CRY2 oligomerization). We always wrap the samples in aluminum foil unless the samples are used for other procedures. Herein, we call these working conditions *the light clearance protocol.* In a conventional room lighting system (i.e. neutral white bulbs), the light intensity of room light measured with 488 nm wavelength at 1.5–2 m away from the bulb is 7–13 mW/mm^2. When we tested our experiment, the room light as well as even deem light or transmitted light from a microscope was able to activate the CRY2 system. If samples are exposed to light for a short time accidentally, let samples recover to a dark state for at least 30 min.
14. This step allows the lasers to warm up for the stable performance and the microscope to equilibrate to the temperature set. If this stabilization time is not allowed long enough, fluctuations in the power of lasers and in the temperatures can occur. Significant differences in the temperatures throughout the equipment can cause focus drifts, affecting the quality of images, whereas the desired field of imaging (x- and y-coordinates) or optimal focal plan (z-coordinate) can be lost.
15. The optical path to the A1 can be selected by pressing the [L100] button on the microscope body.
16. When more than one confocal microscope is connected, we open the Driver Section dialog box and select Nikon Confocal. We recommend running the software after having turned on all the connected equipment to the microscope to avoid any error occurring.
17. If the lasers cannot be interlocked, we recommend checking if the Optical Path is set as the A1.
18. Time-lapse imaging can be done by simply changing the setting in the Time tab here. An appropriate interval time should be determined by running the 1 time loop, which will

show the total duration of time for one loop with your current settings, including the image resolution, scan size, scan speed, line average count, numbers of positions and channels.

19. In replacing *Solution* A to *Solution B* in the wells containing cells, the removal of the Solution A should be followed by immediate application of the Solution B in order to prevent the cells from drying out at any point. Thus, we recommend the complete sequential replacement of each well to be proceeded, one by one. To aid a gentle performance, we recommend removing Solution A with a micropipette instead of using vacuum aspiration.
20. It is important to drop an appropriate amount of immersion oil on the surface of the lens with care. If too much oil is applied, the oil can flow off the surface of the lens when the bottom of a plate touches the oil. This becomes problematic when the oil turns into sticky clots at the unreachable space in the objective, disrupting the movement of the Spring-Loaded Retraction Stopper. On the other hand, too little oil will cause the surface of the lens to dry-over in the middle of the image acquisition process so that the focal plane is lost. The optimal volume of the immersion oil on the surface of the lens is half-oval shaped. If any bubbles are found in it, we recommend cleaning the lens and applying the immersion oil again.
21. If the plate is not firmly fixed on the stage holder, we recommend using tape for firm fixation.
22. Since CRY2 oligomerization can be induced by excitation of CFP, GFP, or YFP with respective 457-, 488-, and 514-nm lasers, we uncheck the checkbox of the Ch2 for the moment in order to avoid photoactivation of the dark samples. In the first performance of the IM-LARIAT system, we recommend acquiring time-lapse images with your Light plate, and plotting graphs of time vs. cluster information. Although 488 nm is the most commonly used wavelength of the laser for CRY2 photoactivation, you can use the Ch2 (514-nm laser) to visualize YFP as well as a source for photoactivation. In other words, photoactivation can be performed at the same time as acquiring the images. Under this co-imaging condition, we usually use a light power density of 0.2 mW/mm^2.
23. The Find Mode allows monitoring the imaging field of view with lower imaging setting values. This allows you to monitor the live image faster than you monitor with the Live button. The Finding Mode also reduces the risk of bleaching the samples.
24. The A1R microscope is equipped with the PFS, which maintains the *z*-coordinate in a narrow range where cells are seen. Once it is on, the focus knob on the microscope is disabled.

With the PFS offset dial, we find a point at the cells in the field of view become blurry and then move the dial toward the opposite direction to find another point at which the cells become blurry. In doing so, we can find an optimal focal plane where the fluorescent proteins have sharp visuals. Pinhole size can be adjusted to increase sharpness, although this makes the powers for lasers become higher, thereby increasing risk of bleaching.

25. In the NIS software, the laser value indicates relative powers. For practical convenience, we recommend preparing a graph for relative power vs. absolute power using an optical power meter in advance. Without adjusting the LUT panel, saturations of any fluorescent signals should be avoided. For amplifying the visual signal or reducing the noise, we can increase the Gain (HV) or decrease the Offset (−10, recommended).
26. The timing between the loading transferrin and acquiring images should be as short and similar as possible in between all testing groups. Thus, we load transferrin on the Dark plate and immediately proceed imaging procedures. For the Light plate, we start from the loading transferrin steps and imaging during the 1 h incubation time of the Dark plate.
27. According to our previous work with CRY2, pulse stimulation works efficiently to activate CRY2 oligomerization while decreasing risks of photo-toxicity or bleaching under long-term photoactivation conditions. Although much longer time intervals also can be used, we usually use LED photoactivation conditions with cycles of 5–10 s stimulation at 1–2 min intervals with a light power density at 6.5–10 mW/mm^2.
28. This step is to give the IM-LARIAT system a recovery phase in dark (Fig. 1). Oligomerized CRY2 proteins have 6–10 min of dissociation kinetics [7, 11–13, 16]. However, photoactivation time, light power, and CRY2 expression level have effects on the time taken for the complete dispersion of clusters. In most experimental conditions, 30 min is enough for their recovery phase.
29. We need to check the expression level of YFP-CRY2 by imaging the samples after fixing to protect samples from the unwanted photoactivation (*see* **Note 19**). It should be avoided making any changes to other settings, including the pinhole size.
30. For the latest version of NIS software (Ver.4.50), the moving ROI detecting analysis is possible with the ROI Changes Over Time tool.
31. When culturing below DIV7, either a 96-well plate or a 24-well plate can be used. However, we recommend using the latter for raising the culture of primary neurons up to

higher DIVs. The density but also the total amount of neurons seem to be important factors for culturing primary neurons. Too low density or few cells give a much sparser distribution where neuronal growth is hindered by the lack of enough signaling, while too many can be another problem; lack of the growth areas when you plan to grow the neuron up to higher DIVs. For imaging up to DIV21, we recommend seeding 3–5 $\times$ 10^4 cells per well in a 24-well plate.

32. We recommend coating plates a day in advance to culturing primary neurons but coating procedures can also be done on the same day as long as the incubation time for coating is at least 1 h.

33. This is a critical step. The tip size should not be too large nor small. We recommend the size of the tip at 1 mm in the inner diameter.

34. This step-wise trypsin inactivation procedure should be carried out slowly and gently as the tissues are fragile. When transferring hippocampi, attention is also required to check if all the tissues have settled down to the bottom of the tube otherwise some of the tissues can be lost.

35. This is a critical step. We recommend setting the speed of the pipette-aid at the second slowest and patiently take the pipette-aid in and out the supernatant 2–3 times without touching the tissues yet. This is to minimize the chance of tissues sticking on the inner surface of the glass pipette where complete trituration is hindered. We then take 1.5 mL supernatant and the tissues within the solution to make sure the tissues are in the narrowest part of the pipette. It should be ensured that the tip is dipped in the remaining solution during aspiration to avoid air being introduced. When releasing the solution in the pipette back into the tube, it should be ensured that the tip is not emerged in the solution, while touching the inner surface of the tube. Repeat this procedure 10 times.

36. Typical yields are 4–6 $\times$ 10^5 cells per embryo. We recommend at least two embryos to be prepared at once even if you are not using them all.

37. For the growth cone analysis, we transfect neurons at DIV2–4 and perform imaging on the following day.

38. For cultured neurons in a 24-well plate at densities of 3–5 $\times$ 10^4 cells/well, we recommend about 500–600 ng of DNA to be transfected with a lipofection condition at a ratio of DNA:PLUS:LTX = 1:1:1. To increase transfection efficiency, we recommend using transfecting media with no serum. For neuron media change, we use 1 volume of fresh media with 2 volumes of the original culture media.

39. For local photoactivation, we usually use 1.5 s stimulation with a 488 nm (or 514 nm) laser at a light power density of 0.2–0.4 mW/mm^2.
40. Imaging Lifeact in live cells can be demanding due to their dynamic movements in the *z*-coordinate. In the ND Acquisition dialog box, the Z tab can be used to acquire images in z-stacks by clicking the Asymmetric Mode with the Steps at 3 and the Range at 1 μm. Later, images can be seen in 3D or at the maximized intensity.

References

1. Nijenhuis W, van Grinsven MMP, Kapitein LC (2020) An optimized toolbox for the optogenetic control of intracellular transport. J Cell Biol 219(4):e201907149. https://doi.org/10.1083/jcb.201907149
2. Zhang J, Schulze KL, Hiesinger PR, Suyama K, Wang S, Fish M, Acar M, Hoskins RA, Bellen HJ, Scott MP (2007) Thirty-one flavors of Drosophila rab proteins. Genetics 176(2): 1307–1322. https://doi.org/10.1534/genetics.106.066761
3. Gillespie EJ, Ho C-LC, Balaji K, Clemens DL, Deng G, Wang YE, Elsaesser HJ, Tamilselvam B, Gargi A, Dixon SD, France B, Chamberlain BT, Blanke SR, Cheng G, de la Torre JC, Brooks DG, Jung ME, Colicelli J, Damoiseaux R, Bradley KA (2013) Selective inhibitor of endosomal trafficking pathways exploited by multiple toxins and viruses. Proc Natl Acad Sci 110(50):E4904. https://doi.org/10.1073/pnas.1302334110
4. Passmore JB, Nijenhuis W, Kapitein LC (2021) From observing to controlling: inducible control of organelle dynamics and interactions. Curr Opin Cell Biol 71:69–76. https://doi.org/10.1016/j.ceb.2021.02.002
5. Wittmann T, Dema A, van Haren J (2020) Lights, cytoskeleton, action: optogenetic control of cell dynamics. Curr Opin Cell Biol 66: 1–10. https://doi.org/10.1016/j.ceb.2020.03.003
6. van Bergeijk P, Adrian M, Hoogenraad CC, Kapitein LC (2015) Optogenetic control of organelle transport and positioning. Nature 518(7537):111–114. https://doi.org/10.1038/nature14128
7. Nguyen MK, Kim CY, Kim JM, Park BO, Lee S, Park H, Heo WD (2016) Optogenetic oligomerization of Rab GTPases regulates intracellular membrane trafficking. Nat Chem Biol 12(6):431–436. https://doi.org/10.1038/nchembio.2064
8. Hutagalung AH, Novick PJ (2011) Role of Rab GTPases in membrane traffic and cell physiology. Physiol Rev 91(1):119–149. https://doi.org/10.1152/physrev.00059.2009
9. Pfeffer SR (2013) Rab GTPase regulation of membrane identity. Curr Opin Cell Biol 25(4): 414–419. https://doi.org/10.1016/j.ceb.2013.04.002
10. Stenmark H (2009) Rab GTPases as coordinators of vesicle traffic. Nat Rev Mol Cell Biol 10(8):513–525. https://doi.org/10.1038/nrm2728
11. Lee S, Park H, Kyung T, Kim NY, Kim S, Kim J, Heo WD (2014) Reversible protein inactivation by optogenetic trapping in cells. Nat Methods 11(6):633–636. https://doi.org/10.1038/nmeth.2940
12. Kennedy MJ, Hughes RM, Peteya LA, Schwartz JW, Ehlers MD, Tucker CL (2010) Rapid blue-light-mediated induction of protein interactions in living cells. Nat Methods 7(12):973–975. https://doi.org/10.1038/nmeth.1524
13. Bugaj LJ, Choksi AT, Mesuda CK, Kane RS, Schaffer DV (2013) Optogenetic protein clustering and signaling activation in mammalian cells. Nat Methods 10(3):249–252. https://doi.org/10.1038/nmeth.2360
14. Heo WD, Meyer T (2003) Switch-of-function mutants based on morphology classification of Ras superfamily small GTPases. Cell 113(3): 315–328
15. Riedl J, Crevenna AH, Kessenbrock K, Yu JH, Neukirchen D, Bista M, Bradke F, Jenne D, Holak TA, Werb Z, Sixt M, Wedlich-Soldner R (2008) Lifeact: a versatile marker to visualize F-actin. Nat Methods 5(7):605–607. https://doi.org/10.1038/nmeth.1220
16. Park H, Kim NY, Lee S, Kim N, Kim J, Heo WD (2017) Optogenetic protein clustering through fluorescent protein tagging and extension of CRY2. Nat Commun 8(1):30. https://doi.org/10.1038/s41467-017-00060-2

Chapter 21

Generating Custom Pooled CRISPR Libraries for Genetic Dissection of Biological Pathways

Daniel R. Gulbranson

Abstract

Genetic screens are a classic approach to dissecting biological pathways including membrane trafficking. Clustered regularly interspaced short palindromic repeats (CRISPR)/Cas9 have enabled the utility of this approach in diploid models, including cultured mammalian cells. Here, we present detailed protocols for generating custom CRISPR libraries. These methods are useful for generating genome-wide libraries for new model organisms that lack an existing genome-wide library, and for generating smaller focused libraries.

Key words CRISPR screen, CRISPR library, Genetic screen, sgRNA cloning, Pooled screen, Custom library

1 Introduction

The genomic age has led to an unprecedented number of candidate disease-modifying genes, some of which may be efficacious drug targets [1–3]. A better understanding of the biological function of the products these genes encode will aid basic research, including the membrane trafficking field, and may contribute to the development of therapies for disease. A classic genetic approach to dissecting biological pathways is the mutagenesis screen, which can determine the subset of genes whose manipulation alters the biological phenotype under investigation. Indeed, genome-wide mutagenesis screens have led to major breakthroughs in diverse fields of biology including genetic inheritance [4], biosynthesis [5], cell-cycle regulation [6], animal behavior [7], embryonic development [8], and the secretory pathway [9]. However, until recently, this approach was limited mainly to haploid model organisms. The discovery and repurposing of CRISPR/Cas9 to enable mammalian genome editing has empowered the use of mutagenesis

Jingshi Shen (ed.), *Membrane Trafficking: Methods and Protocols*,
Methods in Molecular Biology, vol. 2473, https://doi.org/10.1007/978-1-0716-2209-4_21,

screens (CRISPR screens) to dissect complex cellular pathways that are either not present or less complex in haploid model organisms [10–19].

One limitation of genetic screens is the large number of cells required. This is particularly challenging for the study of biological phenomena that occur in post-mitotic cells such as adipocytes and neurons. This limitation can be overcome by performing the initial genetic screen in a cell line, and then validating a subset of hits in a physiologically relevant model, as we have done previously [20, 21]. Because individual gene validation is laborious, an alternative approach is to design a custom CRISPR library composed of selected hits from the initial screen, and then carry out a pooled secondary screen in a relevant cell type [22]. While this strategy is effective at reducing false positives and eliminating hits that do not validate in a relevant cell type, it cannot identify false negatives—genes whose manipulations only manifest phenotypic changes in relevant cell types. To overcome this challenge in models where cell number is limited, custom CRISPR libraries can be used to focus on a subset of genes identified through alternative strategies. For example, genome-wide association studies of Alzheimer's disease (AD) have identified 29 genetic loci—implicating at least 215 possible causal genes—associated with the disease [23, 24]. Although it is difficult to obtain the large number of neurons necessary for a genome-wide CRISPR screen, it is readily feasible to build a CRISPR library focused on this smaller subset of genes. Similarly, one could design custom CRISPR libraries based on expression profiling datasets or interactome datasets. Here, we present a protocol for generating custom pooled CRISPR libraries we have used to perform secondary screens [20, 21], and to build a large-scale CRISPR library targeting genes expressed in cultured rat primary neurons (Fig. 1). This protocol describes the designing and generation of a custom CRISPR library, the production and quantification of lentiviral particles, and the transduction of cells to obtain a mutant cell library.

2 Materials

Use 18.2 MΩ ultra-pure H_2O for all buffers and reactions.

2.1 Designing a Pooled CRISPR Library

1. Computer with internet access.
2. Spreadsheet software, such as Microsoft Excel or R.

2.2 Cloning a Pooled CRISPR Library

1. High-fidelity DNA Polymerase. We use Phusion Hot Start II DNA Polymerase (Thermo Fisher Scientific, F530L).
2. Forward Primer 5′ TAACTTGAAAGTATTTCGATTTCTTG GCTTTATATATCTTGTGGAAAGGACGAAACACCG.

A

Number of sgRNAs with reads	90,415
Number of sgRNAs without reads	6
Mean number of reads/sgRNA	1,126
Median number of reads/sgRNA	1,009
Maximum number of reads/sgRNA	15,842
Standard deviation	607
Quartile 1 (Q1)	734
Quartile 3 (Q3)	1,371
Interquartile range (IQR)	1,214
Bottom whisker (Q1 - 1.5 x IQR)	0
Top whisker (Q3 + 1.5 x IQR)	3,192
Number of sgRNAs above top whisker	3,437
Percent of sgRNA above top whisker	0.7%

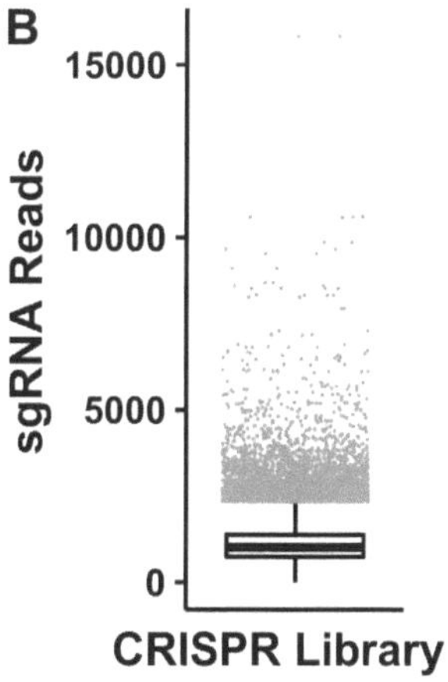

Fig. 1 Abundance of sgRNAs present in a large-scale CRISPR library. sgRNAs were designed and cloned as described here. (**a**) Summary statistics. (**b**) Box plot of sgRNA abundance with individual sgRNAs above the top whisker shown as dots

3. Reverse Primer 5′ ACTTTTTCAAGTTGATAACGGAC TAGCCTTATTTTAACTTGCTATTTCTAGCTCTAAAAC.
4. PCR strip tubes.
5. Tris EDTA (TE) buffer: 10 mM Tris-HCl, pH 8.0, 1 mM EDTA-NaOH.
6. 1% agarose gel: TE buffer containing 1% agarose.
7. Gel electrophoresis apparatus.
8. Gel extraction kit.
9. Lentiviral transfer plasmid for CRISPR library expression, such as lentiCRISPR v2 (Addgene, #52961) (*see* **Note 1**).
10. Esp3l or an isoschizomer such as BsmB1.
11. Alkaline phosphatase.
12. 100 mM dithiothreitol (DTT): dissolve 15.4 mg DTT in 1 mL of 18.2 MΩ ultra-pure water.
13. DNA assembly kit, we use NEBuilder HiFi DNA Assembly Master Mix.
14. Mixed cellulose esters (MCE) filter membrane, 0.025 μm pore size.
15. Electrocompetent Cells. We use Endura ElectroCompetent Cells (Lucigen, #60242).
16. 0.1 cm electroporation cuvettes.
17. 14 mL bacterial culture tube.
18. Large (~25 cm × 25 cm) bioassay Luria-Bertani (LB) agar plates: Mix 24.8 g LB broth media and 12 g agar with 1 L of H_2O. Autoclave and cool to 50 °C. Add 1 mL of a 100 mg/mL ampicillin stock solution. Mix and pour on plates. Allow the plates to cool and gel then store at 4 °C until use.

19. 10-cm LB agar plates: prepare as in **item 18** using 10-cm plates.
20. Electroporator, such as a Bio-Rad Gene Pulser Electroporation System.
21. Plasmid DNA maxiprep kit.

2.3 Generating Lentivirus

1. 293 T cells. (ATCC, #CRL-11268).
2. Cell culture media: Dulbecco's Modified Eagle Medium (DMEM), 4500 mg/mL glucose, 1× GlutaMax, 100 U/mL penicillin-streptomycin, and 10% fetal bovine serum (FBS).
3. Phosphate buffered saline (PBS).
4. 0.25% Trypsin-EDTA solution.
5. Polyethylenimine (PEI) transfection solution: Dissolve 30 mg of PEI, MW 25,000, in 30 mL H_2O by adding 1 M HCl dropwise until the solution becomes clear. Aliquot and store at −20 °C.
6. psPax2 plasmid (Addgene #12260).
7. pCMV-VSV-G plasmid (Addgene #8454).
8. 15-cm cell culture dishes.
9. Ultracentrifuge, such as a Beckman Coulter Optima L-90K Ultracentrifuge.
10. Ultracentrifuge rotor with buckets, we use a SW28 rotor (Beckman Coulter).
11. Ultracentrifuge tubes that fit the ultracentrifuge rotor you are using, we use 25 × 89 mm polypropylene tubes (Beckman Coulter).

2.4 Lentivirus Titration

1. CountBright Absolute Counting Beads (ThermoFisher, #C36950). Dilute the beads fivefold in PBS prior to the experiment.
2. Puromycin stock solution: Dissolve 20 mg puromycin in 10 mL of 18.2 MΩ ultra-pure water.
3. Flow cytometer.
4. Flow cytometer analysis software, such as FlowJo.

3 Methods

3.1 Designing a Pooled CRISPR Library

3.1.1 Designing a Targeted Pooled CRISPR Library from an Existing CRISPR Library

1. Select an existing CRISPR library to use as the basis for the custom targeted library (*see* **Note 2**). Ensure that the library you select is designed against the correct species and uses your desired approach (e.g. knockout, inhibition, or activation).
2. Extract the subset of sgRNAs from the library that target your genes of interest and control sgRNAs.

3. The sgRNA sequence list for some available CRISPR libraries includes the protospacer adjacent motif (PAM) sequence at the 3′ end (NGG for Cas9). If the sgRNA sequence list you are using includes the PAM sequence, delete it.
4. Add a guanine to the 5′ end of all sgRNA target sequences that do not already begin with one. This guanine increases the sgRNA expression from a U6 promoter.
5. Add 5′-TATCTTGTGGAAAGGACGAAACACC-3′ to the 5′ end of the sgRNA target list (*see* **Note 3**).
6. Add 5′-GTTTTAGAGCTAGAAATAGCAAGTTAAAAT-3′ to the 3′ end of the sgRNA target list (*see* **Note 3**).

3.1.2 Designing a New Pooled CRISPR Library

1. Go to the sgRNA design website: https://chopchop.cbu.uib.no/.
2. Choose the desired target gene and species (*see* **Note 4**).
3. Make sure CRISPR/Cas9 is selected under the "Using" dropdown menu.
4. On the "For" dropdown menu, select the desired genetic manipulation (e.g. knockout, inhibition, or activation).
5. Additional options are available under the "General," "Cas9," and "Primers" tabs, but we recommend not deviating from the default settings (*see* **Note 5**).
6. After running the search, copy the desired number of sgRNAs to a spreadsheet.
7. Repeat the process for the remaining genes you wish to include.
8. Delete three bases from the 3′ end of the target sequence (the PAM sequence).
9. Add non-targeting control sgRNAs to your library (*see* **Note 6**).
10. Add a guanine to the 5′ end of all sgRNA target sequences that do not already begin with one.
11. Add 5′-TATCTTGTGGAAAGGACGAAACACC-3′ to the 5′ end of the sgRNA target list (*see* **Note 3**).
12. Add 5′-GTTTTAGAGCTAGAAATAGCAAGTTAAAAT-3′ to the 3′ end of the sgRNA target list (*see* **Note 3**).

3.2 Cloning a Pooled CRISPR Library

3.2.1 DNA Assembly

1. If your library is small (less than ~100 sgRNAs), order individually synthesized oligos and pool them. For a larger library, have the oligo pool synthesized at a commercial company (*see* **Note 7**).
2. Prepare the PCR reaction mixture composition (Table 1) using at least 1 PCR reaction per 10,000 sgRNAs present in the library. Perform a minimum of three PCR reactions.

Table 1
PCR reaction mixture composition

Reagent	Volume (μL)/reaction
GC buffer (5× master mix)	10
dNTPs (5 μM)	2
Primer 1 (50 μM stock)	0.2
Primer 2 (50 μM stock)	0.2
Pool of oligos (Subheading 3.2.1, **step 1**)	1
Phusion	0.5
milliQ H_2O	36.1

3. Run the PCR reaction on a thermocycler with an initial denaturation cycle of 98 °C for 30 s, followed by 5 cycles at 98 °C for 10 s, 55 °C for 30 s, and 72 °C for 30 s, 15 cycles at 98 °C for 10 s, and 72 °C for 1 min, and a final elongation cycle of 72 °C for 10 min.
4. Run the PCR products on a 1% agarose gel and purify them using a gel extraction kit (*see* **Note 8**).
5. Prepare ~1 restriction digestion reaction per 5000 sgRNAs present in the library as follows: 6 μL of 10× digestion buffer, 0.6 μL of 100 mM DTT, 5 μg of lentiCRISPR v2 plasmid, 3 μL of alkaline phosphatase, 3 μL of Esp3l, and sufficient H_2O to bring the total volume to 60 μL.
6. Incubate the restriction digest reaction(s) at 37 °C for 30 min.
7. Run the restriction digestion products on a 1% agarose gel and purify them using a gel extraction kit.
8. Prepare HiFi assembly on ice as follows using at least 1 assembly reaction per 10,000 sgRNAs present in the library: 350 ng digested lentiCRISPR v2 (Subheading 3.2.1, **step** 7), 30 ng PCR product (Subheading 3.2.1, **step 4**), H_2O to a volume of 10 μL, 10 μL HiFi assembly master mix. Include a control reaction in which the PCR product is omitted from the reaction.
9. Incubate the reactions at 50 °C for 1 h.
10. Cool the reactions on ice.
11. If your library contains >20,000 sgRNAs, concentrate the plasmid DNA using isopropanol precipitation prior to electroporation. If your library contains fewer than 20,000 sgRNAs, proceed directly to **step 17**.

12. To concentrate the HiFi assembly using isopropanol precipitation, pool all DNA assembly reactions and add 2 μL of 1 M NaCl and 20 μL of isopropanol for every 20 μL of HiFi assembly reaction.
13. Incubate at room temperature for 15 min and then centrifuge at 13,000 × *g* for 10 min.
14. Gently remove the supernatant and wash using 500 μL of 70% ethanol without disturbing the pellet. If the pellet is disturbed, centrifuge at 13,000 × *g* for 5 min.
15. Remove any residual ethanol and air-dry for 1 min.
16. Resuspend the DNA pellet in TE buffer at 1/5th of the original volume.
17. Float an MCE membrane on ~200 mL of water.
18. Carefully pipette the HiFi assembly onto the floating dialysis membrane. Incubate at room temperature for 15 min. Freeze the HiFi assembly at −20 °C or proceed directly to electroporation.

3.2.2 Electroporation

1. Chill electroporation cuvettes on ice.
2. Thaw electrocompetent cells on ice.
3. Transfer 1.5 μL of the HiFi assembly to each 50 μL aliquot of electrocompetent cells.
4. Transfer 1.5 μL of the negative control reaction to a 50 μL aliquot of electrocompetent cells.
5. Transfer each aliquot of electrocompetent cells to a 0.1 cm electroporation cuvette.
6. Set an electroporator to 1800 V, 600 Ω, and 10 μF.
7. Electroporate a cuvette and immediately add 1 mL of the supplied recovery media. Transfer the cells to a bacterial culture tube.
8. Repeat for all cuvettes.
9. Add an additional 1 mL of recovery media to each bacterial culture tube.
10. Incubate in a shaking incubator for 1 h at 37 °C.
11. Pool all cells transformed with the HiFi assembly together and, separately, all cells transformed with the HiFi negative control reaction. Record the total volume of each cell suspension.
12. Prepare a counting dilution by adding 10 μL of the bacterial suspension to 990 μL of LB. Mix well, then add 10 μL of the dilution to 990 μL of LB. Plate 100 μL of the final dilution onto a 10 cm LB agar plate containing 100 μg/mL ampicillin.
13. Repeat **step 12** for the HiFi negative control reaction.

14. Plate the remaining bacteria on large bioassay LB agar plates containing 100 μg/mL ampicillin. Use a cell spreader to evenly distribute the bacteria across the plates. There is no need to plate the remaining bacteria from the HiFi negative control reaction.
15. Incubate the plates at 30 °C for ~16 h.
16. Count the number of colonies on the 10 cm LB agar plates. You should have at least 10× as many colonies on the sgRNA library plate compared to the HiFi negative control plate.
17. Multiply by the volume (in milliliters) of the total cell suspension in **step 11**, then multiply by 100,000. This is an estimate of the total number of colonies. Proceed only when you have cloned at least 180× the number of sgRNAs present in the library (*see* **Note 9**).
18. Purify plasmid DNA with a plasmid DNA maxiprep kit.

3.3 Generating Lentiviruses

1. 24 h before transfection, passage ~85% confluent 293 T cells at a ratio of 1:2 into six 15-cm dishes containing 30 mL of media.
2. The next morning, add 135 μg cloned CRISPR library (Subheading 3.2.2, **step 18**), 90 μg psPax2, 67 μg pCMV-VSV-G, and 912 μL PEI solution to 5 mL Opti-MEM, mix using a vortex mixer, and incubate at room temperature for 15 min.
3. Add ~1 mL of the mixture dropwise to each 15-cm dish of 293 T cells.
4. Return the cultures to the incubator for 6–12 h.
5. Change media.
6. 48 h after transfection, collect media containing lentiviruses, and replace with fresh media.
7. Centrifuge media at 200 × *g* for 5 min.
8. Collect supernatant, add to ultracentrifuge tubes, and transfer the tubes to rotor buckets.
9. After balancing rotor buckets, centrifuge at 25,000 rpms for 1.5 h at 4 °C with a slow break.
10. When the centrifugation is complete, pour off the supernatant, and resuspend the pellet in 500 μL of PBS.
11. Combine lentiviruses and keep on ice.
12. Collect and concentrate lentiviruses from the 293 T cultures three additional times every 24 h following **steps 6–11**.
13. Combine all lentiviruses, aliquot, and freeze at −80 °C.

3.4 Lentivirus Titration

1. We recommend quantifying the viral titer using the same cell line used in the CRISPR screen (*see* **Note 10**).
2. Wash cells with PBS, treat with 0.25% Trypsin-EDTA solution for ~5 min, and wash cells off the plate using growth media.

3. Centrifuge the cells at 200 × *g* for 5 min, resuspend the cells in growth media at 1 million cells/mL, and plate 1 mL per well of a 12-well plate.
4. Add a serial dilution of lentivirus from 0 to 160 μL of virus solution and incubate overnight in a cell culture incubator.
5. The next day, passage the cells as in **steps 2** and **3**, and plate cells in 6 wells of a 12-well dish at 50,000 cells/well using 1 mL of media per well.
6. The next day, treat 3 wells/condition with 2 μg/mL puromycin for ~3 days, or until there are no surviving cells in the wells that were not transduced with virus but were treated with puromycin.
7. Wash the wells with PBS and treat with 0.5 mL 0.25% Trypsin-EDTA solution for ~5 min. Add 0.5 mL of media and transfer the cells to a flow cytometry tube.
8. Add 25 μL of prediluted CountBright beads to each tube.
9. Vortex and analyze on a flow cytometer (Fig. 2a). Collect data until you have counted ~200 beads.

$$\text{number of cells in a well} = 5000 \times \frac{\text{number of cells counted}}{\text{number of beads counted}}$$

10. To calculate the percent transduction of a given viral treatment volume, divide the cell number in the puromycin treated wells by the mean cell number in the untreated wells that received the same viral titer, and multiply by 100 (Fig. 2b).
11. Calculate the viral titer by choosing a virus volume that gave a transduction of 15–40%. Multiply the percent transduction (expressed as a decimal) by 1,000,000 (the number of cells in the well) and divide by the volume of virus added to the well.

3.5 Transducing a Mutant Cell Library

1. Calculate the volume of lentivirus needed per well by dividing 400,000 [1,000,000 cells at a mean of infection (MOI) of 0.4] by the viral titer calculated in Subheading 3.4 **step 11**.
2. Calculate the minimum number of cells needed by multiplying the number of sgRNAs in your library by 300 (*see* **Note 11**).
3. Calculate the starting number of cells by dividing the minimum number of cells (from **step 2**) by 0.4 (the expected percent transduction, expressed as a decimal).
4. Dissociate cells and plate them in 1 mL of media at 1 million cells/well of a 12-well dish as above.
5. Add lentiviruses to achieve a MOI of ~0.4.
6. The next day, dissociate and pool cells, pellet them at 200 × *g* for 5 min, and re-plate them in ~15× the culture volume.
7. The next day, treat cells with 2 μg/mL puromycin.

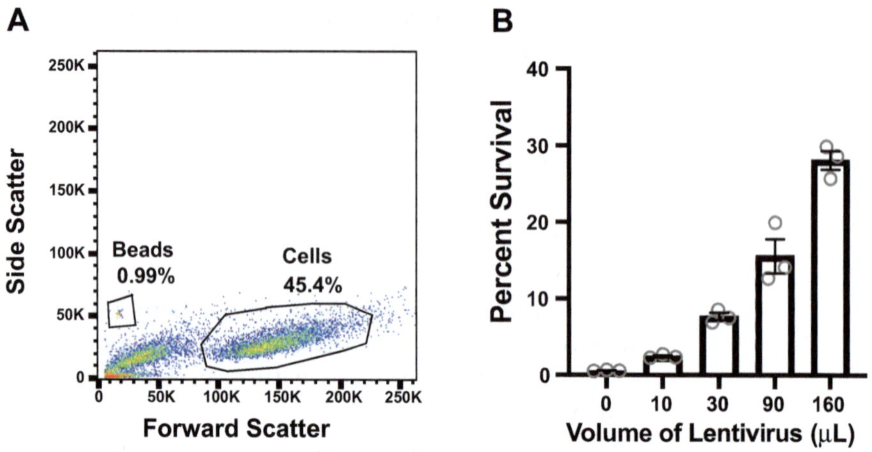

Fig. 2 Increased cell survival of HeLa cells transduced with a lentiCRISPR v2 library after treatment with puromycin. (**a**) Flow cytometry quantification of HeLa cell number after growth in the absence of puromycin. Note the relative size (Forward Scatter) of cells and beads. The bead population can be confirmed by visualizing their fluorescence across any of a wide range of excitation and emission flow cytometry filter channels (not shown). (**b**) HeLa cells were transduced as described in Subheading 3.4 by adding the indicated volumes of lentivirus to one million HeLa cells. 16 h later, cells were passaged, and 40 h after transduction, vehicle or 2 mg/mL puromycin was added to the culture. 3 days later, cells were dissociated and analyzed on a flow cytometer as shown in (**a**) after the addition of counting beads. Cell counts were normalized to the mean vehicle treated cell number. Dots represent different culture wells (technical replicates) and bars are means $\pm$ SEM

8. Continue puromycin selection for 5–7 days and passage when the cells when they reach ~80% confluency.
9. During passages, maintain the complexity of the library by re-plating the minimum number of cells needed (calculated in **step 2**).
10. Freeze aliquots of the mutant cell library during the early passages after puromycin selection. These aliquots can be thawed and used to perform CRISPR screens as shown here using a custom CRISPR library to identify regulators of clathrin mediated endocytosis (Fig. 3) [21].

4 Notes

1. If you use a different transfer vector, you may need to change the cloning strategy described in Subheading 3.1. These changes could include appending different sequences flanking the sgRNA target site and/or performing restriction digest of the transfer vector with a different restriction enzyme.
2. We favor activity-optimized libraries such as Addgene catalogue numbers 1000000096 and 73632.

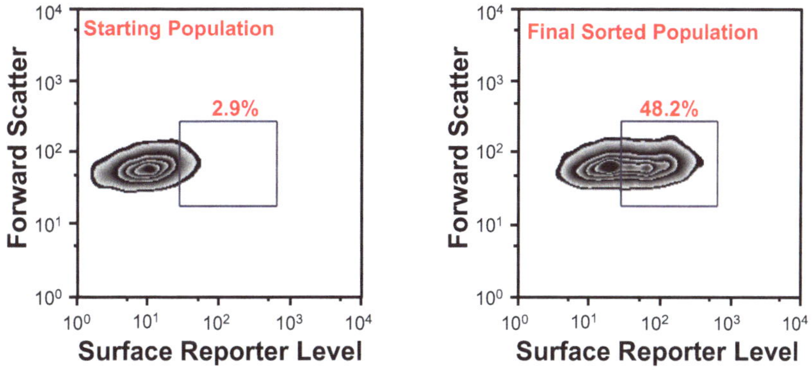

Fig. 3 Enrichment for mutants defective in clathrin mediated endocytosis. Flow cytometry analysis of the starting mutant population (left) and the final sorted population (right) of HeLa cells transduced with a custom CRISPR library designed to identify regulators of clathrin mediated endocytosis (CME) as previously described [21]. A CME cargo reporter was stably expressed in HeLa cells, and cell surface levels of the reporter were determined by labeling cells with anti-HA antibodies and APC-conjugated secondary antibodies. Reprinted from Dev Cell, 50(4), Gulbranson et al. AAGAB Controls AP2 Adaptor Assembly in Clathrin-Mediated Endocytosis, p. 436–446, 2019, with permission from Elsevier

3. This sequence is specific for cloning into lentiCRISPR v2 or lentiCRISPRv2GFP (Addgene catalogue numbers 52961 and 82416, respectively).
4. The online version of CHOPCHOP requires that genes be entered individually. A command line version is also available for bulk analysis (https://bitbucket.org/valenlab/chopchop/src/master/). Alternatively, CRISPick (https://portals.broadinstitute.org/gppx/crispick/public) allows for bulk design of sgRNAs targeting up to 500 genes at a time but it is currently limited to mouse, human, and rat genomes.
5. For a knockout manipulation, settings should be set to target the coding region of a gene, restrict targeting to search exons and immediate flanking regions, have the isoform consensus determined by the intersection, identify sgRNAs whose GC content is 20–80%, require self-complementarity to be below −1, have a 3′ NGG PAM sequence, identify off-targets with up to three mismatches in the protospacer, use the Doench et al. efficiency scoring method [25], and have no 5′ requirements for the sgRNA sequence.
6. It is easiest to use a subset of control sgRNAs from an existing library. If such a library is unavailable, you can use the command line version of CHOPCHOP to design them. Alternative options include using sgRNAs designed against negative control genes or scrambled sequences of the sgRNAs already included in your library. We recommend including a control sgRNA for every experimental sgRNA present in the library, up to a maximum of 1000 control sgRNAs. If fewer than 25% of

experimental sgRNAs are expected to manifest phenotypic changes, the number of negative control sgRNAs can be reduced to 10–25% of the sgRNA library.

7. We have had success with both CustomArray Inc. and Twist Bioscience. Although CustomArray has a higher reported error rate than Twist Bioscience, this has not been a problem for us.
8. Use fresh TE buffer in the gel apparatus. Old buffer may result in degradation of DNA samples.
9. You may need to pool plasmid DNA from multiple cloning reactions to achieve the 180× coverage. For such cases, pool a proportional mass of DNA relative to the library coverage achieved in a given cloning reaction.
10. This approach is optimized for HeLa cells. If another cell line is used, we recommend performing a dose-response curve with puromycin and adjusting the dosage accordingly. It may also be necessary to modify the density of cells during transduction as some cell types are bigger than HeLa cells.
11. This number is referred to as the fold coverage of the library. In general, a higher coverage leads to a greater screen resolution (ability to distinguish true positives from false positives). We view 300-fold coverage of the library as a minimum and advise using a higher coverage if it is feasible.

Acknowledgments

This work was supported by a NIH fellowship award F32AG062039 (to D.G.).

References

1. Edwards SL, Beesley J, French JD, Dunning AM (2013) Beyond GWASs: illuminating the dark road from association to function. Am J Hum Genet 93(5):779–797. https://doi.org/10.1016/j.ajhg.2013.10.012
2. Gallagher MD, Chen-Plotkin AS (2018) The post-GWAS era: from association to function. Am J Hum Genet 102(5):717–730. https://doi.org/10.1016/j.ajhg.2018.04.002
3. Spreafico R, Soriaga LB, Grosse J, Virgin HW, Telenti A (2020) Advances in genomics for drug development. Genes (Basel) 11(8):942. https://doi.org/10.3390/genes11080942
4. Muller HJ (1927) Artificial transmutation of the gene. Science 66(1699):84–87. https://doi.org/10.1126/science.66.1699.84
5. Beadle GW, Tatum EL (1941) Genetic control of biochemical reactions in neurospora. Proc Natl Acad Sci U S A 27(11):499–506. https://doi.org/10.1073/pnas.27.11.499
6. Hartwell LH, Mortimer RK, Culotti J, Culotti M (1973) Genetic control of the cell division cycle in yeast: V. Genetic analysis of cdc mutants. Genetics 74(2):267–286
7. Brenner S (1974) The genetics of Caenorhabditis elegans. Genetics 77(1):71–94
8. Nüsslein-Volhard C, Wieschaus E (1980) Mutations affecting segment number and polarity in Drosophila. Nature 287(5785): 795–801. https://doi.org/10.1038/287795a0
9. Novick P, Field C, Schekman R (1980) Identification of 23 complementation groups required for post-translational events in the yeast secretory pathway. Cell 21(1):205–215.

https://doi.org/10.1016/0092-8674(80)90128-2

10. Ishino Y, Shinagawa H, Makino K, Amemura M, Nakata A (1987) Nucleotide sequence of the iap gene, responsible for alkaline phosphatase isozyme conversion in Escherichia coli, and identification of the gene product. J Bacteriol 169(12):5429–5433. https://doi.org/10.1128/jb.169.12.5429-5433.1987
11. Mojica FJ, Ferrer C, Juez G, Rodríguez-Valera F (1995) Long stretches of short tandem repeats are present in the largest replicons of the Archaea Haloferax mediterranei and Haloferax volcanii and could be involved in replicon partitioning. Mol Microbiol 17(1):85–93. https://doi.org/10.1111/j.1365-2958.1995.mmi_17010085.x
12. Deltcheva E, Chylinski K, Sharma CM, Gonzales K, Chao Y, Pirzada ZA, Eckert MR, Vogel J, Charpentier E (2011) CRISPR RNA maturation by trans-encoded small RNA and host factor RNase III. Nature 471(7340):602–607. https://doi.org/10.1038/nature09886
13. Jinek M, Chylinski K, Fonfara I, Hauer M, Doudna JA, Charpentier E (2012) A programmable dual-RNA-guided DNA endonuclease in adaptive bacterial immunity. Science 337(6096):816–821. https://doi.org/10.1126/science.1225829
14. Gasiunas G, Barrangou R, Horvath P, Siksnys V (2012) Cas9-crRNA ribonucleoprotein complex mediates specific DNA cleavage for adaptive immunity in bacteria. Proc Natl Acad Sci U S A 109(39):E2579–E2586. https://doi.org/10.1073/pnas.1208507109
15. Cong L, Ran FA, Cox D, Lin S, Barretto R, Habib N, Hsu PD, Wu X, Jiang W, Marraffini LA, Zhang F (2013) Multiplex genome engineering using CRISPR/Cas systems. Science 339(6121):819–823. https://doi.org/10.1126/science.1231143
16. Shalem O, Sanjana NE, Hartenian E, Shi X, Scott DA, Mikkelson T, Heckl D, Ebert BL, Root DE, Doench JG, Zhang F (2014) Genome-scale CRISPR-Cas9 knockout screening in human cells. Science 343(6166):84–87. https://doi.org/10.1126/science.1247005
17. Wang T, Wei JJ, Sabatini DM, Lander ES (2014) Genetic screens in human cells using the CRISPR-Cas9 system. Science 343(6166):80–84. https://doi.org/10.1126/science.1246981
18. Parnas O, Jovanovic M, Eisenhaure TM, Herbst RH, Dixit A, Ye CJ, Przybylski D, Platt RJ, Tirosh I, Sanjana NE, Shalem O, Satija R, Raychowdhury R, Mertins P, Carr SA, Zhang F, Hacohen N, Regev A (2015) A genome-wide CRISPR screen in primary immune cells to dissect regulatory networks. Cell 162(3):675–686. https://doi.org/10.1016/j.cell.2015.06.059
19. Menasche BL, Davis EM, Wang S, Ouyang Y, Li S, Yu H, Shen J (2020) PBRM1 and the glycosylphosphatidylinositol biosynthetic pathway promote tumor killing mediated by MHC-unrestricted cytotoxic lymphocytes. Sci Adv 6(48):eabc3243. https://doi.org/10.1126/sciadv.abc3243
20. Gulbranson DR, Davis EM, Demmitt BA, Ouyang Y, Ye Y, Yu H, Shen J (2017) RABIF/MSS4 is a Rab-stabilizing holdase chaperone required for GLUT4 exocytosis. Proc Natl Acad Sci U S A 114(39):E8224–E8233. https://doi.org/10.1073/pnas.1712176114
21. Gulbranson DR, Crisman L, Lee M, Ouyang Y, Menasche BL, Demmitt BA, Wan C, Nomura T, Ye Y, Yu H, Shen J (2019) AAGAB controls AP2 adaptor assembly in Clathrin-mediated endocytosis. Dev Cell 50(4):436–446.e435. https://doi.org/10.1016/j.devcel.2019.06.013
22. Kramer NJ, Haney MS, Morgens DW, Jovičić A, Couthouis J, Li A, Ousey J, Ma R, Bieri G, Tsui CK, Shi Y, Hertz NT, Tessier-Lavigne M, Ichida JK, Bassik MC, Gitler AD (2018) CRISPR-Cas9 screens in human cells and primary neurons identify modifiers of C9ORF72 dipeptide-repeat-protein toxicity. Nat Genet 50(4):603–612. https://doi.org/10.1038/s41588-018-0070-7
23. Jansen IE, Savage JE, Watanabe K, Bryois J, Williams DM, Steinberg S, Sealock J, Karlsson IK, Hägg S, Athanasiu L, Voyle N, Proitsi P, Witoelar A, Stringer S, Aarsland D, Almdahl IS, Andersen F, Bergh S, Bettella F, Bjornsson S, Brækhus A, Bråthen G, de Leeuw C, Desikan RS, Djurovic S, Dumitrescu L, Fladby T, Hohman TJ, Jonsson PV, Kiddle SJ, Rongve A, Saltvedt I, Sando SB, Selbæk G, Shoai M, Skene NG, Snaedal J, Stordal E, Ulstein ID, Wang Y, White LR, Hardy J, Hjerling-Leffler J, Sullivan PF, van der Flier WM, Dobson R, Davis LK, Stefansson H, Stefansson K, Pedersen NL, Ripke S, Andreassen OA, Posthuma D (2019) Genome-wide meta-analysis identifies new loci and functional pathways influencing Alzheimer's disease risk. Nat Genet 51(3):404–413. https://doi.org/10.1038/s41588-018-0311-9
24. Kunkle BW, Grenier-Boley B, Sims R, Bis JC, Damotte V, Naj AC, Boland A, Vronskaya M, van der Lee SJ, Amlie-Wolf A, Bellenguez C, Frizatti A, Chouraki V, Martin ER, Sleegers K,

Badarinarayan N, Jakobsdottir J, Hamilton-Nelson KL, Moreno-Grau S, Olaso R, Raybould R, Chen Y, Kuzma AB, Hiltunen M, Morgan T, Ahmad S, Vardarajan BN, Epelbaum J, Hoffmann P, Boada M, Beecham GW, Garnier JG, Harold D, Fitzpatrick AL, Valladares O, Moutet ML, Gerrish A, Smith AV, Qu L, Bacq D, Denning N, Jian X, Zhao Y, Del Zompo M, Fox NC, Choi SH, Mateo I, Hughes JT, Adams HH, Malamon J, Sanchez-Garcia F, Patel Y, Brody JA, Dombroski BA, MCD N, Daniilidou M, Eiriksdottir G, Mukherjee S, Wallon D, Uphill J, Aspelund T, Cantwell LB, Garzia F, Galimberti D, Hofer E, Butkiewicz M, Fin B, Scarpini E, Sarnowski C, Bush WS, Meslage S, Kornhuber J, White CC, Song Y, Barber RC, Engelborghs S, Sordon S, Voijnovic D, Adams PM, Vandenberghe R, Mayhaus M, Cupples LA, Albert MS, De Deyn PP, Gu W, Himali JJ, Beekly D, Squassina A, Hartmann AM, Orellana A, Blacker D, Rodriguez-Rodriguez-E, Lovestone S, Garcia ME, Doody RS, Munoz-Fernadez C, Sussams R, Lin H, Fairchild TJ, Benito YA, Holmes C, Karamujić-Čomić H, Frosch MP, Thonberg H, Maier W, Roshchupkin G, Ghetti B, Giedraitis V, Kawalia A, Li S, Huebinger RM, Kilander L, Moebus S, Hernández I, Kamboh MI, Brundin R, Turton J, Yang Q, Katz MJ, Concari L, Lord J, Beiser AS, Keene CD, Helisalmi S, Kloszewska I, Kukull WA, Koivisto AM, Lynch A, Tarraga L, Larson EB, Haapasalo A, Lawlor B, Mosley TH, Lipton RB, Solfrizzi V, Gill M, Longstreth WT, Montine TJ, Frisardi V, Diez-Fairen M, Rivadeneira F, Petersen RC, Deramecourt V, Alvarez I, Salani F, Ciaramella A, Boerwinkle E, Reiman EM, Fievet N, Rotter JI, Reisch JS, Hanon O, Cupidi C, Andre Uitterlinden AG, Royall DR, Dufouil C, Maletta RG, de Rojas I, Sano M, Brice A, Cecchetti R, George-Hyslop PS, Ritchie K, Tsolaki M, Tsuang DW, Dubois B, Craig D, Wu CK, Soininen H, Avramidou D, Albin RL, Fratiglioni L, Germanou A, Apostolova LG, Keller L, Koutroumani M, Arnold SE, Panza F, Gkatzima O, Asthana S, Hannequin D, Whitehead P, Atwood CS, Caffarra P, Hampel H, Quintela I, Carracedo Á, Lannfelt L, Rubinsztein DC, Barnes LL, Pasquier F, Frölich L, Barral S, McGuinness B, Beach TG, Johnston JA, Becker JT, Passmore P, Bigio EH, Schott JM, Bird TD, Warren JD, Boeve BF, Lupton MK, Bowen JD, Proitsi P, Boxer A, Powell JF, Burke JR, JSK K, Burns JM, Mancuso M, Buxbaum JD, Bonuccelli U, Cairns NJ, McQuillin A, Cao C, Livingston G, Carlson CS, Bass NJ, Carlsson CM, Hardy J, Carney RM, Bras J, Carrasquillo MM, Guerreiro R, Allen M, Chui HC, Fisher E, Masullo C, Crocco EA, DeCarli C, Bisceglio G, Dick M, Ma L, Duara R, Graff-Radford NR, Evans DA, Hodges A, Faber KM, Scherer M, Fallon KB, Riemenschneider M, Fardo DW, Heun R, Farlow MR, Kölsch H, Ferris S, Leber M, Foroud TM, Heuser I, Galasko DR, Giegling I, Gearing M, Hüll M, Geschwind DH, Gilbert JR, Morris J, Green RC, Mayo K, Growdon JH, Feulner T, Hamilton RL, Harrell LE, Drichel D, Honig LS, Cushion TD, Huentelman MJ, Hollingworth P, Hulette CM, Hyman BT, Marshall R, Jarvik GP, Meggy A, Abner E, Menzies GE, Jin LW, Leonenko G, Real LM, Jun GR, Baldwin CT, Grozeva D, Karydas A, Russo G, Kaye JA, Kim R, Jessen F, Kowall NW, Vellas B, Kramer JH, Vardy E, LaFerla FM, Jöckel KH, Lah JJ, Dichgans M, Leverenz JB, Mann D, Levey AI, Pickering-Brown S, Lieberman AP, Klopp N, Lunetta KL, Wichmann HE, Lyketsos CG, Morgan K, Marson DC, Brown K, Martiniuk F, Medway C, Mash DC, Nöthen MM, Masliah E, Hooper NM, McCormick WC, Daniele A, McCurry SM, Bayer A, McDavid AN, Gallacher J, McKee AC, van den Bussche H, Mesulam M, Brayne C, Miller BL, Riedel-Heller S, Miller CA, Miller JW, Al-Chalabi A, Morris JC, Shaw CE, Myers AJ, Wiltfang J, O'Bryant S, Olichney JM, Alvarez V, Parisi JE, Singleton AB, Paulson HL, Collinge J, Perry WR, Mead S, Peskind E, Cribbs DH, Rossor M, Pierce A, Ryan NS, Poon WW, Nacmias B, Potter H, Sorbi S, Quinn JF, Sacchinelli E, Raj A, Spalletta G, Raskind M, Caltagirone C, Bossù P, Orfei MD, Reisberg B, Clarke R, Reitz C, Smith AD, Ringman JM, Warden D, Roberson ED, Wilcock G, Rogaeva E, Bruni AC, Rosen HJ, Gallo M, Rosenberg RN, Ben-Shlomo Y, Sager MA, Mecocci P, Saykin AJ, Pastor P, Cuccaro ML, Vance JM, Schneider JA, Schneider LS, Slifer S, Seeley WW, Smith AG, Sonnen JA, Spina S, Stern RA, Swerdlow RH, Tang M, Tanzi RE, Trojanowski JQ, Troncoso JC, Van Deerlin VM, Van Eldik LJ, Vinters HV, Vonsattel JP, Weintraub S, Welsh-Bohmer KA, Wilhelmsen KC, Williamson J, Wingo TS, Woltjer RL, Wright CB, Yu CE, Yu L, Saba Y, Pilotto A, Bullido MJ, Peters O, Crane PK, Bennett D, Bosco P, Coto E, Boccardi V, De Jager PL, Lleo A, Warner N, Lopez OL, Ingelsson M, Deloukas P, Cruchaga C, Graff C, Gwilliam R, Fornage M, Goate AM, Sanchez-Juan P, Kehoe PG, Amin N, Ertekin-Taner N, Berr C, Debette S, Love S, Launer LJ, Younkin SG,

Dartigues JF, Corcoran C, Ikram MA, Dickson DW, Nicolas G, Campion D, Tschanz J, Schmidt H, Hakonarson H, Clarimon J, Munger R, Schmidt R, Farrer LA, Van Broeckhoven C, O'Donovan M, Destefano AL, Jones L, Haines JL, Deleuze JF, Owen MJ, Gudnason V, Mayeux R, Escott-Price V, Psaty BM, Ramirez A, Wang LS, Ruiz A, van Duijn CM, Holmans PA, Seshadri S, Williams J, Amouyel P, Schellenberg GD, Lambert JC, Pericak-Vance MA, Consortium ADG, Initiative EAsD, Heart Cf, Consortium ARiGE, Genetic, Environmental Risk in AD/Defining Genetic Pl, Consortium ERfAsD (2019) Genetic meta-analysis of diagnosed Alzheimer's disease identifies new risk loci and implicates Aβ, tau, immunity and lipid processing. Nat Genet 51(3):414–430. https://doi.org/10.1038/s41588-019-0358-2

25. Doench JG, Fusi N, Sullender M, Hegde M, Vaimberg EW, Donovan KF, Smith I, Tothova Z, Wilen C, Orchard R, Virgin HW, Listgarten J, Root DE (2016) Optimized sgRNA design to maximize activity and minimize off-target effects of CRISPR-Cas9. Nat Biotechnol 34(2):184–191. https://doi.org/10.1038/nbt.3437

Chapter 22

Studying Unconventional Secretion of Misfolded Proteins in Cultured Cells and Primary Neurons

Juhyung Lee and Yihong Ye

Abstract

Protein misfolding poses a significant threat to the fitness of eukaryotic cells, particularly for neurons facing environmental stress. To effectively triage and remove defective and unwanted proteins, cells have evolved diverse protein quality control (PQC) mechanisms relying on proteasome- and endolysosome-mediated degradation systems. Defects in PQC functions are linked to various human diseases including many aging-associated neurodegenerative diseases. Misfolding-associated protein secretion (MAPS) is a recently reported PQC mechanism that eliminates misfolded cytosolic proteins by an unconventional secretory pathway using an endo-vesiclular network. This process implicates DNAJC5, a chaperone that escorts misfolded cargos to intracellular vesicles to facilitate their secretion. Cargos of DNAJC5 include Parkinson's and Alzheimer's disease-associated proteins known to undergo cell-to-cell transmission during disease progression. Thus, elucidating how these proteins are secreted may reveal novel therapeutic targets for these diseases. Here we describe a collection of methods used to detect either the basal or induced secretion of misfolded proteins from cell lines and cultured primary neurons.

Key words Misfolding-associated protein secretion (MAPS), Unconventional protein secretion (UPS), Protein quality control (PQC), DNAJC5/CLN4/CSPα, α-Synuclein, Tau, β-Galactosidase/LacZ assay, Primary murine hippocampal neurons, Parkinson's disease, Alzheimer's disease

1 Introduction

Cells release a variety of proteins in response to environmental or internal ques. in either a constant or inducible manner. Besides the classical secretory pathway that relies on a leader sequence (also known as signal sequence or signal peptide) to translocate nascent polypeptides into the endoplasmic reticulum (ER), several other mechanisms exist to export certain proteins lacking leader sequences. These processes are collectively termed unconventional protein secretion (UPS). To reach the cell exterior, cytosolic cargos can either translocate across the plasma membrane directly or first enter an intracellular vesicle compartment before secretion [1, 2].

Jingshi Shen (ed.), *Membrane Trafficking: Methods and Protocols*,
Methods in Molecular Biology, vol. 2473, https://doi.org/10.1007/978-1-0716-2209-4_22,

Interestingly, UPS can be used to selectively export misfolded proteins without causing cell membrane leakage. This process, dubbed as misfolding-associated protein secretion (MAPS), helps eukaryotic cells to alleviate protein misfolding-associated proteotoxic stress and is critical for cell fitness under stress conditions [3]. Intriguingly, substrates of the MAPS pathway include many neurodegenerative disease-associated proteins such as microtubule-associated protein Tau and the pre-synaptic protein α-Synuclein (α-Syn), both of which are known to be present in extracellular space such as cerebrospinal fluid, but how these proteins are released from cells had been unclear until recently [3–6].

To date, an ER-associated deubiquitinase (USP19) and a cytosolic co-chaperone (DNAJC5 also known as CSPα or CLN4) are the only known factors that transport misfolded substrates to an unconventional secretory compartment overlapping with the endo-lysosomal system for releasing into the cell exterior [3, 5]. The level of misfolded proteins undergoing unconventional secretion is usually low, but the secretion capacity can be stimulated to accommodate increased protein misfolding or in response to stress, suggesting that this process may be a back-up PQC mechanism safeguarding the integrity of the cell proteome. The low secretion efficiency makes the MAPS pathway difficult to detect, and thus have hindered our understanding of the underlying mechanism and its biological significance. In this chapter, we report methodology focusing on the detection of the MAPS substrates Tau and α-Syn in conditioned medium (CM) obtained from cell lines and cultured primary neurons.

2 Materials

2.1 Cell Culture, Treatment, and DNA Transfection

1. Complete media: DMEM (high glucose), 10% FBS, 1× Penicillin-Streptomycin.
2. Opti-MEM™ I Reduced Serum Medium (Gibco).
3. TransIT®-LT1 Transfection Reagent (Mirus).
4. Ionomycin calcium salt: Dissolve in DMSO to make 10 mM stock. Store at −20 °C.

2.2 Preparation of Conditioned Media

1. NP40 lysis buffer: 0.5% Nonidet P-40, 50 mM Tris-HCl, pH 7.4, 150 mM NaCl 2 mM $MgCl_2$, 1 mM EDTA. Add 1 mM DTT and a protease inhibitor mixture immediately before use.
2. Benchtop microcentrifuge.
3. 1.5 mL tubes.
4. 4× Laemmli sample buffer.
5. Nitrocellulose Membrane, 0.45 μm.

6. 5% BSA in PBS. 0.02% of Sodium Azide is added.
7. 5% skim milk in PBS.
8. Phosphate buffer saline (PBS).
9. Western Lightning Plus, Enhanced Chemiluminescence Substrate (PerkinElmer).
10. Western Lightning Ultra (PerkinElmer).
11. Western blotting imaging system. We use the ChemiDoc MP Imaging instrument from Bio-Rad.

2.3 LacZ-α-Syn Secretion Experiments

1. 5× lacZ buffer: 40 mM Na_2HPO_4, 60 mM NaH_2PO_4, 10 mM KCl, 1 mM $MgSO_4$, adjust pH to pH 7.0 using either NaOH or HCl.
2. Chlorophenol Red-β-D-galactopyranoside (CPRG): Dissolve in dH_2O to make a 0.4 M solution. Store in −20 °C.
3. LacZ substrate solution (*see* Table 1).
4. 96-well Clear Flat Bottom TC-treated Microplates.
5. Multi-channel pipette.
6. Plate reader. We use the VICTOR3 Multilabel Plate Reader (PerkinElmer).

2.4 Mouse Primary Hippocampal Neuron Culture

2.4.1 Reagents

1. Poly-D-lysine (0.1 mg/mL): Dilute to 50 μg/mL in sterile PBS.
2. MEM (Minimum Essential Medium).
3. Hank's Balanced Salt Solution (HBSS).
4. Laminin (1 mg/mL): Dilute to 2 μg/mL in MEM.
5. Trypsin-EDTA (0.25%) with phenol red.
6. Trypsin inactivation media (10% FBS in MEM).
7. DNase I (20 mg/mL).
8. Neuronal plating media (*see* Table 2).
9. Neuronal culture media (*see* Table 3).
10. Recommended plating media volume (*see* Table 4).

2.4.2 Materials

1. P0-1 mouse pups.
2. Dissection Microscope.
3. 15 mL cornical centrifuge tubes.
4. Cell culture plates or μ-Slide 4 Well Glass Bottom for live imaging.
5. Ice.
6. Petri dishes, 60 × 15 mm.
7. Petri dishes, 100 × 20 mm.

Table 1
lacZ substrate solution

lacZ substrate solution	Working conc.	200 μL (for one reaction)
5× lacZ buffer	1×	40 μL
0.4 M CPRG	500 μM	0.25 μL
β-Mercaptoethanol	0.14% (v/v)	0.28 μL
dH_2O	Up to 200 μL	159.47 μL

Table 2
Neuron plating media (*see* Note 1)

	Stock conc.	Solvent	Working conc.	Dilution
FBS	100%	–	10%	10×
Sodium pyruvate	100 mM	–	1 mM	100×
L-Glutamine	200 mM	H_2O, filtering	2 mM	100×
Primocin	50 mg/mL	–	100 μg/mL	500×
Glucose solution	20%	H_2O, filtering	0.6%	33.33×
MEM	Up to final volume			

Table 3
Neuron culture media

	Stock conc.	Solvent	Working conc.	Dilution
B-27 Supplement (50×)	50×	–	1×	50×
L-Glutamine	200 mM	H_2O, filtering	0.5 mM	400×
Primocin	50 mg/mL		100 μg/mL	500×
Neurobasal medium	Up to final volume			

Table 4
Recommended plating media volume

Plate type	6 well	12 well	24 well
Surface area per well (cm^2)	10	4	2
Plating media volume (μL)	2500	1000	500

8. Fine forceps, straight.
9. Surgical scissors, small.
10. Pipettes.
11. Centrifuges.
12. Cell strainer with a pore size of 70 μm.

2.5 Immuno-cytochemistry in Cultured Hippocampal Neurons

1. Coverslips (#1.5 Thickness).
2. Fixation solution: 4% paraformaldehyde in PBS (v/v).
3. Permeabilization solution: 0.5% Triton X-100 in PBS (v/v).
4. Blocking solution: 0.1% Triton X-100 in PBS (v/v) with 10% fetal bovine serum.
5. Parafilm.
6. Antifade Mounting Medium.
7. Super Frost Plus Micro Slide (1 mm Thickness): Clean surface with methanol or isopropanol before use.
8. Sealant.
9. Imaging microscope.

2.6 Plasmids and Antibodies

See Tables 5 and 6, respectively.

2.7 Production of Lentivirus for Infection

1. 293FT cells.
2. 6 well plates.
3. 3 mL syringes with luer lock tip.
4. Syringe Filter Units, 0.45 μm, PVDF.

3 Methods (*See* Note 3)

3.1 Preparation of Conditioned Media (CM) from HEK293T Cells for Immunoblotting

1. Seed 2.5×10^5 cells to poly-D-lysine coated 12-well plates.
2. When cells reach 40–50% confluency (usually on the next day after seeding), replace medium with 1 mL of fresh DMEM medium. Use the TransIT 293 transfection reagent to transfect cells with 250 ng of the plasmid expressing a MAPS substrate (α-Syn, Tau. or GFP1-10, etc.) together with 250 ng of the plasmid expressing a MAPS stimulator (USP19 or DNAJC5) if necessary.
3. 4 h after the transfection, add 1 mL of fresh medium to each well.
4. 24 h post-transfection, the cell confluency should reach 80–90%. Replace media with 1.5 mL of fresh media to start collecting the secreted proteins.

Table 5
Plasmids (*see* Note 2)

Name	Vector	Promoter	Note
FLAG-GFP (1-10)	pCMV6-XL6	CMV	
EGFP-hTau E14	pRK5	CMV	Addgene #46907, all 14S/P or T/P in human Tau were replaced with E.
EGFP-hαSyn	pEGFP C1	CMV	
FLAG-DNAJC5 WT	pCMV6-XL6	CMV	
FLAG-DNAJC5 ΔJ	pCMV6-XL6	CMV	
lacZ-hαSyn	pcDNA6/V5-His/lacZ	CMV	V5-His was destroyed
FLAG-DNAJC5 WT	FSW	Synapsin	Neuron-specific expression, lentiviral
FLAG-DNAJC5 ΔJ	FSW	Synapsin	Neuron-specific expression, lentiviral
EGFP-hαSyn	FSW	Synapsin	Neuron-specific expression, lentiviral
psPAX2			Addgene #12260, for lentiviral packaging
pVSV-G			Addgene #138479, for lentiviral packaging

5. After 12–16 h, gently tilt the plate back-and-forth a few times and then transfer 1 mL of the CM from each well to a 1.5 mL microtube. Centrifuge at 1000 × *g* for 10 min at 4 °C to remove contaminated cells.
6. Transfer 500 μL of supernatant carefully to a new 1.5 mL microtube and centrifuge at 10,000 × *g* for 30 min at 4 °C to remove cell debris.
7. Take 120 μL (or your desired volume) of supernatant and add 40 μL of 4× laemmli sample buffer. Heat the samples at 95 °C for 5 min. The CM samples are now ready for SDS-PAGE and immunoblotting.
8. For preparation of cell lysates, remove the remaining media from each well and wash the cells with ice-cold PBS twice. Lyse the cells in 200 μL of the NP40 lysis buffer containing a protease inhibitor cocktail and 1 mM DTT for 20 min on ice.
9. Transfer cell lysates to 1.5 mL tubes and centrifuge at 16,000 × *g* for 20 min at 4 °C.

Table 6
Antibodies

Name	Source	Manufacturer	Application	Dilution
α-Syn	Mouse	BD Bioscience (610787)	IB, ICC	1:1000 (IB), 1: 200 (ICC)
Clusterin	Mouse	SantaCruz (sc-5289)	IB	1:2000
DNAJC5	Rabbit	LSbio (LS-C22593)	IB, ICC	1:2000 (IB), 1: 200 (ICC)
GFP (B2)	Mouse	SantaCruz (sc-9996)	IB	1:2000
FLAG	Mouse	Sigma (F1804)	IB, ICC	1:2000 (IB), 1: 500 (ICC)
HSP90	Mouse	SantaCruz (sc-13119)	IB	1:2000
HSC70	Mouse	SantaCruz (sc-7298)	IB	1:2000
Anti-mouse IgG, HRP conjugated		Sigma (A4416)	IB	1:2000
Anti-rabbit IgG, HRP conjugated		Sigma (A6154)	IB	1:2000
IRDye® 680RD Goat anti-Mouse IgG Secondary Antibody	Goat	Thermo (A21058)	IB	1:10,000
IRDye® 800CW Goat anti-Rabbit IgG Secondary Antibody	Goat	Rockland (611-132-003)	IB	1:10,000
anti-Rabbit IgG (H + L) Cross-Adsorbed Secondary Antibody, Alexa Fluor 568	Goat	Thermo (A11011)	ICC	1:500
DAPI (1 mg/mL in dH_2O)		Sigma ('D9542)	ICC	1:1000

10. Transfer cleared cell extracts to new tubes and mix it with 4× laemmli sample buffer at a 3:1 volume ratio. Heat the lysate samples at 95 °C for 5 min before running SDS-PAGE electrophoresis.
11. Separate proteins in the medium and lysate samples by SDS-PAGE and perform immunoblotting analysis (*see* Subheading 3.2).

3.2 Immunoblotting

1. After transfer proteins from SDS-PAGE gels to a nitocellular membrane, incubate the membrane with 5% skim milk in PBS for 30 min at room temperature.

2. Wash the membrane with PBS for 10 min three times.
3. Incubate the membrane in PBS containing 5% BSA and primary antibodies with appropriate dilution factors (*see* Table 6) in 4 °C for 12–16 h. (*See* **Note 4**).
4. Wash the membrane with PBS for 10 min three times.
5. Incubate the membrane in PBS with 5% Skim milk and HRP-conjugated secondary antibodies (*see* Table 6) for 1 h at room temperature (*see* **Note 5**).
6. Wash the membrane with PBS for 10 min three times.
7. Develop membrane using Western Lightning ECL-Plus solution. If the signal is weak, use the more sensitive ECL substrate Western Lightning Ultra instead, and scan with an imager (Fig. 1, *see* **Note 6**).

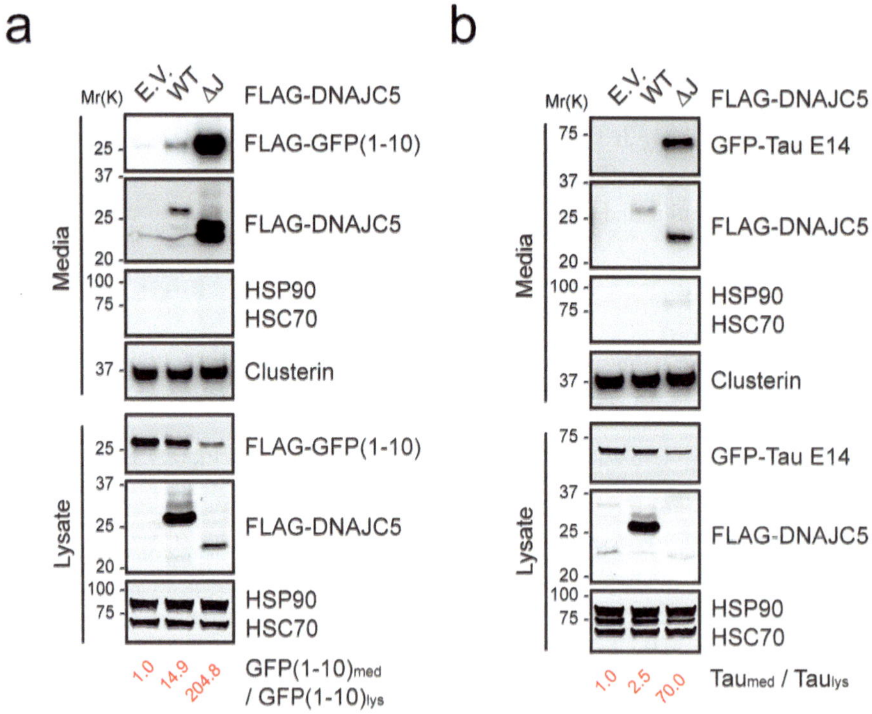

Fig. 1 Secretion of MAPS substrates from HEK293T cell. HEK293T cells were transfected with MAPS substrates such as FLAG-GFP1-10 (**a**) or FLAG-Tau E14 (**b**) together with either FLAG-tagged DNAJC5 WT or the hyper-active DNAJC5 mutant ΔJ to induce MAPS. At 24 h post-transfection, the medium was replaced with fresh medium. After 16 h incubation, the CM were harvested and analyzed by immunoblotting as indicated. Note that DNAJC5 itself is also secreted along with its cargoes. The levels of secretion were quantified as the ratio of the secreted protein in the medium to that in the lysate

3.3 LacZ-α-Syn Secretion Assay (See Note 7)

1. Prepare CM and lysate samples from HEK293T cells expressing LacZ-α-Syn following the steps 1-9 in Subheading 3.1. Also prepare a CM sample from HEK293T cells with no LacZ expression ($Blank_{med}$) for background subtraction.
2. Place 100 μL of CM sample into a 96-well microplate. Each condition is usually analyzed in triplicate.
3. Dilute each lysate sample at 1:100 in the same lysis buffer and place 100 μL of the diluted samples into a 96-well plate. Use the lysis buffer as the background control ($Blank_{lys}$). Each sample should be analyzed in triplicate.
4. Add 200 μL of substrate solution to each well using a multichannel pipette (300 μL reactant in total). Avoid introducing bubbles.
5. Monitor the absorbance at A_{570} immediately using a microplate reader. Make sure the readings are within the linear range (*see* **Note 8**) (Fig. 2a, b).
6. For quantification, calculate the ratio of LacZ-α-Syn in media and in lysate after subtracting the corresponding background as follows:

$$\text{LacZ-}\alpha\text{-Syn secretion} = \left(\text{LacZ-}\alpha\text{-Syn}_{med} - \text{Blank}_{med}\right) / \left(\text{LacZ-}\alpha\text{-Syn}_{lys} - \text{Blank}_{lys}\right)$$

7. Normalize the values and perform statistical analysis (Fig. 2c).

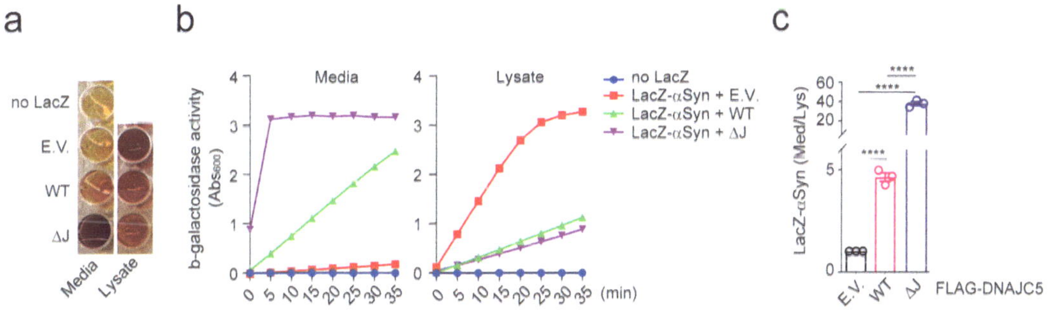

Fig. 2 The LacZ-α-Syn secretion assay. (**a**) Examples of the LacZ-α-Syn reaction. CM and lysate samples from HEK293T cells stably expressing LacZ-α-Synuclein alone or together with the indicated DNAJC5 variants were assayed with the CPRG substrate (*see* Subheading 3.3). The β-galactosidase activity converts CPRG (yellow) to CPR (red). (**b**) Kinetics of CPR production as a result of β-galactosidase activity. The reactants for the CM and lysates were monitored using a plate reader for 35 min with 5 min intervals. Note that the enzymatic activity in the medium from ΔJ-expressing cells is saturated within 5 min. The optimal time point to compare this set of samples should be within 5 min. (**c**) The LacZ-α-Syn assay reveals an increased MAPS-stimulating activity for the DNAJC5 ΔJ mutant. The ratio of Abs_{570} in media/lysates was calculated and normalized to the control ($n = 3$). ****$p < 0.0001$

3.4 Preparation of Mouse Primary Hippocampal Neurons (See Note 9)

3.4.1 Coating and Preparation of Cell Culture Plates

1. Dilute poly-D-lysine solution with sterile dH_2O to a final concentration of 50 μg/mL.
2. Cover the wells of a culture plate with 50 μg/mL poly-D-lysine solution and incubate the plate for 1 h or longer at room temperature.
3. Aspirate the poly-D-lysine solution. Wash the wells two times with sterile dH_2O. Aspirate the wells to remove all remaining liquids. The poly-D-lysine solution can be reused 3–4 times. Store at 4 °C until use.
4. The coated plates can be wrap-sealed with parafilm or plastic wrap and stored at 2–8 °C for up to 1 month.
5. Start the below remaining steps on the day of mouse hippocampal neuron isolation. Cover the wells of a poly-D-lysine coated plate with the laminin media.
6. Incubate the plate at room temperature for 1 h.
7. Aspirate the laminin media while tilting the plate and remove all remaining liquids.

3.4.2 Dissection of Mouse Hippocampi

Before starting, dissecting tools should be autoclaved or disinfected with 70% ethanol.

1. Anesthetize P0 or P1 pups by putting them on crushed ice for 2–5 min. Use thin materials such as latex gloves to avoid freezing damage to the pup skins.
2. Wash pups with ice-cold PBS and transfer them one-by-one to a 100 × 20 mm petri dish.
3. Decapitate each pup at the head/neck junction using a pair of small surgical scissors.
4. Place the head in a new 60 × 15 mm petri dish containing ice-cold HBSS.
5. Under a dissecting microscope, peel back the two halves of the skull using fine forceps.
6. Remove the whole brain from the skull using forceps and place it in a 60 × 15 mm petri dish containing ice-cold HBSS. Keep the dish on ice. Repeat **steps 3–6** for the remaining heads.
7. Cut off and discard the brainstem and cerebellum using forceps.
8. Separate the brain hemisphere following the median longitudinal fissure using forceps.
9. Using fine forceps, peel off the meninges from each hemisphere. Open up the brain to reveal the midsagittal side.
10. Cut the hippocampi, the darker and c-shaped area using fine forceps, and collect them in a 15 mL cornical tube containing ice-cold HBSS.

3.4.3 Dissociation and Culture of Mouse Hippocampal Neurons

1. Ensure all tissues are at the bottom of the tube and remove HBSS carefully.
2. Wash the tissues with 5 mL of MEM twice.
3. Decant the solution and add 3 mL of 0.25% trypsin containing 30 μL of DNase I (100×, 0.2 mg/mL) and incubate the samples at 37 °C for 10 min.
4. Remove trypsin and incubate the tissues with 5 mL of warmed trypsin-inactivating media at 37 °C for 5 min.
5. Wash the tissues with MEM three times.
6. After aspirating MEM, add 1 mL of MEM and 10 μL of DNase I. Gently triturate the tissues with a 1 mL pipette tip (avoid bubbles). Typically 10–15 strokes are sufficient to dissociate the tissues into single cells. If the solution is sticky due to extraction of genomic DNA, add another 5–10 μL of DNase I to the solution.
7. Centrifuge at 500 × *g* at 4 °C for 5 min.
8. Aspirate MEM and add 1 mL of plating media. Carefully resuspend the cells using pipetting.
9. Transfer the cells into a 50 mL cornical tube containing plating media.
10. Pass the cell solution through a 70 μm strainer and collect single cells to a new 50 mL conical tube.
11. Plate the neurons on the prepared culture plates. Incubate cells at 37 °C and 5% CO_2. Mark the cells as days in vitro (DIV) 0 (*see* **Note 1**).
12. At DIV 1, change the medium with fresh-made neuron culture medium.

Note: For media exchange throughout the culture, do not expose neurons to air for more than a few seconds. Replace media one well at a time.

3.5 Collecting Conditioned Medium from Cultured Hippocampal Neurons

1. To detect secretion of endogenous α-Syn from neurons, start at **step 3** after changing the medium with 2 mL of fresh neuron culture meda at DIV 5. At DIV 4, change medium with 1 mL of fresh neuron culture medium and add 0.5 mL of lentivirus-containing medium (*see* Subheading 3.7) to infect cells with GFP-α-Syn-expressing virus. If necessary, add 0.5 mL of viral medium for the expression of DNAJC5 to stimulate MAPS (*see* **Note 10**).
2. After 24 h (DIV 5), change medium with 2 mL of fresh-made neuron culture medium.

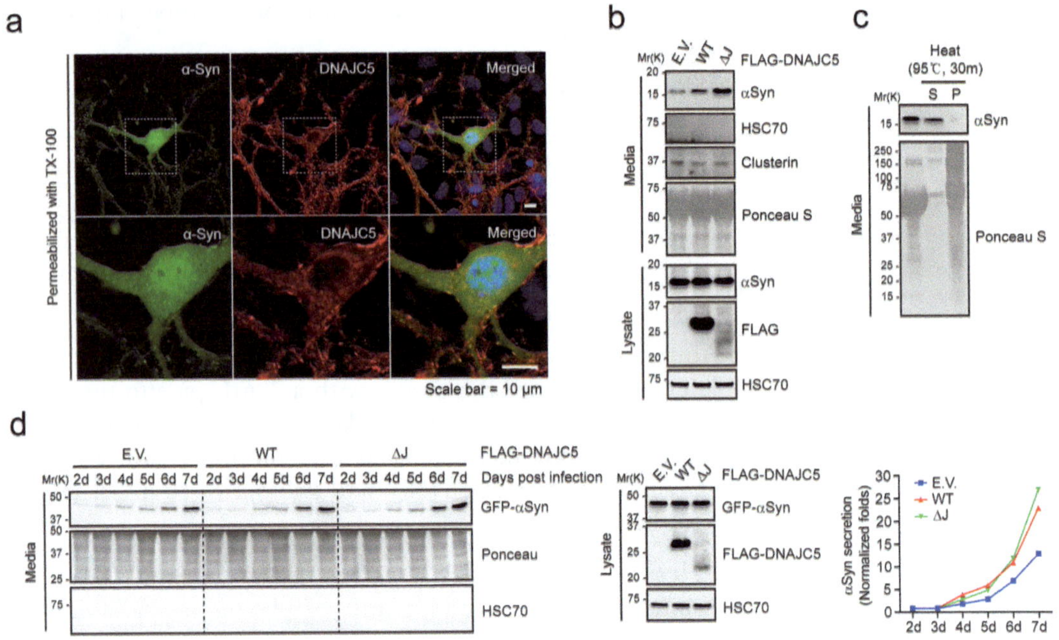

Fig. 3 α-Syn secretion from mouse primary hippocampal neurons. (**a**) Primary mouse hippocampal neurons were immunostained with antibodies against α-Syn and DNAJC5, respectively. Note that subpopulation of α-Syn (green) is localized to intracellular vesicular structures decorated by DNAJC5 (Red) (*See* enlarged views in the lower panels). (**b**) Endogenous α-Syn secretion. Neurons were infected with lentiviruses expressing either FLAG-DNAJC5 or the ΔJ mutant as indicated on DIV 3, and the CM were collected between DIV 13 and 14. The samples were analyzed by immunoblotting as indicated. (**c**) The secreted α-Syn is heat-resistant. The CM obtained from (**b**) was heated at 95 °C for 30 min, centrifuged and fractionated into a soluble (S) and a pellet (P) fraction. These fractions were analyzed by immunoblotting. Note that serum albumin is exclusively enriched in the pellet fraction due to heat-induced aggregation. (**d**) Neurons were infected with lentiviruses expressing GFP-α-Syn either alone or together with the indicated DNAJC5 variants on DIV3. The medium were collected between 2 (DIV5) and 7 (DIV10) days post-infection. The CM and lysate samples were analyzed by immunoblotting as indicated. The quantification graph shows the ratio of GFP-α-Syn in medium and lysate, indicating that expression of either wildtype DNAJC5 and the ΔJ mutant increases the α-Syn secretion efficiency compared to the control

3. Between DIV 6 (2 days post-infection) and DIV 14, harvest 100 μL of CM from the culture and prepare media samples as described in **steps 5–7** of Subheading 3.1 (*see* **Note 11**).
4. After harvesting the last CM at DIV 14, lyse the neurons as described in **steps 8–10** of Subheading 3.1.
5. Analyze samples by immunoblotting (*see* Subheading 3.2) (*see* **Note 12**) (Fig. 3b–d).

3.6 Immunocytochemistry in Cultured Hippocampal Neurons

1. Place No. 1.5 coverslips into a 12-well plate using sterile forceps and cover the coverslips with poly-D-lysine and Lamin solutions sequentially as described in Subheading 3.4.1. Alternatively, a 4-well imaging chamber can be used.

2. Add fresh primary hippocampal neurons into the prepared plate or imaging chamber (*see* **Notes 1** and **13**).
3. On the assay day, wash neurons with ice-cold PBS once and add a fixation solution.
4. Incubate the cells at room temperature for 10 min.
5. Wash the cells with PBS three times.
6. Add 1 mL of permeabilization solution to neurons and incubate the samples at room temperature for 10 min (*see* **Note 14**).
7. Wash cells with PBS three times.
8. Add 1 mL of blocking solution to neurons and incubate at room temperature for 30 min.
9. Place parafilm on a flat plastic tray and fix it using tapes. When taping it, gently stretch the parafilm to avoid wrinkles.
10. Prepare primary antibody solution (60 μL per coverslip) using the blocking solution. *See* Table 6 for the dilution factor of each antibody.
11. Place 50 μL of antibody solutions as a drop on the parafilm using a pipette.
12. Take out coverslips from the plate carefully using forceps, and place them inversely on the antibody solution drops. The side where neurons were on should contact the solution. Release coverslips carefully to avoid introducing bubbles.
13. Prepare a container with a lid, place a piece of wet paper towel at the bottom, and place the sample tray in it. Close the lid firmly to prevent evaporation of antibody solutions during incubation.
14. Incubate the cells at 4 °C overnight.
15. Transfer the coverslips to a new 12 well plate (Now the cell should face upward) and wash cells with ice-cold PBS three times.
16. Prepare secondary antibody solutions using the blocking solution with appropriate dilution factors as described in Table 6. Add DAPI to stain the nuclei if necessary.
17. Incubate the coverslips with the secondary antibody solutions at room temperature for 1 h, as described in **steps 11–13**.
18. Wash the coverslips three times as described in **step 15**.
19. Prepare slide glasses and drop mounting medium on it.
20. Using forceps, place the coverslips inversely on the mounting medium.
21. Absorb the extra mounting medium using paper wipes.

22. Seal the rims of the coverslips with a sealant (e.g. nail polish) (about 1 h). The sample slides can be stored at 4 °C for a month or at −20 °C for several months.
23. Observe the slides with an imaging microscope (Fig. 3a).

3.7 Small Scale Production of Lentivirus-Containing Medium for Infection

1. Seed 7.5 × 10^5 of 293FT cells into a 6 well plate and incubate at 37 °C with 5% CO_2.
2. Transfect cells when 293FT cells reach ~50% confluence as follows. Change medium with 2 mL of fresh medium. For each transfection reaction, add 100 μL of OPTI-MEM to a 1.5 mL microtube. Add 6.8 μg pVSV-G, 10.1 μg psPAX2, and 13 μg FSW plasmids (*see* **Note 15**). Add 5.4 μL of TransIT293 reagent to another 100 μL of OPTI-MEM tube.
3. Let the solutions to sit at room temperature for 5 min and then mix the DNA solution with the TransIT solution.
4. Incubate at room temperature for 20 min before adding the transfection mixtures (200 μL/well) to cells.
5. After 24 h, change the medium with 2.5 mL of fresh medium.
6. Collect the entire medium 72 h post transfection using a pipette.
7. Centrifuge at 1500 × *g* at 4 ° C for 10 min to pellet cell debris.
8. Pass the supernatant through a 3 mL syringe installed with a syringe filter unit (0.45 μm, PVDF) to further remove cell debris. Collect the virus-containing medium to a new 5 mL tube. Do not push the piston hard to avoid introducing bubbles.
9. The virus-containing medium can be stored in single-use size aliquots at 4 °C for 2 weeks or at −80 °C for several months. Avoid freeze/thaw cycles (No more than 2 cycles).

3.8 Stimulating MAPS by Increasing Intracellular Ca^{2+} or Serum Add-Back (See Note 16)

1. To elevate intracellular Ca^{2+} ion concentration, treat cultured cell lines or primary neurons with 5–10 μM of ionomycin calcium salt solution for 2–4 h. Use minimum volume of medium (e.g. 600 μL/well in 12 well plates) when treating cells to achieve higher concentration of secreted proteins in the short collection time window.
2. For serum add-back conditions, wash cells with serum-free DMEM and then replenish with 2 mL of serum-free medium 24 h post-transfection. Starve the cells for 16 h.
3. After serum starvation, change the medium with serum-containing complete media and incubate for 4 h. Minimize the medium volume to achieve higher concentration of secreted proteins.

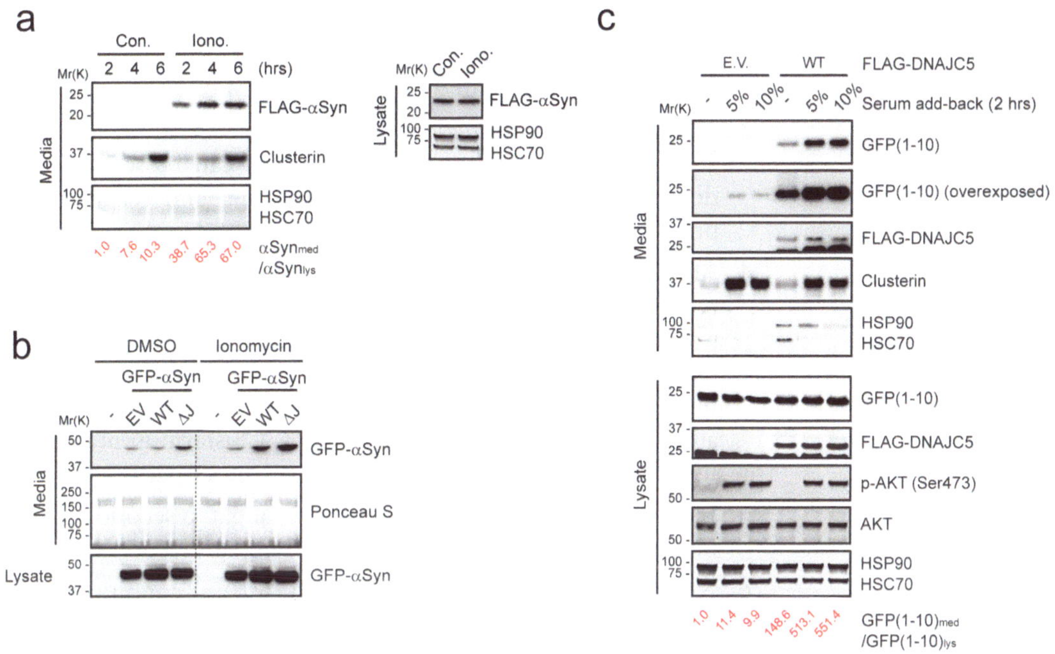

Fig. 4 Ca^{2+} or serum-induced secretion of MAPS cargoes. (**a**) HEK293T cells transfected with FLAG-α-Syn were treated with 10 μM of Ionomycin for the indicated time points. The CM were analyzed by immunoblotting as indicated. Note that the secretion of Clusterin, a conventional ER-Golgi cargoes is not affected by the treatment. (**b**) Primary hippocampal neurons expressing EGFP-α-Syn and FLAG-DNAJC5 at DIV 14 were treated with 5 μM of Ionomycin for 4 h. The conditioned media were harvested and analyzed by immunoblotting. (**c**) Serum-induced secretion of GFP(1-10). HEK293T cells transfected with GFP1-10 and FLAG-DNAJC5 were starved by replacing the culture medium with serum-free medium. After 16 h of starvation, the medium was changed back to serum-containing medium. After 2 h of incubation, the conditioned medium was analyzed by immunoblotting. Note that HSP90 and HSC70, exosomal markers, were secreted before serum-add back but disappeared when serum is added. Clusterin secretion is also affected by serum starvation and add back but it does not respond to DNAJC5 overexpression

4. For both stimulations, harvest CM and prepare the lysates as in Subheading 3.1. Analyze the samples by immunoblotting as in Subheading 3.2 (Fig. 4).

4 Notes

1. Always prepare an extra amount of medium for hippocampal cell culture based on number of wells used. In general, two hippocampi obtained from each brain hemisphere of one pup can be seeded to a 6 cm^2-area for immunoblotting or 12 cm^2 for immunostaining. *See* Table 4 for the surface area of different cell culture plates. For example, four hippocampi are required to cover 3 wells of a 12-well plate for immunoblotting.

2. Purities of plasmids are critical for high expression of a transgene while preventing cytotoxicity. Specifically, lentiviral plasmids should be prepared using Midi- or Maxi-prep kit.
3. All the abovementioned protocols are specifically tailored for 12 well dish-size cultures. The protocol can be scaled up or down by modifying the medium volume and reagent level proportionally.
4. All primary antibody solutions in BSA-containing PBS can be reused several times if sodium azide is added. Store the used primary antibodies at 4 °C. HSP90 and HSC70 are used as exosomal secretion markers and also as a control for cell membrane damage. For cell lysages, immunoblotting with these antibodies serves as a loading control. Clusterin in medium is used as an indicator of conventional secretory pathway [3].
5. In general, fluorescence- and HRP-conjugated secondary antibodies are used for lysate and media samples, respectively.
6. No exosomal markers were detected in basal condition using this method [3].
7. This β-Galactosidase Assay (CPRG) provides an easy, rapid, and sensitive method for determining the amounts of LacZ-α-Syn secreted from cells.
8. The enzymatic experiment should be conducted within the detection linear range. To this end, the absorbance can be recorded over a period of time to determine plateau phase of the reaction (Fig. 2b). If necessary, medium or lysate samples can be further diluted to avoid signal saturation.
9. Since the composition of cell types in the hippocampus is relatively simple (Pyramidal neurons and interneurons) compared to other brain regions [7], we choose hippocampal culture as a system to study MAPS phenotypes in primary neurons.
10. In general, transgene expression can be detected by immunoblotting or imaging from 2 days post-infection. Use of CMV promoter for gene expression in neurons is also possible [8]. To this end, a pLenti CMV vector (e.g., Addgene #17447) can be used in this case, and the procedure for virus production is the same as Subheading 3.7.
11. The hippocampal culture can be maintained up to 3 weeks if culture medium is replenished once every week. Therefore, CM can also be collected at latter time points when neurons are more mature.
12. Since α-Syn is an intrinsically misfolding-prone protein and heat-resistant, the CM can be boiled before the addition of the 4× Lamaelli buffer. After heating at 95 °C for 10 min, endogenous α-Syn remains in the supernatant and this method

reduces immunoblotting noise and thus increases detection sensitivity. By contrast, GFP-α-Syn forms aggregates after heating probably due to the GFP tag (unpublished observation). The onset of endogenous α-Syn expression in primary hippocampal neurons is approximately on DIV 6 [9], thus CM can be collected after DIV 6.

13. For live-cell imaging, neurons can be cultured on glass-bottom chambers (e.g. μ-Slide 4 Well Glass Bottom, Ibidi) after PDL-laminin coating.
14. To detect synapse-localized α-Syn, saponin solution (0.2% Saponin and 10% FBS in PBS) can be used for permeabilization, blocking, and antibody incubation instead.
15. Including either EGFP/FSW or mCherry/FSW plasmids in transfection is highly recommended. A fraction of the resulting viruses will express these fluorescence reporters when applied to neurons, and thus can be used to verify transduction efficiency. Otherwise, you can manually determine the titer of the viruses before use in each infection.
16. Secretion of MAPS cargoes can be regulated by environmental ques. such as increase in intracellular Ca^{2+} concentration (Fig. 4a, b) and serum add-back after starvation (Fig. 4c). To increase Ca^{2+} ion concentration, ionomycin Calcium salt can be used to treated cells for 2–4 h. It has been reported that serum starvation condition inhibits MAPS [6], while it activates exosomal secretion [10–12]. When the starvation is lifted by adding serum, a robust increase in MAPS secretion is observed.

References

1. Rabouille C (2017) Pathways of unconventional protein secretion. Trends Cell Biol 27(3):230–240. https://doi.org/10.1016/j.tcb.2016.11.007
2. Zhang M, Liu L, Lin X, Wang Y, Li Y, Guo Q, Li S, Sun Y, Tao X, Zhang D, Lv X, Zheng L, Ge L (2020) A translocation pathway for vesicle-mediated unconventional protein secretion. Cell 181(3):637–652.e615. https://doi.org/10.1016/j.cell.2020.03.031
3. Lee JG, Takahama S, Zhang G, Tomarev SI, Ye Y (2016) Unconventional secretion of misfolded proteins promotes adaptation to proteasome dysfunction in mammalian cells. Nat Cell Biol 18(7):765–776. https://doi.org/10.1038/ncb3372
4. Fontaine SN, Zheng D, Sabbagh JJ, Martin MD, Chaput D, Darling A, Trotter JH, Stothert AR, Nordhues BA, Lussier A, Baker J, Shelton L, Kahn M, Blair LJ, Stevens SM Jr, Dickey CA (2016) DnaJ/Hsc70 chaperone complexes control the extracellular release of neurodegenerative-associated proteins. EMBO J 35(14):1537–1549. https://doi.org/10.15252/embj.201593489
5. Xu Y, Cui L, Dibello A, Wang L, Lee J, Saidi L, Lee JG, Ye Y (2018) DNAJC5 facilitates USP19-dependent unconventional secretion of misfolded cytosolic proteins. Cell Discov 4:11. https://doi.org/10.1038/s41421-018-0012-7
6. Lee J, Xu Y, Zhang T, Cui L, Saidi L, Ye Y (2018) Secretion of misfolded cytosolic proteins from mammalian cells is independent of chaperone-mediated autophagy. J Biol Chem 293(37):14359–14370. https://doi.org/10.1074/jbc.RA118.003660
7. Kaech S, Banker G (2006) Culturing hippocampal neurons. Nat Protoc 1(5):2406–2415. https://doi.org/10.1038/nprot.2006.356

8. Yaguchi M, Ohashi Y, Tsubota T, Sato A, Koyano KW, Wang N, Miyashita Y (2013) Characterization of the properties of seven promoters in the motor cortex of rats and monkeys after lentiviral vector-mediated gene transfer. Human Gene Ther Methods 24(6):333–344. https://doi.org/10.1089/hgtb.2012.238
9. Murphy DD, Rueter SM, Trojanowski JQ, Lee VM (2000) Synucleins are developmentally expressed, and alpha-synuclein regulates the size of the presynaptic vesicular pool in primary hippocampal neurons. J Neurosci 20(9): 3214–3220. https://doi.org/10.1523/jneurosci.20-09-03214.2000
10. Aubertin K, Silva AK, Luciani N, Espinosa A, Djemat A, Charue D, Gallet F, Blanc-Brude O, Wilhelm C (2016) Massive release of extracellular vesicles from cancer cells after photodynamic treatment or chemotherapy. Sci Rep 6:35376. https://doi.org/10.1038/srep35376
11. Sun L, Wang HX, Zhu XJ, Wu PH, Chen WQ, Zou P, Li QB, Chen ZC (2014) Serum deprivation elevates the levels of microvesicles with different size distributions and selectively enriched proteins in human myeloma cells in vitro. Acta Pharmacol Sin 35(3):381–393. https://doi.org/10.1038/aps.2013.166
12. Taverna S, Ghersi G, Ginestra A, Rigogliuso S, Pecorella S, Alaimo G, Saladino F, Dolo V, Dell'Era P, Pavan A, Pizzolanti G, Mignatti P, Presta M, Vittorelli ML (2003) Shedding of membrane vesicles mediates fibroblast growth factor-2 release from cells. J Biol Chem 278(51):51911–51919. https://doi.org/10.1074/jbc.M304192200

Chapter 23

Single-Molecule Optical Tweezers Study of Protein–Membrane Interactions

Lu Ma, Jinghua Ge, Xin Bian, and Yongli Zhang

Abstract

Numerous proteins directly or indirectly bind membranes to exert their roles in a wide variety of biological processes. Such membrane binding often occurs in the presence of an external mechanical force. It remains challenging to quantify these interactions using traditional experimental approaches based on a large number of molecules, due to ensemble averaging or the lack of mechanical force. Here we described a new single-molecule approach based on high-resolution optical tweezers to characterize protein–membrane interactions. A single membrane binding protein is attached to the lipid bilayer coated on a silica bead via a flexible polypeptide linker, tethered to another bead via a long DNA handle, and pulled away from the bilayer using optical tweezers. Dynamic protein binding and unbinding is detected by the corresponding changes in the extension of the protein-DNA tether with high spatiotemporal resolution, which reveals the membrane binding affinity, kinetics, and intermediates. We demonstrated the method using C2 domains of extended synaptotagmin 2 (E-Syt2) with a detailed protocol. The method can be widely applied to investigate complex protein–membrane interactions under well-controlled experimental conditions.

Key words Optical tweezers, Protein–membrane interactions, Extended synaptotagmins, Supported bilayer, C2 domain

1 Introduction

Protein-membrane binding plays vital roles in a wide variety of biological processes, such as cell signaling and motility [1–3], membrane trafficking [4–6], lipid metabolism and transport [7–10], and membrane protein folding [11–13]. These membrane binding proteins often contain one or more evolutionarily conserved domains that recognize specific lipids, often in cooperation with other ligands [14]. These domains include C1 domains that bind to diacylglycerols, C2 domains that preferentially bind negatively charged lipids, such as phosphatidylserine (PS) and

Lu Ma and Jinghua Ge contributed equally with all other contributors.

Jingshi Shen (ed.), *Membrane Trafficking: Methods and Protocols*,
Methods in Molecular Biology, vol. 2473, https://doi.org/10.1007/978-1-0716-2209-4_23,

phosphatidylinositol (4,5)-bisphosphate (PI(4,5)P_2), in either Ca^{2+}-dependent or Ca^{2+}-independent manner [15], and pleckstrin homology (PH) domains that specifically bind lipids containing PI groups. Together, these lipid-binding domains allow their host proteins to target different membranes and regulate their conformations or enzymatic activities.

Although many methods have been developed to examine protein-membrane binding, it is often challenging to quantify the interactions, especially in the presence of mechanical force [1, 3, 16]. We recently developed a single-molecule manipulation approach based on optical tweezers and validated the approach by measuring the membrane binding affinities and kinetics of a few representative C2 domains [17]. Compared with other methods, the new approach is especially advantageous in dissecting the role of mechanical force in complex protein–membrane interactions. In addition, the approach overcomes the issues of protein aggregation or membrane bridging found in traditional protein-membrane binding assays [18].

We described our single-molecule assay using the C2 domains isolated from extended synaptotagmins (E-Syts) [16, 19]. Three E-Syts (E-Syt1, E-Syt2, and E-Syt3) are resident in the endoplasmic reticulum (ER) membrane in mammalian cells [20, 21]. They contain an N-terminal hairpin membrane anchor, a synaptotagmin-like mitochondrial lipid-binding protein (SMP) domain, and tandem C2 domains (five in E-Syt1 and three in E-Syt2 and E-Syt3). E-Syts bind the inner leaflet of the plasma membrane enriched with negatively charged lipids in trans via their tandem C2 domains. The binding draw the ER and the plasma membrane close to induce their lipid exchange via the SMP domain when triggered by an elevation of Ca^{2+} level in the cell [7, 10, 21]. Here we focus on the C2 domains in E-Syt2, i.e., C2A, C2B, and C2C. Structural studies reveal that E-Syt2 C2A and C2B associated with each other (Fig. 1) and link to the C-terminal C2C via a long disordered polypeptide [19, 22]. We describe our protocols to study interactions of the isolated C2AB domain pair and the C2C domain with membranes. Using this method, the force, energy, and kinetics associated with the protein binding process are determined, which help reveal their biological roles.

2 Materials

2.1 Preparation of E-Syt2 C2AB and C2C Protein Constructs

2.1.1 Plasmid Modifications

1. T4 DNA ligase.
2. DH5α competent cells.
3. SOC medium: 20 g/L tryptone, 5 g/L yeast extract, 4.8 g/L $MgSO_4$, 3.603 g/L dextrose, 0.5 g/L NaCl, 0.186 g/L KCl.

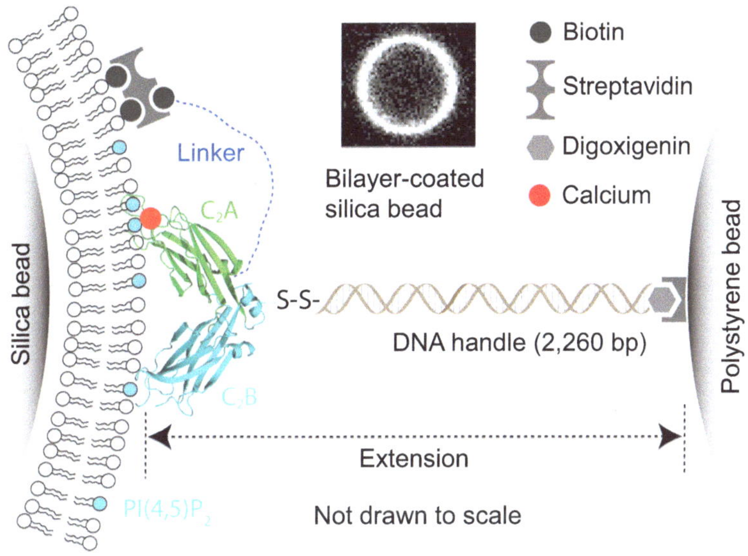

Fig. 1 Experimental setup to study protein-membrane binding using optical tweezers. A single protein represented by the E-Syt C2AB domain is attached to the bilayer supported on a silica bead at the N-terminus of the protein via a flexible polypeptide linker and pulled from its C-terminus via a 2260 bp DNA handle. The DNA handle is crosslinked to the protein through a disulfide bond at one end and attached to an anti-digoxigenin-coated polystyrene bead at the other end. The force applied to the single protein is controlled by moving one of the optical traps and the extension of the protein-DNA tether is detected. The figure is reproduced from our earlier work with permission [17]

4. High-fidelity DNA polymerase.
5. Plasmid purification kit.

2.1.2 Bacterium Growth and Harvest

1. BL21-CodonPlus (DE3)-RIPL chemically competent cells.
2. LB medium:10 g/L tryptone, 5 g/L yeast extract, 10 g/L NaCl.
3. SB medium:35 g/L tryptone, 20 g/L yeast extract, 5 g/L NaCl.
4. 1 M IPTG (Isopropyl-1-thio-β-D-galactopyranoside) stock solution (1000×, 2.38 g/10 mL). Filter the solution with 0.22 μm syringe filter.
5. 50 mg/mL kanamycin stock solution (1000×). 100 mg/mL ampicillin stock solution (1000×). Filter with 0.22 μm syringe filter.
6. LB Agar: 25 g LB broth and 15 g Agar per 1 L of ddH$_2$O. Autoclave for 20 min, and cool to approximately 55 °C. Pour into petri dishes.

2.1.3 Protein Purification

1. Protease inhibitor cocktail tablet, EDTA free.
2. Nickel beads: Ni Sepharose 6 Fast Flow.
3. Centrifuge filter with molecular cutoff of 10 kDa.
4. 0.5 M TCEP stock solution: Weight 1.43 g of TCEP·HCl (Tris (2-carboxyethyl) phosphine hydrochloride), add 8 mL of H_2O, adjust pH to 7.4 using KOH, and make up to 10 mL with water. Store 0.5 mL aliquots at −20 °C.
5. 1 M Tris store buffer: Weight 60.6 g of Tris power, dissolve in 400 mL of H_2O, adjust pH to 8.0, and then make up to 500 mL with water. Store at 4 °C.
6. 5 M NaCl stock solution: Weight 292.2 g of NaCl, dissolve in 1 L of H_2O. Store at room temperature.
7. 2 M imidazole stock solution: Weight 112.2 g of imidazole, dissolve in 400 mL of H_2O, adjust pH to 7.4 using HCl, and add up to 500 mL with water. Cover with aluminum film and store at 4 °C.
8. Lysis buffer: 25 mM Tris–HCL, pH 8.0, 300 mM NaCl, 0.5 mM TCEP, 10 mM imidazole, 1× EDTA-free protease inhibitor cocktail. Take 25 mL of 1 M HEPES, 60 mL of 5 M NaCl, 1 mL of 0.5 M TCEP, 5 mL of 2 M imidazole stock solutions, EDTA-free protease inhibitor cocktail, respectively. Add up to 1 L with water.
9. Washing buffer I: 25 mM Tris–HCL, pH 8.0, 300 mM NaCl, 0.5 mM TCEP, 20 mM imidazole, 1× EDTA-free protease inhibitor cocktail. Same recipe as in Lysis buffer in **item 8**, except using 10 mL of imidazole instead of 5 mL.
10. Elution buffer: 25 mM Tris–HCL, pH 8.0, 300 mM NaCl, 0.5 mM TCEP, 400 mM imidazole, 1× EDTA-free protease inhibitor cocktail. Same recipe as in Lysis buffer in **item 8**, except using 200 mL of imidazole instead of 10 mL.
11. Gel filtration buffer: 25 mM Tris–HCL, pH 8.0, 100 mM NaCl, 0.5 mM TCEP. Same recipe as in Lysis buffer in **item 8**, except using 20 mL of NaCl instead of 60 mL and no imidazole and EDTA-free protease inhibitor cocktail.
12. 4–12% Bis-Tris precast protein gels.
13. 4× SDS loading buffer.
14. SDS-PAGE Coomassie staining solution: 0.1–0.2 g of Coomassie Brilliant Blue G-250, 225 mL of methanol, 225 mL of H_2O, 50 mL of glacial acetic acid.
15. SDS-PAGE Coomassie de-staining solution: 150 mL of methanol, 300 mL of ddH_2O, 50 mL of glacial acetic acid.

16. MOPS SDS running buffer: 50 mM MOPS, 50 mM Tris Base, 0.1% SDS, 1 mM EDTA, pH 7.7.
17. SUMO protease (a recombinant fragment of ULP1).

2.1.4 Protein Biotinylation

1. BirA biotin-protein ligase [1 mg/mL (BirA500)].
2. BiomixA (10× concentration: 0.5 M bicine buffer, pH 8.3).
3. BiomixB (10× concentration: 100 mM ATP, 100 mM MgOAc, 500 μM d-biotin).
4. BIO200 (d-biotin; 10× concentration: 500 μM).

2.2 Preparation of DNA Handles

1. Taq DNA polymerase with reaction buffer.
2. dNTP mix (10 mM).
3. PCR primer 1: Dual-digoxigenin modified oligo (Dual-Dig-A TCATCCAAGGCTGAGCCTGCAGGTCT).
4. PCR primer 2: Thiol modified oligo. (thiol-AAATCGACGCTCAAGTCAGAGGTGGCG).
5. Template plasmid: pUC19 plasmid with a modified multi-cloning site [23–25].

2.3 Crosslinking

1. Buffer A: 0.1 M sodium phosphate buffer, pH 5.8, 400 mM NaCl.
2. Buffer C: 0.1 M sodium phosphate buffer, pH 8.5, 400 mM NaCl.
3. 10 mM 2,2′-dithiodipyridine disulfide (DTDP) stock solution.
4. Micro Bio-spin 6 columns (Bio-Rad), similar desalting columns also work.
5. 0.5 M pH 7 TCEP (Tris(2-carboxyethyl) phosphine hydrochloride) stock solution.
6. PCR purification kit.

2.4 Membrane Coating on Silica Beads

1. Lipids stock solution: 25 mg/mL POPC (1-palmitoyl 2-oleoyl-glycero-3-phosphocholine, 850457P), 10 mg/mL DOPS (1,2-dioleoyl-sn-glycero-3-phospho-L-serine (sodium salt), 840035P), 1 mg/mL Brain PI(4,5)P_2 (L-α-phosphatidylinositol-4,5-bisphosphate (Brain, Porcine) (ammonium salt), 840046X), 0.5 mg/mL DSPE-PEG(2000) Biotin (1,2-distearoyl-sn-glycero-3-phosphoethanolamine-N-[biotinyl(polyethylene glycol)-2000] (ammonium salt), 880129P), 0.01 mg/mL DSPE-PEG(2000) Biotin. Lipids POPC, DOPS, and DSPE-PEG (2000) Biotin are dissolved in chloroform, and PI(4,5)P_2 is dissolved in 70% chloroform and 30% methanol. All lipids are from Avanti Polar Lipids.
2. Thermal mixer with 1.5 mL microtube block.

2.5 Single-Molecule Protein-Membrane Binding Assay

2.5.1 Home-Built Dual-Trap High-Resolution Optical Tweezers

Our dual-trap optical tweezer setup is home-built and calibrated as previously described [26, 27]. The tweezer setup is built on an optical table in an acoustically isolated room to minimize environmental noises and vibrations. Briefly, a 1064 nm laser (CW, Nd: YVO4 laser with a maximum power of 5 W) beam is expanded ~5 fold in diameter by two telescopes. Between the telescopes, the laser beam is split into two orthogonally polarized beams by a polarizing beam splitter. One beam is reflected by a mirror attached to a nano-positioning stage, which is used to move the optical trap in the sample plane. The two beams are then combined by another polarizing beam splitter and focused by a water immersion 60× objective (NA = 1.2) to form two optical traps inside the central channel of the microfluidic flow cell. The outgoing laser beams are collected by a second identical objective, split again by polarization, and projected onto two position-sensitive detectors to detect bead positions using back-focal-plane interferometry [28, 29].

2.5.2 Optical Trapping Assay

1. HEPES buffer: 200 mM NaCl, 25 mM HEPES, pH 7.4 (*see* **Note 1**).
2. 1 M EGTA: Weight 3.8035 g of EGTA, add 5 mL of H_2O, adjust pH to 8 using concentrated NaOH and the total volume to 10 mL using H_2O.
3. 1 M $CaCl_2$: Weight 1.1098 g of anhydrous $CaCl_2$ and dissolve it in 10 mL of H_2O.
4. Catalase from bovine liver: Aqueous suspension, 10,000–40,000 units/mL.
5. Glucose oxidase from *Aspergillus niger*: 50 mg/mL, ≥100 units/mg.
6. 400 mg/mL glucose stock solution: Weight 20 g of glucose and completely dissolve it in 50 mL of ddH_2O. Store 1 mL aliquots at −20 °C.
7. Streptavidin (1 mg/mL in PBS buffer).
8. HEPES buffer with oxygen scavenging system: Dissolve 75 μL of glucose, 6 μL of glucose oxidase (final concentration ≥ 10 units/mL), and 4.7 μL of catalase (final concentration 15.7–62.7 units/mL) in 3 mL of HEPES buffer.
9. 100 μM Ca^{2+} buffer in the main channel of the microfluidic chamber: HEPES buffer with oxygen scavenging system, 500 μM EGTA and 600 μM $CaCl_2$.
10. Polystyrene beads: 1% w/v anti-digoxigenin antibody-coated polystyrene particles (2.1 μm, called DIG beads), streptavidin-coated polystyrene particles (1.8 μm). We use beads from Spherotech, IL.

11. Silica beads: 10% w/v non-functionalized silica microspheres aqueous suspensions with a mean diameter of 2.06 μm. We use silica beads from Bangs Laboratories, Inc., with a part number SS04N.
12. Sodium azide solution: 0.02% sodium azide solution in ddH_2O filtered with 0.22 μm syringe filter.

3 Methods

3.1 Experimental Setup and Protein Construct

Our experimental setup and protein construct are shown in Fig. 1. A single E-Syt2 C2AB or C2C construct is attached to the supported bilayer at the N-terminus of the protein and pulled from its C-terminus via a 2260 bp DNA handle [30–33]. An Avi Tag sequence (GLNDIFEAQKIEWHE) is added to the N-terminus of the protein, which is recognized by the BirA biotin-ligase to catalyze biotinylation of the lysine residue in the Avi Tag sequence in the presence of free biotin. The supported bilayer contains 0.03 mol% biotin-PEG-DSPE, which allows E-Syt2 C2AB or C2C protein to bind to the bilayer through biotin–streptavidin interactions. Intrinsic cysteine residues in the E-Syt2 proteins are mutated to glycine or alanine, and a unique cysteine residue is added to the C-terminus of the protein constructs, which enables the proteins to crosslink with the thiol-containing 2260 bp DNA handle. The other end of the DNA handle is labeled with two digoxigenin moieties, which bind to anti-digoxigenin antibody-coated polystyrene bead.

3.2 Membrane Coating of Silica Beads (Fig. 2)

1. Mix appropriate quantities of lipids based on a target lipid composition in a 15 mL glass tube, with a total lipid amount of 250 nmol (Fig. 2a).
2. Completely dry the lipids under a gentle nitrogen flow followed by evaporation in vacuum for 1 h. The lipids will form a thin, transparent, and uniform film on the interior glass surface (Fig. 2b).
3. Hydrate the dried lipids by adding 500 μL of HEPES buffer. Resuspend the lipids by pipetting up and down the lipid solution (Fig. 2c, d). The suspended vesicle solution should appear cloudy.
4. Transfer the vesicle solution to a 1.5 mL microtube and sonicate the vesicle solution until the solution became clear (Fig. 2e). This step generates small unilamellar vesicles (SUVs).
5. Centrifuge the vesicle solution at 15,000 × *g* for 5 min to precipitate any large vesicles or lipid aggregates (Fig. 2f). Collect the SUVs in the supernatant.

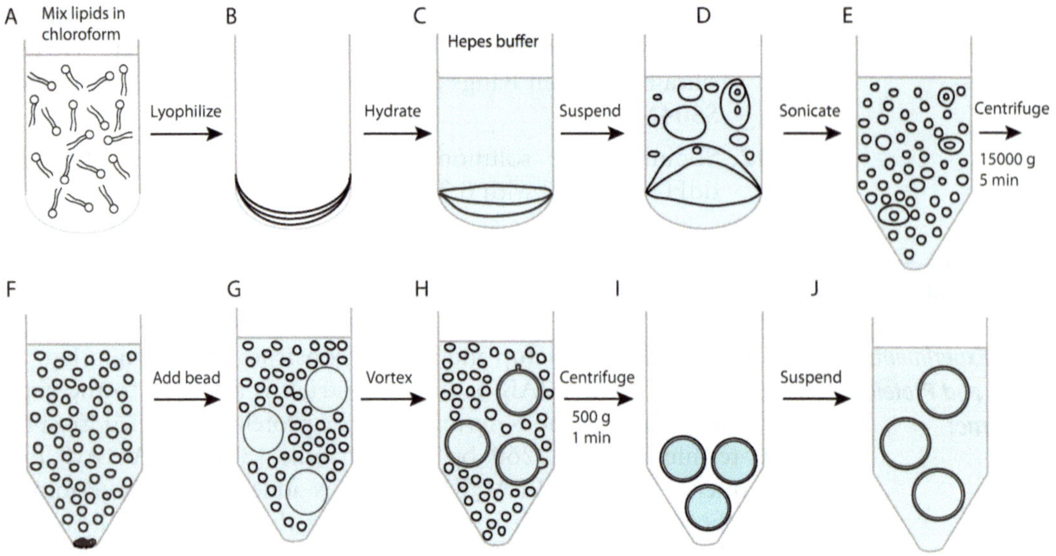

Fig. 2 Experimental procedures to prepare the membrane-coated silica beads required for the single-molecule experiment. The figure is reproduced from our earlier work with permission [17]

6. Resuspend the silica beads and add 70 μL into a 1.5 mL microtube. Wash the beads using the HEPES buffer by vortexing and then centrifuging the beads at 500 × *g* for 1 min. Transfer the resuspended silica beads to the SUV solution. SUVs spontaneously collapse onto silica surfaces to form supported bilayers (Fig. 2g).
7. Vortex the beads at 1500 rpm at 37 °C for 45 min to strip away excessive membranes from bead surfaces using a thermal mixer. Membrane coating at a higher temperature yields supported bilayers on the silica surfaces with a high membrane tension when the protein-membrane binding assay is conducted at room temperature (Fig. 2h) [17, 34], which prevents the supported bilayers from detaching from bead surfaces in the presence of high pulling force.
8. Centrifuge the bead solution at room temperature and 500 × *g* for 1 min to precipitate the beads and then remove the supernatant (Fig. 2i).
9. Add 1 mL of suspension buffer and resuspend the beads by pipetting up and down the bead solution. Repeat **steps 8** and **9** twice to remove residual vesicles in the bead solution (Fig. 2j, *see* **Note 2**).

3.3 E-Syt2 Plasmid Constructs

The gene of human E-Syt2 C2AB domain (a.a. 343–646) is inserted into the pET-SUMO vector containing a SUMO tag, a 6xHis tag, and an Avi tag at the N-terminus. The gene of the C2C domain (a.a. 744–893) is inserted into the pETDuet-1 vector containing a 6xHis tag and an Avi tag at the N-terminus.

1. Synthesize PCR primers to amplify the genes corresponding to the C2AB domain or C2C domain of E-Syt2.
2. Prepare the PCR reaction: Mix 5 μL of 10× PCR reaction buffer, 5–50 ng of template, 1 μL of 10 μM forward oligonucleotide primer, 1 μL of 10 μM reverse oligonucleotide primer, 1 μL of 10 mM dNTP, and ddH_2O to a final volume of 50 μL. Then add 1 μL of high-fidelity DNA polymerase (2.5 U/μL).
3. Perform the PCR reaction using the following cycling parameters: Initial denaturation at 95 °C for 3 min, 30 cycles of denaturation at 95 °C for 30 s, annealing at 55 °C for 1 min, and extension at 68 °C for 2 min, followed by a final extension for 5 min at 68 °C and cooling to 12 °C.
4. Add 1 μL BamHI and 1 μL NotI restriction enzymes (both 10 U/μL) to the PCR product purified by gel extraction. Gently and thoroughly mix the reaction mixture by pipetting the solution up and down several times. Spin down the PCR product in a microtube for 1 min and immediately incubate the reaction at 37 °C for 1 h to digest.
5. Purify the digested PCR product using a PCR purification kit and ligate the DNA into the pET-SUMO vector for C2AB or the pETDuet-1 vector for C2C using the T4 DNA ligase. Purify the plasmid using a plasmid purification kit.
6. Gently thaw 50 μL of DH5α competent cells on ice.
7. Transfer 10 μL of the DNA ligation product to the competent cells, mix the DNA and cells by gently swirling the transformation reaction, and incubate the cells on ice for 30 min.
8. Put the cells in a 42 °C water bath for 45 s and then quickly place them on ice for 2 min (Heat shock).
9. Add 0.5 mL of S.O.C broth preheated to 42 °C and incubate the transformation reaction at 37 °C for 1 h while shaking at 225–250 rpm.
10. Plate the appropriate volume of the transformation reaction on agar plates containing kanamycin for the pET-SUMO vector and ampicillin for the pETDuet-1 vector.
11. Pick five colonies from the agar plate and inoculate each colony in a culture of 5 mL of LB medium containing the appropriate selective antibiotic. Incubate the cells at 37 °C for 12–16 h while shaking at 225–250 rpm.
12. Harvest the bacterial cells by centrifugation at >6800 × *g* in a cooled microcentrifuge for 5 min.
13. Extract plasmids.
14. Sequence and confirm the protein genes in the extracted plasmids.

3.4 Protein Expression and Purification

1. Transform 5–10 ng of protein-encoding plasmids (1–5 μL) to BL21-CodonPlus (DE3)-RIPL chemically competent cells and mix by gently stirring with the pipette tip.
2. Incubate on ice for 30 min.
3. Heat shock the cells at 42 °C for 30 s without shaking.
4. Immediately transfer the tubes to ice and wait for 2 min.
5. Add 200 μL of S.O.C. medium at room temperature and shake at 37 °C with 225–250 rpm for 1 h.
6. Spread 100 μL of transformed cells evenly on a selective LB plate containing 100 μg/mL ampicillin or 50 μg/mL kanamycin prewarmed to 37 °C.
7. Incubate at 37 °C overnight.
8. Pick a single colony from the plate and inoculate the cells in 100–200 mL of LB with 100 μg/mL ampicillin or 50 μg/mL kanamycin. Grow the cells at 37 °C overnight while shaking at 225–250 rpm.
9. Transfer the precultured cells to 1 L of LB medium containing 100 μg/mL ampicillin or 50 μg/mL kanamycin and continue to grow the culture for about 2 h until the absorbance of the cell culture at 600 nm reaches 0.6–0.8 (middle logarithmic phase).
10. Induce expression of the protein by adding IPTG to a final concentration of 0.5 mM.
11. Inoculate the cell culture at 18 °C for 12–16 h. Harvest the cells by centrifuging the culture at 4500 × *g* for 15 min at 4 °C.
12. Proceed to protein purification or store the cells at −20 °C for future protein purification.
13. Thaw the cell pellet for 15 min on ice. Resuspend the cells in 30 mL of lysis buffer containing the protease inhibitor cocktail.
14. Disrupt cells using ultrasonication.
15. Centrifuge the cell suspension at 17,000 × *g* for 1 h at 4 °C and keep the clarified supernatant.
16. Wash Nickel beads using the lysis buffer 3 times. To wash the Nickel beads, add the lysis buffer to suspend the beads and then precipitate the beads by centrifugation at 500 × *g* for 3 min.
17. Add the supernatant from **step 15** to the Nickel beads (500 μL of bead/1 L *E. coli* culture) and nutate the bead solution at 4 °C for 2 h.
18. Centrifuge the Nickel beads at 500 × *g* for 3 min at 4 °C and remove the supernatant.

19. Wash the beads with 10 mL of washing buffer 3 times to remove nonspecifically bound proteins.
20. Elute the protein from the Nickel beads using the elution buffer.
21. For E-Syt2 C2AB domain, use SUMO protease to remove the SUMO tag. Re-purify the protein with Nickel beads.
22. Concentrate the purified protein to 500 μL using Centrifugal filters.
23. Further purify the protein by size exclusion chromatography using Superdex 200 Increase 10/300 GL in the gel filtration buffer.
24. Run SDS-PAGE to check the purity and concentration of the purified protein.

3.5 Protein Biotinylation

1. Add the biotinylation buffer (a final 1× concentration) containing BiomixA, BiomixB, d-biotin, and BirA to the purified C2 protein. If the protein solution contains glycerol, remove it by passing through the Micro Bio-spin 6 column.
2. Incubate at 4 °C overnight.
3. Pass spin column to remove biotin if needed.
4. Test the biotinylation levels of E-Syts using a SDS gel.

3.6 Protein and DNA Handle Crosslinking (Fig. 1)

1. Run PCR to produce the DNA handles.
2. Purify DNA handles using a PCR purification kit.
3. Add TCEP to 2 mM final concentration and incubate for 30 min to reduce the DNA handles.
4. Remove free TCEP by passing through the Micro Bio-spin 6 column once.
5. Add DTDP to 1 mM final concentration to activate the thiol group in the DNA handle and incubate for 30 min.
6. Remove free DTDP and change buffer from Buffer A to Buffer C using a PCR purification spin column.
7. Measure the DNA concentration.
8. Reduce E-Syt2 C2AB or C2C protein with 2 mM TCEP and incubate for 30 min at room temperature. Remove free TCEP by passing through the Micro Bio-spin 6 columns twice and then change buffer to Buffer C.
9. Measure the concentration of E-Syt2 C2AB or C2C protein.
10. Crosslink the E-Syt2 C2AB or C2 protein to the DNA handle: Mix the DTDP-treated DNA handle with the E-Syt2 C2AB protein at a molar ratio of 1:100 and incubate the mixture at 4 °C overnight.

3.7 Protein-Membrane Binding Assay Based on Optical Tweezers

3.7.1 Measure the Force-Extension Curves

1. Wash all three microfluidic channels with the HEPES buffer.
2. Mix the crosslinked protein-DNA sample containing ~2 ng of DNA with 20 μL of DIG beads and incubate at room temperature for 15 min to allow the DNA to bind to the DIG beads. Dilute the beads with 1 mL of HEPES buffer and transfer the bead solution to a 1 mL syringe.
3. Add 5 μL of streptavidin (1 mg/mL) to 50 μL of membrane-coated beads (*see* **Note 3**). Incubate at room temperature for 15 min.
4. Dilute the membrane-coated beads with 1 mL of HEPES buffer.
5. Centrifuge the bead solution at 500 × *g* at room temperature for 1 min to precipitate the beads and remove the supernatant (including free streptavidin).
6. Repeat **steps 4** and **5** twice to remove residual streptavidin in the bead solution.
7. Dilute the membrane-coated beads into 1 mL of HEPES buffer. Transfer the bead solution to a 1 mL syringe.
8. Connect the syringes to the top and bottom channels of the microfluidic chamber.
9. Transfer 3 mL of HEPES buffer containing the oxygen scavenging system to a 10 mL syringe and connect the syringe to the central channel. The pressure in the 10 mL syringe is remotely controlled via a computer.
10. Inject the beads to the top and bottom channels manually. Trap a single C2 domain-DNA-bound DIG bead and a single membrane-coated silica bead in each optical trap.
11. Calibrate the optical traps to determine their stiffnesses. The stiffness of each optical trap is determined by measuring the power-spectrum density distribution of the Brownian motion of the trapped bead at 80 kH and fitting the distribution with the Lorentzian distribution [23].
12. Bring the two beads close to form a protein-DNA tether between the two trapped beads [35].
13. Pull the C2 domain at a trap separation speed of 10 nm/s to observe dissociation of the C2 domain from the supported bilayer and its subsequent unfolding at a higher force range.
14. Test more single C2 proteins by repeating **steps 12** and **13**.

Representative FECs obtained by pulling a single E-Syt2 C2AB are shown in Fig. 1c in reference [17].

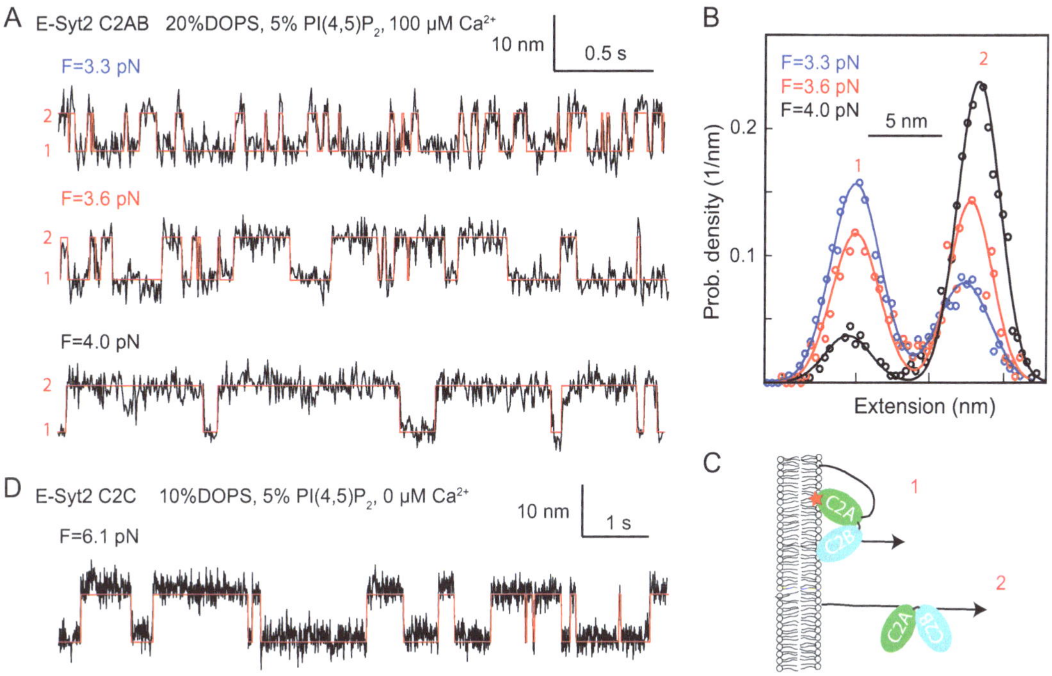

Fig. 3 Reversible protein binding to the supported bilayer detected by optical tweezers at constant trap separation or mean force. (**a**) Time-dependent extension trajectories of E-Syt2 C2AB at different forces. The red lines are idealized transitions derived from hidden-Markov modeling. The red numbers indicate the corresponding states illustrated in **c**. (**b**) Probability density distributions of the extensions (symbols) shown in (**a**) and their best-fits by a sum of two Gaussian functions (curves). (**c**) Schematic diagram of membrane-bound and unbound states derived from the extension trajectories. (**d**) Time-dependent extension trajectory of E-Syt2 C2C domain at a constant force. Note that E-Syt2 C2AB and E-Syt2 C2C binds to the membrane in a Ca^{2+}-dependent and Ca^{2+}-independent manner, respectively. The figure is reproduced from our earlier work with permission [17]

3.7.2 Measure Dynamic Membrane Binding and Unbinding of a Single E-Syt2 C2 Domain Under an Equilibrium Condition at Constant Trap Separations

To better resolve reversible membrane binding and unbinding of a single C2 domain, pull the protein to ~4 pN and hold it at a constant trap separation or constant mean force to observe the C2 transition with an extended period of time (Fig. 3). Then the trap separation is increased or decreased stepwise to scan the reversible transition in the full force range. This experiment generates time-dependent extension trajectories at different constant mean forces, which are used to derive the unbinding probabilities and binding and unbinding rates as a function of force (Fig. 4). These measurements are fit with a theoretical model to determine the binding affinity and binding and unbinding rates at zero force.

1. Pull a single E-Syt2 C2 domain to a force that shows reversible binding and unbinding transitions, hold the molecule at the trap separation, and record the time-dependent transitions for an extended period of time so that more than 100 C2 binding and unbinding events are detected.

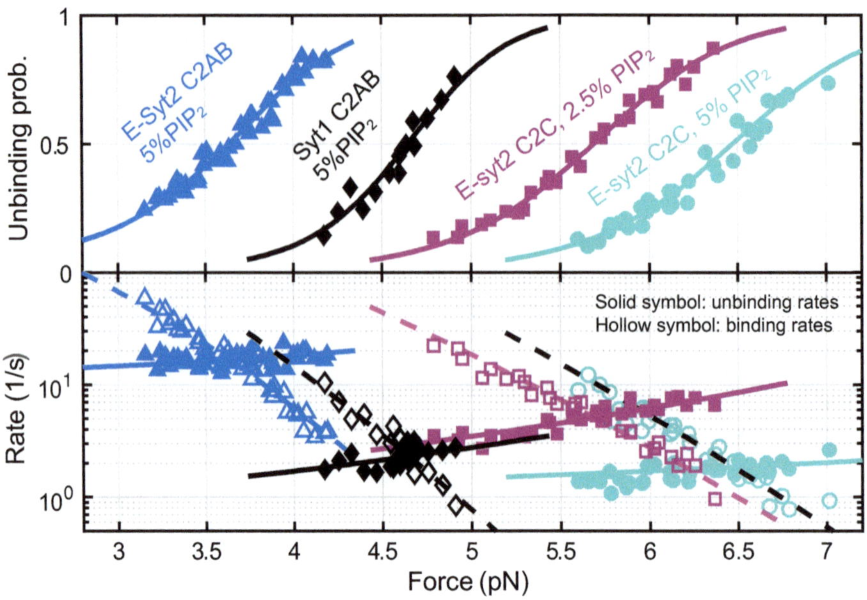

Fig. 4 Force-dependent unbinding probabilities (top), transition rates (bottom), and their best model fits (solid and dashed curves). Unbinding probabilities and rates are indicated by solid symbols, while binding rates are shown by hollow symbols. The figure is reproduced from our earlier work with permission [17]

2. Increase or decrease the trap separation in a stepwise manner and repeat **step 1** until the reversible transitions in their full force range are collected.
3. Repeat **steps 1** and **2** at different Ca^{2+} concentrations.
4. Pull the E-Syt2 C2 domain to high force to completely unfold the C2 domain and record the corresponding FEC.

3.8 Data Analysis

3.8.1 Hidden-Markov Modeling

The force-extension curves reveal all state transitions of the protein induced by the pulling force, including membrane binding/ unbinding and folding/unfolding of the protein. The curves are equivalent to a phase diagram of the protein in the force-extension coordinate, which helps to confirm that a single protein is being pulled and protein-membrane binding occurs in the presence of an intact protein. The pulling site on the protein should be carefully chosen such that the protein structure is not altered during the force-induced reversible membrane binding and unbinding.

1. Identify the time regions with different constant trap separations in the measured time-dependent extension, force, and trap separation trajectories. Determine the starting and ending time points of each region and their corresponding mean trap separation.
2. Calculate probability density functions of the extension and force trajectories corresponding to constant trap separation,

fit the probability density functions with a sum of two Gaussian functions, and determine the average extensions and forces of the membrane-bound and -unbound states and their probabilities [35].

3. Perform two-state hidden-Markov modeling (HMM) for each extension trajectory at constant trap separation using the calculated average state extensions as input parameters [35, 36]. The modeling maximizes the likelihood of observing the measured extension trajectory by optimizing parameters in the hidden-Markov model with an iterative optimization method based on an expectation-maximization algorithm or a gradient search [36, 37]. The HMM reveals the optimized model parameters, including probabilities and average extensions of both states and their associated transition rates.
4. Calculate the idealized extension transition based on the best-fit parameters using the Viterbi algorithm [37]. The quality of the hidden-Markov modeling can be confirmed by a good match of the measured and idealized extension transitions (Fig. 3).
5. Plot unbinding probabilities and binding and unbinding rates as a function of force (Fig. 4).

3.8.2 Nonlinear Model Fitting to Determine Binding Affinity and Kinetics

The binding energy is determined by the mechanical work required to pull the protein away from the membrane, which is equilibrium force (the force corresponding to an unbinding probability of 0.5) multiplied by the corresponding extension change. Note that the mechanical work not only dissociates the protein from the membrane, but also stretches the polypeptide linker joining the protein to the membrane to the equilibrium force. The stretching energy can be calculated based on the worm-like chain model of the semiflexible polypeptide and should be subtracted from the mechanical work to give the membrane binding energy. The binding and unbinding rates as a function of force are described by the Kramers rate equation. To facilitate the binding energy and rate calculations, we simultaneously fit the force-dependent unbinding probabilities, rates, and extension changes with the work and rate formula and derive the membrane binding energy and rates at zero force as best-fit parameters [38] (Fig. 4).

4 Notes

1. To test Ca^{2+}-dependent membrane binding, the HEPES buffer, but not the PBS buffer, is used, as Ca^{2+} induces precipitation in the PBS buffer.

2. Fresh membrane-coated beads should be used, which cannot be stored for more than 3 days at 4 °C.
3. Streptavidin is bound to the membrane-coated beads in the presence of slightly higher molar amount than that of the accessible biotin on the membranes.

Acknowledgments

This work was supported by NIH grant R35GM131714, R01GM093341, and R01GM120193 to Y. Z. and by the National Natural Science Foundation of China grant 31770812 to L. M.

References

1. Das DK, Feng Y, Mallis RJ, Li X, Keskin DB, Hussey RE, Brady SK, Wang JH, Wagner G, Reinherz EL, Lang MJ (2015) Force-dependent transition in the T-cell receptor beta-subunit allosterically regulates peptide discrimination and pMHC bond lifetime. Proc Natl Acad Sci U S A 112:1517–1522
2. Cho WH, Stahelin RV (2005) Membrane-protein interactions in cell signaling and membrane trafficking. Annu Rev Biophys Biomol Struct 34:119–151
3. Wang XF, Ha T (2013) Defining single molecular forces required to activate integrin and notch signaling. Science 340:991–994
4. Brunger AT, Choi UB, Lai Y, Leitz J, White KI, Zhou Q (2019) The pre-synaptic fusion machinery. Curr Opin Struct Biol 54:179–188
5. Brose N, Brunger A, Cafiso D, Chapman ER, Diao J, Hughson FM, Jackson MB, Jahn R, Lindau M, Ma C, Rizo J, Shin YK, Sollner TH, Tamm L, Yoon TY, Zhang YL (2019) Synaptic vesicle fusion: today and beyond. Nat Struct Mol Biol 26:663–668
6. Baker RW, Hughson FM (2016) Chaperoning SNARE assembly and disassembly. Nat Rev Mol Cell Biol 17:465–479
7. Yu HJ, Liu YH, Gulbranson DR, Paine A, Rathore SS, Shen JS (2016) Extended synaptotagmins are Ca^{2+}-dependent lipid transfer proteins at membrane contact sites. Proc Natl Acad Sci U S A 113:4362–4367
8. Saheki Y, De Camilli P (2017) Endoplasmic reticulum-plasma membrane contact sites. Annu Rev Biochem 86:659–684
9. Reinisch KM, De Camilli P (2016) SMP-domain proteins at membrane contact sites: structure and function. Biochim Biophys Acta 1861:924–927
10. Bian X, Zhang Z, Xiong QC, De Camilli P, Lin CX (2019) A programmable DNA-origami platform for studying lipid transfer between bilayers. Nat Chem Biol 15:830
11. Yu H, Siewny MGW, Edwards DT, Sanders AW, Perkins TT (2017) Hidden dynamics in the unfolding of individual bacteriorhodopsin proteins. Science 355:945–949
12. Choi HK, Min D, Kang H, Shon MJ, Rah SH, Kim HC, Jeong H, Choi HJ, Bowie JU, Yoon TY (2019) Watching helical membrane proteins fold reveals a common N-to-C-terminal folding pathway. Science 366:1150–1156
13. Popot JL, Engelman DM (2016) Membranes do not tell proteins how to fold. Biochemistry 55:5–18
14. Lemmon MA (2008) Membrane recognition by phospholipid-binding domains. Nat Rev Mol Cell Biol 9:99–111
15. Corbalan-Garcia S, Gomez-Fernandez JC (2014) Signaling through C2 domains: more than one lipid target. Biochim Biophys Acta 1838:1536–1547
16. Saheki Y, De Camilli P (2017) The extended-synaptotagmins. Biochim Biophys Acta 1864: 1490–1493
17. Ma L, Cai Y, Li Y, Jiao J, Wu Z, O'Shaughnessy B, De Camilli P, Karatekin E, Zhang YL (2017) Single-molecule force spectroscopy of protein-membrane interactions. eLife 6:e30493
18. Perez-Lara A, Thapa A, Nyenhuis SB, Nyenhuis DA, Halder P, Tietzel M, Tittmann K, Cafiso DS, Jahn R (2016) $PtdInsP_2$ and PtdSer cooperate to trap synaptotagmin-1 to the plasma membrane in the presence of calcium. eLife 5:e15886

19. Schauder CM, Wu XD, Saheki Y, Narayanaswamy P, Torta F, Wenk MR, De Camilli P, Reinisch KM (2014) Structure of a lipid-bound extended synaptotagmin indicates a role in lipid transfer. Nature 510:552–555
20. Min SW, Chang WP, Sudhof TC (2007) E-Syts, a family of membranous Ca^{2+}-sensor proteins with multiple C_2 domains. Proc Natl Acad Sci U S A 104:3823–3828
21. Giordano F, Saheki Y, Idevall-Hagren O, Colombo SF, Pirruccello M, Milosevic I, Gracheva EO, Bagriantsev SN, Borgese N, De Camilli P (2013) PI(4,5)P_2-dependent and Ca^{2+}-regulated ER-PM interactions mediated by the extended synaptotagmins. Cell 153: 1494–1509
22. Xu JJ, Bacaj T, Zhou A, Tomchick DR, Sudhof TC, Rizo J (2014) Structure and Ca^{2+}-binding properties of the tandem C2 domains of E-Syt2. Structure 22:269–280
23. Zhang YL, Sirinakis G, Gundersen G, Xi ZQ, Gao Y (2012) DNA translocation of ATP-dependent chromatin remodelling factors revealed by high-resolution optical tweezers. Methods Enzymol 513:3–28
24. Zhang YL, Smith CL, Saha A, Grill SW, Mihardja S, Smith SB, Cairns BR, Peterson CL, Bustamante C (2006) DNA translocation and loop formation mechanism of chromatin remodeling by SWI/SNF and RSC. Mol Cell 24:559–568
25. Ma L, Jiao J, Zhang YL (2019) Single-molecule optical tweezers study of regulated SNARE assembly. Methods Mol Biol 1860: 95–114
26. Sirinakis G, Ren YX, Gao Y, Xi ZQ, Zhang YL (2012) Combined and versatile high-resolution optical tweezers and single-molecule fluorescence microscopy. Rev Sci Instrum 83: 093708
27. Moffitt JR, Chemla YR, Izhaky D, Bustamante C (2006) Differential detection of dual traps improves the spatial resolution of optical tweezers. Proc Natl Acad Sci U S A 103:9006–9011
28. Gittes F, Schmidt CF (1998) Interference model for back-focal-plane displacement detection in optical tweezers. Opt Lett 23:7–9
29. Zhang XM, Ma L, Zhang YL (2013) High-resolution optical tweezers for single-molecule manipulation. Yale J Biol Med 86:367–383
30. Cecconi C, Shank EA, Bustamante C, Marqusee S (2005) Direct observation of the three-state folding of a single protein molecule. Science 309:2057–2060
31. Cecconi C, Shank EA, Marqusee S, Bustamante C (2011) DNA molecular handles for single-molecule protein-folding studies by optical tweezers. DNA Nanotechnol Method Protoc 749:255–271
32. Gao Y, Zorman S, Gundersen G, Xi ZQ, Ma L, Sirinakis G, Rothman JE, Zhang YL (2012) Single reconstituted neuronal SNARE complexes zipper in three distinct stages. Science 337:1340–1343
33. Gao Y, Sirinakis G, Zhang YL (2011) Highly anisotropic stability and folding kinetics of a single coiled coil protein under mechanical tension. J Am Chem Soc 133:12749–12757
34. Petrache HI, Dodd SW, Brown MF (2000) Area per lipid and acyl length distributions in fluid phosphatidylcholines determined by ^2H NMR spectroscopy. Biophys J 79:3172–3192
35. Jiao JY, Rebane AA, Ma L, Zhang YL (2017) Single-molecule protein folding experiments using high-resolution optical tweezers. Methods Mol Biol 1486:357–390
36. Zhang YL, Jiao J, Rebane AA (2016) Hidden Markov modeling with detailed balance and its application to single protein folding. Biophys J 111:2110–2124
37. Rabiner LR (1989) A tutorial on hidden Markov-models and selected applications in speech recognition. Proc IEEE 77:257–286
38. Rebane AA, Ma L, Zhang YL (2016) Structure-based derivation of protein folding intermediates and energies from optical tweezers. Biophys J 110:441–454

Chapter 24

Generation of Hybrid Extracellular Vesicles by Fusion with Functionalized Liposomes

Max Piffoux, Amanda Karine Andriola Silva, Florence Gazeau, and David Tareste

Abstract

Extracellular vesicles (EVs) and liposomes are natural and synthetic drug delivery systems, respectively, with their own advantages and limitations. EV/liposome fusion allows the generation of hybrid EVs that benefit from both the versatility of liposomes (tunable lipid and protein composition, surface functionalization, lumen loading, etc.) and the functionality of EVs (natural targeting properties, low immunogenicity, anti-inflammatory properties, etc.). Here, we describe the methods to (1) produce EVs and liposomes, (2) induce and monitor their fusion, and (3) purify the obtained hybrid EVs.

Key words Extracellular vesicles, Liposomes, Membrane fusion, Polyethylene glycol, Ultracentrifugation, Fluorescence resonance energy transfer

1 Introduction

EVs are relatively new actors in the drug delivery field, which display interesting properties such as good immune tolerance [1], immunomodulation [2], enhanced capacity to cross biological barriers [3], and natural targeting properties [4]. However, EVs are relatively difficult to load, especially with hydrophilic molecules, and to modify in a controlled way [5, 6]. On the contrary, liposomes are well-known drug delivery systems whose composition can be easily tuned both in their membrane and lumen [7]. However, liposomes have limited abilities to cross biological membranes to deliver drugs, and are recognized by the immune system as foreign bodies [8]. We have developed the EV/liposome fusion method to create new natural/artificial hybrid vectors that benefit from both vectors' specific properties. We recently published a proof of concept study describing that the method is versatile enough to be used with different types of EVs and liposomes, leads to complete fusion between EVs and liposomes, and can be

Jingshi Shen (ed.), *Membrane Trafficking: Methods and Protocols*,
Methods in Molecular Biology, vol. 2473, https://doi.org/10.1007/978-1-0716-2209-4_24,

controlled by tuning the reaction settings [9]. The method allows the transfer of lipophilic and hydrophilic molecules from liposomes to EVs, while keeping the intrinsic cargoes of EVs intact. Surface functionalization of EVs allows modification of EV–cell interaction properties. Hybrid EVs also display enhanced drug delivery properties compared to commercial liposomes [9, 10]. Several liposomal formulations have received market approval, including Caelyx®, Evacet® and Lipodox® (doxorubicin), DaunoXome® (daunorubicin), Onivyde® (irinotecan), DepoCyt® (cytarabine), Marqibo® (vincristine), AmBisome® and Abelcet® (amphotericin B) [11]. EV-liposome fusion may represent a strategy of interest in order to render commercial liposomes more "cell-friendly." Besides, we anticipate that the pharmacokinetics and pharmacodynamics properties of commercial liposomes may be modulated by introducing cellular lipids and proteins into their core structures. For example, EV fusion with commercial liposomes may improve accumulation at the site of interest and reduce biodistribution to off-target organs. Fusion with mesenchymal stem/stromal cells EVs (MSC-EVs) may for instance direct liposome-carried drugs to injured and inflammatory tissues, which exhibit enhanced uptake of MSC-EVs due to up-regulated integrin expression levels [12].

We describe below the EVs and liposomes production methods, the polyethylene glycol (PEG)-mediated fusion process used to generate hybrid EVs (Fig. 1), the fluorescence resonance energy transfer (FRET)-based lipid mixing assay to monitor the formation of hybrid EVs, and the density gradient ultracentrifugation method used to purify hybrid EVs (Fig. 2).

2 Materials

Prepare all aqueous solutions using molecular biology grade reagents and 18.2 MΩ ultra-pure water. Filter with sterile 0.22 μm polyethersulfone (PES) membranes.

2.1 Liposome Formation

1. Synthetic lipids in chloroform solutions with a purity >99%: 1-palmitoyl-2-oleoyl-sn-glycero-3-phosphocholine (POPC, M_w = 760.1 g/mol, C = 10 mg/mL); 1,2-dioleoyl-sn-glycero-3-phosphoethanolamine (DOPE, M_w = 744.0 g/mol, C = 10 mg/mL); 1,2-dioleoyl-sn-glycero-3-phospho-L-serine-N-(7-nitro-2-1,3-benzoxadiazol-4-yl) (NBD-PS, M_w = 985.2 g/mol, C = 1 mg/mL); 1,2-dioleoyl-sn-glycero-3-phosphoethanolamine-N-(lissamine rhodamine B sulfonyl) (Rho-PE, M_w = 1301.7 g/mol, C = 1 mg/mL). Store at −20 °C (*see* **Note 1**).
2. Analytical grade ethanol and chloroform. Store at room temperature.

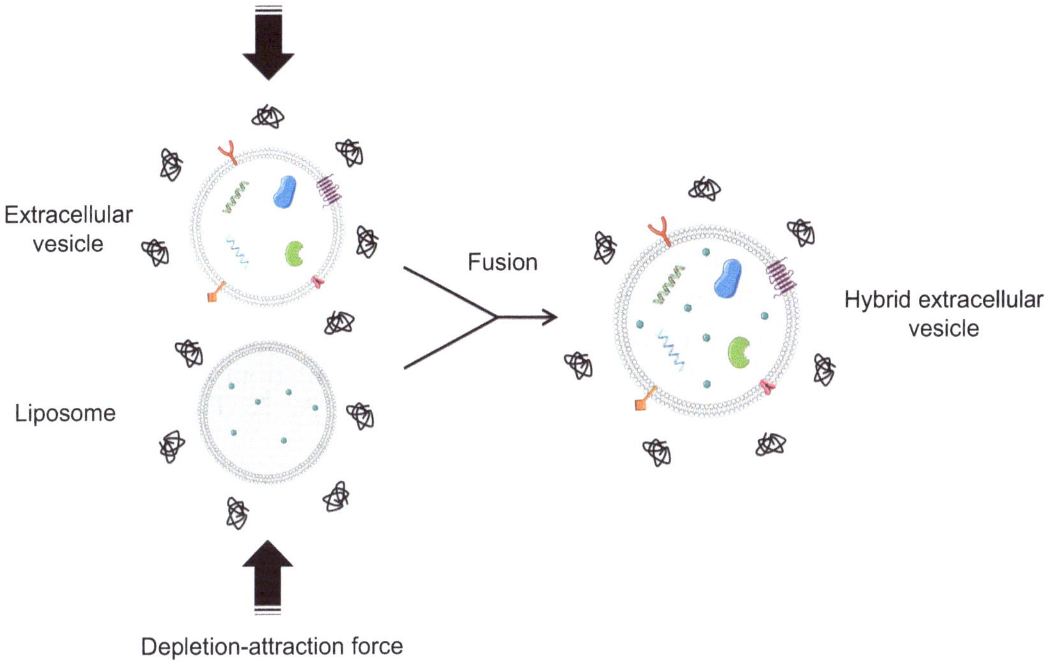

Fig. 1 Polyethylene glycol (PEG)-mediated fusion between EVs and liposomes to generate hybrid EVs. PEG (black random coil) produces its main fusogenic effect via an attractive depletion force resulting from its exclusion from the space between EVs and liposomes. The attractive force comes from the increase of the osmotic pressure of PEG in the bulk region surrounding two approaching EVs and liposomes compared to the gap region between their membrane surfaces

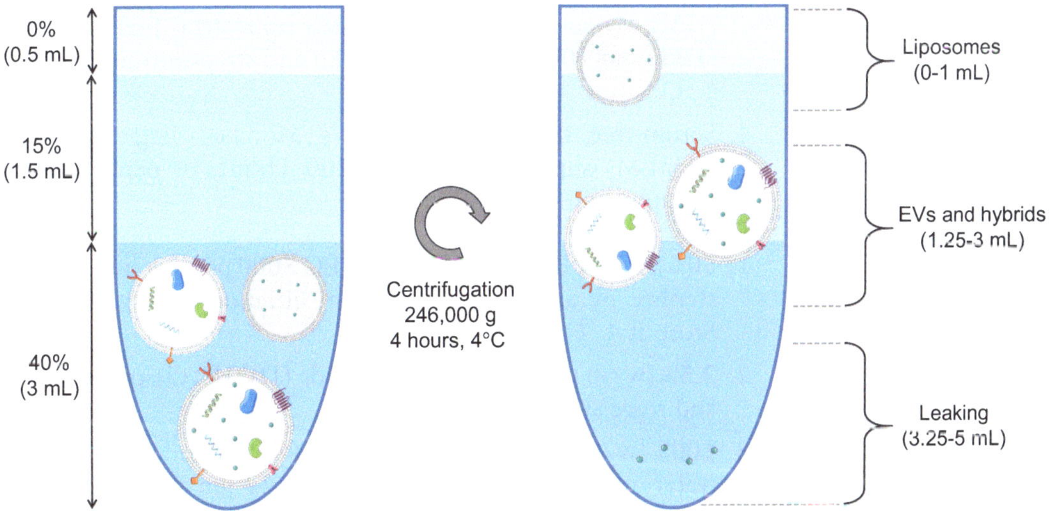

Fig. 2 Purification of hybrid EVs by density gradient ultracentrifugation. Unfused liposomes (if there are any) are located at the 0–15% Nycodenz interface. Unfused EVs (if there are any) are found at the 15–40% Nycodenz interface. Hybrid EVs, depending on the EV/liposome molar ratio are located in the 15% Nycodenz layer toward the top if they are mostly made of liposomes (e.g. 1/9 EV/liposome molar ratio) and toward the bottom if they are mostly made of EVs (e.g. 9/1 EV/liposome molar ratio)

3. Sterile Phosphate Buffer Saline (PBS). Store at 4 °C.
4. Round-bottom borosilicate glass tube.
5. Argon gas (99.9992% overall purity).
6. Oil-free vacuum pump.
7. Liposome extruder. We use the Mini-Extruder Set from Avanti Polar Lipids (Cat#610023), which includes 100 nm pore size polycarbonate membranes.
8. Ultracentrifuge with appropriate rotors.
9. 5 mL transparent centrifuge tubes. We use Open-Top Thinwall Ultra-Clear tubes (13 × 51 mm) from Beckman Coulter (Cat#344057), which fit into a SW 55 Ti swinging-bucket rotor (Beckman Coulter).
10. 32.4 mL polypropylene centrifuge tubes. We use OptiSeal Polypropylene tubes (26 × 77 mm) from Beckman Coulter (Cat#361625), which fit into a SW 32 Ti swinging-bucket rotor (Beckman Coulter).
11. Nanoparticle tracking analysis system. We use the Nanosight LM10-HS (NanoSight, UK) equipped with a 532 nm laser.

2.2 Cell Culture

1. Mammalian cells: human umbilical vein endothelial cells (HUVEC) or murine mesenchymal stem cell line (MSC) C3H/10 T1/2, Clone 8 (ATCC®) (*see* **Note 2**).
2. 150 cm^2 cell culture flasks.
3. Complete medium: Dulbecco's Modified Eagle Medium (DMEM) supplemented with 10% (v/v) fetal bovine serum (FBS) and 100 U/mL of penicillin and streptomycin. Store at 4 °C.
4. Serum-free medium: Dulbecco's Modified Eagle Medium (DMEM) supplemented with 100 U/mL of penicillin and streptomycin. Store at 4 °C.

2.3 Hybrid EVs Formation and Purification

1. 50% (w/v) PEG 8000: dissolve PEG 8000 powder in water and sterilize by autoclaving (0.22 μm filtration is very difficult). Store at 4 °C.
2. 2.5% (w/v) n-dodecyl β-D-maltoside (DDM): dissolve in water and store at room temperature.
3. 100% (w/v) Nycodenz solution: dissolve Nycodenz powder in water and sterilize by autoclaving (0.22 μm filtration is very difficult). Store at 4 °C (*see* **Note 3**).
4. 80% (w/v) Nycodenz stock solution: mix 24 mL of 100% (w/v) Nycodenz solution with 6 mL of water. Store at 4 °C.
5. 15% (w/v) Nycodenz solution: mix 7.5 mL of 100% (w/v) Nycodenz solution with 17.5 mL of water and 25 mL of 1× PBS. Store at 4 °C.

6. Pipet tips: use pipet tips with a large opening (about 1.5 mm in diameter) for pipetting PEG 8000 and Nycodenz solutions (*see* **Note 4**).
7. Mixer/incubator with evaporation control. We use the ThermoMixer C equipped with SmartBlock and ThermoTop (Eppendorf).
8. Flat bottom 96-well white polystyrene plate.
9. Fluorescence plate reader.

3 Methods

3.1 EV Production by the Starvation Method (See Note 5)

1. Culture cells at 37 °C and 5% CO_2 in complete medium until they reach 80% confluence.
2. Switch to serum-free medium and culture for 3 days.
3. Isolate EVs (microvesicles and exosomes) from the serum-free culture medium by differential ultracentrifugation (*see* **Note 6**).
4. Eliminate cell debris by centrifugation at 2000 × *g* for 10 min in 50 mL conical tubes.
5. Ultracentrifuge the supernatant at 10,000 × *g* for 20 min at 4 °C using 32.4 mL polypropylene tubes and recover microvesicles (MVs) from the pellet.
6. Ultracentrifuge the supernatant at 100,000 × *g* for 1 h at 4 °C using 32.4 mL polypropylene tubes and recover exosomes (Exos) from the pellet.
7. Alternatively, isolate the total population of EVs (containing both MVs and Exos) by a single ultracentrifugation step at 100,000 × *g* for 1 h at 4 °C (i.e. without the 10,000 × *g* step).
8. Resuspend EVs in a small volume of PBS (typically 0.5 mL, leading to 1–6 × 10^{12} EVs/mL).
9. Snap-freeze EVs in liquid nitrogen and store them at −80 °C or in liquid nitrogen for up to 6 months.

3.2 Liposomes Production by the Extrusion Method (See Note 7)

1. Rinse a glass tube sequentially with water (3 times), ethanol (3 times), and chloroform (3 times).
2. Add to the glass tube 1 μmol (total lipid content) of the appropriate lipid mix (*see* **Note 8**) prepared from chloroform solutions of each lipid component. Add 53.2 μL of POPC (70 mol%) and 22.3 μL of DOPE (30 mol%) for EV-liposome fusion without FRET-based lipid mixing measurements. Add 50.9 μL of POPC (67 mol%), 22.3 μL of DOPE (30 mol%), 14.8 μL of NBD-PS (1.5 mol%), and 19.5 μL of Rho-PE (1.5 mol%) for EV-liposome fusion checked by the FRET-based lipid mixing assay.

3. Dry the lipid mix for 10 min under a gentle stream of argon, followed by 2 h under vacuum to remove all trace of organic solvent.
4. Resuspend the dried lipid film with 1 mL of PBS to obtain a final lipid concentration of 1 mM. To produce cargo-loaded liposomes, use instead a solution of cargo in PBS at the desired final concentration to resuspend the lipid film (*see* **Note 9**).
5. Vortex the solution vigorously for 1 h at room temperature.
6. Freeze the obtained multilamellar liposomes in liquid nitrogen (~30 s) and thaw them immediately in a 40 °C water bath (~5 min). Repeat this sequence 7 times to generate unilamellar liposomes.
7. Homogenize the unilamellar liposomes by extrusion through a 100 nm polycarbonate membrane using a liposome extruder. Perform at least 21 passages (*see* **Note 10**).
8. In case of cargo-loaded liposomes, separate liposomes from non-encapsulated cargoes by flotation in a Nycodenz density gradient.
9. Dilute the liposome solution with PBS to a final volume of 1.5 mL and add 1.5 mL of 80% Nycodenz solution to obtain a 40% Nycodenz mixed solution.
10. Put this 40% Nycodenz liposome mixture at the bottom of a 5 mL transparent centrifuge tube.
11. Add very gently 1.5 mL of 15% Nycodenz solution on top of the 40% Nycodenz layer (*see* **Note 11**).
12. Put a third layer consisting of 0.5 mL PBS solution on top of the 15% Nycodenz layer.
13. Ultracentrifuge the gradient at 246,000 × *g* for 4 h at 4 °C with acceleration and deceleration profiles at the slowest settings to preserve the gradient.
14. Recover 0.5 mL of liposomes by gently pipetting within the third layer of the gradient and dilute with 0.5 mL of PBS, leading to about 5–7 × 10^{12} liposomes/mL.
15. Store the liposomes on ice and protected from light.

3.3 EVs and Liposomes Characterization by Nanoparticle Tracking Analysis (NTA)

1. Before measurements, dilute EVs, and liposomes, typically 3500× and 6000× with PBS (confirmed to be particle-free), respectively, to obtain a concentration of about 10^9 particles/mL.
2. Determine the actual EVs and liposomes size distribution and concentration using an NTA system (*see* **Note 12**).
3. For each sample, record 10 movies of 12 s each using predefined camera level 16.
4. Analyze data with the NTA Analytical Software suite.

3.4 Generation and Purification of Hybrid EVs

1. Mix EVs and unlabeled liposomes at an EV/liposome molar ratio of 1:1 (2×10^{11} particles total) (*see* **Note 13**) in the presence of 20% PEG 8000 (*see* **Note 14** and Fig. 1). Typically, add 30 μL of EVs, 20 μL of liposomes, 40 μL of 50% PEG 8000, and 10 μL of PBS in a 0.5 mL microtube.
2. Incubate the tube at 40 °C for 2 h in a mixer/incubator with evaporation control. Alternatively, seal the 0.5 mL microtube with Parafilm (to prevent evaporation and ensure impermeability) and place it in a 40 °C water bath for 2 h.
3. Stop the fusion reaction by diluting the reaction mix with 0.4 mL of PBS at 4 °C.
4. Purify hybrid EVs in a three-step density gradient consisting of 40%, 15%, and 0% Nycodenz solutions. First, add 1 mL of PBS to the 0.5 mL reaction solution and mix with 1.5 mL of 80% Nycodenz solution to obtain a 40% Nycodenz mixed solution.
5. Put this 40% Nycodenz EV-liposome mixture at the bottom of a 5 mL transparent centrifuge tube.
6. Add very gently 1.5 mL of 15% Nycodenz solution on top of the 40% Nycodenz layer (*see* **Note 12**).
7. Put a third layer consisting of 0.5 mL PBS solution on top of the 15% Nycodenz layer using the same technique.
8. Ultracentrifuge the gradient at 246,000 × *g* for 4 h at 4 °C with acceleration and deceleration profiles at the slowest settings to preserve the gradient.
9. Collect fractions of 250 μL from the top of the gradient (*see* **Note 15** and Fig. 2).
10. (Optional) To remove Nycodenz, dilute the collected fractions 100× with PBS, followed by ultracentrifugation at 100,000 × *g* for 1 h at 4 °C using 32.4 mL polypropylene tubes. Recover hybrid EVs from the pellet by resuspension with 250 μL of PBS (*see* **Note 16**).

3.5 Monitoring Hybrid EV Formation Using a FRET-Based Lipid Mixing Assay

1. Mix EVs and fluorescent liposomes in a 96-well plate maintained at 40 °C in a fluorescence plate reader. Typically, add 30 μL of EVs, 20 μL of liposomes, 40 μL of 50% PEG 8000, and 10 μL of PBS to obtain an EV/liposome molar ratio of 1:1 (2×10^{11} particles total) and 20% PEG 8000 (*see* **Note 17**).
2. Monitor lipid mixing for 2 h at 40 °C by following the increase of NBD fluorescence intensity (excitation at 460 nm, emission at 535 nm, cutoff at 530 nm) over time (*see* **Note 18**).
3. Stop the fusion reaction after 2 h by adding 10 μL of 2.5% DDM, which solubilizes all liposomes and thus gives F_{max}(NBD), the NBD fluorescence intensity at infinite dilution.

4. Normalize the fusion curves using the following equation that gives the percentage of NBD fluorescence increase at time t, % F(NBD, t) (*see* **Note 19**):

$$\%F(\text{NBD}, t) = [F(\text{NBD}, t) - F_{\min}(\text{NBD})]/[F_{\max}(\text{NBD}) - F_{\min}(\text{NBD})]$$

where $F_{\min}$(NBD) is the lowest NBD fluorescence value from all time points.

4 Notes

1. Lipids are stored at −20 °C in glass bottles with polytetrafluoroethylene (PTFE) lined caps. They should be taken out of the freezer 10 min before being pipetted in order to avoid water condensation inside the bottle at the opening. Pipet lipids with a Hamilton gastight microsyringe previously rinsed with ethanol and then chloroform. After use, replace air in the glass bottles with pure argon gas (through the application of a very gentle stream of argon). Close the bottles and seal with Parafilm. These steps prevent oxidation of lipids and maintain their functionality for 3–6 months when stored at −20 °C.
2. Other parental cell types can be used to produce EVs, with adapted culture medium (e.g. human mesenchymal stem cells, HeLa cells, or THP-1 cells).
3. To obtain a volume "V" of 100% (w/v) Nycodenz, add a volume "V/2" of water to the powder. Nycodenz dissolution can be facilitated by sonication in an ultrasonic bath heated at 50 °C.
4. Nycodenz (80% and 100%) and PEG (50%) solutions are very viscous and need to be pipetted very slowly in order to reach the equilibrium level. Pipetting is facilitated by the use of large opening tips, keeping the pipet vertical during the aspiration and release steps. Make these large openings tips by cutting the extremity of conventional 200 μL or 1 mL tips with a sterile cutter. Autoclave to sterilize.
5. To produce EVs, we have used the starvation method which, to the best of our knowledge, works well with all cell types. The EV production yield obtained by this method depends on the cell type but typically ranges from 1000 to 30,000 EVs/cell, as quantified by NTA. Typically, 40 flasks of 150 cm^2 are used in our lab to produce 0.5–3 × 10^{12} EVs, which are sufficient for 5–30 experiments. Other EV production methods (e.g. hollow fiber bioreactor) [13] may be used for EV-liposome fusion if they ensure a good purity (low protein contamination from the culture media and parental cells) to ensure the accuracy of the EV/liposome ratio measured by NTA.

6. Although complete separation between small EVs (exosomes) and large EVs (microvesicles) is difficult to obtain (and to confirm), it seems that the EV-liposome fusion method is efficient for both exosomes and microvesicles irrespective of whether they are separated by sequential ultracentrifugation. EV isolation is usually performed from large volumes of media (liter scale). Therefore, the use of large volume ultracentrifuge tubes is highly recommended to limit the number of time-consuming ultracentrifuge runs.
7. Other liposome production methods may be used for EV-liposome fusion although we have not tested them in producing hybrid EVs. For example, the sonication method prepares small, highly fusogenic liposomes, while the detergent-assisted method allows for reconstitution of membrane proteins and preparation of several samples at once (in a flow dialysis chamber) [14].
8. The lipid composition of liposomes can be tuned in order to obtain hybrid EVs with specific properties [9]. For example, PEGylated lipids were shown to increase the circulation time of hybrid EVs [15]. Lipid composition can also modulate the efficiency of fusion and thus the extent of EV modifications (e.g. membrane functionalization, cargo loading, etc.). Of note, cationic lipids increase fusion with EVs, whereas anionic or PEGylated lipids reduce fusion with EVs. In order to maintain fusion efficiency *if less fusogenic lipids are used*, PEG concentration and the duration of the fusion reaction can be increased.
9. Liposomes can be loaded with various compounds. As a proof of concept, we have successfully formed hybrid EVs containing 2 μM of Sulforhodamine B [9]. Loading efficiency, defined as the proportion of cargo transferred from the liposomes to the EVs, depends on the cargo‘s physical properties. For example, lipophilic molecules are easily transferred with a loading efficiency of more than 90%, whereas the transfer of hydrophilic molecules can vary from 20% to 60% [9]. Transfer of small interfering RNA (siRNA), iron oxide nanoparticles, or adeno-associated virus (AAV) vectors is also possible (unpublished data).
10. Before liposome extrusion, the remaining bubbles should be removed from the syringe and all parts of the extruder should be tightly screwed and carefully wiped. Potential leakage should be first tested with PBS to avoid content loss during extrusion. Dead volume is usually about 50 μL. After use, the extruder should be cleaned from remaining liposomes using PBS and then water solutions.

11. To limit hydrodynamic perturbations when gradients are overlaid, the 15% Nycodenz solution should be deposited on the tube wall, about 1–2 mm above the 40% Nycodenz surface, with the tube inclined at 75°.
12. The number of EVs seen by NTA is highly dependent on the purity of the sample. In our case, EVs are produced by cell starvation in serum-free culture medium, therefore limiting contamination by protein aggregates. With this EV generation method, the purity ratio is about 2×10^{10} particles/μg of proteins, corresponding to a pure sample [16]. Other techniques such as multi-angle dynamic light scattering (MADLS) or tunable resistive pulse sensing (TRPS) can also be used to characterize the concentration and size distribution of EVs and liposomes [17].
13. The EV/liposome molar ratio can also be set at 9:1 (1.8×10^{11} EVs and 2×10^{10} liposomes) or 1:9 (2×10^{10} EVs and 1.8×10^{11} liposomes) to control the hybrid EVs composition and properties. As an example, a liposomal formulation of a high-cost drug will benefit from a high EV/liposome ratio to ensure high drug loading efficiency. On the contrary, an easy to produce liposomal formulation should be used in excess to ensure that most EVs fuse with a liposome.
14. Although fusion can be triggered with other PEG molecules [9], best results are obtained using PEG with a molecular weight of 8000 kDa.
15. Fraction collection is more precise when using small pipet tips. We recommend using 100 μL pipet tips to recover the 250 μL layers in 3 × 83 μL.
16. Purified hybrid EVs can be used for further characterization, functional tests, or drug delivery experiments. They can be kept at 4 °C for 2 days, or be snap-frozen in liquid nitrogen and stored at −80 °C or in liquid nitrogen for up to 6 months, without any significant cargo loss (<10%) if the cargo is lipophilic, but they may lose some of the cargo if the cargo is hydrophilic [9].
17. To control the timing of EV-liposome fusion, the reaction mix is done by adding the EV-containing droplet and the liposome-containing droplet in two different places at the bottom of the well. The PEG solution is then added on the wall of the well so that it can fall and mix with the two other droplets when the plate is shaken at the beginning of the experiment. Shaking can be done by hand on the bench or by the plate reader in a controlled manner just before the fluorescence measurement begins. Because the PEG solution is relatively thick, a >5 s vigorous shaking is recommended.

18. Lipid mixing during EV-liposome fusion is monitored by FRET [18]. Donor fluorescent liposomes contain an equal molar ratio (1.5%) of the fluorescent lipids NBD-PS and Rho-PE. Upon excitation of NBD, energy is transferred to Rhodamine in a FRET process, which is strongly dependent on the distance between the two fluorophores. Therefore, dilution of the NBD-PS and Rho-PE lipids within the membrane of acceptor non-fluorescent EVs (when donor liposomes and acceptor EVs fuse) results in an increase of the NBD fluorescence.
19. $\%F(\mathrm{NBD}, t)$ is related to the average number of fusion events per donor liposome (rounds of fusion) as previously described [19] and confirmed in our setup [9]:

$$\text{Rounds of fusion} = \left(0.49666 \times \exp^{(0.036031 \times X)}\right) - \left(0.50597 \times \exp^{(-0.053946 \times X)}\right)$$

where $X = \%F(\mathrm{NBD}, t)$.

Acknowledgments

This work was supported by the ITMO-Inserm Plan Cancer 2014-2019, the IdEx Université de Paris ANR-18-IDEX-0001, and the Région Île-de-France under the convention SESAME 2019—IVETh (n°EX047011). M.P. received a PhD fellowship from Servier Laboratories, and funds by the PhD Program Frontières du Vivant (FdV)—Cursus Bettencourt, and the École de l'Inserm Liliane Bettencourt.

References

1. Zhu X, Badawi M, Pomeroy S et al (2017) Comprehensive toxicity and immunogenicity studies reveal minimal effects in mice following sustained dosing of extracellular vesicles derived from HEK293T cells. J Extracell Vesicles 6:1324730
2. Xie M, Xiong W, She Z, Wen Z, Abdirahman AS, Wan W, Wen C (2020) Immunoregulatory effects of stem cell-derived extracellular vesicles on immune cells. Front Immunol 11:13
3. Elsharkasy OM, Nordin JZ, Hagey DW, de Jong OG, Schiffelers RM, Andaloussi SE, Vader P (2020) Extracellular vesicles as drug delivery systems: why and how? Adv Drug Deliv Rev 159:332–343
4. Murphy DE, de Jong OG, Brouwer M, Wood MJ, Lavieu G, Schiffelers RM, Vader P (2019) Extracellular vesicle-based therapeutics: natural versus engineered targeting and trafficking. Exp Mol Med 51:1–12
5. Kooijmans SAA, Stremersch S, Braeckmans K, de Smedt SC, Hendrix A, Wood MJA, Schiffelers RM, Raemdonck K, Vader P (2013) Electroporation-induced siRNA precipitation obscures the efficiency of siRNA loading into extracellular vesicles. J Control Release 172:229–238
6. Vader P, Breakefield XO, Wood MJA (2014) Extracellular vesicles: emerging targets for cancer therapy. Trends Mol Med 20:385–393

7. Palazzolo S, Hadla M, Spena CR et al (2019) Proof-of-concept multistage biomimetic liposomal dna origami nanosystem for the remote loading of doxorubicin. ACS Med Chem Lett 10:517–521
8. Inglut CT, Sorrin AJ, Kuruppu T, Vig S, Cicalo J, Ahmad H, Huang H-C (2020) Immunological and toxicological considerations for the design of liposomes. Nano 10(2):190. https://doi.org/10.3390/nano10020190
9. Piffoux M, Silva AKA, Wilhelm C, Gazeau F, Tareste D (2018) Modification of extracellular vesicles by fusion with liposomes for the design of personalized biogenic drug delivery systems. ACS Nano 12:6830–6842
10. Rayamajhi S, Nguyen TDT, Marasini R, Aryal S (2019) Macrophage-derived exosome-mimetic hybrid vesicles for tumor targeted drug delivery. Acta Biomater 94:482–494. https://doi.org/10.1016/j.actbio.2019.05.054
11. Farjadian F, Ghasemi A, Gohari O, Roointan A, Karimi M, Hamblin MR (2019) Nanopharmaceuticals and nanomedicines currently on the market: challenges and opportunities. Nanomedicine 14:93–126
12. van Balkom BW, Gremmels H, Giebel B, Lim SK (2018) Proteomic signature of mesenchymal stromal cell-derived small extracellular vesicles. Proteomics 19(1–2):e1800163
13. Piffoux M, Nicolás-Boluda A, Mulens-Arias V, Richard S, Rahmi G, Gazeau F, Wilhelm C, Silva AKA (2019) Extracellular vesicles for personalized medicine: the input of physically triggered production, loading and theranostic properties. Adv Drug Deliv Rev 138:247–258
14. Ji H, Coleman J, Yang R, Melia TJ, Rothman JE, Tareste D (2010) Protein determinants of SNARE-mediated lipid mixing. Biophys J 99: 553–560
15. Kooijmans SAA, Fliervoet LAL, van der Meel R, Fens MHAM, Heijnen HFG, van Bergen-En-Henegouwen PMP, Vader P, Schiffelers RM (2016) PEGylated and targeted extracellular vesicles display enhanced cell specificity and circulation time. J Control Release 224:77–85
16. Webber J, Clayton A (2013) How pure are your vesicles? J Extracell Vesicles. https://doi.org/10.3402/jev.v2i0.19861
17. Vogel R, Savage J, Muzard J et al (2021) Measuring particle concentration of multimodal synthetic reference materials and extracellular vesicles with orthogonal techniques: who is up to the challenge? J Extracell Vesicles 10:e12052
18. Scott BL, Van Komen JS, Liu S, Weber T, Melia TJ, McNew JA (2003) Liposome fusion assay to monitor intracellular membrane fusion machines. Methods Enzymol 372:274–300
19. Parlati F, Weber T, McNew JA, Westermann B, Söllner TH, Rothman JE (1999) Rapid and efficient fusion of phospholipid vesicles by the alpha-helical core of a SNARE complex in the absence of an N-terminal regulatory domain. Proc Natl Acad Sci U S A 96:12565–12570

Chapter 25

Quantitative Measurement of Extracellular Vesicle Content Delivery Within Acceptor Cells

Émeline Bonsergent, Shéryl Bui, and Grégory Lavieu

Abstract

Extracellular vesicles (EVs), including exosomes and microvesicles, are thought to transport bioactive molecules from donor to acceptor cells. Although EV uptake has been qualitatively assessed through subcellular imaging, EV content delivery has been rarely addressed due to a lack of adequate methods. Here we present a sensitive bulk assay to quantitatively measure EV uptake and content delivery in mammalian cell. In this assay, EVs containing a NanoLuc luciferase-tagged cargo are mixed with unlabeled acceptor cells. Cell fractionation separates membrane and cytosolic fractions, and luciferase activity is measured within each fraction to determine the percentage of cytosolic release. This assay can be used to further decipher cellular and molecular mechanisms that regulate the EV delivery process or to quantitatively test specific pairs of donor–acceptor cells.

Key words Extracellular vesicles, Exosomes, Microvesicles, Uptake, Content release, Fusion

1 Introduction

Extracellular vesicles (EVs) are emerging vectors in intercellular cellular communication. They transfer biomolecules such as proteins, lipid, and nucleotides from donor to acceptor cells [1–3]. While EVs are thought to contribute to many physiological functions and diseases, the EV delivery process remains mysterious. Cellular aspects of the process are only superficially apprehended, and the molecular machinery that governs the EV content release remains a black box [4]. Previous studies assessed EV uptake, mainly relying on fluorescent labeling of EV membrane or cargo and tracking their fates in acceptor cells using fluorescence-activated cell sorting (FACS) or subcellular imaging [5]. These methods, however, do not allow to assess EV content delivery.

We developed a cargo-based approach in which a generic EV cargo, Hsp70, is tagged with the NanoLuc luciferase (NLuc) to track its fate within acceptor cells [6]. Briefly, we generated donor

Jingshi Shen (ed.), *Membrane Trafficking: Methods and Protocols*,
Methods in Molecular Biology, vol. 2473, https://doi.org/10.1007/978-1-0716-2209-4_25,

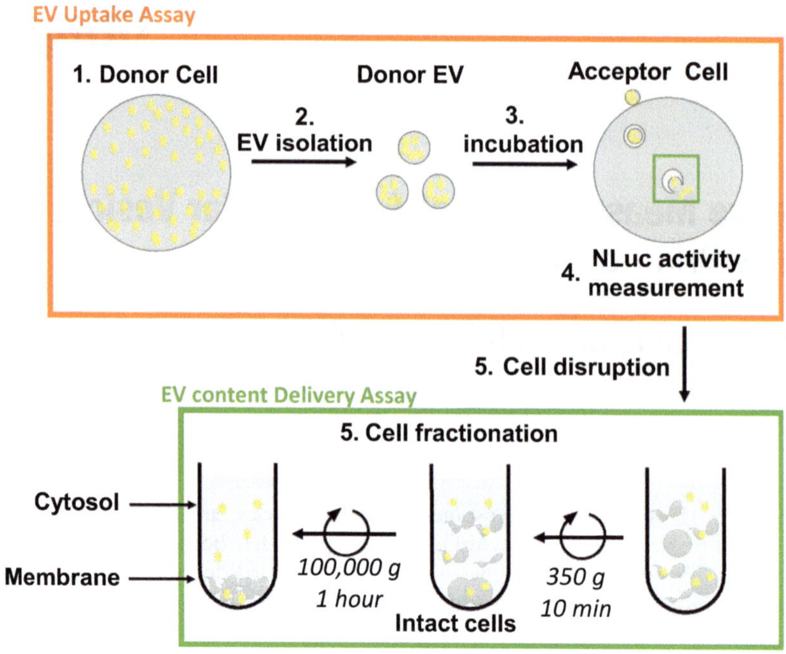

Fig. 1 EV Uptake and EV Content Delivery assays. Donor cells expressing Nluc-EV cargo release EVs (stage 1) that are isolated (stage 2). EVs are incubated on unlabeled acceptor cells (stage 3). After EV removal, luminescence can be measured within the acceptor cells (stage 4) to quantify the amount EV that (1) are docked at the plasma membrane, (2) are internalized within endo/lysosomal compartments, (3) have release their Nluc-cargo within the cytosol of the acceptor cells (EV Uptake assay in orange). Alternatively, acceptor cells can be further processed for mechanical disruption (stage 5) and cell fractionation (stage 6) to separate cytosolic and membrane fractions, in which Nluc-activity can be measured to determine the percentage of cytosolic release (EV content delivery in green)

cells stably expressing NLuc-Hsp70, isolated EVs by sequential centrifugation. Here we start right after the EV isolation, which has been extensively described [7]. Then we assessed the topology of the cargo through a proteinase protection assay, incubated NLuc-Hsp70 positive EVs with unlabeled acceptor cells, and eventually measured luminescence within the membrane and cytosolic fractions of the acceptor cells (Fig. 1). This method is flexible and adaptable to many cellular models and cargoes.

2 Materials

2.1 Common Materials

1. Donor cell lines expressing the EV cargo of interest tagged with the NLuc reporter. In our studies, we use NLuc-Hsp70, a model for EV cytosolic cargos, and NLuc-CD63, a model for EV membrane cargos. To generate new plasmids, a NLuc template can be purchased from Promega. We use HeLa cells as both the donor and acceptor cells, but the current protocol can be adapted to any cargo or cell of interest.

2. Phosphate Buffered Saline (PBS): 138 mM NaCl, 8 mM Na_2HPO_4, 3 mM KCl, 1.5 mM KH_2PO_4, pH 7.4.
3. Triton X-100 solution: PBS containing 50 μM of Triton-X100. Keep at 4 °C.
4. MicroBCA kit for protein quantification.
5. Plate Reader/Luminometer.
6. 96-well opaque white plates and clear-bottom white plates.

2.2 Protease Protection Assay

1. Proteinase K stock solution: 1 mg/mL proteinase K in ultra-pure water. Weight 1 mg of lyophilized proteinase K and dissolve it in 1 mL of ultra-pure water. Store single-use aliquots at −80 °C.
2. PBS PI: PBS containing 1× protease inhibitor cocktail. Reconstitute a protease inhibitor cocktail to a 1× solution in PBS. Use fresh dilution for each experiment.
3. Triton 1% solution: PBS PI containing 5 μM of Triton X-100. Dilute Triton X-100 stock solution with PBS PI to reach a 1% Triton X-100 solution. Use fresh dilution for each experiment.
4. Nonlytic NLuc reagent: we use the Nano-Glo™ Live Cell assay reagent from Promega.

2.3 Uptake and Content Delivery Assays

1. Fetal bovine serum (FBS)-free or EV-depleted cell culture media.
2. Lytic NLuc reagent: we use the Nano-Glo® Luciferase Assay System from Promega.
3. EDTA solution: PBS with 0.5 mM EDTA. Dilute EDTA in PBS to reach a concentration of 0.5 mM. Store at 4 °C.
4. 30 G needles adapted on 1 mL syringes.
5. Micro-tip sonicator.

3 Methods

3.1 Assessment of Cargo's Topology

Treat isolated EVs (*see* **Note 1**) emanating from donor cells expressing putative NLuc-tagged EV cargo with a protease with or without detergents. In the absence of detergents, a cargo inside EVs is expected to be protected from protease degradation.

1. Prepare four aliquots (1 μg equivalent proteins into 10 μL of PBS) for each EV preparation to test the following conditions, with 2–3 technical replicates per condition. Sample A: EVs alone. Sample B: EVs + Triton X-100, to rule out effect of Triton X-100 on NLuc activities (*see* **Note 2**). Sample C: EVs + proteinase K, to assess the amount of unprotected cargo. Sample D: EVs + Triton X-100 + proteinase K, to assess

Table 1
Summary of sample preparation to perform protease protection assay on NLuc EV cargo

Conditions	A: EVs alone	B: Triton	C: Protease K	D: Triton + Prot. K
Sample (μL)	10	10	10	10
Triton 1× solution (μL)	–	10	–	10
PBS PPI solution (μL)	90	80	40	30
Proteinase K solution at 0.1 mg/mL (μL)	–	–	50	50

the total amount of EV-associated NLuc-cargo (samples preparation is summarized in Table 1 and detailed in the following **steps 2–4**). All samples are loaded on a 96 opaque white well-plate and must be kept on ice, except when mentioned otherwise.

2. Add 10 μL of Triton 1% solution for samples B and D. Add 10 μL of PBS for samples A and C.
3. Dilute proteinase K stock solution (1 mg/mL) within ice-cold PBS PI to 100 μg/mL (*see* **Note 3**).
4. Add 50 μL of the diluted proteinase K solution to samples C and D. The final volume of each well should be 100 μL, supplemented with ice-cold PBS PI when required.
5. Incubate at room temperature for 5 h (*see* **Note 2**).
6. Add 20 μL of nonlytic NLuc reagent to each well.
7. Read NLuc activities immediately using a luminometer.
8. Compare the readings of samples C and D to determine whether the EV cargo is inside the vesicles. A cargo located inside the vesicle is expected to be proteinase K resistant in absence of detergent.

3.2 EV Uptake Assay

Incubate EV containing NLuc-cargo with acceptor cells. The EV uptake assay measures the NLuc activity that remains associated with acceptor cells, after donor EVs washout. EV uptake does not discriminate between EVs docked at the cell surface, EVs internalized within the endo/lysosomal compartments, or EV cargo delivered within the cytosol of acceptor cells.

1. Seed acceptor cells in a clear-bottom white plate to reach around 80–90% confluency by the end of the assay (day 1 for short kinetics or day 2 for long kinetics) (*see* **Note 4**).
2. 24 h later, replace the media with FBS-free media or EV-depleted media according to acceptor cell compatibility/sensitivity. If not possible, experiments can be performed in complete media but FBS-derived EVs may interfere with the NLuc-positive EVs.

3. Add EV input (1 μg equivalent protein into 10 μL of PBS) in each well and incubate for the desired time. We highly recommend performing kinetics (from 1 to 24 h) and dose responses in initial attempts.
4. Wash 3 times with ice-cold PBS.
5. Add 50 μL of PBS to each well.
6. Add EV input in empty wells (replicates recommended). Dilute EV input in ice-cold PBS to reach a total of 50 μL.
7. If using the Nano-Glo® Luciferase Assay System (Lytic NLuc kit), dilute 1 μL of NLuc-substrate in 50 μL of buffer for each well, and add it to the top of each sample.
8. Read NLuc activities with a luminometer.
9. Subtract backgrounds using signals from acceptor cells not receiving any labeled EVs. EV uptake can be quantified as follows:

$$\text{Uptake (\%input)} = \frac{(\text{acceptor cells signal} - \text{background}) \times 100}{\text{input signal} - \text{background}}$$

3.3 EV Content Release Assay

After incubation with NLuc-EVs, mechanically disrupt acceptor cells to separate membrane and cytosolic fractions. This allows us to acutely measure EV cargo delivery into the cytosol of acceptor cells.

1. Seed acceptor cells to reach around 80–90% confluency by the end of the assay (day 1 for short kinetics or day 2 for long kinetics). Seed at least 100,000 cells in a 24 well plate to collect enough cells for the cell fractionation step.
2. Proceed as for the EV uptake assay describe in Subheading 3.2 until **step 6**.
3. After PBS wash, add the EDTA solution to adherent cells (300 μL per well for a 24-well plate) and incubate at room temperature for 10 min to detach the cells.
4. Collect the detached cells in 1.5 mL microtubes and centrifuge at 500 × *g* for 10 min 4 °C. Cells in suspension can be directly collected using low-speed centrifugation. From this step, all samples should remain on ice or at 4 °C.
5. Remove the supernatant and resuspend the cell pellet in 200 μL of PBS PI.
6. Vortex the sample for 5 min.
7. Pass each sample through a 30 G needle attached to a 1 mL syringe for 5 times.

8. Sonicate the samples (5 s at 30% duty cycle, output control level 3 if using the Ultrasonic Processor of Thomas scientific) (*see* **Note 5**).
9. Centrifuge the samples at 350 × *g* for 10 min at 4 °C to collect the remaining intact cells.
10. Transfer the supernatant to a new microtube and resuspend the pellet (intact cells) in 200 μL of PBS PI and store on ice to assess the percentage of intact cells.
11. Centrifuge the supernatant at 100,000 × *g* for 1 h at 4 °C to separate membrane fractions (pellet) from cytosolic fractions (supernatant). Resuspend the membrane fraction in 200 μL of PBS PI.
12. Transfer each sample to a 96-well white plate to measure luminescence activities within each fraction and initial EV input.
13. Calculate the percentage of cytosolic release using the following formula:

$$\text{Cytosolic release }(\%) = \frac{\text{Cytosol signal} \times 100}{(\text{Membrane signal} + \text{Cytosol signal})}$$

4 Notes

1. Here we start right after the EV isolation, which has been extensively described [7]. We highly recommend to assess the quality of the EV samples, following MISEV recommendations [8], especially to quantify the protein concentration of EV samples.
2. Ensure that the NLuc activity within the EV input is at least 1000× above the background. If not, we recommend increasing the input of donor cells. For HeLa cells expressing NLuc-Hsp70, we recommend starting with at least 200×10^6 cells and collecting conditioned media after 24 h, when isolating EVs through sequential ultracentrifugation.
3. For the protease protection assay, we noticed that the amount and type of detergent affect the NLuc activity [6]. We therefore highly recommend systematically assessing the effects of detergents. In our hands, 0.1% Triton X-100 works well. Similarly, the concentration, time, and temperature of incubation with the proteinase K need optimization, and the parameters mentioned above are only indicative. Our indicated protocol should be considered as a starting point and may benefit from optimization.

4. We mostly used white-wall plates with clear bottoms, which enable us to visually check the cells during the culture. A white tape is sticked to the bottom just before reading the luminescence on the plate reader, to avoid any leakage of luminescence signal. The EV uptake efficiency is low, around 1% in our cell model. We highly recommend separating the samples for measurements of EV uptake and EV input, which shows low and high NLuc activities, respectively, to avoid cross-contamination between adjacent wells.
5. The cell disruption/fractionation procedure needs to be set up or optimized with each sonicator device. We recommend starting by assessing the behaviors of several cargo models (e.g., Hsp70 as a soluble cargo and CD63 as a membrane cargo) in different fractions using western blotting for endogenous proteins or luminescent signals for NLuc-tagged cargoes [6]. Intact cell samples should be used to calculate the overall EV uptake (sum of the 3 fractions) and to estimate cell fractionation efficiency.

Acknowledgments

This work is supported by ANR-18-IDEX-0001, IdEx Université de Paris, ANR-19-CE18-002001, and ANR-20-CE15-002102. E.B. is supported by a PhD fellowship from La Ligue contre le cancer. S.B. is supported by a PhD fellowship from PSLUniversity.

References

1. Valadi H, Ekström K, Bossios A et al (2007) Exosome-mediated transfer of mRNAs and microRNAs is a novel mechanism of genetic exchange between cells. Nat Cell Biol 9:654–659
2. Balaj L, Lessard R, Dai L et al (2011) Tumour microvesicles contain retrotransposon elements and amplified oncogene sequences. Nat Commun 2:180
3. Skog J, Würdinger T, van Rijn S et al (2008) Glioblastoma microvesicles transport RNA and proteins that promote tumour growth and provide diagnostic biomarkers. Nat Cell Biol 10: 1470–1476
4. Mathieu M, Martin-Jaular L, Lavieu G et al (2019) Specificities of secretion and uptake of exosomes and other extracellular vesicles for cell-to-cell communication. Nat Cell Biol 21:9–17
5. Costa Verdera H, Gitz-Francois JJ, Schiffelers RM et al (2017) Cellular uptake of extracellular vesicles is mediated by clathrin-independent endocytosis and macropinocytosis. J Control Release 266:100–108
6. Bonsergent E, Grisard E, Buchrieser J et al (2021) Quantitative characterization of extracellular vesicle uptake and content delivery within mammalian cells. Nat Commun 12:1864
7. Théry C, Clayton A, Amigorena S et al (2006) Isolation and characterization of exosomes from cell culture supernatants. Curr Protoc Cell Biol 30:3.22.1–3.22.29
8. Théry C, Witwer KW, Aikawa E et al (2018) Minimal information for studies of extracellular vesicles 2018 (MISEV2018): a position statement of the International Society for Extracellular Vesicles and update of the MISEV2014 guidelines. J Extracell Vesicles 7:1535750

INDEX

Jingshi Shen (ed.), *Membrane Trafficking: Methods and Protocols*, Methods in Molecular Biology, vol. 2473, https://doi.org/10.1007/978-1-0716-2209-4,

MIX
Papier aus verantwortungsvollen Quellen
Paper from responsible sources
FSC® C105338

If you have any concerns about our products,
you can contact us on
ProductSafety@springernature.com

In case Publisher is established outside the EU,
the EU authorized representative is:
Springer Nature Customer Service Center GmbH
Europaplatz 3, 69115 Heidelberg, Germany

Printed by Libri Plureos GmbH
in Hamburg, Germany